Fishery Water Treatment

Editor:

Bruno Augusto Amato Borges

Delve Publishing

www.delvepublishing.com

Fishery Water Treatment

Editor: **Bruno Augusto Amato Borges**

708 3rd Avenue, 6th Floor

New York, NY 10017

United States of America

www.delvepublishing.com

ISBN: 978-1-68095-844-7

Library of Congress Control Number: 2016951597

About the Editor

Bruno Augusto Amato Borges

Aquaculture Engineer Florianópolis, Brazil

Bruno Augusto is Aquaculture Engineer from Federal University of Santa Catarina. His focus is on writing projects and development of new technologies in aquaculture. He is currently project service provider, consultant and develops research in the field of Biofloc Technology (BFT) for freshwater fish farming.

Contents

List of Contributors

Chen-Lin Soo
Department of Chemistry, Faculty of Resource Science and, Technology, Universiti Malaysia Sarawak, 94300 Kota Samarahan, Sarawak, Malaysia

Teck-Yee Ling
Department of Chemistry, Faculty of Resource Science and, Technology, Universiti Malaysia Sarawak, 94300 Kota Samarahan, Sarawak, Malaysia

Nyanti Lee
Department of Aquatic Science, Faculty of Resource Science, and Technology, Universiti Malaysia Sarawak, 94300 Kota Samarahan, Sarawak, Malaysia

Kasing Apun
Department of Molecular Biology, Faculty of Resource Science, and Technology, Universiti Malaysia Sarawak, 94300 Kota Samarahan, Sarawak, Malaysia

Aazami J
Department of Environment, Faculty of Natural Resources, Tarbiat Modares University, Tehran, Iran

Esmaili-Sari A
Department of Environment, Faculty of Natural Resources, Tarbiat Modares University, Tehran, Iran

Abdoli A
Department of Biodiversity and Ecosystem Management, Environmental Research Institute, Shahid Beheshti University, Tehran, Iran.

Sohrabi H
Department of Forestry, Faculty of Natural Resources, Tarbiat Modares University, Tehran, Iran.

Van den Brink PJ
Department of Aquatic Ecology and Water Quality Management, Wageningen University, Wageningen University and Research Centre, Wageningen, The Netherlands.
Alterra, Wageningen University and Research Centre, Wageningen, The Netherlands.

Goran Bakalar
University of Zadar, Zadar, Croatia.
International Maritime Technology Consultancy, Split, Croatia

John J. Cullen
Department of Oceanography, Dalhousie University, Halifax, Nova Scotia, Canada B3H 4R2

Hugh L. MacIntyre
Department of Oceanography, Dalhousie University, Halifax, Nova Scotia, Canada B3H 4R2

Innocent Chirisa
Department of Rural and Urban Planning, University of Zimbabwe, Harare, Zimbabwe

Elmond Bandauko
Department of Rural and Urban Planning, University of Zimbabwe, Harare, Zimbabwe

Abraham Matamanda
Centre for Applied Social Sciences, University of Zimbabwe, Harare, Zimbabwe

Gladys Mandisvika
Department of Rural and Urban Planning, University of Zimbabwe, Harare, Zimbabwe

Erick Butler
School of Engineering and Computer Science, West Texas A&M University, Canyon, TX 79016, USA

Yung-Tse Hung
Department of Civil and Environmental Engineering, Cleveland State University, Cleveland, OH 44115, USA

Mohammed Suleiman Al Ahmad
Department of Civil and Environmental Engineering, Cleveland State University, Cleveland, OH 44115, USA

Ruth Yu-Li Yeh
Department of Materials Science and Engineering, National Dong Hwa University, Hualien, 97401, Taiwan

Robert Lian-Huey Liu
Department of Materials Science and Engineering, National Dong Hwa University, Hualien, 97401, Taiwan

Yen-Pei Fu
Chemical Engineering Department, Ming Hsin University of Science and Technology, Hsin-chu 30401, Taiwan

Tianlong Zheng
Department of Environmental Engineering, University of Science and Technology Beijing, 30 Xueyuan Road, Haidian District, Beijing 100083, China

Juan Wang
Department of Environmental Engineering, University of Science and Technology Beijing, 30 Xueyuan Road, Haidian District, Beijing 100083, China

Qunhui Wang
Department of Environmental Engineering, University of Science and Technology Beijing, 30 Xueyuan Road, Haidian District, Beijing 100083, China

Huimin Meng
Laboratory for Corrosion-Erosion and Surface Technology, Corrosion and Protection Center, University of Science and Technology Beijing, 30 Xueyuan Road, Haidian District, Beijing 100083, China

Lihong Wang
Department of Environmental Engineering, University of Science and Technology Beijing, 30 Xueyuan Road, Haidian District, Beijing 100083, China
Department of Architectural Engineering, Handan Polytechnic College, 141 Zhuhe Road, Handan 056001, Hebei, China

Ashok Kumar Chopra
Department of Zoology and Environmental Sciences, Gurukula Kangri University, Haridwar 249404, Uttarakhand, India

Arun Kumar Sharma
Department of Zoology and Environmental Sciences, Gurukula Kangri University, Haridwar 249404, Uttarakhand, India

N. V. Pradeep
Department of Civil Engineering, Angadi Institute of Technology and Management, Belgaum, India

S. Anupama
Department of Civil Engineering, M. V. J. College of Engineering, Bangalore, India

K. Navya
Department of Biotechnology, Ballari Institute of Technology and Management, Bellary, India

H. N. Shalini
Department of Biotechnology, Ballari Institute of Technology and Management, Bellary, India

M. Idris
Department of Civil Engineering, University Visvesvaraya, College of Engineering, Bangalore University, Bangalore, India

U. S. Hampannavar
Angadi Institute of Technology and Management, Belgaum,India
Patcharawalai Sriyasak
Faculty of Fisheries Technology and Aquatic Resources, Maejo University, Chiang Mai, Thailand
Unit for Social and Environmental Research, Chiang Mai University, Chiang Mai, Thailand

Chanagun Chitmanat
Faculty of Fisheries Technology and Aquatic Resources, Maejo University, Chiang Mai, Thailand

Niwooti Whangchai
Faculty of Fisheries Technology and Aquatic Resources, Maejo University, Chiang Mai, Thailand

Jongkon Promya
Faculty of Fisheries Technology and Aquatic Resources, Maejo University, Chiang Mai, Thailand

Louis Lebel
Unit for Social and Environmental Research, Chiang Mai University, Chiang Mai, Thailand

Magnus Land
Mistra EviEM, Stockholm Environment Institute, 10451 Stockholm, Sweden

Wilhelm Granéli
Department of Biology, Lund University, 22362 Lund, Sweden.
Everglades Wetland Research Park, Florida Gulf Coast University, 4940 Bayshore Drive, Naples 34112, FL, USA
Department of Bioscience, Aarhus University, 8600 Silkeborg, Denmark.

Anders Grimvall
Swedish Institute for the Marine Environment, Box 260, 405 30 Gothenburg, Sweden.

Carl Christian Hoffmann
Department of Bioscience, Aarhus University, 8600 Silkeborg, Denmark.

William J. Mitsch
Everglades Wetland Research Park, Florida Gulf Coast University, 4940 Bayshore Drive, Naples 34112, FL, USA

Karin S. Tonderski
Department of Physics, Chemistry and Biology, Linköping University, 581 83 Linköping, Sweden.

Jos T. A. Verhoeven
Department of Biology, Utrecht University, 3584 CA Utrecht, Netherlands.

H. J. De Lange
Alterra, Wageningen UR, 6700 Wageningen, The Netherlands

M. P. C. P. Paulissen
Alterra, Wageningen UR, 6700 Wageningen, The Netherlands

Panneerselvam Karthikeyan
Centre of Advanced Study in Marine Biology, Faculty of Marine Sciences, Annamalai University, Parangipettai 608 502, Tamil Nadu, India

Kuppusamy Manimaran
Centre of Advanced Study in Marine Biology, Faculty of Marine Sciences, Annamalai University, Parangipettai 608 502, Tamil Nadu, India

Pitchai Sampathkumar
Centre of Advanced Study in Marine Biology, Faculty of Marine Sciences, Annamalai University, Parangipettai 608 502, Tamil Nadu, India

Lakshmanan Rameshkumar
Centre of Advanced Study in Marine Biology, Faculty of Marine Sciences, Annamalai University, Parangipettai 608 502, Tamil Nadu, India

Anja Henneberg
Animal Physiological Ecology, *University of Tübingen*

Rita Triebskorn
Animal Physiological Ecology, *University of Tübingen*

Rajasri Yadavalli
Department of Biotechnology, Sreenidhi Institute of Science and Technology, Hyderabad, Andhra Pradesh 501301, India

Goutham Rao Venkata Naga Heggers
Department of Biotechnology, Sreenidhi Institute of Science and Technology, Hyderabad, Andhra Pradesh 501301, India

Hidayah Manan
Institute of Tropical Aquaculture, Universiti Malaysia Terengganu, 21030 Kuala Nerus, Terengganu, Malaysia

Julia Hwei Zhong Moh
Institute of Tropical Aquaculture, Universiti Malaysia Terengganu, 21030 Kuala Nerus, Terengganu, Malaysia

Nor Azman Kasan
School of Fisheries and Aquaculture Sciences, Universiti Malaysia Terengganu, 21030 Kuala Nerus, Terengganu, Malaysia

Suhaimi Suratman
Institute of Oceanography and Environment, Universiti Malaysia Terengganu, 21030 Kuala Nerus, Terengganu, Malaysia

Mhd Ikhwanuddin
Institute of Tropical Aquaculture, Universiti Malaysia Terengganu, 21030 Kuala Nerus, Terengganu, Malaysia

Hefni Effendi
Center for Environmental Research, Bogor Agricultural University (IPB), Dramaga, Bogor 16680, Indonesia
Department of Aquatic Resources Management, Faculty of Fisheries and Marine Sciences, Bogor Agricultural University (IPB), Dramaga, Bogor 16680, Indonesia

Sri Wahyuningsih
Graduate Program of Aquatic Resources Management, Bogor Agricultural University (IPB), Dramaga, Bogor 16680, Indonesia

Yusli Wardiatno
Department of Aquatic Resources Management, Faculty of Fisheries and Marine Sciences, Bogor Agricultural University (IPB), Dramaga, Bogor 16680, Indonesia

Jason J. Danaher
Pentair Aquatic Ecosystems, 2395 Apopka Boulevard, Apopka, FL 32703, USA

Jeremy M. Pickens
Department of Horticulture, Auburn University, 101 Funchess Hall, Auburn, AL 36849, USA

Jeffrey L. Sibley
Department of Horticulture, Auburn University, 101 Funchess Hall, Auburn, AL 36849, USA

Jesse A. Chappell
School of Fisheries, Aquaculture and Aquatic Sciences, 203 Swingle Hall, Auburn, AL 36849, USA

Terrill R. Hanson
School of Fisheries, Aquaculture and Aquatic Sciences, 203 Swingle Hall, Auburn, AL 36849, USA

Claude E. Boyd
School of Fisheries, Aquaculture and Aquatic Sciences, 203 Swingle Hall, Auburn, AL 36849, USA

Preface

With the increase in population and the growing demand for aquatic food chain of fish see the need to continue increasing their production. The use of natural resources should be managed in a way that does not compromise the environment. One of the main impacts caused by the activity, have the excessive use of water and wastewater generation by cultivation and processing industry.

The treatment of effluents from fishery production began to be discussed recently due to increased concern with environmental issues, so there are many possibilities of innovation and development of new technologies that improve the current processes. In addition to the concern about the treatment of effluents generated by the fishery production, we must also get quality analysis of natural water bodies to understand the actual situation with the aim to reduce pollution and manage the areas so that this precious resource is still available for future generations.

Three-quarters of the Earth's surface is covered by water. It seems unbelievable that the world is about to face a crisis of water supply, but there's no point in having plenty of this resource if it cannot be used by the population. Much of the freshwater reserves in the world are already infected and the few healthy reserves are not being managed responsibly. I see that this issue is extremely delicate and should be discussed increasingly seeking improvements in the use of water within the fishery production.

One of the main trends in wastewater treatment and reuse of water is the Aquaponic culture, whose principle is the use of effluent generated by fish farming for the production of vegetables. In addition to this we have the cultivation method in Biofloc technology (BFT), in which there is no renewal of water in the system generating zero effluent.

The book sought a general approach regarding pollution, waste generation and treatment techniques already in use. Some techniques reported in the book are not of fishery water treatment but has the potential to be applied, so I had an interest in adding.

My main intention in this book is to present the various possibilities in the use and treatment of water by fishery production, seeking always the responsible use and proper management of this resource, which is essential for life.

Editor
Bruno Augusto Amato Borges
Aquaculture Engineer
Florianópolis, Brazil

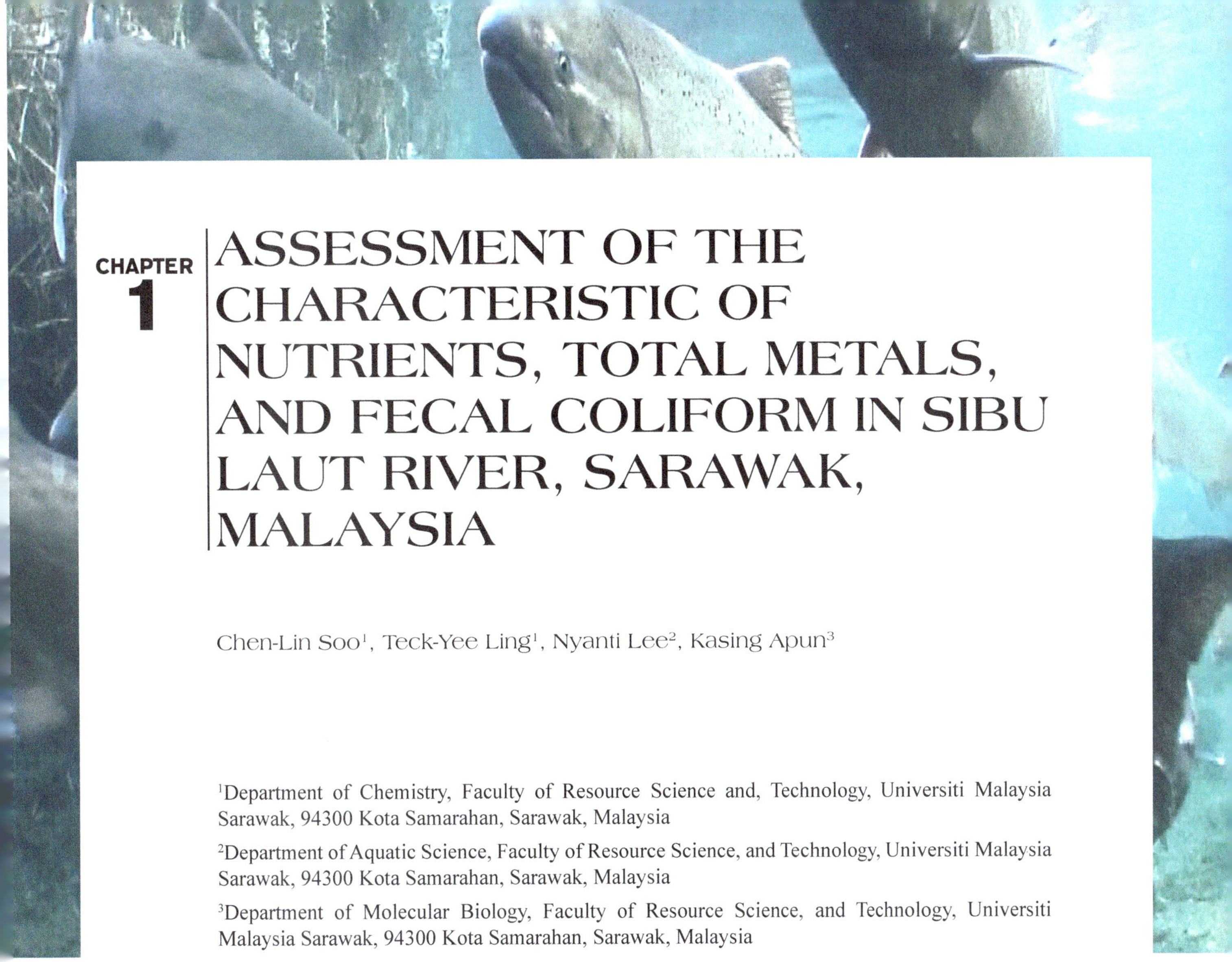

CHAPTER 1

ASSESSMENT OF THE CHARACTERISTIC OF NUTRIENTS, TOTAL METALS, AND FECAL COLIFORM IN SIBU LAUT RIVER, SARAWAK, MALAYSIA

Chen-Lin Soo[1], Teck-Yee Ling[1], Nyanti Lee[2], Kasing Apun[3]

[1]Department of Chemistry, Faculty of Resource Science and, Technology, Universiti Malaysia Sarawak, 94300 Kota Samarahan, Sarawak, Malaysia

[2]Department of Aquatic Science, Faculty of Resource Science, and Technology, Universiti Malaysia Sarawak, 94300 Kota Samarahan, Sarawak, Malaysia

[3]Department of Molecular Biology, Faculty of Resource Science, and Technology, Universiti Malaysia Sarawak, 94300 Kota Samarahan, Sarawak, Malaysia

ABSTRACT

The concentrations of nutrients (nitrogen and phosphorus), total metals, and fecal coliform (FC) coupling with chlorophyll-*a* (chl-*a*), 5-day biochemical oxygen demand (BOD_5) and other general environmental parameters were evaluated at the sub-surface and near-bottom water columns of 13 stations in the Sibu Laut River during low and high slack waters. The results indicated that inorganic nitrogen (mainly nitrate) was the primary form of nitrogen whereas organic phosphorus was the major form of phosphorus. The abundance of total heavy metals in Sibu Laut River and its tributaries was in the order of Pb < Cu < Zn < Cd. Fecal coliform concentration was relatively low along Sibu Laut River. The shrimp farm effluents contributed a substantial amount of chl-*a*, BOD_5, nutrients, and FC to the receiving creek except for total metals. Nevertheless, the influence was merely noticeable in the intake creek and amended rapidly along Selang Sibu River and brought minimal effects on the Sibu Laut River. Besides, the domestic sewage effluents from villages nearby also contributed a substantial amount of pollutants.

INTRODUCTION

Deterioration in water quality is one of the greatest concerns associated with population growth and accompanying land-use changes (Ribeiro and Kjerfve 2002; Davis and Koop 2006; Ham et al. 2009; Whitall et al. 2010; Hadibarata et al. 2012). Eutrophication, metals contamination, and bacterial pollution are among the major problems and stresses of the marine environment and have been receiving global attention and interest (Malik and Ahmad 2002; Parnell 2003; Jonas and Millward 2010; Reopanichkul et al. 2010; Shin et al. 2012).

Eutrophication can be defined as the process of increasing nutrients of an ecosystem that causes changes to its nutritional status (Nixon 1995). Most frequently, this has usually been characterized by increased supplies of nutrients in particular nitrogen and phosphorus to that system, although the supply of excessive decomposable organic carbon leads to eutrophication as well (Young et al. 1999; Pinckney et al. 2001; Howarth and Marino 2006). The discharge of excessive nutrients from urban and industrial wastewaters, or rural and agricultural runoff contributes to the enrichment of inorganic and organic material in marine waters. Eutrophication can lead

Citation: Chen-Lin Soo, Teck-Yee Ling, Nyanti Lee, and Kasing Apun, Assessment of the characteristic of nutrients, total metals, and fecal coliform in Sibu Laut River, Sarawak, Malaysia, DOI 10.1007/s13201-014-0205-7.

to the detrimental changes of the structure and function of both living organisms and non-living components in an ecosystem (Paez-Osuna et al. 1998; Bricker et al.1999; Smith et al. 1999; Meyer-Reil and Koster 2000; Parnell 2003; Del-Pilar-Ruso et al. 2009; Reopanichkul et al. 2010). Often, the "bloom" or greatly increase of aquatic vegetation or phytoplankton and algal in a water body is the first response to the increased levels of nutrients or the eutrophication (Pan and Subba Rao 1997; Smith et al. 1999). Hence, chl-*a* concentration which provides a measure of phytoplankton biomass often used as an indicator of eutrophication and studied along with nutrients (Ho et al. 2008; Xu et al. 2008; Madhu et al.2010; Lefebvre et al. 2011).

Besides nutrients, excessive amounts of heavy metals have been introduced to estuarine and coastal environments as a result of intense industrial development and anthropogenic activities (Fianko et al. 2007; Nduka and Orisakwe 2009; Sundaray et al. 2012; Zhang et al. 2012; Lacerda et al. 2013). Heavy metals are deemed to be dangerous pollutants and have raised environmental concern due to their toxicity, persistence and non-degradability, bioaccumulation in organisms, and ultimately their effect on human health (Burgos and Rainbow 2001; Kress et al. 2004; Otero et al. 2005; Kurun et al. 2007; Du Laing et al. 2008). Special attentions have been paid to Cd, Cu, Pb, Zn and many other heavy metals.

The discharge of domestic and industrial wastewaters with the presence of pathogenic microorganisms in the adjacent water frequently threatened the microbiological quality of the water (Aslan-Yilmaz et al. 2004; Nduka and Orisakwe 2009; Dorsey et al. 2012). The microbiological risks depend on the bacteria concentration in water and on the water usage such as drinking, recreational activities, irrigation, and shellfish harvesting purposes. The microbiological quality of water is classically monitored by searching for and quantifying bacterial indicators of fecal contamination. Total coliform, fecal coliform, fecal streptococci and enterococci are among the most frequently used microbial indicators of fecal pollution (Malik and Ahmad 2002; Cabral and Marques 2006; Djuikom et al. 2006; Kovačić et al. 2011). Although not usually harmful themselves, their presence may indicate the possibility of pathogen presence (Bonadonna et al. 2002; Dorsey et al. 2012).

Sibu Laut River is surrounded by various villages, shrimp farms and tourism's activities make it vulnerable to human activities and susceptible to water quality degradation. Furthermore, Sibu Laut River, which forms the western boundary of the Kuching Wetlands National Park, a wetland classified as a sensitive area also renders it of high research value. What happened to Sibu Laut River and its tributaries now? Is Sibu Laut River suffering from eutrophication, metals contamination or bacterial pollution? The level and the variability in the water quality over neap–spring tidal cycles in this river are yet to be explored in detail. Hence, physicochemical and major water quality data were collected and were employed to determine the water quality characteristics and current health status of the Sibu Laut River and its tributaries. The data presented here provide a snapshot of hydrodynamic conditions and water quality of the Sibu Laut River and its tributaries experienced during high and low tides.

MATERIALS AND METHODS

Study Area and Sampling Regime

The Sibu Laut River is in the northwest region of Kuching, Sarawak, Malaysia (Fig. 1). The area experiences a humid tropical climate with an annual rainfall of 3,600–4,000 mm. The region is not directly exposed to the northeast monsoon. Rainfall is the lowest during June and July, and reaches a peak in December and January. Temperature ranges from 19 to 36 °C receives only 5 h of sunshine per day on average (Malaysian Meteorological Department 2012). This river is one of the boundaries of the Kuching Wetlands National Park at which the international importance of the wetland has been recognized when Malaysia designated the park as the RAMSAR site in November 2005 (Ramsar Site No., 1568), which is also the first Ramsar site in Sarawak and the fifth in Malaysia (Wetlands International 2013). The river drains the mountainous region at the upper reaches of the wetland and the mangrove swamp within the delta on the estuary with tidal ranged from 0.0 to 4.7 m. Selang Sibu River and Loba Kilong River are the left and right tributary at the downstream of the Sibu Laut River.

Samplings were divided into two stages. Stage one focused on the eutrophication study while stage two concentrated on heavy metals and bacteriological water quality. Table 1 shows the details of the sampling cruises. Four samplings were conducted to obtain the data of nutrients in 2010 while two samplings were conducted for total heavy metals and bacteriological water quality study in 2011. A total of 13 stations was selected along Sibu Laut River and tributaries (Fig. 2). The total river length studied was approximately 8 km. The locations were chosen mainly due to their proximity to the suspected domestic pollution sources and aquaculture activities; and some relatively unpolluted sites as a control. The distance between stations was ranged from 1 to 2 km. The first station was at the most downstream of the river. Station 2 was in the Loba Kilong River near to the village of Sibu Laut, just before its conjunction with Sibu Laut River. Station 3, 4, and 5 were along the Selang Sibu River. Station 5 was the closest site to the discharge point of a shrimp farm (Sea Horse Marine Life Sdn. Bhd.)

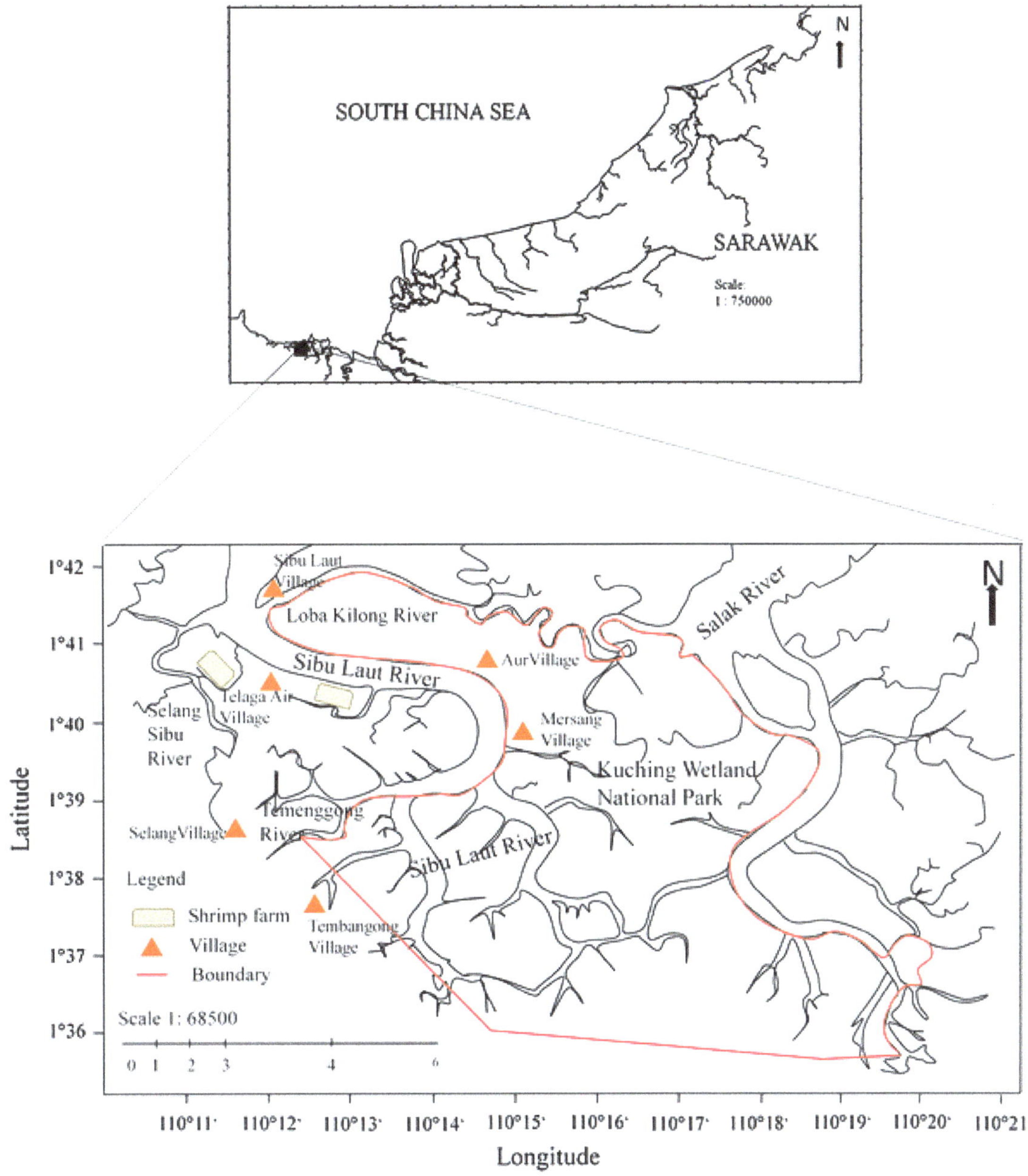

Figure. 1: Study area.

These stations are intended to monitor the possible impacts of the discharges of wastewater and the influences to the tributary and main river body. Stations 6, 7, 8, and 9 were along the main river body of Sibu Laut. Station 6 was chosen to observe the possible influences from the surrounding activities of the Telaga Air Town and the public jetty. Station 7 was close to another shrimp farm belonging to CCK Pelita Telaga Air Sdn. Bhd. Station 8 and 9 were near to the village of Aur and Mersang, respectively. Station 10 and 11 were at the right upstream while station 13 was at the left upper reach of the Sibu Laut River near to Tembangong Village. Station 12 was at the Temenggong River, which is the left upper tributary of the Sibu Laut River.

Table 1: Details of the sampling cruises

Sampling	Date	Tide	Water column	Parameters
1st trip	25.2.10	High	Near-bottom	General environmental parameters, nutrients, chl-*a*, BOD_5
2nd trip	25.3.10	High	Sub-surface	
3rd trip	29.4.10	Low	Near-bottom	
4th trip	18.5.10	Low	Sub-surface	
5th trip	16.6.11	Low	Sub-surface and Near-bottom	General environmental parameters, total heavy metals, and fecal coliform
6th trip	27.7.11	High	Sub-surface and Near-bottom	

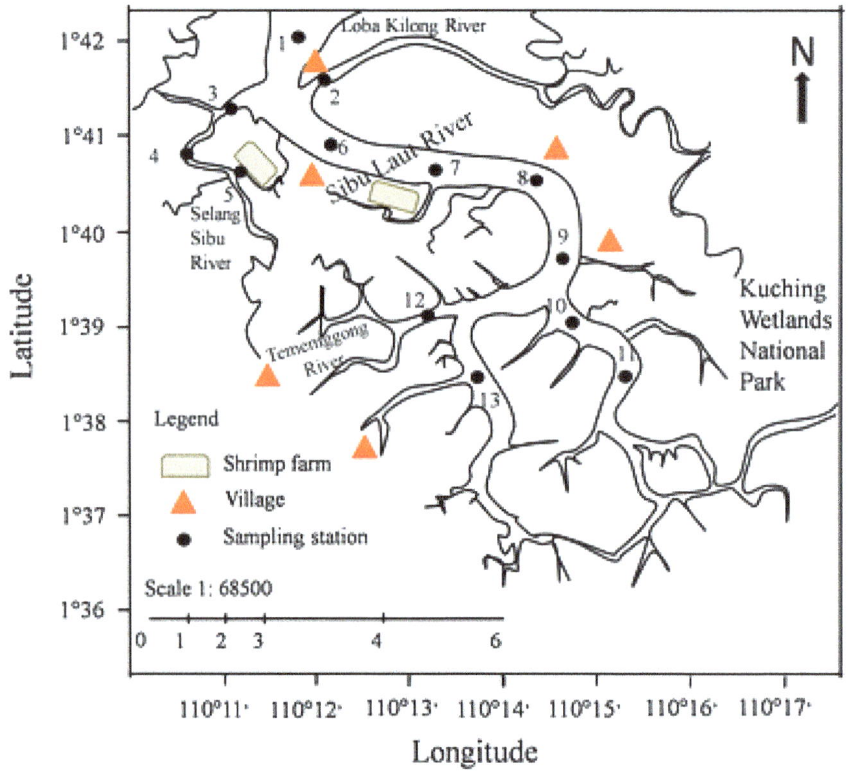

Figure. 2: Sampling stations along the Sibu Laut River and its principal tributaries.

Sampling Techniques

Portable Global Positioning System (GPSMAP® 76CSx, GARMIN) was used to determine the actual coordinates of the sampling stations and to re-confirm the location of the stations during subsequent sampling periods. Hydrological parameters including salinity, pH, temperature, dissolved oxygen (DO), turbidity, and transparency were measured using a refractometer (MR100ATC, Milwaukee), a pH meter with temperature probe (HI 8424, Hanna), a DO meter (HI 9142, Hanna), a turbidity meter (HI 93703, Hanna), and a secchi disc with measuring tape, respectively. Temperature, pH, DO, total dissolved solids (TDS), and conductivity were measured using a multi-parameter water quality meter (PCD650, Eutech) for samplings conducted in stage two samplings in the year 2011.

In stage one sampling for nutrients study, three replicate water samples were taken from the sub-surface and near-bottom water columns using a 2-l Wildco® Van Dorn water sampler. The water sample was taken in the middle of the river with approximately 1 m below the surface water and 1 m above the river bed after confirmed the river width and depth using a rangefinder (Elite 1500, Bushnell) and a depth sounder (PS-7, Hondex). Table 2 summarizes the river width and depth throughout the study period. The concentrations of chlorophyll-*a* (chl-*a*), total suspended solids (TSS), 5-day biochemical oxygen demand (BOD_5), total nitrogen (TN), ammonium nitrogen (NH_4^+-N), nitrite nitrogen (NO_2^--N), nitrate nitrogen (NO_3^--N), organic nitrogen (Org-N), total phosphorus (TP), inorganic phosphorus (IP), and organic phosphorus (OP) were studied. In the study of total heavy metals and bacteriological water quality, three replicate water samples were also taken from the sub-surface and near-bottom water columns for the determination of TSS, total heavy metals (Zn, Cu, Cd, and Pb) and fecal coliform (FC) bacteria. All sampling bottles were acid-washed, cleaned and dried before use. Sampling bottles for FC bacteria analysis were pre-sterilized with 70 % v/v ethanol before use. Water samples were placed in ice box and transported to the laboratory for immediate analysis (Parsons et al.1984; APHA 2005).

Table 2: Summary of the physical characteristic of Sibu Laut River pooled from the six samplings in year 2010 and year 2011

Station	Width (m)		Depth (m)	
	Low tide	High tide	Low tide	High tide
1	520.3 ± 67.0	557.3 ± 30.9	17.9 ± 1.5	22.0 ± 2.8
2	176.5 ± 26.3	184.0 ± 30.7	4.9 ± 0.9	7.0 ± 0.6
3	211.3 ± 30.3	241.0 ± 76.5	3.0 ± 1.2	5.3 ± 0.9
4	62.3 ± 9.6	78.9 ± 26.2	8.2 ± 1.7	8.1 ± 1.5
5	13.6 ± 3.6	17.4 ± 0.4	2.0 ± 0.5	3.3 ± 0.5
6	521.8 ± 27.3	517.8 ± 37.9	16.6 ± 0.7	16.7 ± 1.6

7	639.3 ± 20.6	661.6 ± 24.9	13.0 ± 0.6	14.3 ± 0.9
8	527.7 ± 34.1	545.9 ± 18.9	20.8 ± 1.4	23.5 ± 0.8
9	541.1 ± 12.4	549.4 ± 31.1	16.4 ± 1.2	18.3 ± 1.7
10	418.4 ± 5.5	414.1 ± 25.2	12.1 ± 3.4	11.8 ± 3.5
11	308.7 ± 17.8	336.4 ± 55.4	11.8 ± 0.6	11.3 ± 0.9
12	233.5 ± 79.1	240.9 ± 75.8	7.4 ± 1.9	9.0 ± 2.7
13	173.0 ± 113.5	176.8 ± 125.3	8.7 ± 2.6	8.1 ± 0.6

Analytical Techniques

All analyses were done according to standard procedures for water analysis in triplicate unless otherwise mentioned (Parsons et al. 1984; APHA 2005). Concentrations of chl-*a* were determined from adequate samples filtered through 0.7 μm Whatman GF/F filters and extracted for 24 h using 90 % (v/v) acetone. Five-day biochemical oxygen demand was determined as the difference between initial and 5-day dissolved oxygen content, after incubation of the sample in the dark at 20 °C. Total suspended solids were assayed as the difference between initial and final weight of the 0.7 μm Whatman GF/F filter paper, after filtration of an adequate sample volume and drying at 105 °C.

The filtered water samples were then used to determine nitrogen and phosphorus concentrations. Inorganic phosphorus was determined by the colorimetric ascorbic acid method while TP was determined as IP after potassium persulfate digestion of samples. Organic phosphorus was calculated as the difference between TP and IP. Ammonium nitrogen and NO_2^--N were determined by phenate method and colorimetric diazotization method, respectively. Nitrate nitrogen was determined by the ultraviolet spectrophotometric method while TN was determined as NO_3^--N after potassium persulfate digestion of the sample. Organic nitrogen was calculated as the difference between TN and inorganic nitrogen (NH_4^+-N, NO_2^--N, and NO_3^--N). A five-point calibration curve was constructed for each element. Blank and standard solutions were treated in a similar way as sample.

The unfiltered water samples, acidified to pH < 2 using ultrapure nitric acid (Merck) were used to determine total heavy metals (Zn, Pb, Cu, and Cd) concentrations. Total heavy metals were determined by Flame Atomic Absorption spectrophotometer (iCE 3000 Series AA, Thermo Fisher Scientific) after nitric acid digestion of samples. A three-point calibration curve was constructed for each element. Blank and standard solutions were treated in a similar way as sample. The detection limits for Zn, Pb, Cu, and Cd are 0.0033, 0.013, 0.0045, and 0.0028 mg l^{-1}, respectively.

A membrane filtration method was used to enumerate FC bacteria as colony-forming unit per 100 ml of sample, cfu/100 ml. Duplicate water samples were filtered through a 0.45 μm Advantec mixed cellulose ester membrane. Filters were then placed on the m-FC agar (Merck) without rosolic acids. Fecal coliform counts were performed as the number of blue colonies developed on m-FC agar plate after 24 h incubation at 44.5 °C.

Statistical Analysis

Hierarchical cluster analysis, HCA was used to investigate the grouping of the sampling stations along the Sibu Laut River and its tributaries. Ward's method using squared Euclidean distances as a measure of similarity was used. The dendrogram was used to interpret the result of the cluster analysis. Principal component analysis (PCA) was conducted to characterize the loadings of all physicochemical parameters for each of the PCs which were obtained. The PCA was carried out with factors having eigenvectors higher than one (Kaiser criterion). The quality of data for PCA analysis was confirmed with Kaiser–Meyer–Olkin (KMO) measure of sampling adequacy test. The rotation of PCs was made by varimax-normalized algorithm. All the statistical analyses were carried out by the Statistical Software for Social Sciences (SPSS Version 19.0, SPSS Inc. 1995).

RESULTS AND DISCUSSION

General Environmental Variables

Figure 3 summarizes the general environmental parameters along the Sibu Laut River and tributaries. All thirteen stations along Sibu Laut River and tributaries were determined to be tidally influenced based on the observed salinity values. Salinity was more than 20 ppt even at the most upstream of the river in the present study. Salinity can reach up to 33 ppt at stations near river mouth. Salinity gradient was evident along the river while near-bottom waters were more saline than the sub-surface waters, resulting in weak horizontal and vertical stratification. The mean upstream to downstream salinity difference measured was around 2–3 ppt while the mean sub-surface to near-bottom salinity difference measured was approximately 2 ppt. Malaysia is a tropic country where the temperature is uniform throughout the year (Malaysian Meteorological Department 2012). The temperature was relatively consistent (≈29.9 °C) along the river throughout the study periods. Generally, near-bottom waters were slightly cooler than the sub-surface waters, resulting in weak vertical stratification. This is probably attributable to the solar heating of sub-surface waters (Jones et al. 2008). The pH of the Sibu Laut River was well within range, which near to 7.8 on average. The pH value was increased towards downstream with the lowest pH value of 7.0 at the most upstream sites and up to pH value of 8.9 at the downstream. Near-bottom waters were generally more alkaline than the sub-surface waters. The pH value was mainly influenced by tides as pH was higher during high tide. It is known that DO concentrations above 5 mg l^{-1} are rarely harmful to aquatic life, whereas concentrations below 2 mg l^{-1}

are almost always detrimental and are considered a hypoxia threshold (Diaz and Rosenberg 1995). In sub-surface water samples collected in the present study, 62 % (n = 52) showed DO concentrations more than 5 mg l^{-1} while 71 % (n = 52) of near-bottom water samples contained more than 5 mg l^{-1} of the DO. Dissolved oxygen was generally well aerated along the river. However, lower concentration of DO was observed at station 2 in Loba Kilong River and stations 3–5 along Selang Sibu River in low tide condition.

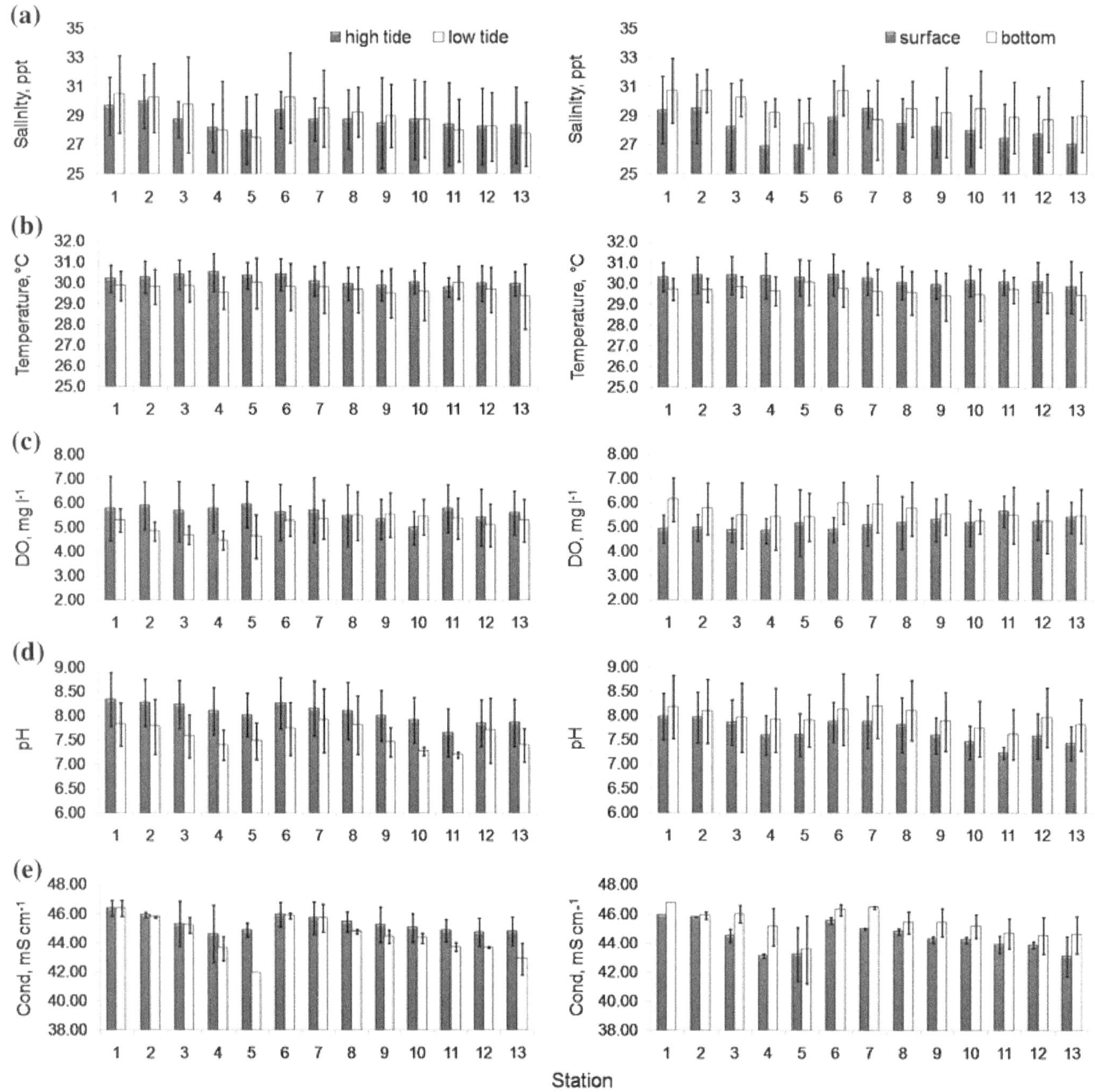

Figure. 3: Hydrological parameters of **a** salinity (ppt), **b** temperature (°C), **c** DO (mg l^{-1}), **d** pH, and**e** conductivity (mS cm^{-1}) at 13 sampling stations along Sibu Laut River and tributaries.

Total suspended solids and turbidity values were steadily increased from the upstream to downstream (Fig. 4). Besides the clear gradient along the river, TDS and turbidity in near-bottom water column were higher than the sub-surface water, most apparent at the downstream and the middle of the river. Total suspended solids and turbidity were also higher during low tide and the variations were larger at downstream and the middle of the main river but smaller at the upstream. Transparency was relatively high in the Sibu Laut River with the lowest transparency observed at station 5 (0.56 m) during low tide. The highest averaged transparency was observed at station 7 (>2 m). Transparency was higher during high tide at all stations and the variation was most apparent in the middle of the river and upstream. TDS and conductivity exhibited a similar increment trend towards downstream. The lowest value of TDS and conductivity was observed at station 5 during low tide and at station 4 during high tide. Generally, TDS and conductivity were higher during high tide and in near-bottom water column at all stations. TDS and conductivity concentrations are associated with ocean influence and surface water runoff from the shrimp farm effluent. As salinity increases, TDS and conductivity also increase. Shin et al.

(2012) demonstrated similar distribution of TDS but the authors attributed the low TDS concentration at upstream to surface water runoff and eliminated the possibility of ocean influence as their sampling was conducted at low tide.

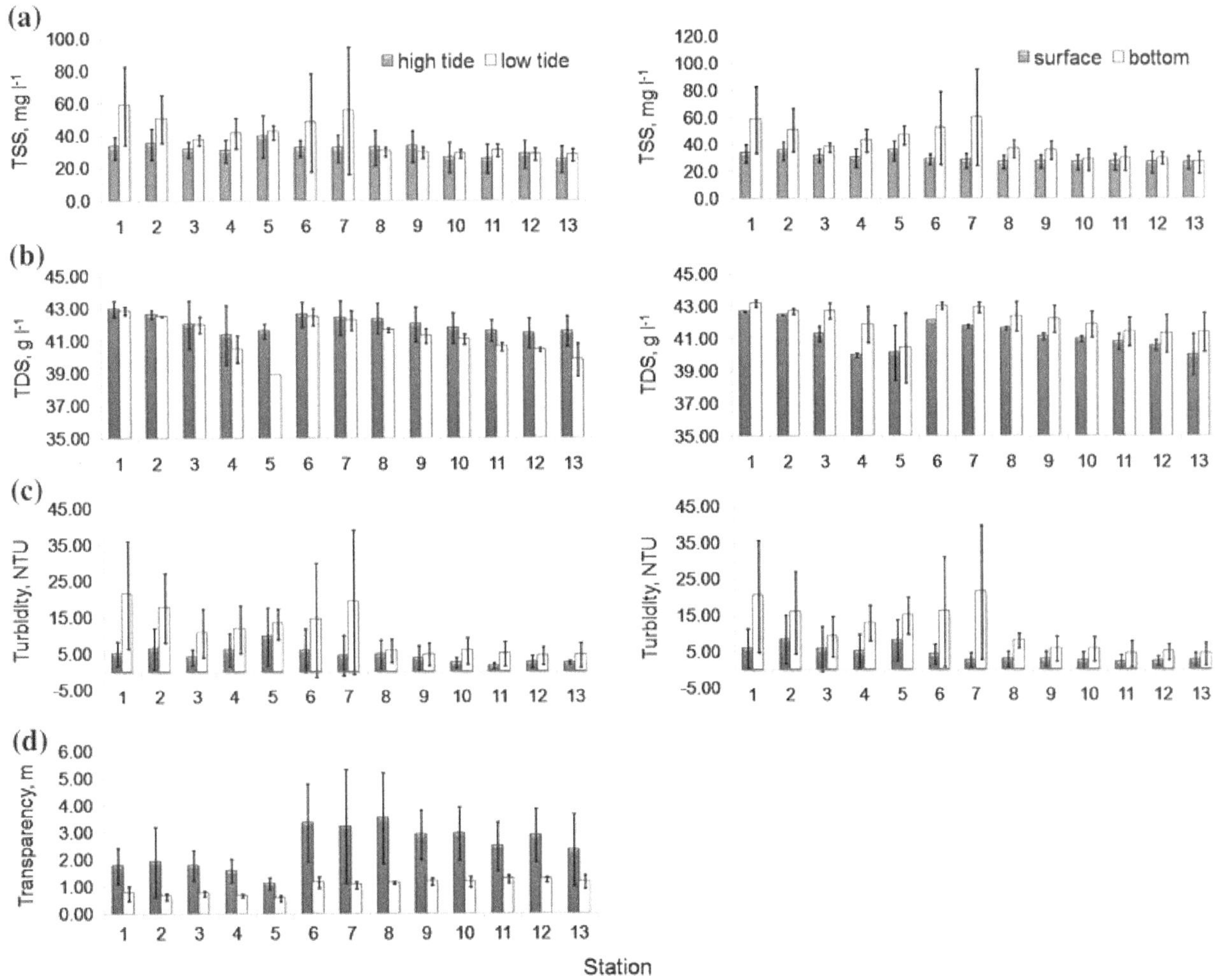

Figure. 4: Mean concentrations of **a** TSS (mg l^{-1}), **b** TDS (g l^{-1}), **c** turbidity (NTU), and **d**transparency (m) at 13 sampling stations along Sibu Laut River and tributaries.

Chlorophyll-*a* (chl-*a*)

Chlorophyll-*a* content of sub-surface waters samples collected in the present study ranged from 0.207 ± 0.095 to 6.144 ± 1.367 mg m^{-3}, with 50 % of the samples measured <1 mg m^{-3}, 42.3 % of the samples measured between 1 and 3 mg m^{-3}, and 7.7 % of the samples measured more than 5 mg m^{-3} (Fig. 5). Near-bottom chl-*a* ranged from 0.259 ± 0.097 to 9.680 ± 1.900 mg m^{-3}, with 69.2 % of the samples measured <1 mg m^{-3}, 26.9 % of the samples measured between 1 and 3 mg m^{-3}, and 3.9 % of the samples measured more than 5 mg m^{-3}. On average, the distribution of chl-*a* was generally low at all stations (0.637 ± 0.117 to 0.992 ± 0.487 mg m^{-3}) except station 5 which exhibited a peak of 5.921 ± 3.260 mg m^{-3}. The chl-*a* density at station 5 was about 9.5 times higher than the other stations during low tide, but only around 4.5 times higher than the other stations during high tide. Nonetheless, the high chl-*a* concentration rapidly decreased along the Selang Sibu River down to 0.919 ± 0.379 mg m^{-3} at the conjunction with Sibu Laut River. The observation was in good agreement with the study of Thomas et al. (2010) and Trott and Alongi (2000). The decrease of the chl-*a* may be due to the dilution or high grazing rate of zooplankton on phytoplankton as suggested by La Rosa et al. (2002). Chlorophyll-*a* data in the present study were comparable to the study conducted in Pearl River Estuary, South China as chl-*a* concentration in the estuary was 0.8–5.6 mg m^{-3} in ebb tide, and 0.8–7.8 mg m^{-3} in flood tide (Huang et al. 2003). The present results are also similar with the surface water chl-*a* concentration (0.28–8.29 mg m^{-3}) observed in the coastal waters of Kalpakkam in the study of Satpathy et al. (2010).

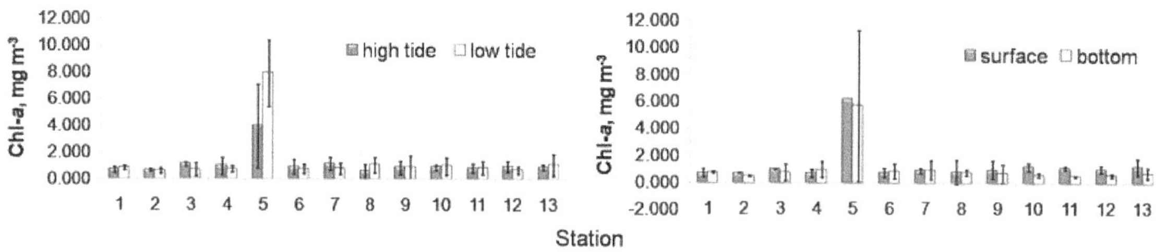

Figure. 5: Mean concentrations of chl-a (mg m^{-3}) at 13 sampling stations along Sibu Laut River and tributaries.

Generally, chl-*a* level less than 3.4 mg m^{-3} is indicative of the oligotrophic condition in tidal and fresh waters; ranging from 3.0 to 7.4 mg m^{-3} is indicative of mesotrophic condition, and chl-*a* >6.7 mg m^{-3} is indicative of eutrophic condition in the Chesapeake Bay (USEPA 2003). Molvaer et al. (1997) stated that for salinity above 20 ppt, chl-*a* concentration <2 mg m^{-3} is considered as Class I or 'very good'. On the other hand, Smith et al. (1999) equated oligotrophic, mesotrophic, eutrophic, and hypertrophic status in marine systems to mean chl-*a*concentrations <1 mg m^{-3}, ranging from 1 to 3 mg m^{-3}, ranging from 3 to 5 mg m^{-3}, and more than 5 mg m^{-3}, respectively. When compared to the concentrations of chl-*a* in Sibu Laut River with the above standard and guideline, the chl-*a* concentration in the Sibu Laut River can be defined as 'very good' as the chl-*a* value was lower than 2 mg m^{-3} except station 5 which was characterized as eutrophic or even hypertrophic status. Although chl-*a* concentration at station 5 showed signs of eutrophication, but the problem was localized as chl-*a* concentration was decreased rapidly along Selang Sibu River.

Five-day Biochemical Oxygen Demand (BOD_5)

Figure 6 shows that the sub-surface water of the Sibu Laut River and its tributaries was generally low in BOD_5, with 69 % (*n* = 26) of the samples contained <1 mg l^{-1}, 27 % (*n* = 26) of the samples ranged between 1 and 2 mg l^{-1}, and 4 % (*n* = 26) of the samples had BOD_5concentration more than 2 mg l^{-1}. In near-bottom water column, 42.3 % (*n* = 26) showed BOD_5concentration <1 mg l^{-1}, 42.3 % (*n* = 26) ranged between 1 and 2 mg l^{-1}, and the remaining 15.4 % (*n* = 26) of the samples had more than 2 mg l^{-1}. The highest concentration of BOD_5 was found at station 5 which in the receiving creek of the wastewater discharged from a shrimp farm. Besides, stations 2, 6, and 9 also contained higher BOD_5 values. The high BOD_5 values in those sites indicated that the organic matter being input to the river. Generally, BOD_5 was higher in near-bottom (≈1.34 mg l^{-1}) water column compared to sub-surface (≈0.81 mg l^{-1}); and during high tide (≈1.40 mg l^{-1}) compared to low tide (≈0.75 mg l^{-1}). The result was in contrast to the study of Yin and Harrison (2007) where the BOD_5 was about 1 mg l^{-1} near the bottom and more than 1.5 mg l^{-1} at the surface water at Victoria Harbour, respectively. Nevertheless, the authors also showed that bottom BOD_5 may be higher as phytoplankton cell sinks to the bottom layer.

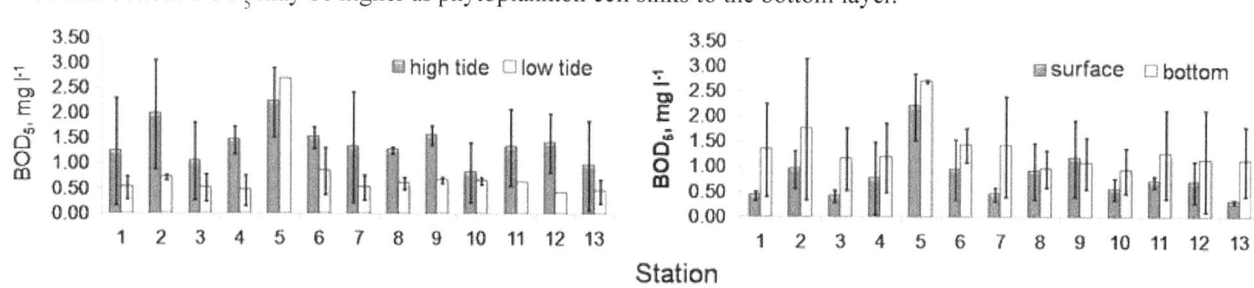

Figure. 6: Mean concentrations of BOD_5 (mg l^{-1}) at 13 sampling stations along Sibu Laut River and tributaries.,

Elevated concentrations of BOD_5 were found near to shrimp farm, villages and jetty in the present study, yet, the BOD_5 concentration was comparatively lower than other studies with sewage and anthropogenic influences. Shin et al. (2012) demonstrated that BOD_5 concentration was higher at upriver near combined sewer overflow and sewage treatment plant effluents in the Hackensack River Estuary with an average of BOD_5 fluctuated between 3.5 and 12.8 mg l^{-1}. Organic pollution is even more severe in Lower Sebou polluted by Gharb plain wastes with the means BOD_5 vary from 19 to 76 mg l^{-1} (Bennasser et al. 1997). Extremely high BOD_5concentration was also found in Chetumal Bay which registered a mean BOD_5 of 32.26 mg l^{-1}(Ortiz-Hernandez and Saenz-Morales 1999). Similar results with the present study was found in the study of Song et al. (2011) who reported that the BOD_5 concentration ranged from 0.807 to 7.557 mg l^{-1} near the industrial and urban area in a tributary of the Pearl Estuary—Beijiang River. In Malaysia, the mean values of BOD_5 were in the range of 0.26 to 2.95 mg l^{-1} in the Chini Lake (Hamzah and Hattasrul 2008) and from 2.36 to 15.37 mg l^{-1} in Langat River from 1995 to 2002 (Juahir et al. 2011).

Dissolved Phosphorus

Figure 7 shows the distribution of phosphorus along Sibu Laut River and tributaries. The highest TP concentration was observed at station 5 in the Selang Sibu River, which steadily reduced along the river from 0.047 ± 0.041 to 0.010 ± 0.009 mg l^{-1} at station 3. Station 2 which was located in Loba Kilong

River also showed elevated phosphorus concentration with an average of 0.022 ± 0.009 mg l^{-1}. Total phosphorus was lower in the middle and upstream sites, ranging from 0.007 ± 0.003 to 0.011 ± 0.006 mg l^{-1} except stations 6, 9, and 10 (0.014 ± 0.012 to 0.018 ± 0.013 mg l^{-1}). Organic phosphorus constituted the major proportion of the composition of phosphorus on the river basin, contributing 57.5–88.4 % of the TP. The distribution of OP followed the trend of TP where higher concentrations of OP were found at stations 2, 4, and 5. Inorganic phosphorus was generally low at all stations, contributing only 11.6–42.5 % of the TP. Nevertheless, station 5 which was near to the discharge point of the shrimp farm exhibited a peak of 0.014 ± 0.019 mg IP l^{-1}.

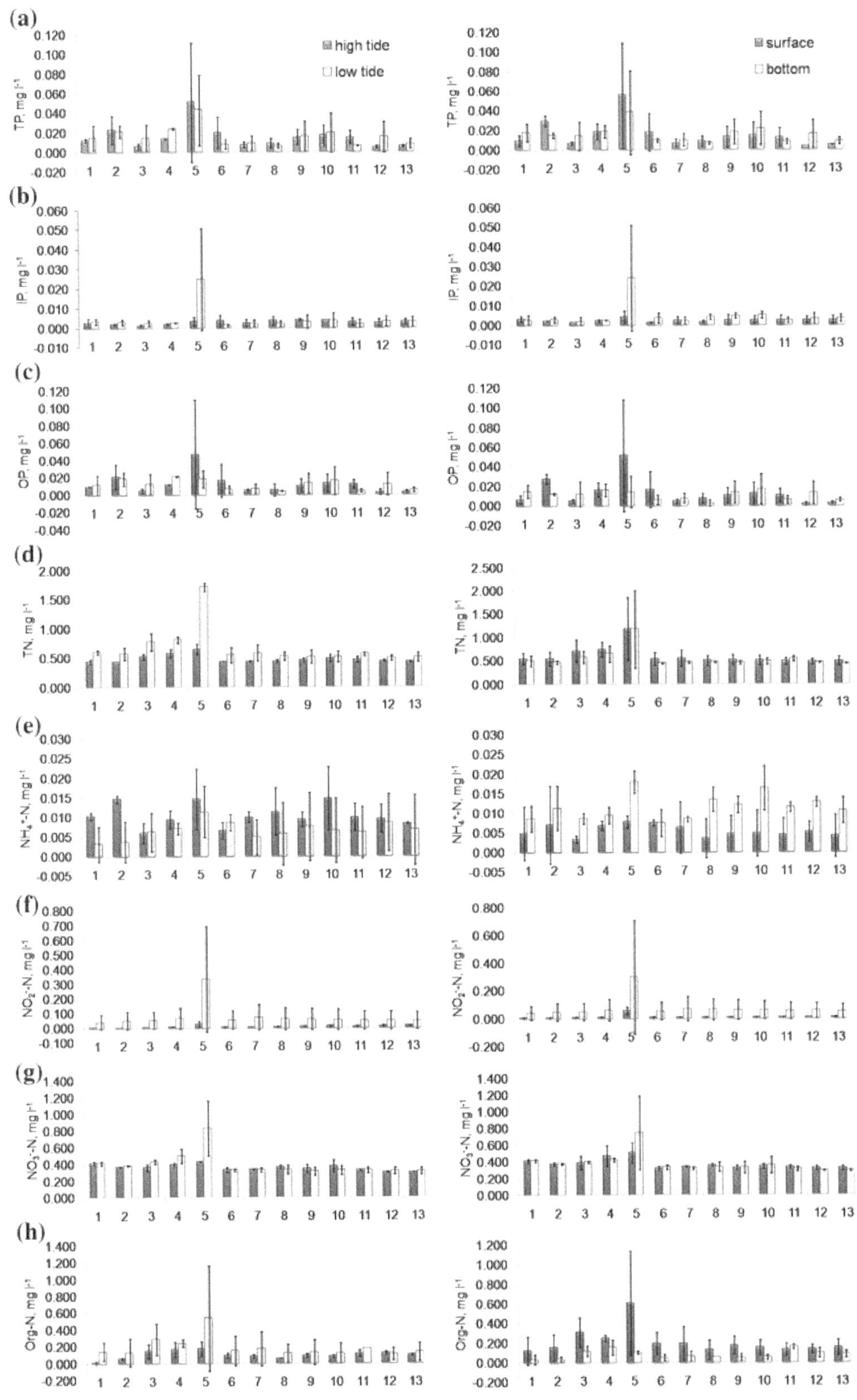

Figure. 7: Mean concentrations (mg l^{-1}) of **a** TP, **b** IP, **c** OP, **d** TN, **e** NH_4^+-N, **f** NO_2^--N, **g** NO_3^--N, and **h** Org-N at 13 sampling stations along Sibu Laut River and tributaries.

The Sibu Laut River contained extremely low concentrations of IP with an average of 0.003 mg l^{-1}. The result was comparable to the study of Troussellier et al. (2004) who reported the orthophosphate concentration was 0.03 μmol l^{-1} (≈0.001 mg l^{-1}) in the Senegal River Estuary. On the other hand, the concentration of IP in the present study was considerably lower than a Pearl River estuarine system of Hong Kong (Huang et al. 2003) and Kalpakkam Estuary in India (Satpathy et al. 2010) with an IP range of 0.007–0.017 mg l^{-1} and 0.05–1.96 μmol l^{-1} (≈0.002–0.061 mg l^{-1}), respectively.

Averages of phosphorus concentration exhibited horizontal gradients, which were most pronounced near shrimp farm discharge point on Selang Sibu River. The phosphorus concentrations decreased with distance from the shrimp farm discharge point from 0.047 ± 0.041 to 0.010 ± 0.009 mg l^{-1}, 0.014 ± 0.019 to 0.002 ± 0.001 mg l^{-1}, and 0.033 ± 0.040 to 0.009 ± 0.008 mg l^{-1} for TP, IP, and OP, respectively. This was in good agreement with results from Wu et al. (1994) where the impacts of marine fish farming activities on water quality were localized and did not appear to extend beyond a distance of 1–1.5 km from the fish rafts. Trott and Alongi (2000) suggested that the shrimp farm effluent was dissipated by tides and assimilated by the estuarine food web. Tidal mangrove estuaries have some capacity to process the intermittent inputs of pond-derived nutrients at least over short spatial and temporal scales. Thomas et al. (2010) and Ribeiro and Kjerfve (2002) also reported similar water quality improvement with distance. The dilution factor along Selang Sibu River was varied among phosphorus species. Inorganic phosphorus was dissipated faster than TP and OP. The dilution factor for IP (≈7) was relatively higher than TP (≈2.5) and OP (≈2.1) from station 5 to station 4. Inorganic phosphorus was rapidly assimilated within the 1 km distance from station 5 to station 4; but no further decrement was observed from station 4 to station 3. Total phosphorus and OP were continued to decrease from station 4 to station 3 with a lower rate approximately 1.8–1.9 times. It is hypothesized that though phytoplankton is able to assimilate various types of phosphorus, but is preferred IP rather than OP (Cotner and Wetzel1992), hence IP was decreased faster than OP.

In addition to phosphorus enrichment associated with shrimp farm effluents, Sibu Laut River is also susceptible to the domestic sewage effluent. This can be seen from the results that measurements of phosphorus revealed considerable enrichment at station 2 besides of station 5. In the present study, sub-surface TP and OP at station 2 were 2.9 times and 3.4 times higher than the other stations along Sibu Laut River; whereas IP concentration was similar to the other stations. The elevated phosphorus concentrations were confined to sub-surface water as near-bottom phosphorus at station 2 was similar or even lower than other stations. Since station 2 is near to the Sibu Laut Village, the high mean phosphorus concentration indicated that wastewater discharged from the village was likely to contribute to the high phosphorus observed as phosphorus is an important component of domestic detergents. Ling et al. (2010a) reported high loading of reactive phosphorus into rivers in Kuching, Sarawak, from household wastewater, especially low income residential area which are attributable to more usage of phosphorus detergent. Besides, as station 2 was in the Loba Kilong River near the downstream, the intrusion of the seawater also helps to dilute the phosphorus concentration in the near-bottom water column.

Higher concentrations of phosphorus in the sub-surface water of station 2 and station 5 revealed that the sub-surface phosphorus runoff from shrimp farm and domestic sewage effluent is entering the area. Nevertheless, higher near-bottom IP concentration was observed at station 5. Sedimentation and re-suspension may play a major role in the distribution of phosphorus in the near-bottom water column (Pomeroy et al. 1965). The high load of phosphorus from a shrimp farm effluent may sorb to particulate matter and settle on the sediment in the intake creek of shrimp farm effluents (Sondergaard et al. 2001). There are no noticeable trends of the vertical profiles of phosphorus at other stations.

A mean TP concentration <0.010 mg l^{-1} is characterized as oligotrophic status, ranging from 0.010 to 0.030 mg l^{-1} is mesotrophic status, ranging from 0.030 to 0.040 mg l^{-1} is eutrophic status, and more than 0.040 mg l^{-1} is hypertrophic status (Smith et al. 1999). Total phosphorus concentration is low in the Sibu Laut River as 48 % of the samples (n = 52) contained <0.010 mg l^{-1} and 42 % of the samples (n = 52) ranged from 0.010 to 0.030 mg l^{-1}. There are only 6 and 4 % of the water samples characterized as eutrophic and hypertrophic status. On the other hand, IP in all the samples collected in this study complied with the limit of 75 μg PO_4 l^{-1}(≈0.024 mg l^{-1}) according to the Malaysia Marine Water Quality Criteria and Standard except a water sample collected from station 5 during low tide (0.043 ± 0.001 mg l^{-1}).

Dissolved Nitrogen

The distribution of nitrogen was relatively consistent along Sibu Laut River and tributaries, ranging from 0.466 ± 0.063 to 0.697 ± 0.144 mg l^{-1} except station 5 which exhibited a peak of 1.177 ± 0.618 mg l^{-1}. In addition, concentrations of NH_4^+-N, NO_2^--N, NO_3^--N, and Org-N were also the highest at station 5 and steadily reduced along Selang Sibu River. Inorganic nitrogen (InOrg-N) constituted the major proportion of the composition of nitrogen in the river basin, contributing 75.6 % of TN. Nitrate nitrogen was the major composition among the InOrg-N, contributing about 67.0 %, followed by NO_2^--N (7.1 %), and NH_4^+-N (1.5 %). Ammonium nitrogen was low in the study area, ranging from 0.006 ± 0.003 to 0.013 ± 0.006 mg l^{-1}. Nitrite nitrogen was relatively consistent along the river, ranging from 0.024 ± 0.035 to 0.040 ± 0.062 mg l^{-1} when excluding station 5 (0.179 ± 0.273 mg l^{-1}). Concentration of NO_3^--N was found higher at the river mouth (0.415 ± 0.018 mg l^{-1}) while Org-N exhibited a different trend as concentrations were higher at the upper reaches of the river and lower at downstream sites.

The present study showed that the influence of the shrimp farm was reduced along Selang Sibu Laut as indicated by the decreasing nitrogen concentrations as similar to phosphorus distribution described previously. The nitrogen concentrations decreased with distance from the shrimp farm discharge point from 1.177 ± 0.618 to 0.647 ± 0.174 mg l^{-1} of TN, 0.013 ± 0.006 to 0.006 ± 0.003 mg l^{-1} of NH_4^+-N, 0.179 ± 0.273 to 0.030 ± 0.043 mg l^{-1} of NO_2^--N, 0.630 ± 0.298 to 0.396 ± 0.047 mg l^{-1} of NO_3^--N, and 0.355 ± 0.421 to 0.216 ± 0.143 mg l^{-1} of Org-N, respectively. The dilution factor from station 5 to station 4 was relatively similar among TN, NH_4^+-N, NO_3^--N, and Org-N, ranging from 1.40 to 1.76 times. Nitrite nitrogen was the fastest dissipated nitrogen species with a dilution factor of 4.84 times. A dilution factor of about 40 times for nitrogen components was reported from the head of channels to the Santos Bay (Braga et al. 2000). The authors attributed the dilution to assimilation, physical mixture and other removal processes that present in the system. Nitrogen was rapidly dissipated within the 1 km distance from station 5 to station 4; but no further apparent decrement was observed from station 4 to station 3 as the nitrogen was decreased at an extremely slow rate. It is hypothesized that NO_2^--N as an unstable intermediate form of NH_4^+-N and NO_3^--N will tend to reduce to NH_4^+-N or oxidize to NO_3^--N forms, hence the dilution factor of NO_2^--N was the highest. Organic nitrogen was the second fastest nitrogen species that dissipated other than NO_2^--N. Seitzinger and Sanders (1997) suggested that in estuaries with residence times on the order of weeks to months, such as Delaware Bay, river inputs of the biologically available portion of dissolved organic nitrogen (DON) are first utilized within the estuary. In contrast, in estuaries with residence times of less than a week, such as New York Bay, a portion of the biologically available DON may be utilized first within the estuary, with the remainder exported and utilized in continental shelf waters.

There was noticeable variation in the nitrogen concentration explained by differences in tidal stage. In general, the concentrations of TN, NO_2^--N, and Org-N were found higher during low tide in most observations. Higher NO_3^--N concentration was also observed at station 3 to station 5 on Selang Sibu River during low tide. A similar observation was reported by Ribeiro and Kjerfve (2002) where the Bay water quality was best at high tide at the time of maximum tidal dilution. On the other hand, NH_4^+-N exhibited different trends where higher concentration was observed during high tide at most of the stations. Marginally high sub-surface TN and Org-N concentrations were noticed at all stations except station 11. The largest variety of sub-surface and near-bottom Org-N was observed at station 5 indicated that the sub-surface water in the intake creek is influenced by the shrimp farm effluents rich in Org-N. Contrary to TN and Org-N concentrations, near-bottom NH_4^+-N and NO_2^--N concentrations were higher than sub-surface water. Sub-surface and near-bottom concentrations of NO_3^--N were relatively similar, except station 5 where near-bottom NO_3^--N was higher than the sub-surface. Accumulation of high amounts of nutrients in sediments was evident from the high concentration of the overlying water discharged from a shrimp farm (Shahidul et al. 2004). The presence of higher nitrogen in near-bottom water samples could be due to either re-suspension of fine sediments by currents or to the disturbance of the bottom during sampling (Balls et al.1994).

Smith et al. (1999) equated oligotrophic, mesotrophic, eutrophic, and hypertrophic status in marine systems to mean TN concentrations <0.26 mg l^{-1}, ranging from 0.260 to 0.350 mg l^{-1}, ranging from 0.350 to 0.400 mg l^{-1}, and more than 0.400 mg l^{-1}, respectively. Sibu Laut River and its tributaries are characterized as hypertrophic status as the whole system contained more than 0.400 mg TN l^{-1}. Comparisons of the results of the present study with the Malaysia Marine Water Quality Criteria and Standard (Department of Environment 2010) indicated that NH_4^+-N in Sibu Laut River complied with the guideline value set by the Department of Environment, Malaysia. All the NH_4^+-N concentration observed in the present study was lower than 70 µg $l^{-1}$of un-ionized ammonia (≈0.054 mg l^{-1} of un-ionized ammonia nitrogen). Nitrite nitrogen complied with the guideline value of 55 µg NO_2 l^{-1} (≈0.017 mg l^{-1}) at most of the stations. However, near-bottom NO_2^--N concentration in the Sibu Laut River exceeded the guideline value during low tide with NO_2^--N concentration ranged from 0.077 to 0.587 mg l^{-1}. On the other hand, NO_2^--N concentrations on three out of four water samples at station 5 were violated the guideline value. Nitrate nitrogen violated the guideline value of 60 µg NO_3 l^{-1}(≈0.014 mg l^{-1}) at all stations throughout the study periods. Nitrate is thermodynamically the most stable form of InOrg-N in well-oxygenated waters. The reference level of 0.014 mg l^{-1} for nitrate may be too low because nitrate is relatively non-toxic to aquatic organisms (Wickins1976; Rubin and Elmaraghy 1977). No formal water quality criteria and standard exist for nitrate in estuarine waters. The Canadian Council of Ministers of the Environment (2003) has recommended nitrate concentration ranging 2.9–3.6 mg l^{-1} to protect freshwater and marine life; whereas Camargo et al. (2005) proposed a maximum level of 2 mg l^{-1} for the protection of sensitive aquatic animals. What is more, the nitrate concentration for USA federal maximum level for drinking water is 10 mg l^{-1} (USEPA 2009). Accordingly, nitrate concentrations in the Sibu Laut River are still under the recommended value as nitrate concentration was below 2 mg l^{-1} even at station 5.

Total Heavy Metals (Zn, Pb, Cu, and Cd)

Figure 8 shows the distributions of total Zn, Pb, and Cu along Sibu Laut River and tributaries. The distribution of total Cd was not shown as the concentration of total Cd was below the detection limit at all stations throughout the study period. Lead appears to be the most abundant metal measured in the present study. The distribution of total Pb was relatively consistent along the river and tributaries when the data was pooled from the 13 stations during high and low tides, with an average of 0.0277 mg l^{-1}. Total Pb in this study was lower than the values quoted by Shin et al. (2012) in the study of total metal accumulation in the Hackensack River, Northern New Jersey (0.0349–0.2139 mg Pb l^{-1}) but was higher than the reported value for Potiguar Basin (≤0.002 mg l^{-1}) and Thames Estuary (≤0.008 mg l^{-1}) by Lacerda et al. (2013) and Pope and Langston (2011), respectively. Copper is the second most abundant metal in the river and is distributed geochemically in a similar manner to total Pb, with an average of 0.0160 mg l^{-1}. Total Cu concentration was within the range of Thames Estuaries (0.0097–0.2346 mg l^{-1}) and Hackensack River (0.0102–0.0363 mg l^{-1}) and higher than the reported values (≈0.002 mg l^{-1}) for Potiguar Basin. The concentration of total Zn was comparatively more fluctuated along Sibu Laut River and tributaries. The lowest concentration of total Zn was found in downstream (0.0024 mg l^{-1}) while the highest concentration of total Zn was observed at station 9 (0.0119 ± 0.0097 mg l^{-1}). Other than that, total Zn were relatively consistent on the main river with an average of 0.0079 mg l^{-1}. Total Zn was lower than the highest values reported in the Hackensack River (0.0312–0.0849 mg l^{-1}) and Thames Estuary (0.0330–0.0773 mg l^{-1}). Cadmium was the least metal measured in the present study at which it is below the detection limit at all stations throughout the study periods. Total Cd concentration is similar with Potiguar Basin, which is also below detection level and is lower than values reported in the Hackensack River (0.0042–0.0328 mg l^{-1}). Nduka and Orisakwe (2009) reported high level of Cd on the surface water of Aja-Etan and Ijala (≈1.33 mg l^{-1}) while high Pb was found on the surface water of Ifie-kporo and Ekpan (≈0.95 mg l^{-1}).

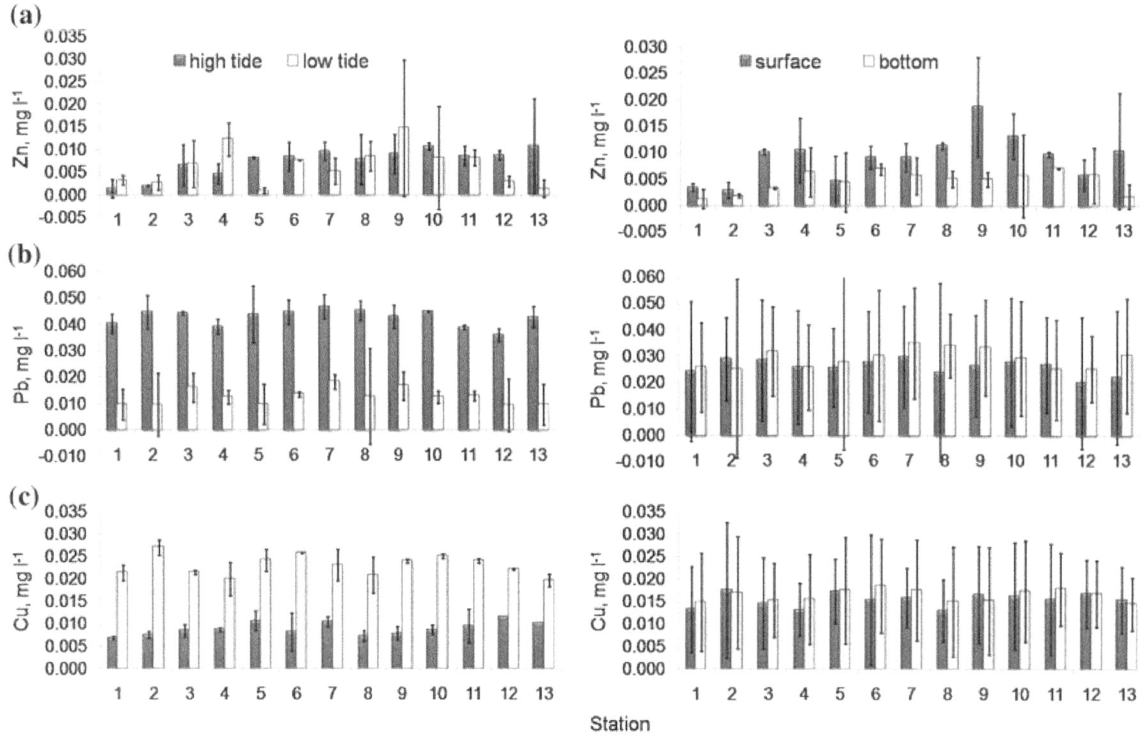

Figure. 8: Mean concentrations (mg l^{-1}) of **a** Zn, **b** Pb, and **c** Cu at 13 sampling stations along Sibu Laut River and tributaries.

Total heavy metals in the intake creek of shrimp farm effluent (station 5) were within ambient level with other stations in the Sibu Laut River. The results showed that the shrimp farm effluent had no major impact on heavy metals pollution in the receiving creek. Heavy metals are not major pollutants in shrimp farm effluents. In many cases, the study of heavy metal related to shrimp farm is more concerned on the accumulation of heavy metals in shrimps as it may adversely affect the health of the final consumption user (Amaraneni 2006; Kurun et al. 2007).

The main natural sources of metals into the aquatic system are the weathering of soils and rocks and precipitation; or from anthropogenic activities such as industrial, domestic, and urban sewage runoffs (Rippey 1982; Caccia and Millero 2003; Du Laing et al. 2008; Nduka and Orisakwe 2009; Sundaray et al. 2012). Elevated concentrations of total Zn were observed at station 9. These concentration peaks may represent events of re-suspension or the presence of Zn-enriched particles as suggested by Lacerda et al. (2013). In contrast, the total metal concentrations of Cu and Pb in the river showed no particular trend and this indicated that the source might be diffused. The source of total Pb on Sibu Laut River could be due to the anthropogenic activities from adjacent villages as domestic sewage discharge and seepage from waste disposal sites could contribute greatly to the Pb concentration (Simeonov et al. 2003; Fianko et al. 2007). Soil leaching is the main factor that contributes to the Cu concentration as suggested by Simeonov et al. (2003). Besides, Cu released from the antifouling paints used on boat hulls could also contribute to the Cu concentration (Caccia and Millero 2003; Whitall et al.2010).

Sub-surface Zn were generally higher than near-bottom water, whereas Pb and Cu were not notably different between sub-surface and near-bottom water column. Lourino-Cabana et al. (2011) attributed the highly variable of Zn concentration in the Deûle River to re-suspension of polluted sediment and biological activities. In contrary, metal concentrations in surface and bottom waters in Potiguar Basin, Northeastern Brazil under the influence of the Guamare outfalls showed no difference, suggesting completing mixing of the water column in this relatively (<20 m) shallow waters (Lacerda et al. 2013). On the other hand, the present study showed that Zn and Pb were higher during high tide while Cu was considerably higher during low tide. Hatje (2003) demonstrated that concentrations of particulate trace metals, in general, tended to be higher at flood than ebb tide in Port Jackson Estuary, Australia. The higher concentration of Cu during low tide in the present study might indicate the contribution of Cu from the upstream sites.

No formal water quality criteria exist for total heavy metals in brackish waters. Department of Environment in Malaysia has constructed a system for the criteria and standard for metals in marine waters but was limited to dissolved metals. Nevertheless, the dissolved Zn, Pb, Cu, and Cd complied with the guideline value set by the Department of Environment, Malaysia as all the dissolved metals measured in the present study were below detection limits (data not shown).

Bacteriological water quality (fecal coliform bacteria)

Figure 9 shows that, on average, FC was lower in the main river body ranging from 1 ± 1 cfu/100 ml to 8 ± 9 cfu/100 ml. The highest value of FC was observed at station 9, followed by station 12. Station 2 at Loba Kilong River also contained relatively higher FC with a mean of 8 ± 7 cfu/100 ml. Fecal coliform was found high, but steadily decreased along Selang Sibu River from 20 ± 24 cfu/100 ml at station 5 to 13 ± 15 cfu/100 ml at station 3. Changes in concentrations in the downstream direction reflect the attenuation of FC through sedimentation, dilution, and die-off. The results were consistent with Ortiz-Hernandez and Saenz-Morales (1999) who demonstrated that the densities of FC were the highest in discharge samples and decreased with increasing distance from the discharge.

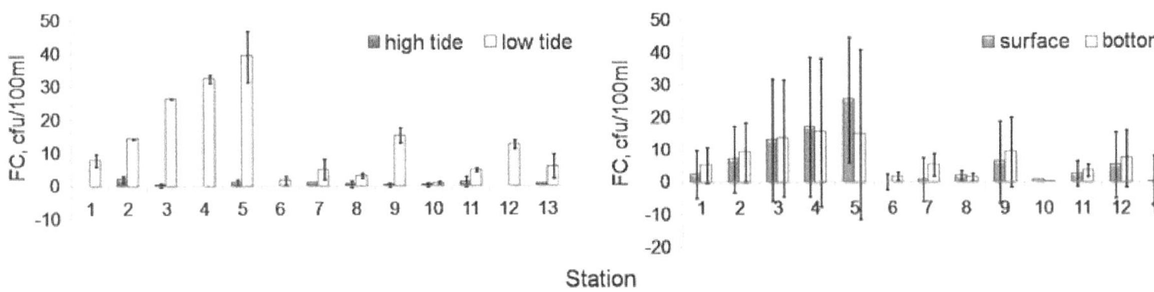

Figure. 9: Mean concentrations of FC (cfu/100 ml) at 13 sampling stations along Sibu Laut River and tributaries.

At low tide condition, FC was found high in Selang Sibu River especially at station 5 which is near to the discharge point of shrimp farm as illustrated in Fig. 8. Fecal coliform was steadily decreased from 39 ± 8 cfu/100 ml at station 5 to 26 ± 0 cfu/100 ml at station 3 in the sub-surface water column. Besides, FC was also found high at stations 2, 9, and 12 in both sub-surface and near-bottom water columns with an average of 14 cfu/100 ml. In addition, FC was found higher in the near-bottom water column at stations 1, 10, and 13 with an average of 10 cfu/100 ml. Nonetheless, FC was low at the rest of the stations during low tide, ranging from BDL to 6 ± 4 cfu/100 ml. Contrary to low tide condition, FC was generally low during high tide as most of the stations contained <2 cfu/100 ml on average. The highest value of FC during high tide was found in near-bottom water column at station 2. Fecal coliform was not notably different between sub-surface and near-bottom water columns. Nevertheless, FC was drastically higher during low tide at most of the stations. A similar observation was described by Mallin et al. (1999) who noted an increase in the abundance of FC at or near low tide in tidal creeks that received polluted water from feeder creeks in North Carolina. Bordalo (2003) attributed the dramatic improvement in microbiological water quality to the increase in salinity during high tide in coastal beaches of Portugal.

Occasionally, slightly elevated concentration of FC (>10 cfu/100 ml) was found at stations 1, 2, 9, and 12 in the present study. Fecal coliform is typically found in fecal material originating from human or other warm-blooded animals. One of the most common FC sources is the non-point sources and uncontrolled discharge of sewage from densely populated urban and touristic areas (Aslan-Yilmaz et al. 2004; Ling et al. 2010b). Station 2 and station 12 received untreated waste discharge from Sibu Laut and Temenggong Village. On top of that, station 1 at downstream received polluted water from Selang Sibu River and Loba Kilong River that contained shrimp farm discharge and domestic waste. It can be concluded that the untreated sewage discharge from the adjacent villages may cause the higher counts of FC at the area. However, those villages are not large enough and generate large enough sewage to contaminate the waters in a noteworthy way when compared to other studies. Mallin et al. (2009) demonstrated that frequent excessive FC concentrations were found particularly at some of the most urbanized sites in Burnt Mill Creek (955 cfu/100 ml) while periodic high counts occurred in some rural (536 cfu/100 ml) and suburban (493 cfu/100 ml) locations. Besides, the geometric mean concentration of FC bacteria was 790 cfu/100 ml in Tama River, an urban river flowing through densely populated areas of Tokyo (Ham et al. 2009). Even more, Troussellier et al. (2004) delineated that high TTC abundances (33,000 cfu/100 ml) were recorded in oceanic water near the large village of Guet N'dar where around 15000 fishermen live with their family.

The present mean FC value was comparatively lower when comparing to the FC abundances reported in other riverine, estuarine or coastal areas; the Sibu Laut River is far from being among the polluted sites. The mean concentration of FC bacteria in Sibu Laut River was complied with water quality standards established for FC (<100 cfu/100 ml) by the World Health Organization (WHO 1998) or the European microbiological standards for bathing waters (Barrell et al. 2000). In fact, despite the influence of the shrimp farm wastewater, the contamination level of the Selang Sibu River appeared not so high; with the highest observed value of 50 ± 12 cfu/100 ml in sub-surface water at station 5 during low tide.

Hierarchical Cluster Analysis

Hierarchical cluster analysis was applied to detect similarities between the sampling stations along the Sibu Laut River and its tributaries using the sub-surface and near-bottom water quality data collected during low and high tide in 2010 and 2011, respectively. A total of 19 variables and 14 variables collected in 2010 and 2011 were first subjected to HCA, respectively. The 19 variables include the general environmental parameters, nutrients, chl-*a*, TSS, and BOD_5 collected in 2010 while the 14 variables include the general environmental parameters, total metals, and fecal coliform bacteria collected in 2011. The dendrograms showing the cluster of the sampling stations along the Sibu Laut River and its tributaries are presented in Fig. 10. The result showed that water quality was grouped into three and two groups along the Sibu Laut River and its tributaries in 2010 and 2011, respectively.

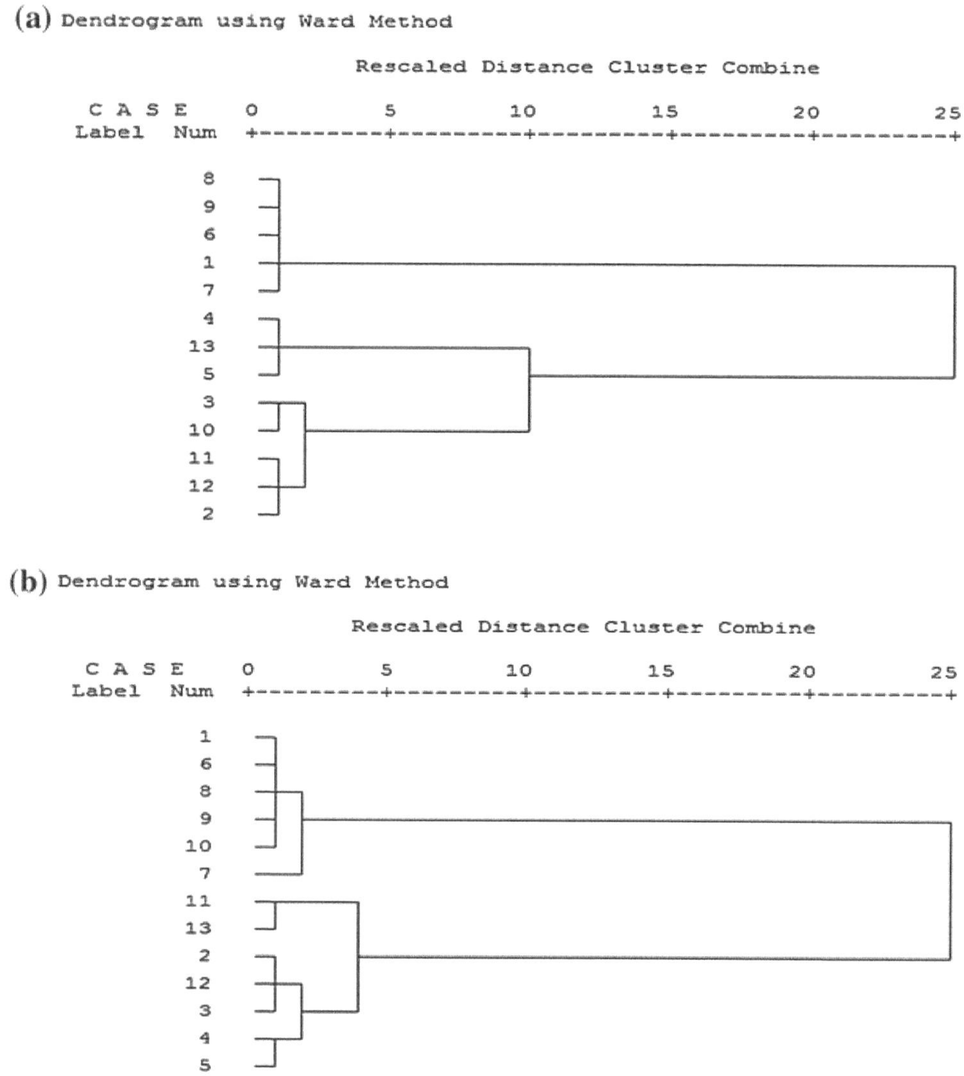

Figure. 10: Cluster of sampling stations along the Sibu Laut River and its tributaries based on the water quality data collected in **a** 2010 and **b** 2011, respectively.

In 2010, cluster 1 of the water quality consisted of stations 8, 9, 6, 1, and 7 which were all located at main river. Station 4 and station 5 which were located near to a shrimp farm at Selang Sibu River; and station 13 which was located at the left upper reach of the Sibu Laut River near to Tembangong Village were grouped into cluster 2. Lastly, cluster 3 consisted of sampling stations 3, 10, 11, 12, and 2, which are mainly located on the tributaries of the Sibu Laut River. On the other hand, water quality collected in 2011 was grouped into two major clusters which were the main stem of the Sibu Laut River and the tributaries of the river. The cluster analysis further confirmed that the effluents from the shrimp farm near Selang Sibu River consisted relatively higher nutrients compared to total metals and fecal coliform bacteria; hence, the stations near the shrimp farm were dissimilar with the other stations located on the main stem and other tributaries.

Principal Component Analysis

Principal component analysis was conducted to characterize the loadings of 19 and 14 variables in the water quality of the Sibu Laut River and its tributaries in 2010 and 2011, respectively. The KMO measure of sampling adequacy obtained was 0.445 and 0.760 for water samples in 2010 and 2011, respectively. Six PCs were obtained for the water quality in 2010 (Table 3); and three PCs were obtained for the water quality in 2011 (Table 4), with eigenvalues larger than 1, explaining 86.045 and 75.544 % of the total variance, respectively.

Table 3: Loadings of environmental variables on the first six varimax-rotated PCs for water quality data collected from Sibu Laut River and its tributaries in 2010

Rotated component matrix[a]						
Parameter	**Component**					
	1	2	3	4	5	6
IP	0.937					

NO_3^--N	0.878					
NO_2^--N	0.840					
Chl-a	0.818					0.365
TN	0.811		−0.428			
pH		0.900				
DO		0.822				
BOD_5	0.392	0.767				
Temperature		−0.763	−0.435			
Org-N			−0.819			
NH_4^+-N		0.348	0.727			
Salinity			0.637		0.434	
Width				0.907		
Depth				0.905		
Transparency		0.412		0.451	−0.416	
TSS					0.934	
Turbidity					0.835	
OP						0.976
TP	0.402					0.883
Initial eigen-value	5.814	3.517	2.593	1.782	1.497	1.145
% of variance	30.598	18.513	13.647	9.381	7.878	6.028
Cumulative %	30.598	49.111	62.758	72.139	80.017	86.045

[a]Rotation converged in 7 iterations

Table 4: Loadings of environmental variables on the first three varimax-rotated PCs for water quality data collected from Sibu Laut River and its tributaries in 2011

Rotated component matrix[a]			
Parameter	**Component**		
	1	**2**	**3**
Total Cu	−0.941		
Total Pb	0.929		
Temperature	0.923		
Transparency	0.808		0.332
Turbidity	−0.658	0.639	
FC	−0.575		−0.546
Salinity		0.802	
Conductivity	0.365	0.780	0.398
TDS	0.380	0.766	0.403
TSS	−0.326	0.750	
pH	0.565	0.592	
Total Zn		−0.522	0.305
Width			0.931
Depth			0.875
DO		0.396	0.578
Initial eigen-value	5.654	3.828	1.850
% of variance	37.696	25.518	12.330
Cumulative %	37.696	63.213	75.544

[a]Rotation converged in 5 iterations

The first PC for water quality in 2010 has high positive loadings for the IP, NO_3^--N, NO_2^--N, chl-*a*, TN, BOD_5, and TP, accounted for 30.598 % of total variance. The second PC accounted for 18.513 % of the total variance and positively loaded with pH, DO, BOD_5, NH_4^+-N, and transparency; but negatively loaded with temperature. The third PC which contributed 13.647 % of the total variance showed positive loadings on NH_4^+-N and salinity, and negative loadings on TN, Org-N, and temperature. Positive loading of width, depth, and transparency was observed in PC 4 which accounted for 9.381 % of total variance. The PC 5 and PC 6 contributed 7.878 and 6.028 % of the total variance, respectively. The PC 5 was found to be positively loaded with salinity, TSS, and turbidity; and negatively loaded with transparency. Positive loading was observed in chl-*a*, TP, and OP in the last PC.

In the present study, PCA showed positive loadings on nutrients, phytoplankton, and organic matter indicated that those enrichments on the Sibu Laut River were derived from the same sources which were mainly regulated by the shrimp farm effluent and riverine runoff of domestic sewage. The positive loading of chlorophyll and nutrients indicated that high chl-*a*concentration may be due to the availability of higher concentration of nutrients which favor their growth. The BOD_5 and NH_4^+-N were positively loaded in PC 2 indicated that they are possibly derived from the same sources particulary the shrimp farm. The oxidation of NH_4^+-N contributes to high BOD_5, and the process is strongly regulated by pH, DO, and temperature as BOD_5 and NH_4^+-N are positively loaded with pH and DO, but negatively loaded with temperature. Positive loading on salinity and negative loadings of TN and Org-N were observed in the present study. In other word, TN and Org-N concentrations decreased as salinity increased which may be the result of mixing those nutrients with high salinity coastal water as depicted by Whitall et al. (2010). On the other hand, the positive loadings on salinity, TSS, and turbidity indicated that suspended solids were high in the downstream due to the movement of water in and out of the estuary by tides, causing fine particles to mix and be stirred up.

In 2011, PC 1 accounted for 37.696 % of the total variance, showing positive loadings on total Pb, temperature, transparency, conductivity, TDS, and pH; but negatively loaded with total Cu, turbidity, FC, and TSS. Positive loadings on turbidity, salinity, conductivity, TDS, TSS, pH, and DO; but negatively loaded with total Zn was observed on PC 2 which contributed 25.518 % of the total variance. The contribution of PC 3 to the total variance was 12.330 % and positively loaded with transparency, conductivity, TDS, total Zn, depth, width and DO; but has a negative loading on FC.

Positive loadings on pH, salinity, conductivity, TDS, and negative loadings on total Cu and total Zn were observed in the PCA. This phenomenon indicated that the most important sources of these metals could be the riverine and estuarine discharge, or the anthropogenic sources in the upstream. The decrease in metals concentration is probably the result of the dilution by seawater (Norton et al. 1984) or the precipitation of metals in the tidal area with higher pH (Grande et al. 1999). Positive loadings on temperature, conductivity, TDS, total Pb, and total Zn; and negative loading of FC were observed in the PCA. This agrees well with the previous studies that increase in temperature (Chigbu et al. 2004; Noble et al. 2004), salinity (Hamzah et al. 2011; Yayintaş et al. 2007), and heavy metals (Malik and Ahmad 2002; Jana and Bhattacharya 1988) were harmful to FC survival in natural environments. Besides, the negative loadings on FC, turbidity and TSS are not surprising. Transport of FC via re-suspension of the sediments through the storm or tidal actions into the overlying water column is an important mechanism in the aquatic environment (Brauwere et al. 2011; Ouattara et al. 2011) as FC has longer survival in association with sediment particles (Ishii et al. 2007; Lee et al. 2006).

CONCLUSIONS

Sibu Laut River is a highly dynamic, tidally dominated and well-oxygenated river with relatively consistent temperature. The pH value of the water is well within range and demonstrates an increasing trend from upstream to downstream. Total suspended solids and turbidity are low in sub-surface waters; show increasing values from upstream to downstream, but may be high at the intersection between freshwater and seawater. TDS and conductivity concentrations show changes with river miles, with higher concentration were found in near-bottom water column and in high tide condition. The spatial distribution pattern of the parameters studied indicates substantial inputs of contaminants possibly from shrimp farms as well as domestic sewage from villages nearby Sibu Laut River. The higher elemental concentrations which included chl-*a*, BOD_5, nitrogen, phosphorus, and FC were associated with the stations near the intake creek of shrimp farm effluent in Selang Sibu River, and sewage discharge points of the villages of Sibu Laut, Telaga Air, and Temenggong. Nevertheless, the shrimp farm effluent had no major influence on heavy metals pollution. Sibu Laut River was characterized as oligotrophic and/or mesotrophic status according to chl-*a* and TP inputs, but was characterized as hypertrophic status according to the TN input. Nitrate concentrations in the Sibu Laut River violated the national standard, while occasional violation of NO_2^--N concentration was observed in near-bottom water column during low tide. Other water quality criteria in the Sibu Laut River are well within range and complied with the international and national standard.

ACKNOWLEDGMENTS

The authors appreciate the financial support provided by the Malaysian Ministry of Science, Technology and Innovation (MOSTI) through E-Science Grant; 06-01-09-SF0026, and Ministry of Higher Education (MOHE) through FRGS Grant; FRGS/07(02)/749/2010(35), and the facilities provided by Universiti Malaysia Sarawak.

REFERENCES

1. Amaraneni SR (2006) Distribution of pesticides, PAHs and heavy metals in prawn ponds near Kolleru Lake Wetland, India. Environ Int 32(3):294–302
2. APHA (2005) Standard methods for the examination of water and wastewater, 21st edn. APHA, AWWA, and WEF, USA
3. Aslan-Yilmaz A, Okus E, Ovez S (2004) Bacteriological indicators of anthropogenic impact prior to and during the recovery of water quality in an extremely polluted estuary, Golden Horn, Turkey. Mar Pollut Bull 49(11–

12):951–958

4. Balls PW, Laslett RE, Price NB (1994) Nutrient and trace metal distributions over a complete semi-diurnal tidal cycle in the Forth Estuary, Scotland. Neth J Sea Res 33(1):1–17

5. Barrell RA, Hunter PR, Nichols G (2000) Microbiological standards for water and their relationship to health risk. Commun Dis Public Health 3(1):8–13

6. Bennasser L, Fekhaoui M, Benoit-Guyod JL, Merlin G (1997) Influence of tide on water quality of lower Sebou polluted by Gharb plain wastes (Morocco). Water Res 31(4):859–867

7. Bonadonna L, Briancesco R, Coccia AM, Semproni M, Stewardson D (2002) Occurrence of potential bacterial pathogens in coastal areas of the Adriatic Sea. Environ Monit Assess 77:31–49

8. Bordalo AA (2003) Microbiological water quality in urban coastal beaches: the influence of water dynamics and optimization of the sampling strategy. Water Res 37(13):3233–3241

9. Braga ES, Bonetti CVDH, Burone L, Filho JB (2000) Eutrophication and bacterial pollution caused by industrial and domestic wastes at the Baixada Santista estuarine system—Brazil. Mar Pollut Bull 40(2):165–173

10. Bricker SB, Clement CG, DE Pirhalla, Orlando SP, Farrow DRG (1999) National Estuarine Eutrophication Assessment: effects of nutrient enrichment in the Nation's Estuaries. NOAA, National Ocean Service, Special Projects Office and the National Centers for Coastal Ocean Science. Silver Spring, MD

11. Burgos MG, Rainbow PS (2001) Availability of cadmium and zinc from sewage sludge to the flounder, *Platichthys flesus*, via a marine food chain. Mar Environ Res 51(5):417–439

12. Cabral JP, Marques C (2006) Faecal coliform bacteria in Febros River (Northwest Portugal): temporal variation, correlation with water parameters, and species identification. Environ Monit Assess 118(1–3):21–36

13. Caccia VG, Millero FJ (2003) The distribution and seasonal variation of dissolved trace metals in Florida Bay and adjacent waters. Aquat Geochem 9(1):111–144

14. Camargo JA, Alonso A, Salamanca A (2005) Nitrate toxicity to aquatic animals: a review with new data for freshwater invertebrates. Chemosphere 58:1255–1267

15. Canadian Council of Ministers of the Environment (2003) Canadian Water Quality Guidelines for the Protection of Aquatic Life: Nitrate Ion. CCMI, Winnipeg

16. Chigbu P, Gordon S, Strange T (2004) Influence of inter-annual variations in climatic factors on fecal coliform levels in Mississippi Sound. Water Res 38(20):4341–4352

17. Cotner JBJ, Wetzel RG (1992) Uptake of dissolved inorganic and organic phosphorus compounds by phytoplankton and bacterioplankton. Limnol Oceanogr 37(2):232–243

18. Davis JR, Koop K (2006) Eutrophication in Australian rivers, reservoirs and estuaries—a southern hemisphere perspective on the science and its implications. Hydrobiologia 559(1):23–76

19. De Brauwere A, De Brye B, Servais P, Passerat J, Deleersnijder E (2011) Modelling *Escherichia coli* concentrations in the tidal Scheldt River and estuary. Water Res 45(9):2724–2738

20. Del-Pilar-Ruso Y, De-la-Ossa-Carretero JA, Loya-Fernández A, Ferrero-Vicente LM, Giménez-Casalduero F, Sánchez-Lizaso JL (2009) Assessment of soft-bottom polychaeta assemblage affected by a spatial confluence of impacts: sewage and brine discharges. Mar Pollut Bull 58(5):776–782

21. Department of Environment (2010) Malaysia marine water quality criteria and standard. DOE, Malaysia. http://www.doe.gov.my/portal/water-marine-marine-island-marine-water/malaysia-interim-marine-water-quality-standard/ Accessed 20 Oct 2012

22. Diaz RJ, Rosenberg R (1995) Marine benthic hypoxia: a review of its ecological effects and the behavioural responses of benthic macrofauna. Oceanogr Mar Biol 33:245–303

23. Djuikom E, Njine T, Nola M, Sikati V, Jugnia LB (2006) Microbiological water quality of the Mfoundi River watershed at Yaoundé, Cameroon, as inferred from indicator bacteria of fecal contamination. Environ Monit Assess 122(1–3):171–183

24. Dorsey JH, Carmona-Galindo VD, Leary C, Huh J, Valdez J (2012) An assessment of fecal indicator and other bacteria from an urbanized coastal lagoon in the City of Los Angeles, California, USA. Environ Monit Assess 185(3):2647–2669

25. Du Laing G, Moortel AVD, Lesage E, Tack FMG, Verloo MG (2008) Chapter 11: factors affecting metal accumulation, mobility and availability in intertidal wetlands of the Scheldt Estuary (Belgium). In: Vymazal J (ed) Wastewater treatment, plant dynamics and management. Springer Science + Business Media B.V, Netherlands, pp 121–133

26. Fianko JR, Osae S, Adomako D, Adotey DK, Serfor-Armah Y (2007) Assessment of heavy metal pollution of the Iture Estuary in the central region of Ghana. Environ Monit Assess 131(1–3):467–473

27. Grande JA, Borrego J, Morales JA (1999) A study of heavy metal pollution in the Tinto-Odiel Estuary in Southwestern Spain using factor analysis. Environ Geol 39(10):1095–1101

28. Hadibarata T, Abdullah F, Yusoff ARM, Ismail R, Azman S, Adnan N (2012) Correlation study between land use, water quality, and heavy metals (Cd, Pb, and Zn) content in water and green lipped mussels *Perna viridis* (Linnaeus.) at the Johor Strait. Water Air Soil Poll 223(6):3125–3136

29. Ham YS, Kobori H, Takasago M (2009) Effects of combined sewer overflow and stormwater on indicator bacteria concentrations in the Tama River due to the high population density of Tokyo metropolitan area. Environ Monit Assess 152(1–4):459–468

30. Hamzah A, Hattasrul Y (2008) Water quality and bacterial study in Tasik Chini, Pahang. In: Sengupta M, Dalwani

R (eds) Proceedings of Taal 2007: The 12th World Lake Conference. ILEC, Japan, pp 184–189

31. Hamzah A, Kipli SH, Ismail SR, Una R, Sarmani S (2011) Microbiological study in coastal water of Port Dickson, Malaysia. Sains Malays 40(2):93–99
32. Hatje V (2003) Particulate trace metal and major element distributions over consecutive tidal cycles in Port Jackson Estuary, Australia. Environ Geol 44:231–239
33. Ho AYT, Xu J, Yin K, Yuan X, He L, Jiang Y, Lee JHW et al (2008) Seasonal and spatial dynamics of nutrients and phytoplankton biomass in Victoria Harbour and its vicinity before and after sewage abatement. Mar Pollut Bull 57(6–12):313–324
34. Howarth RW, Marino R (2006) Nitrogen as the limiting nutrient for eutrophication in coastal marine ecosystems: evolving views over three decades. Limnol Oceanogr 51(1, part 2):364–376
35. Huang XP, Huang LM, Yue WZ (2003) The characteristics of nutrients and eutrophication in the Pearl River Estuary, South China. Mar Pollut Bull 47(1–6):30–36
36. Ishii S, Hansen DL, Hicks RE, Sadowsky MJ (2007) Beach sand and sediments are temporal sinks and sources of *Escherichia coli* in Lake Superior. Environ Sci Technol 41(7):2203–2209
37. Jana S, Bhattacharya DN (1988) Effect of heavy metals on growth population of a fecal coliform bacterium *Escherichia coli* in aquatic environment. Water Air Soil Poll 38(20):251–254
38. Jonas PJC, Millward GE (2010) Metals and nutrients in the Severn Estuary and Bristol Channel: contemporary inputs and distributions. Mar Pollut Bull 61(1–3):52–67
39. Jones RC, Kelso DP, Schaeffer E (2008) Spatial and seasonal patterns in water quality in an embayment-mainstem reach of the tidal freshwater Potomac River, USA: a multiyear study. Environ Monit Assess 147(1–3):351–375
40. Juahir H, Zain SM, Yusoff MK, Hanidza TIT, Armi ASM, Toriman ME, Mokhtar M (2011) Spatial water quality assessment of Langat River Basin (Malaysia) using environmetric techniques. Environ Monit Assess 173(1–4):625–641
41. Kovačić A, Tripković I, Galov A, Zitko T (2011) Distribution of microbiological indicators of fecal pollution in the riverine substrates. Environ Monit Assess 172(1–4):623–630
42. Kress N, Herut B, Galil BS (2004) Sewage sludge impact on sediment quality and benthic assemblages off the Mediterranean coast of Israel—a long-term study. Mar Environ Res 57(3):213–233
43. Kurun A, Balkis H, Balkis N (2007) Accumulations of total metal in dominant shrimp species (*Palaemon adspersus*, *Palaemon serratus*, Parapenaeus longirostris) and bottom surface sediments obtained from the northern inner shelf of the Sea of Marmara. Environ Monit Assess 135(1–3):353–367
44. La Rosa T, Mirto S, Favaloro E, Savona B, Sarà G, Danovaro R, Mazzola A (2002) Impact on the water column biogeochemistry of a Mediterranean mussel and fish farm. Water Res 36(3):713–721
45. Lacerda LD, Campos RC, Santelli RE (2013) Metals in water, sediments, and biota of an offshore oil exploration area in the Potiguar Basin, Northeastern Brazil. Environ Monit Assess 185(5):4427–4447
46. Lee CM, Lin TY, Lin CC, Kohbodi GA, Bhatt A, Lee R, Jay JA (2006) Persistence of fecal indicator bacteria in Santa Monica Bay Beach sediments. Water Res 40(14):2593–2602
47. Lefebvre A, Guiselin N, Barbet F, Artigas FL (2011) Long-term hydrological and phytoplankton monitoring (1992–2007) of three potentially eutrophic systems in the Eastern English Channel and the Southern Bight of the North Sea. ICES J Mar Sci 68(10):2029–2043
48. Ling TY, Siew TF, Lee N (2010a) Quantifying pollutants from household wastewater in Kuching, Malaysia. World Appl Sci J 8(4):449–456
49. Ling TY, Srikaran R, Kho CP, Lee N (2010b) Organic matter, nutrients and trace metals of Serin River. World Appl Sci J 8(4):496–502
50. Lourino-Cabana B, Billon G, Magnier A, Prygiel E, Baeyens W, Prygiel J, Mikkelsen O, Ouddane B (2011) Evidence of highly dynamic geochemical behaviour of zinc in the Deûle River (Northern France). J Environ Monit 13:2124–2133
51. Madhu NV, Balachandran KK, Martin GD, Jyothibabu R, Thottathil SD, Nair M, Joseph T et al (2010) Short-term variability of water quality and its implications on phytoplankton production in a tropical estuary (Cochin Backwaters-India). Environ Monit Assess 170(1–4):287–300
52. Malaysian Meteorological Department (2012) General climate of Malaysia. MMD Malaysia.http://www.met.gov.my/index.php?option=com_content&task=view&id=75&Itemid=1089Accessed 20 Oct 2012
53. Malik A, Ahmad M (2002) Seasonal variation in bacterial flora of the wastewater and soil in the vicinity of industrial area. Environ Monit Assess 73(3):263–273
54. Mallin MA, Esham EC, Williams KE, Nearhoof JE (1999) Tidal stage variability of fecal coliform and chlorophyll a concentrations in coastal creeks. Mar Pollut Bull 38(5):414–422
55. Mallin MA, Johnson VL, Ensign SH (2009) Comparative impacts of stormwater runoff on water quality of an urban, a suburban, and a rural stream. Environ Monit Assess 159(1–4):475–491
56. Meyer-Reil LA, Koster M (2000) Eutrophication of marine waters: effects on benthic microbial communities. Mar Pollut Bull 41(1–6):255–263
57. Molvaer J, Knutzen J, Magnusson J, Rygg B, Skei J, Sorensen J (1997). Environmental Quality Classification in Fjords and Coastal Areas. Statens Forurensningstilsyn TA-1467, Norway
58. Nduka JK, Orisakwe OE (2009) Effect of effluents from warri refinery petrochemical company WRPC on water and soil qualities of "contiguous host" and "impacted on communities" of Delta State, Nigeria. Open Environ Pollut Toxicol J 1(1):11–17

59. Nixon SW (1995) Costal marine eutrophication—a definition, social causes, and future concerns. Ophelia 41:199–219
60. Noble RT, Lee IM, Schiff KC (2004) Inactivation of indicator micro-organisms from various sources of faecal contamination in seawater and freshwater. J Appl Microbiol 96(3):464–472
61. Norton MG, Rowlatt SM, Nunny RS (1984) Sewage sludge dumping and contamination of Liverpool Bay sediments. Estuar Coast Shelf Sci 19(1):69–87
62. Ortiz-Hernandez MC, Saenz-Morales R (1999) Effects of organic material and distribution of fecal coliforms in Chetumal Bay, Quintana Roo, México. Environ Monit Assess 55:423–434
63. Otero XL, Vidal-Torrado P, Calvo de Anta RM, Macías F (2005) Trace elements in biodeposits and sediments from mussel culture in the Ria de Arousa (Galicia, NW Spain). Environ pollut (Barking, Essex: 1987) 136(1):119–134
64. Ouattara NK, Passerat J, Servais P (2011) Faecal contamination of water and sediment in the rivers of the Scheldt drainage network. Environ Monit Assess 183(1–4):243–257
65. Paez-Osuna F, Guerrero-Galvan SR, Ruiz-Fernandez AC (1998) The environmental impact of shrimp aquaculture and the coastal pollution in Mexico. Mar Pollut Bull 36(1):65–75
66. Pan Y, Subba Rao DV (1997) Impacts of domestic sewage effluent on phytoplankton from Bedford Basin, Eastern Canada. Mar Pollut Bull 34(12):1001–1005
67. Parnell PE (2003) The effects of sewage discharge on water quality and phytoplankton of Hawaiian coastal waters. Mar Environ Res 55:293–311
68. Parsons TR, Maita Y, Lalli CM (1984) A manual of chemical and biological methods for seawater analysis. Pergamon Press, Oxford
69. Pinckney JL, Paerl HW, Tester P, Richardson TL (2001) The role of nutrient loading and eutrophication in estuarine ecology. Environ Health Persp 109(5):699–706
70. Pomeroy RL, Smith EE, Grant CM (1965) The exchange of phosphate between estuarine water and sediments. Limnol Oceanogr X(2):167–172
71. Pope ND, Langston WJ (2011) Sources, distribution and temporal variability of trace metals in the Thames Estuary. Hydrobiologia 672(1):49–68
72. Reopanichkul P, Carter RW, Worachananant S, Crossland CJ (2010) Wastewater discharge degrades coastal waters and reef communities in Southern Thailand. Mar Environ Res 69:287–296
73. Ribeiro C, Kjerfve B (2002) Anthropogenic influence on the water quality in Guanabara Bay, Rio de Janeiro, Brazil. Reg Environ Change 3(1–3):13–19
74. Rippey B (1982) Sediment-water interactions of Cu, Zn and Pb discharged from a domestic waste water source into a Bay of Lough Neagh, Northern Ireland. Environ Pollut B 3:199–214
75. Rubin AJ, Elmaraghy GA (1977) Studies on the toxicity of ammonia, nitrate and their mixtures to guppy fry. Water Res 11:927–935
76. Satpathy KK, Mohanty AK, Natesan U, Prasad MVR, Sarkar SK (2010) Seasonal variation in physicochemical properties of coastal waters of Kalpakkam, East Coast of India with special emphasis on nutrients. Environ Monit Assess 164(1–4):153–171
77. Seitzinger S, Sanders R (1997) Contribution of dissolved organic nitrogen from rivers to estuarine eutrophication. Mar Ecol-Prog Ser 159:1–12
78. Shahidul IM, Jahangir SM, Yamamoto T, Abdul WM, Tanaka M (2004) Water and sediment quality, partial mass budget and effluent N loading in coastal brackish water shrimp farms in Bangladesh. Mar Pollut Bull 48(5–6):471–485
79. Shin JY, Artigas F, Hobble C, Lee YS (2012) Assessment of anthropogenic influences on surface water quality in urban estuary, Northern New Jersey: multivariate approach. Environ Monit Assess 185(3):2777–2794
80. Simeonov V, Stratis JA, Samara C, Zachariadis G, Voutsa D, Anthemidis A, Sofoniou M et al (2003) Assessment of the surface water quality in Northern Greece. Water Res 37(17):4119–4124
81. Smith VH, Tilman GD, Nekola JC (1999) Eutrophication: impacts of excess nutrient inputs on freshwater, marine, and terrestrial ecosystems. Environ Pollut (Barking, Essex: 1987) 100(1-3):179–196
82. Sondergaard M, Jensen PJ, Jeppesen E (2001) Retention and internal loading of phosphorus in shallow, eutrophic lakes. Sci World J 1:427–442
83. Song MW, Huang P, Li F, Zhang H, Xie KZ, Wang XH, He GX (2011) Water quality of a tributary of the Pearl River, the Beijiang, Southern China: implications from multivariate statistical analyses. Environ Monit Assess 172(1–4):589–603
84. Sundaray SK, Nayak BB, Kanungo TK, Bhatta D (2012) Dynamics and quantification of dissolved heavy metals in the Mahanadi River estuarine system, India. Environ Monit Assess 184(2):1157–1179
85. Thomas Y, Courties C, Helwe YE, Herbland A, Lemonnier H (2010) Spatial and temporal extension of eutrophication associated with shrimp farm wastewater discharges in the New Caledonia Lagoon. Mar Pollut Bull 61(7–12):387–398
86. Trott LA, Alongi DM (2000) The impact of shrimp pond effluent on water quality and phytoplankton biomass in a tropical mangrove estuary. Mar Pollut Bull 40(11):947–951
87. Troussellier M, Got P, Bouvy M, M'Boup M, Arfi R, Lebihan F, Monfort P et al (2004) Water quality and health status of the Senegal River Estuary. Mar Pollut Bull 48(9–10):852–862
88. USEPA (2003). Ambient Water Quality Criteria for Dissolved Oxygen, Water Clarity and Chlorophyll a for the Chesapeake Bay and Its Tidal Tributaries. EPA 903-R-03-002. EPA, United States
89. USEPA (2009). National recommended water quality criteria. USEPA.http://water.epa.gov/scitech/swguidance/standards/current/. Accessed 20 October 2012

90. Wetlands International (2013). Ramsar sites information service. wetlands international.http://ramsar.wetlands.org/ Accessed 3 March 2014

91. Whitall D, Hively WD, Leight AK, Hapeman CJ, McConnell LL, Fisher T, Rice CP et al (2010) Pollutant fate and spatio-temporal variability in the Choptank River Estuary: factors influencing water quality. Sci Total Environ 408(9):2096–2108

92. WHO (1998). Guidelines for safe recreational-water environments. WHO/EOS/98. WHO, Geneva

93. Wickins JF (1976) The tolerance of warm-water prawns to recirculated water. Aquaculture 9:19–37

94. Wu RSS, Lam KS, MacKay DW, Lau TC, Yam V (1994) Impact of marine fish farming on water quality and bottom sediment: a case study in the sub-tropical environment. Mar Environ Res 38(2):115–145

95. Xu J, Ho AYT, Yin K, Yuan X, Anderson DM, Lee JHW, Harrison PJ (2008) Temporal and spatial variations in nutrient stoichiometry and regulation of phytoplankton biomass in Hong Kong waters: influence of the Pearl River outflow and sewage inputs. Mar Pollut Bull 57(6–12):335–348

96. Yayintaş OT, Yilmaz S, Türkoğlu M, Colakoğlu FA, Cakir F (2007) Seasonal variation of some heavy metal pollution with environmental and microbiological parameters in sub-basin of Kocabas Stream (Biga, Canakkale, Turkey) by ICP-AES. Environ Monit Assess 134(1–3):321–331

97. Yin K, Harrison PJ (2007) Influence of the Pearl River Estuary and vertical mixing in Victoria Harbor on water quality in relation to eutrophication impacts in Hong Kong waters. Mar Pollut Bull 54(6):646–656

98. Young K, Morse GK, Scrimshaw MD, Kinniburgh JH, MacLeod CL, Lester JN (1999) The relation between phosphorus and eutrophication in the Thames catchment, UK. Sci Total Environ 228:157–183

99. Zhang D, Zhang X, Tian L, Ye F, Huang X, Zeng Y, Fan M (2012) Seasonal and spatial dynamics of trace elements in water and sediment from Pearl River Estuary, South China. Environ Earth Sci 68(4):1053–1063

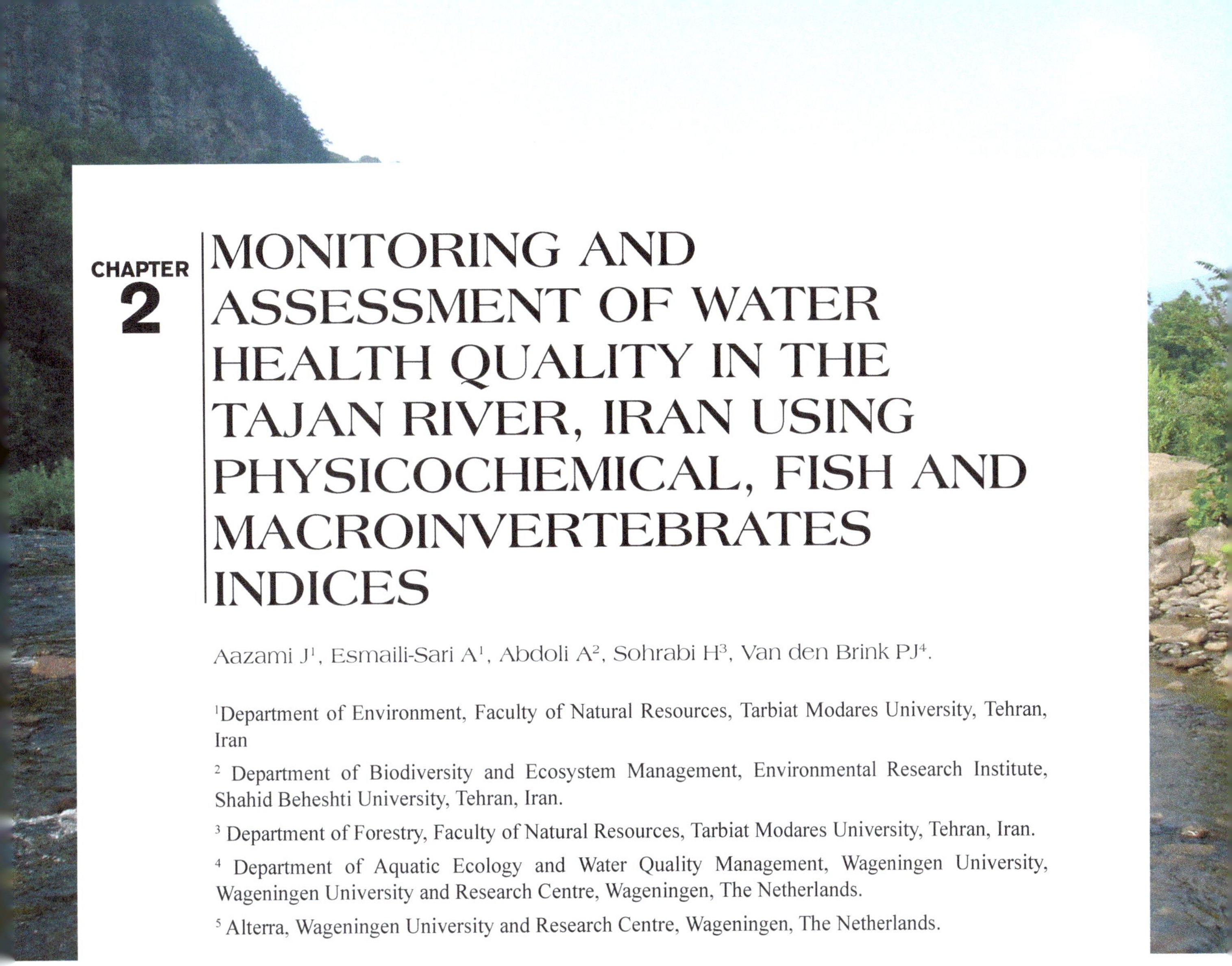

CHAPTER 2

MONITORING AND ASSESSMENT OF WATER HEALTH QUALITY IN THE TAJAN RIVER, IRAN USING PHYSICOCHEMICAL, FISH AND MACROINVERTEBRATES INDICES

Aazami J[1], Esmaili-Sari A[1], Abdoli A[2], Sohrabi H[3], Van den Brink PJ[4].

[1]Department of Environment, Faculty of Natural Resources, Tarbiat Modares University, Tehran, Iran

[2] Department of Biodiversity and Ecosystem Management, Environmental Research Institute, Shahid Beheshti University, Tehran, Iran.

[3] Department of Forestry, Faculty of Natural Resources, Tarbiat Modares University, Tehran, Iran.

[4] Department of Aquatic Ecology and Water Quality Management, Wageningen University, Wageningen University and Research Centre, Wageningen, The Netherlands.

[5] Alterra, Wageningen University and Research Centre, Wageningen, The Netherlands.

ABSTRACT

Background

Nowadays, aquatic organisms are used as bio-indicators to assess ecological water quality in western regions, but have hardly been used in an Iranian context. We, therefore, evaluated the suitability of several indices to assess the water quality for an Iranian case study.

Methods

Measured data on biotic (fish and macroinvertebrates) and abiotic elements (28 physicochemical and habitat parameters), were used to calculate six indices for assessment of water quality and the impact of human activities in the Tajan river, Iran. GIS, uni- and multivariate statistics were used to assess the correlations between biological and environmental endpoints.

Results

The results showed that ecological condition and water quality were reduced from up- to downstream. The reduced water quality was revealed by the biotic indices better than the abiotic ones which were linked to a variety of ecological water quality scales.

Conclusion

The fish index showed a strong relationship with long-term database of physicochemical parameters (12 years (94%)), whereas macroinvertebrates index is more correlated with short-term data (76%). Meanwhile, the biotic and abiotic elements in this study were also classified well by PCA. Pulp and wood plants and sand mining are indicated to have the most negative effects on the river ecosystem.

Citation: Aazami, J et al. J Environ, Monitoring and assessment of water health quality in the Tajan River, Iran using physicochemical, fish and macroinvertebrates indices, doi:10.1186/s40201-015-0186-y

INTRODUCTION

One of the greatest environmental challenges of this century is to sustain natural biological structural and functional attributes of aquatic ecosystems, rivers in particular. This goal requires that we know the condition of these dynamic systems and how they are being affected by specific factors and forces [1]. Nowadays, we can easily see that there are many pollutants in the environment due to anthropogenic activities. The destruction of natural habitats and the presence of environmental pollutants may affect the ecological balance of every ecosystem [2]. Among various ecosystems in the world, rivers which cross different areas such as agriculture and industry are the most threatened and affected by anthropogenic activities [3]. In developing countries such as Iran, water pollution is a common and widespread problem. Therefore, water resource management of rivers is of great importance and especially essential for semiarid and developing countries [4]. Assessment and classification of ecological water quality using indices-based approaches can help the conservation and management of rivers. The measurement of physicochemical parameters is usually time-consuming, cost-intensive and also dependent on special instruments. However, physicochemical parameters can only show water quality at the moment of measurement and these can change over time. Nowadays, indicators based on the presence or absence aquatic organisms have been developed to assess water quality and for the classification of ecological status. Norris and Thoms suggested that the effects on biota are usually the final point of environmental degradation and pollution of rivers and thus are an important indication of ecosystem health [5].

Many living organisms (e.g. small mammals, fish, aquatic plants, algae, invertebrates) can be used to assess ecological water quality. Fish encompass different trophic levels, have a long life cycle, and high mobility, and can herewith be used to integrate the effects of habitat change and environmental pollution over a long period [6]. Macroinvertebrates are used for bioassessment because they are relatively easily sampled and are a very biodiverse group of species inhabiting waters that is contaminated to a different extent (from clean to highly polluted) [7-9]. They are important for the cycling of organic matter and provide food resources for higher trophic levels. The fluctuation of macroinvertebrate richness in the aquatic environment may result in the change of the ecosystem function. Moreover, the relatively low mobility and long life cycles of macroinvertebrates ensure that the presence of a given taxon reflects the past conditions. Many previous studies have shown the importance of biotic indices in the world, especially in Europe [10].

However, there are only a few studies using biotic indices in Asian countries such as Iran. The water quality monitoring programs in Iran are mainly based on the determination of some physicochemical parameters and water quality indices have generally not yet been use as a tool for the assessment and management of river ecosystems. Meanwhile, scientific efforts have often focused on improving freshwater resources that are of economic, cultural or spiritual importance. Unfortunately, most of these efforts have proceeded without documentation of the relative successes and failures of individual activities. Even though success is noted, there is often a lack of biotic data to identify specific results or endpoints for the river management activity [1].

In this study, we tried to employ the most widely used biotic and physicochemical indices to classify the ecological water quality in one of the most important Iranian rivers, the Tajan River. This river was chosen as a pilot river from the 115 rivers in the north of Iran because of having a good water-flow, discharge regime, catchment area, Accessibility and environmental condition [11]. Also, there are many similarities such as environmental landscape, climate and land uses between this river and European rivers. Besides, it is possible to select some stations as references because of being away of human activities.

The goals of this study were to determine and classify the ecological water quality of Tajan River based on different indices of water quality and to evaluate their performance, to zone the water quality based on these indices and GIS (Geographic Information System), to assess the effects of human land uses on the river and to compare the results between up- and down-stream parts of the river. This is the first study that compares biotic and physicochemical indices, as well as uses fish species as a bio indicator in an Iranian River.

MATERIALS AND METHODS

Study Area

The Tajan River is one of the best freshwater ecosystems in the Mazandaran province located in the north of Iran. The river is 140 km long and originates from forested mountains and continues through different land uses, especially the agricultural areas of the coastal plain, where rice is extensively cultivated. The river eventually terminates in the Caspian Sea, the biggest land-locked aquatic ecosystem in the world (Figure 1). There are different land-uses including agriculture, aquaculture, dam construction, sand mining and industrial activities in the river [11,12]. It is chosen as a pilot river from the 115 rivers in north of Iran [13] because of having a good water-flow, discharge regime, catchment area and environmental condition. The Tajan River is divided to tow parts (Upstream and Downstream) by an old big dam (Shahid-Rajaie Dam, 1993). Water, macroinvertebrates and fish were sampled at 17 sites in September 2013. Eight sites are located in the upstream part and 9 sites in the downstream part. Also, five sites were selected as reference sites where there was no or slight pollution expected. Site selection was based on land use, accessibility and anthropogenic activities. Sites were subjected to non-point (i.e. agricultural runoff) and point influents (i.e. fish pond).

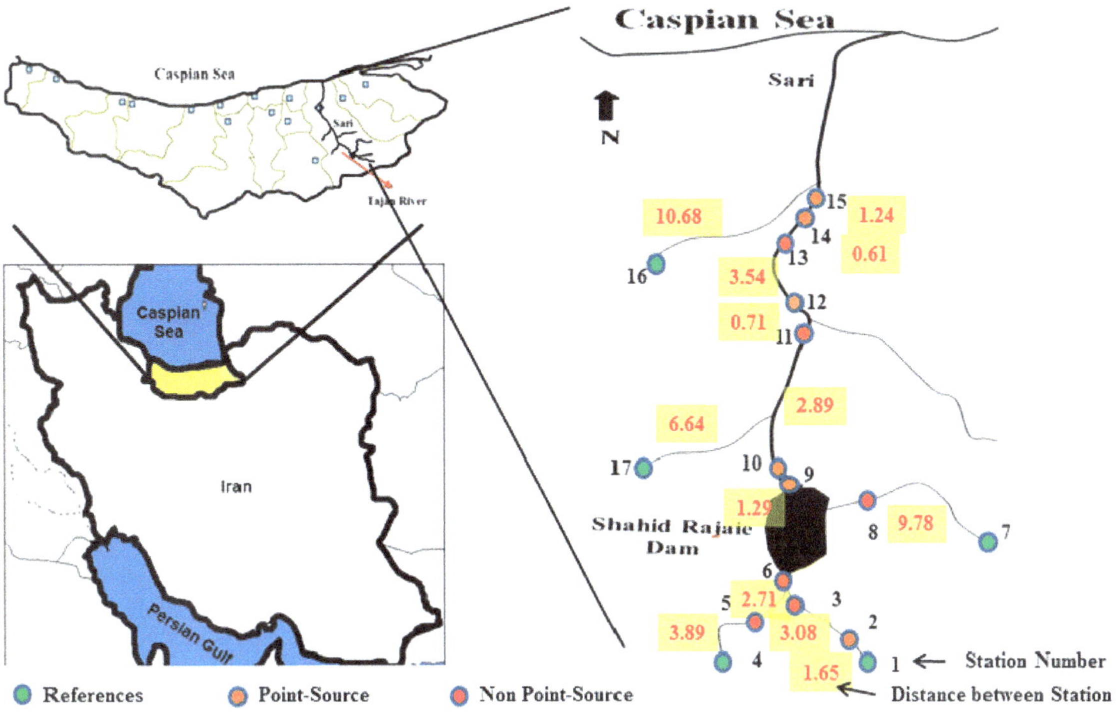

Figure 1: Sites sampled in this case study.

Physicochemical and Habitat Parameters

Width, length, depth, altitude (m), dissolved oxygen (mg/l), pH, water and air temperature (°C), conductivity (µS/cm), turbidity (NTU) and nutrients (NO3-N, NO2-N, NH4-N, and PO4-P, mg/l) were measured in situ by using Portable multi-parameter water analyser and UV–vis Spectrophotometry 8000. Biochemical Oxygen Demand (BOD5, mg/l) and Total Suspended Solids (TSS, mg/l) were determined according to APHA [14] with three replicate samples being measured in the laboratory. Some stations in this study overlapped with the stations of Iranian Water Resources Management (IWRM). Therefore, the accuracy of the collected data of physicochemical parameters was checked with the data of IWRM which had collected monthly data for twelve years. Average water flow (m/sec) was calculated by calculating the average of the recordings available from the IWRM for the same period as our sampling. According to the Field Manual for Water Quality Monitoring, the National Sanitation Foundation Water Quality Index (NSFWQI) surveyed 142 sites representing a wide range of locations at the local, state, and national level, using about 35 water quality tests which outcomes were combined in an index [15]. Nine parameters (dissolved oxygen, faecal coliform, biochemical oxygen demand, pH, nitrates, total phosphate, temperature, turbidity, and total solids) were chosen and some were judged to be more important than others, so the values were combined by calculating a weighted mean, based on the method described by Nikoo [16]. For this, field measurements were converted to index values using a questionnaire in which respondents were asked to estimate the level of water quality (0 through 100) corresponding to the field measurements (e.g., pH 2–12). The curves were then averaged and are assumed to represent the best professional judgment of the respondents. When the test results were not available for all 9 measurements, we preserved the relative weight for each factor and scale the total so that the range remains 0 to 100 [17]. The $IRWQI_{sc}$ index is a modification of NSFWQI based on the local condition in Iran. Habitat assessment was performed using 10 factors assessed by four experts and the Rapid Bioassessment Protocol (RBP) was used for river habitat assessment by visual observations at each site [18]. The range of each habitat parameters was from 0 (very perturbed) to 20 (pristine) and included Epifaunal Substrate/Available Cover, Embeddedness (Embed), Velocity/Depth Regime, Sediment Deposition, Channel Flow Status, Channel Alteration, Frequency of Riffles (Freq), Bank Stability, Vegetative Protection, Distance of References (DisRef), and Riparian Vegetative Zone Width. Finally, an average of the ranges of the parameters resulted in an overall RBP index.

Biological Elements

A surber sampler (30*30 cm and 250 micron mesh) was used for collecting macroinvertebrates based on the RBP [18]. Three replicates were collected on each site and all the three replicates were composited as one sample. Benthic macroinvertebrates were preserved in 4% formaldehyde solution before being sorted, identified and counted to family level in laboratory. Electrofishing (200–300 V) with Blank net 6 mm based of FAME Protocol was used to sample fish [19]. Most of the sampled fish were identified, measured and released in situ. Some fish had to be identified further and were fixed in 10% formalin and transferred to the laboratory. Finally, the species and weight of each fish were recorded. RBP metrics were calculated for each sample and only mean site values are reported. In next step, mean values for each site were compared to references sites to generate a score for that site and classification. In Table 1, the classification of introduced indices and their colours are provided [20-25].

Table 1: Indices used to classify water and habitat quality

	Range	**<15**	**15 – 29.9**	**30 – 44.9**	**45 - 55**	**55.1 - 70**	**70.1 - 85**	**85 ≤**
IRWQI$_{sc}$	Quality	Very Polluted	Polluted	Enough Polluted	Moderate	Enough Good	Good	Excellent
	Colour	Violet	Red	Orange	Yellow	Olive-green	Green	Blue
NSFWQI	Range			0-25	25-50	50-70	70-90	90-100
	Quality			Very bad	Bad	Medium	Good	Excellent
MMIF	Range			0.00–0.29	0.30–0.49	0.50–0.69	0.70–0.89	0.90–1.00
	Quality			Bad	Poor	Moderate	Good	Excellent
Karr	% Index			<40	40-59	60-74	75-87	88-100
	Quality			Very poor	Poor	Fair	Good	Excellent
	Colour			Red	Orange	Yellow	Green	Blue
RBP	% Index				<60	74-60	89-75	≥90
	Quality				Poor	Marginal	Good	Excellent
BMWP /ASPT	Range				4 >	4-5	5-6	6 <
	Quality				Polluted	Moderate	Doubtful	Excellent
	Colour				Red	Yellow	Green	Blue

Physicochemical and Biological indices of Water Quality

Physical habitat quality (type, stability, availability, etc.) and water quality are reflected by the diversity of stream communities. Based on these relationships, environmental quality of aquatic systems can be described or categorized using integrated approaches that incorporate an evaluation of the physical, chemical and biological components. The guidelines associated with the RBP provide systematic approaches for identifying habitat quality and biotic integrity of river systems [26-28].

The NSFWQI is one of the first water quality indices that aggregate some water quality parameters through a weighted arithmetic mean function [15]. The Index is a usual water quality index method that was developed by paying great rigor in selecting parameters, developing a common scale and assigning weights. The water quality data are recorded and transferred to a weighting curve chart, where a numerical value of Qi is obtained. The mathematical expression for NSF WQI is given by [29]:

$$NSFWQI = \sum_{k=1}^{n} WiQi$$

Q_i=sub-index for each water quality parameter;

W_i=weight associated with each water quality parameter;

n=number of water quality parameters.

Also, the Ministry of Energy of Iran developed a local index, IRWQI$_{sc}$ (Iranian Water Quality Index for Surface Water Resource-Conventional Parameters), for the assessment of water quality [23,30]. This index is calculated by:

$$IRWQI = \sum_{1}^{n}(IiWi)^{1/a}$$

$$a = \sum_{1}^{n} Wi$$

Ii=sub-index for each water quality parameter;

Wi=weight associated with each water quality parameter;

n=number of water quality parameters;

a=the sum of the weight.

Both indices are based on physicochemical parameters that are evaluated in this study. For assessment and classification of water quality by macroinvertebrate indices, we used the ratio of Biological Monitoring Working Party score to Average Score per Taxon (BMWP/ASPT) and Multimetric Macroinvertebrates Index Flanders (MMIF) [21]. These two indices are numerical expressions used to assess water quality based on the presence and, in many cases, the diversity of taxa with known environmental-pollution tolerances according to the following equation:

$$BMWP/ASPT = \left(\sum B.n\right)/N$$

Whereas:

B=the value for each species;

n=abundance of each species;

N=total number of species.

Subsequently, ecological water quality is assessed with Karr Biotic Index of Fish (KBI). KBI was firstly invented by Karr to study the River Trent basin in Champaign, then applied by other researchers, and is nowadays used all over the world and well documented by Karr and Ruaro [22,31]. The index is designed to assess the present the status of the community using twelve fish community parameters. These parameters can be roughly grouped into two sets: ecological factors (including number of individuals in samples, the proportion of omnivore individuals, proportion of insectivorous cyprinids, proportion of top carnivorous, proportion of individuals with disease, tumours, fin damage, and other anomalies) and species composition and richness (including number of species, presence of intolerant species, species richness and composition of darters, suckers and sunfish (except green sunfish) and proportion of green sunfish and hybrid individuals) [22]. Habitat condition is classified based on US EPA Rapid Bioassessment Protocol (RBP) [18].

Statistical Analysis

Statistical analysis of the resulting data was performed using Excel and SPSS software version 19 (licensed by Tarbiat Modares University, Iran) and Canoco version 5 (licensed by Wageningen University, The Netherlands). Normality of data was checked by ShapiroWilk test. Because the data were not normally distributed, Mann–Whitney U test were used to assess the significance of the difference between up- and downstream values of physicochemical parameters. Cluster analysis was used for the classification of the stations on the basis of the indices and Casewise analysis was used for assessing the correlation between the classifications (with using SPSS). Principal component analysis (PCA) was used to analyse the correlations between abiotic (habitat and physicochemical) parameters and biotic (macroinvertebrates and fish) community. Before analysis, all the macroinvertebrates were divided into 5 groups based on the pollution tolerance owned by the dominant species in each family (Very sensitive, Sensitive, Neutral, Resistant and Very resistant) that was provided in Appendix B of RBP, EPA [18].

RESULTS AND DISCUSSION

For irrigating of the vast plain area of downstream rice farms, the dam valves are opened from spring to late summer and closed in early September, with a minimal flow at all time of 1 m^3/s. The rains start in early autumn and therefore the best time for sampling is September in accordance with local condition and some previous studies [32]. The main land-use in the upstream areas is not similar to that in the downstream areas, so no similar anthropogenic impact may be expected. Furthermore, there is a larger downstream flow, and herewith also a larger move of pollutants, most of the year because of the opening of the dam valves and irrigation. Habitat parameters show significant differences between up- and downstream sites (see Figure 1 for better understanding). Some land uses such as mining that affect the rivers' physical status are less development upstream compared to downstream. In the downstream part, there are many mining industries that have altered the physical status of the river. Humans have channelized, diverted, drained and filled streams because of the dredging of sand. In the upstream part of the Tajan River, the higher number of residents has no major impact on the physical structure of habitat. Also, the accessibility of the upstream part of the river is lower than downstreams. Therefore, there are more pristine habitats in upstream. Padmalal et al. showed that in the past 3–4 decades, rivers in the densely populated areas of the world are subjected to immense pressures due to various kinds of human interventions, among which indiscriminating mining for construction-grade sand from alluvial reaches is among the most disastrous one [32].

We identified 2426 fish individuals covering 17 species, 5 families and 7639 individuals of macroinvertebrates covering 45 geneses or families. Zoning of ecological river condition was performed with GIS based on the result of indices (Table 2 and Figure 2). All indices showed water quality reductions from upstream to downstream: the references stations (1,4,7,16,17) are indicated to have a good water quality (almost unpolluted) and in the most downstream stations, water quality was classified as moderate to polluted (14,15). We tried to get an even distribution of the stations along the river so we could analysis the effects of land uses on study's parameters, too. Human land use, development and urbanization can alter water flow and degrade stream habitat and biotic conditions through, for example, draining agricultural fields, channelizing streams and increasing sedimentation. The first station is located in an almost pristine area, with hardy any agriculture activities and covered with natural forest. All indices showed that the station was unpolluted. Upstream of the second station, there was a fish pond that probably affected the natural fish community, which is expressed by the low KBI value (green, Figure 2). Other indices, however, did not show a marked response. Because of the limitation to captured fisheries production, aquaculture has been developed worldwide especially in developing countries such as Iran. Aquaculture may have significant impacts on the environment and natural resources, and a number of concerns have been expressed by environmental activists and scientists [33,34]. Raczyńska et al. [35] reported that little attention is paid to the changes in water quality resulting from the inflow of effluent discharged from fish breeding ponds. They, however, report that the aquaculture effluent discharge from the carp breeding ponds had a significant impact on the physicochemical parameters of the river water. They observed an increase in concentrations of the organic and biogenic compounds immediately after the inflow of the polluted water from the ponds [35]. However, more studies are needed in order to understand the decrease in fish biodiversity observed at station 2. Downstream of the fish ponds, rice farms were located. In station 3, physical habitat parameters were altered probably because of the influence of traffic of agriculture. Therefore, the habitat index (RBP) descended from excellent to good levels (Figure 2). Also, the values of macroinvertebrates indices were reduced but no differences were observed for the KBI index. Agriculture may increase nutrient levels, erosion, and turbidity due to the use of fertilizers, planting and harvesting. Especially rice farms may be important for the water quality because they require a lot of water for cultivation. Many scientists studied the effect of agriculture as a non-point pollutant on river basin and biota [36,37]. Traditional agriculture in the research area uses much water and the effluent of the farms often runs directly into the river. The indices did not show many changes for station 6 that is located after of accident tow stream and only the value of the MMIF index decreased. Between station 7 (reference station of the second stretch of the river) and 8, there are large farms and all indices may have decreased in station 8 because of the polluted discharge water from farming. This study is the first one reporting the abundance of fish population in two streams of the Tajan River, as there were 10 fish (*Barbus lacerta*) in station 7. *B. lacerta* is one of the most sensitive species to anthropogenic stress, similar to *Salmo trutta* [38]. *S. trutta* as an indicator species showing a high sensitivity to a variety of human pressures (e.g. water pollution, habitat degradation). Normally, it inhabits headwaters with high oxygen saturation, steep slope, fast flow, suitable temperature regimes and adequate food and due to anthropogenic influences, the brown trout was eliminated from many original habitats in Iran [39]. The comparison of indices in the 8 stations in the upstream catchment showed a comparable decrease in index values between the two streams of the river. A big dam is located between the eight upstream stations and station 9 which changed most of the parameters. Results show that values for pH, TSS, BOD and nutrients were significantly higher in the downstream area compared to the upstream area. The NSFWQI index showed a significant change due to the dam, although the local index ($IRWQI_{SC}$) did not change. This result is similar to other studies on the effect of a dam on physicochemical parameters [40]. The dam altered the diversity and density of macroinvertebrates, but had no effects on the macroinvertebrates indices, because after the dam the total density and total biomass increased, while taxa richness tended to decrease, which is in line with the findings of others [41]. Native fish assemblages have experienced profound changes as a consequence of the dam construction. We documented the decline of native fish and the rise of non-native fish which become the major component of the ichthyofauna in the downstream stations. RBP showed a reduction in the physical structure of the habitat in the downstream part. After the dam, there is a fish pond which may be responsible for the strong reduction in the physical habitat quality as shown by RBP (Table 2). The fish pond are situated very close to the river and also to the dam, and takes in clean water and releases the same amount of effluent water to the downstream part of the river. Downstream of the fish pond there are rice farms and small streams like the Zellem stream entering the main river. In the Zellem stream there were many sand mining and sandblast sites which polluted the water in the Tajan River. The stream is small and is located between station 11 and 12 and a number of indices show a marked decrease between these stations (Table 2). Station 13 and 14 were chosen to show the effect of sand mining which was expressed by difference in levels of $IRWQI_{SC}$. Sand mining increased TSS and negatively affected the physicochemical parameters, herewith affecting some sensitive species such as Salmonids. The most notable changes occurred in this study at station 15 which is located after a wood factory, because the wastewaters of this factory were directly discharged into river. Walker et al. [42] reported that the studies conducted in the early 1990s downstream of pulp and paper mills in Canada, found a number of changes in the physiological, biochemical and reproductive responses of wild fish and macroinvertebrates. The study assessed changes in the taste or odours of fish as a result of exposure to pulp and paper effluent as a measure of effects on the usability of fisheries resources by humans and their conclusion on the effects of the factory on fish and habitat parameters confirms our results [42]. Thompson et al. [43] also described the pulp and paper making industry as the third most water intensive industry in terms of freshwater withdrawal in the world, after the primary metals and the chemical industries. There is a large variation in the quality of the wastewaters from pulping and papermaking operations. This is due to the diversity of processes and the chemicals used [43]. Unfortunately, there has not been any exhaustive study on the effects of the plant's wastewater on the ecology of the Tajan River in Iran. Effluents of the plant were released without or with a preliminary treatment into the river, and hence the colour and odour of river's water were changed into brown and strong distinctive odour. There were large differences between station 15 and the others stations, especially the references stations with the same elevation (16, 17). pH and DO decreased and water temperature, BOD and nutrient increased in station 15. The diversity of animals decreased and some exotic species were present (i.e. *Rasborinae pseudorasbora parva*), while the sensitive species were absent. We suggest that more studies on the environmental effects of the paper plant on the ecology of the river are conducted. After station 15, there is a big city, Sari which is the capital of Mazandarn province. A major component of the human use of aquatic systems is the construction, maintenance, and use of roads that occur as part of human infrastructure and the road/stream interface is one of the main pathways for sediment to reach waterways. Stream crossings, often culverts, can alter in-stream sediment accumulation and the geomorphology of a stream. The effects of sedimentation on macroinvertebrates and fish have been well documented [1], as well as of traffic, excavating bottom sand from the river and changing of the riparian zone , which likely caused the obviously decrease in the habitat parameters at the station.

Table 2: The value of the calculated indices for each station

Station	Longitude	Latitude	KBI	BMWP/ASPT	MMIF	NSFWQI	IRWQIsc	RBP
1	712608	4004824	99	8.9	0.95	81.49	87.99	100
2	712059	4005260	75.03	7.5	0.95	70.91	74.06	95
3	709238	4006751	76.28	6.6	0.95	66.97	68.39	88.5
4	708189	4001145	81.48	8.2	0.95	81.74	81.26	100
5	707769	4006595	69.24	6.7	0.95	61.62	62.57	82
6	706166	4010260	75.94	5.7	0.85	59.6	54.84	79
7	726606	4007091	100	7.8	1	69.37	89.39	97
8	707556	4013824	70.95	5.8	0.85	63.59	72.06	73
9	700152	4014338	49.05	7.8	0.8	67.95	80.29	61
10	699857	4015151	52.03	5.2	0.8	59.67	62.2	58.5
11	695882	4025071	56.21	5.3	0.75	59.72	64.38	57.5
12	695429	4026319	51.96	5.5	0.7	52.02	52.36	44.5
13	689155	4037624	60.86	4.2	0.7	52.47	50.9	46.5

14	688694	4038040	52.08	4.2	0.65	53	46.82	38
15	686889	4039978	56.61	3.6	0.55	44.51	39.16	34.5
16	685112	4033640	90.53	7.6	0.95	67.84	86.57	96
17	689592	4023526	83.46	8.2	0.95	66.14	88.64	97

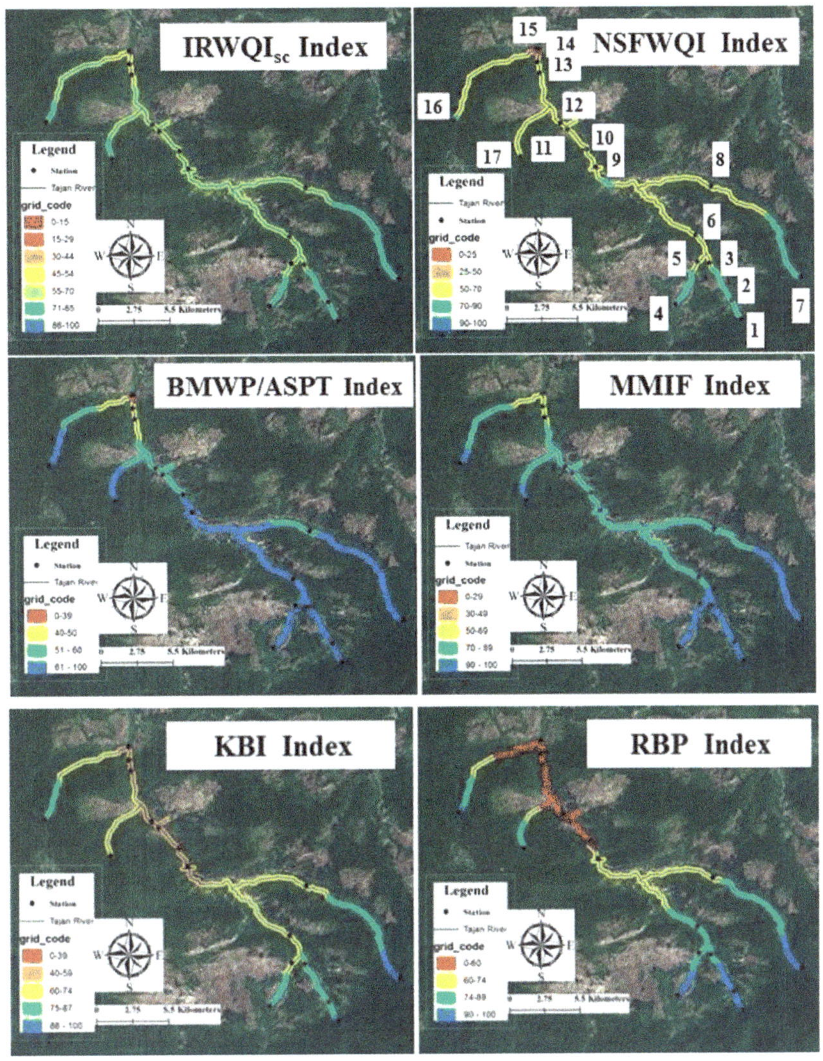

Figure 2: Water quality classification of the Tajan River by different indices and GIS.

Figure 3 shows the classification of the stations of the Tajan River based on biotic indices (KBI, MMIF and BMWP/ASPT) and Figure 4 shows it based on abiotic indices (IRWQIsc, NSFWQI). The dendrogram of the biotic indices divides all stations into two major groups. All references stations are present in the first group and others stations are in the other group. The stations 12, 13, 14 and 15 are then subdivided in a separate group and these stations were located downstream with heavy human activities (Figure 3). The classification based on the abiotic indices shows the impact of land uses on river ecosystems very well. The classification groups the references stations together with station 9 which is located after dam because the dam improved the water quality by reducing sediment deposition, temperature go down and other physicochemical parameters partly improve. The results indicate also the percentage of group membership. On the basis of the results, KBI (fish index) could predict 61% of the changes in physicochemical parameters using our sampling data and 94% of these changes when using the 12 years of data. The prediction ability based on macroinvertebrates indices (MMIF and BMWP/ASPT) were 76 and 71% for our sampling data and 12 years of data, respectively. Fish are at the top of the food chain and could, therefore, show the effects of differences in water pollution at longer time scales than macroinvertebrates, explaining the higher predictive value of the fish index compared to the macroinvertebrate indices. Macroinvertebrate indices had a better correlation with physicochemical parameters in our sampling time, compared to the fish one, probably because of their shorter life cycle, greater dispersal ability, and lower position in the food chain than fish.

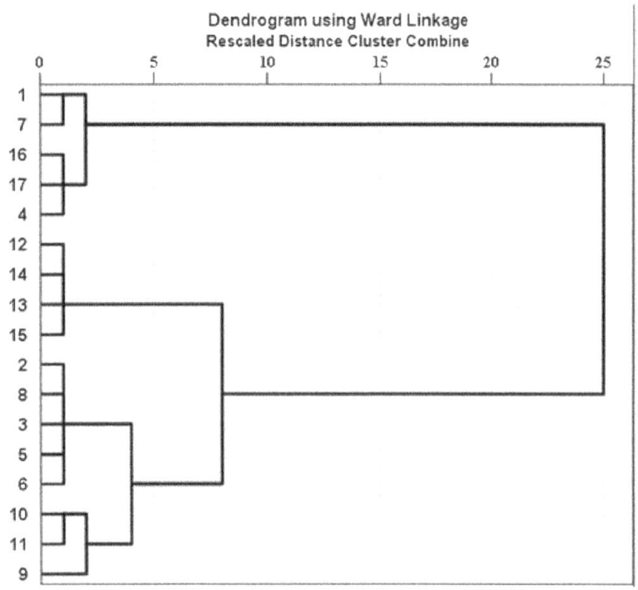

Figure 3: Classification of stations on based biotic indices.

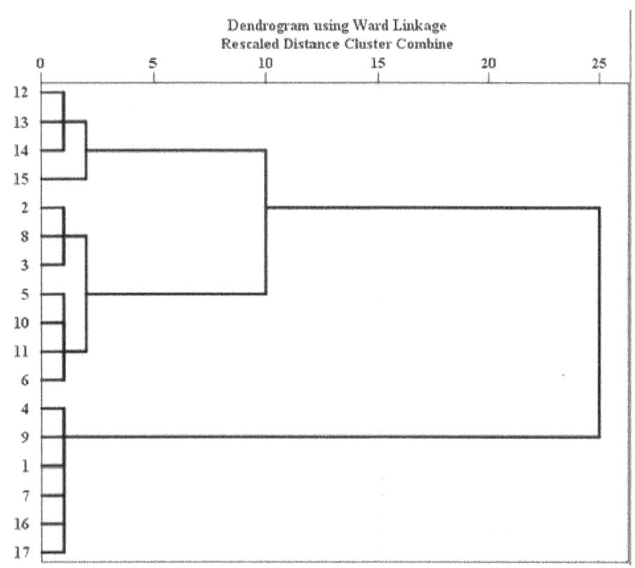

Figure 4: Classification of stations on based abiotic indices.

With the increase of the habitat parameters located at the left side of the diagram, the abundance of the very sensitive and sensitive fish species also increase (Figure 5). DO is the only physicochemical parameters that has a positive correlation with the abundance of the very sensitive and sensitive species. The other physicochemical parameters have negative correlations with both groups (Figure 5). Also the stations with a good status, such as 9, are located on the left side of the diagram. The arrow of neutral species indicates that this group was not strongly correlated with any of the parameters. In other words, the abundance of this group is not correlated with pollution and cannot be used as an indicator. On the right side there are the polluted stations and the groups of resistant and very resistant species. The arrows of the resistant species, pH and Ammonium point into the same direction, which means these two parameters correlate with the abundance of resistant species and less with the others parameters. The abundance of the very resistant species increases with most of the studied physicochemical parameters (except DO) and decrease of habitats parameters. In the upper right quadrant the highly polluted stations are located, with the placement of station 15 being extreme because it receives high pollution from a pulp and wood plant. The result of the PCA analysis complements was similar to the result of bio indicators and classification of stations.

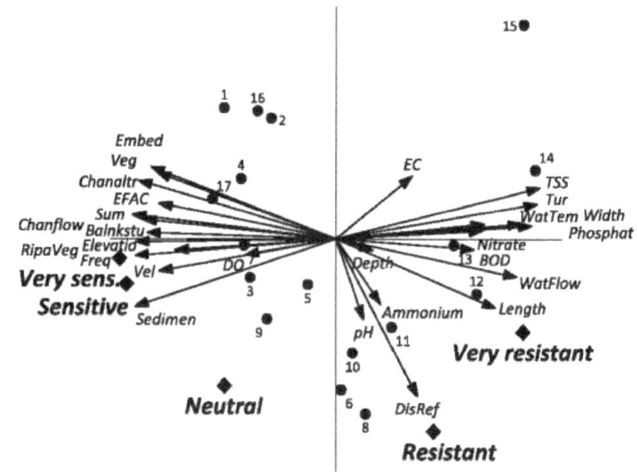

Figure 5: The correlations between abiotic parameters and macroinvertebrates sensitivity groups (PCA, Canoco 5).

Of the variation in group abundances 75% is displayed on the horizontal axis and another 17% on the vertical one.

Also, the result of the PCA analysis on the fish data and abiotic parameters shows that the placement of two fish species (i.e. *B. lacerta* and *S. trutta*) are different from the others (Figure 6). *B. lacerta* and *S. trutta* are correlate positively with high DO and habitat related parameters, while the opposite is true for most other species (Figure 6). On the other hand, the two species are indicated to have a negative correlation with most polluted related physicochemical parameters and therefore may be suitable to indicate clean water. As the physicochemical parameters except DO increase, the presence of the sensitive species decrease which is in accordance with Bengar et al. [44]. Habitat related parameters and DO had a negative correlation with most of the other fish species in the Tajan River and most fish species increase in abundance in the stations at the right side of the diagram (such as station 11, 12).

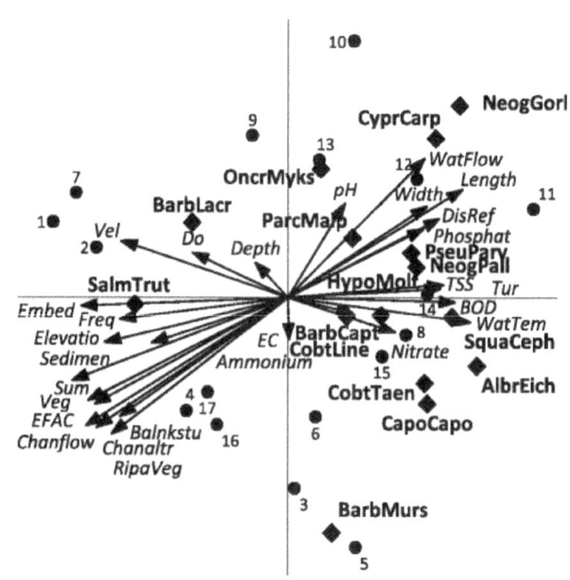

Figure 6: The relationship between abiotic parameters and fish species (PCA, Canoco 5). Of the variation in group abundances 31% is displayed on the horizontal axis and another 23% on the vertical one.

CONCLUSION

This study provides an assessment and comparison of biotic and abiotic indices-based approaches for the Tajan River. As well as Yazdian et al. concluded, we cannot claim that all of the indices would work in other regions as well [45], because of the different range in biodiversity, physicochemical parameters and land-uses. Actually, we have to use, modify and develop some native indices for Iranian ecosystems that they can be used for a rapid assessment of environmental health condition. A comparison between the indices shows that the classification based on biotic indices can show the long-term environmental condition better than those based on abiotic indices. We suggest that macroinvertebrates and fish can be used as indicators of water pollution with having the advantages of low cost, easy identification and it provides a better reflection of water quality than using physicochemical parameters alone. We also suggest to develop and modify the biotic indices for research and management of Iranian rivers, since Iran is located in the mid-dry area where water resources management is particularly urgent and important.

ACKNOWLEDGEMENTS

The manuscript is part of the PhD thesis of Jaber Aazami and is supported by Tarbiat Modares University, Iran. He would never have been able to do this investigation without the guidance of his committee members, help from friends, and support from his family.

REFERENCES

1. Ogren SA. Using indicators of biotic integrity for assessment of stream condition. Michigan Tech. 2014;7(5):10–9.
2. Begon M, Townsend CR, Harper JL. Ecology: from individuals to ecosystems. London, UK: Blackwell Publishing; 2009.
3. Leprieur F, Beauchard O, Blanchet S, Oberdorff T, Brosse S. Fish invasions in the world's river systems: when natural processes are blurred by human activities. PLoS biology. 2008;6(2), e28.
4. Kivaisi AK. The potential for constructed wetlands for wastewater treatment and reuse in developing countries: a review. Ecological Engineering. 2001;16(4):545–60.
5. Norris RH, Thoms MC. What is river health? Freshwater Biology. 1999;41(2):197–209.
6. Mathuriau C, Silva NM, Lyons J, Rivera LMM. Fish and Macroinvertebrates as Freshwater Ecosystem Bioindicators in Mexico: Current State and Perspectives. Water Resources in Mexico. USA: Springer; 2011. p. 251–61.
7. Cheimonopoulou MT, Bobori DC, Theocharopoulos I, Lazaridou M. Assessing Ecological Water Quality with Macroinvertebrates and Fish: a case study from a small Mediterranean river. Environmental management. 2011;47(2):279–90.
8. Li F, Cai Q, Fu X, Liu J. Construction of habitat suitability models (HSMs) for benthic macroinvertebrate and their applications to instream environmental flows: A case study in Xiangxi River of Three Gorges Reservior region. China Progress in Natural Science. 2009;19(3):359–67.
9. Schultz R, Dibble E. Effects of invasive macrophytes on freshwater fish and macroinvertebrate communities: the role of invasive plant traits. Hydrobiologia. 2012;84(1):1–14.
10. Ogleni N, Topal B. Water Quality Assessment of the Mudurnu River, Turkey. Using Biotic Indices Water Resour Manage. 2011;25(10):2487–508.
11. Ahmadi M, Khorasani N, Rafiee G. Investigation of pollution sources and water quality of Tajan River. J Nat Environ. 2011;63(4):327–17.
12. Shokri M, Rossaro B, Rahmani H. Response of macroinvertebrate communities to anthropogenic pressures in Tajan River (Iran). Biologia. 2014;69(10):1395–409.
13. Ahmadi-Mamaqani Y, Khorasani N, Talebi K, Hashemi SH, Rafiee G, Bahadori-Khosroshahi F. Diazinon Fate and Toxicity in the Tajan River (Iran) Ecosystem. Environmental Engineering Science. 2011;28(12):859–68.
14. Eaton AD, Franson MAH. Standard methods for the examination of water & waste water. USA: American Public Health Association (APHA); 2005.
15. Bharti NK. Water quality indices used for surface water vulnerability assessment. Int J Environ Sci. 2011;2(1).
16. Nikoo MR, Kerachian R, Malakpour-Estalaki S, Bashi-Azghadi SN, Azimi-Ghadikolaee MM. A probabilistic water quality index for river water quality assessment: a case study. Environmental monitoring and assessment. 2011;181(1–4):465–78.
17. Mitchell M, Stapp W. Field Manual for Water Quality Monitoring–An Environmental Education Program for Schools. Kendall: Hunt Publishing Company; 2000.
18. Water USEPAOo, Barbour MT, Gerritsen J, Synder BD, Stribling JB. Rapid bioassessment protocols for use in wadeable streams and rivers: periphyton, benthic macroinvertebrates, and fish. USA: EPA Publishing; 1999.
19. Consortium F. Manual for the application of the European Fish Index–EFI; A fish-based method to assess the ecological status of European rivers in support of the Water Framework Directive. Wat Fram Direc. 2005;10(1):12–9.
20. Extence C, Bates A, Forbes W, Barham P. Biologically based water quality management. Environmental pollution. 1987;45(3):221–36.
21. Gabriels W, Lock K, De Pauw N, Goethals PL. Multimetric Macroinvertebrate Index Flanders (MMIF) for biological assessment of rivers and lakes in Flanders (Belgium). Limnologica-Ecology and Management of Inland Waters. 2010;40(3):199–207.
22. Karr JR. Assessment of biotic integrity using fish communities. Fisheries. 1981;6(6):21–7.
23. Management IWR. Iranian Water Quality Index for Surface Water Resource-Conventional Parameters "IRWQI". Tehran: IWRM; 2013. p. 5–14.

24. Said A, Stevens DK, Sehlke G. An innovative index for evaluating water quality in streams. Environmental management. 2004;34(3):406–14.

25. Mali L, Boaf N. Environmental condition of rivers and streams in the ovens catchment. Freshwater Sci. 2003;909(1):10–65.

26. Barbour MT, Gerritsen J, Snyder BD, Stribling JB. Rapid bioassessment protocols for use in streams and river: Pryphyton, Benthic Macroinvertebrates and fish. 2nd edition. Washington D.C: USEPA; 1999:841-B-99-002. 408p.

27. Barbour MT, Stribling JB, Verdonschot PF. The multihabitat approach of USEPA's rapid bioassessment protocols: Benthic macroinvertebrates. Limnetica. 2006;25(3):839–50.

28. Winger PV, Lasier PJ, Bogenrieder KJ. Combined use of rapid bioassessment protocols and sediment quality triad to assess stream quality. Environmental monitoring and assessment. 2005;100(1–3):267–95.

29. Tyagi S, Sharma B, Singh P, Dobhal R. Water quality assessment in terms of water quality index. American Journal of Water Resources. 2013;1(3):34–8.

30. Nabizadeh R, Amin MV, Alimohammadi M, Naddafi K, Mahvi AH, Yousefzadeh S. Development of innovative computer software to facilitate the setup and computation of water quality index. Journal of Environmental Health Science and Engineering. 2013;10(1):32.

31. Ruaro R, Gubiani ÉA. A scientometric assessment of 30 years of the Index of Biotic Integrity in aquatic ecosystems: applications and main flaws. Ecological Indicators. 2013;29:105–10.

32. Padmalal D, Maya K, Sreebha S, Sreeja R. Environmental effects of river sand mining: a case from the river catchments of Vembanad lake, Southwest coast of India. Environmental geology. 2008;54(4):879–89.

33. Saremi A, Saremi K, Sadeghi M, Sedghi H. The effect of aquaculture effluents on water quality parameters of Haraz River. Iranian Journal of Fisheries Sciences. 2013;12(2):445–53.

34. Rempel LL, Gill G, Fisheries Do, Oceans W, MB. Freshwater Inst. Bioassessment of streams along the Mackenzie River Valley, Canada, using the reference condition approach: biological, habitat, landscape and climate data. Winnipeg, MB (Canada): DFO; 2011.

35. Raczyńska M, Machula S, Choiński A, Sobkowiak L. Influence of the fish pond aquaculture effluent discharge on abiotic environmental factors of selected rivers in Northwest Poland. Acta Ecologica Sinica. 2012;32(3):160–4.

36. McCarthy KA, Johnson HM. Effect of agricultural practices on hydrology and water chemistry in a small irrigated catchment, Yakima River basin. Washington: US Geological Survey; 2009.

37. Smakhtin V. Environmental water needs and impacts of irrigated agriculture in river basins. A Framework for a New Research Program IWMI Working Paper. 2002;42

38. Abdoli A, Naderi M. Biodiversity of fishes of the southern basin of the Caspian Sea. Tehran: Abzian Scientific Publication; 2009.

39. Mostafavi H, Pletterbauer F, Coad BW, Mahini AS, Schinegger R, Unfer G, et al. Predicting presence and absence of trout (Salmo trutta) in Iran. Limnologica - Ecology and Management of Inland Waters. 2014;46:1–8.

40. Nnaji J, Uzairu A, Harrison G, Balarabe M. Effect of pollution on the physico-chemical parameters of water and sediments of river Galma, Zaria, Nigeria. Research Journal of Environmental and Earth Sciences. 2011;3(4):314–20.

41. Benítez-Mora A, Camargo JA. Ecological responses of aquatic macrophytes and benthic macroinvertebrates to dams in the Henares River Basin (Central Spain). Hydrobiologia. 2014;728(1):167–78.

42. Walker SL, Hedley K, Porter E. Pulp and paper environmental effects monitoring in Canada: An overview. Water Quality Research Journal of Canada. 2002;37(1):7–19.

43. Thompson G, Swain J, Kay M, Forster C. The treatment of pulp and paper mill effluent: a review. Bioresource technology. 2001;77(3):275–86.

44. Banagar GR, Kiabi BH, Homayoonnezhad I, Piri I, Amirian P. Biodiversity of Fish Species in Haraz River (An Ecological Approach). World Applied Sciences Journal. 2008;5(1):05–11.

45. Yazdian H, Jaafarzadeh N, Zahraie B. Relationship between benthic macroinvertebrate bio-indices and physicochemical parameters of water: a tool for water resources managers. Journal of Environmental Health Science and Engineering. 2014;12(1):30.

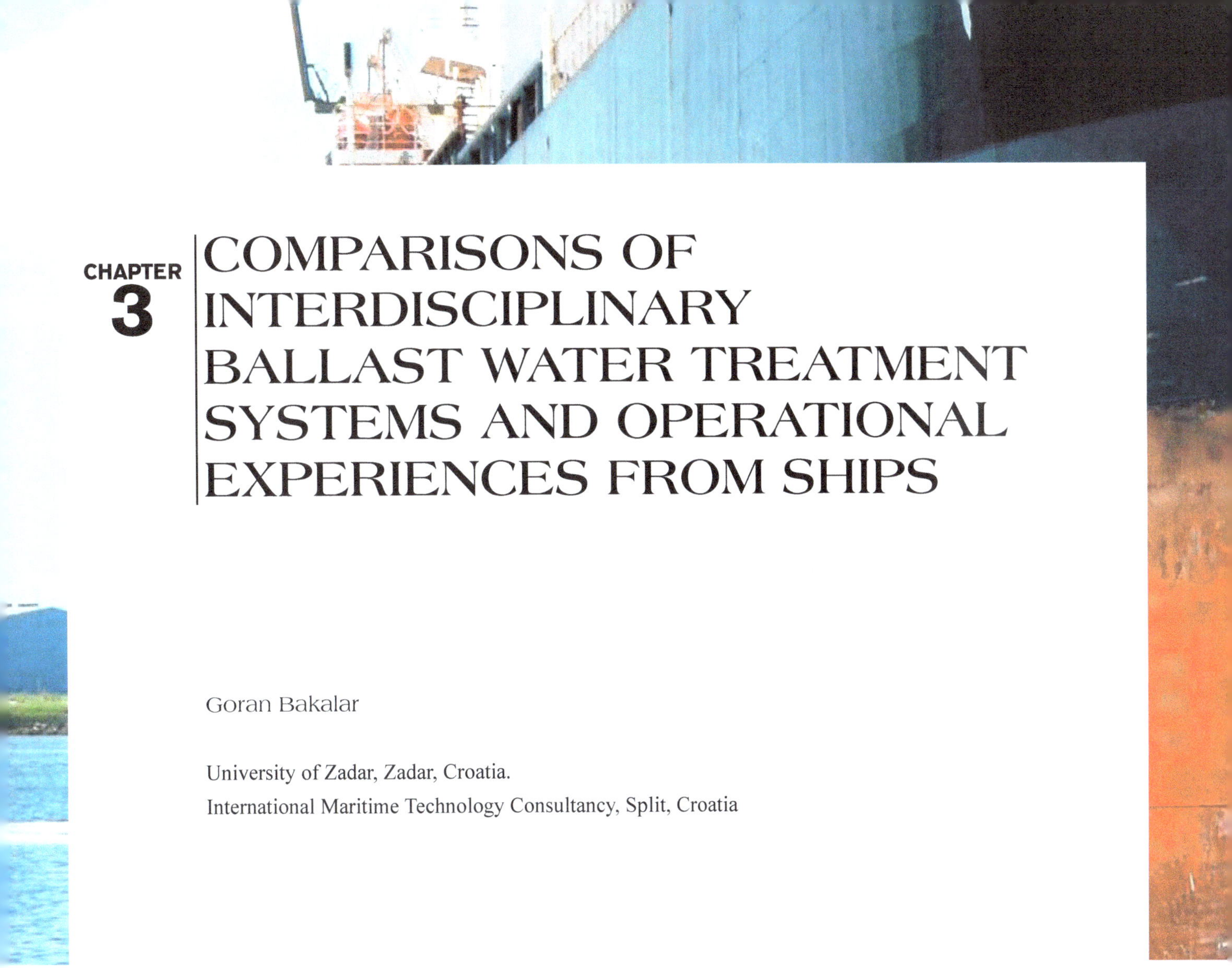

CHAPTER 3

COMPARISONS OF INTERDISCIPLINARY BALLAST WATER TREATMENT SYSTEMS AND OPERATIONAL EXPERIENCES FROM SHIPS

Goran Bakalar

University of Zadar, Zadar, Croatia.

International Maritime Technology Consultancy, Split, Croatia

ABSTRACT

There are high functioning and low functioning ballast water treatment systems on board ships. In this study, five systems were analysed so as to methodically examine the operational difficulties for ship crew members while giving important consideration to sustainable environment practices. Multi-criteria analysis, a questionnaire, survey and interviews were used as the research method so as to ascertain and corroborate existing problems on board ships, and the reliability of the systems was calculated. The co-insistency, maintenance and the efficiency of the systems, were shown as being the major problem as there are no systems for tracking ship ballast operations from land. The treatment system that used oxidants was, through multi criteria analysis, evaluated as being the best and was ranked first. However, the survey results showed that the ship's crew had serious problems with this system which difficult to solve during the ship's operations with cargo. The deoxygenation system was the most appropriate according to ballast water treatment criteria in the port or at sea. The treatment system which used electrolysis with oxidant was better in terms of efficacy and the treatment system electrolysis with ultra violet light was better in terms of the criterion environment pollution footprint. During further research, it was shown that 7 % of the surveyed crew members had major problems with operating ballast water treatment systems, including the system which was ranked first through multi criteria analysis. They by-passed these systems while continuing to ballast or de-ballast. It was calculated that of the total time needed for the ballast water treatment system operation, 9 % of this time was used for repairs or maintenance of the systems. Some examples are changing a used UV bulb, cleaning the filter or controlling the amount of oxidant which would be discharged into the sea. A conclusion was made and solution was suggested. The study results emphasised taking action in the interest of protecting the natural world, with particular attention being given to environmental protection to support human life.

Citation: Goran Bakalar, Comparisons of interdisciplinary ballast water treatment systems and operational experiences from ships, doi: 10.1186/s40064-016-1916-z.

BACKGROUND

When large ships offload their cargo, they flood ballast tanks with seawater for stability on the return trip. Ballast water is widely used on ships to ensure manoeuvrability and stability when cargoes are unloaded (La Carbona et al. 2010). Without proper ballast, the ship's centre of gravity is too high and can result in their capsizing. At the cargo loading port, the ballast water is pumped into the harbour in order to load more cargo. When ballast water is pumped overboard it may carry with it aquatic nuisance species that can seriously harm indigenous species in ports and coastal waters. These nuisance species may include unwanted bacteria such as Escherichia coli or Vibrio cholerae and animals such as zebra mussels. In addition, the ballast tanks are often contaminated with invasive species of animals, plants, and bacteria from the previous port. The BWMC (Ballast Water Management Convention) 2004 protects the sea environment through regulations regarding future ballast water treatment. Regulations and standards established by the IMO (International Maritime Organization) require treatment limits to be met independently and calls for vessels to carry out monitoring themselves (Albert et al. 2013). Ballast water will have to be treated once this convention comes into effect (Gregg et al. 2009).

Many technologies have been developed for ballast water treatment and fall into two broad categories: in-line and in-tank. In-line systems operate so that water is pumped from the sea to the ballast tanks routed through a processing system which kills organisms before they reach the ballast tank. In-tank systems kill the organisms after the ballast tanks are filled with seawater from the port. These are reasonably considered batch process systems. These systems treat water in the ballast tanks rather than during intake or discharge. Treatment of the ballast water can occur en route between ballast water operations and this is the principal advantage of in-tank systems. Different technologies have been suggested for ballast water treatment, such as disinfection with chlorine (Simpson 2001), injecting chemicals (La Carbona et al. 2010), adding biocides to ballast water for the neutralisation of harmful microorganisms (Chelossi and Faimali 2006), sterilisation with ozone (Perrins et al. 2006), filtration with UV (ultraviolet) light (Sutherland et al. 2013), exposure to electrochemical charge or electro-ionization (Aliotta et al.2001), exposure to heat (Mountfort et al. 2001), and sonication (Gavand et al. 2007). Some suggested systems use acids, while others sterilisation with hydrogen peroxide (Smit et al.2008). The deoxygenation system (Browning 2001) kills organisms in the ballast tanks by creating a water environment (inert gas + CO_2) which is low in oxygen, higher than normal in CO_2, and with a lower than normal seawater pH. Some systems employ multiple methods to increase effectiveness. The 2 × filtration + peracetic acid system utilises the method of passing the ballast water through the filters and adding peracetic acid. The filtration + electrolysis + ultraviolet light system provides the method of killing plants, bacteria and animals passing the ballast water through the filter and exposing them to electrolysis and ultraviolet light (Sutherland et al. 2013). The electrolysis system conducts sterilisation passing the ballast water through the filter with electrolysis (Rigby and Taylor 2001). The filtration with electrolysis, electrochlorination and waste control system uses the electrolysis method, injecting chlorine to induct a electrochemical reaction and removing the waste depending on the water quality. For this system, current density is an important parameter effecting the production of total residual chlorine in ballast or brackish water. Low current densities can avoid the production of harmful chlorine species (Lacasa et al. 2013).

A complete listing of the companies, country of origin, type of treatment, current status of testing and comments are available (Lloyd's 2014). Numerous BWTS (Ballast Water Treatment Systems) have been type approved by flag administrations to meet the requirements of the IMO ballast water convention (Lloyd's 2014; EPA Environmental Protection Agency 2011). At the moment, many ships are already equipped with ballast water treatment systems. Not all of them have treatment systems with optimal operational performances. When shipping company owners plan to buy ballast water treatment systems, they can choose between similar treatment systems with regard to the needs of their fleet. Since there is no automatic performance control to alert port authorities, they can choose cheaper and smaller systems with lower treatment costs when considering the load rate of ballast water in m^3/h.

The objective of this study is to assess the performance of various ballast water treatment systems from the point of view of ship crew members. Real, first-hand experience can verify the impact that individuals can have on the environment. The aim of this research is to reduce the negative impact that ship operations have on the environment. The overall aim of the study is to develop processes that will lead to complete environmental sustainability in the future.

METHODS

The materials used in this research were different ballast water treatment systems and their operational processes. Methods used in this research were multi-criteria analysis, a questionnaire, survey and reliability study.

The predominant methods of killing plants, bacteria and animals in ballast water treatment systems require passing the ballast water through filters (Matheickal et al. 2001). Filters in the systems, the maintenance of the UV generator and efficacy were imperative for this research. Efficacy means whether the system performance met the IMO guidelines of not discharging the prescribed quantities of living phytoplankton, zooplankton, microbial organisms and residue of used oxidant into the water.

Multi-criteria analysis

Multi-criteria analysis can be defined as a decision-making model that consists of a set of solutions (variants to rank or sort by the decision-maker), a set of criteria (quantitative and qualitative, economic and ecological, using different measures) and the set value (score) of each variant for each criteria (Hajkowicz and Collins 2007). The PROMETHEE (Preference Ranking Organization Method for Enrichment Evaluations) method was designed to analyse criteria parameters including an alternative choice of specific criteria. In PROMETHEE I partial ranking, $\varphi+\varphi+$ shows how much one criterion (a) prefers the other criteria (b):

$$\varphi^{+}(a) = \left(\frac{1}{n-1}\right)\sum_{b\neq a}\pi(a,b) \quad (1)$$

The other option φ^{-} shows the weakness of one criterion against the other criteria:

$$\varphi^{-}(a) = \left(\frac{1}{n-1}\right)\sum_{b\neq a}\pi(b,a) \quad (2)$$

"*a*" would prefer "*b*" only if:

$$\varphi^{+}(a) \geq \varphi^{+}(b) \quad (3)$$

and also:

$$\varphi^{-}(a) \leq \cdots \varphi^{-}(b) \quad (4)$$

In PROMETHEE II complete ranking, the ranking criteria flow in opposite directions because of the different views which are not touching each other (that is presenting incomparable and different criteria which uses different measures: e.g. economic, environmental, and social); this means that:

"*a*" would prefer "*b*" only if:

$$\varphi(a) > \cdots \varphi(b) \quad (5)$$

The PROMETHEE II method was particularly well-fitted to this research as it could make decisions from incomparable criteria. PROMETHEE II has been used with success to solve many problems (Behzadian et al. 2009). It was suggested that the PROMETHEE II method be used in this paper since this method is based on partial aggregation and possible decisions are compared to each other in pairs and ranked. It was then possible to select the best decision. Possible decisions (solutions) in this research were different ballast water treatment systems with common criteria (parameters) as alternative criteria was not set as an option. Compared to the well-established ELECTRE decision making method (also based on partial aggregation), the PROMETHEE II method was easier to use and able to get more robust results (Al-Shemmeria et al. 1997; Thaillandier and Stinckwich 2011).

Criteria

The purpose of these analyses was to determine which ballast water treatment system had optimal performance from the ship crew members' the point of view and as good as possible a performance from an ecological aspect. Five different effective possible decisions (alternatives) with type approvals were chosen to be ranked using the PROMETHEE II multi-criteria decision making method:

- 2 × filtration + peracetic acid.
- Filtration + electrolysis + UV generator.
- Filtration + electrolysis + electrochlorination + waste control.
- Electrolysis + oxidant.
- Deoxygenation.

The importance of ecological protection was dealt with in three of the five criteria parameters. All criteria alternatives were analysed according to the following criteria parameters, as shown in Table 1.

Table 1: Parameters data (Lloyd's 2014)

Ballast water treatment system	Environment pollution footprint 200/2000 m3/h	Efficacy	G8 or G9	Treatment in port or on the sea during the voyage	Number of methods
2 × filtration + peracetic acid	4.3/-	4	G9	Port	2
Filtration + electrolysis + UV	0.6/3	4	G9	Port	2
Filtration + electrolysis + electrochlorination + waste control	8.7/12.4	4	G9	Port	3
Electrolysis + oxidant	4.5/11	5	G9	Port	2
Deoxygenation	5/9	1	G8	Sea	2

Environmental pollution in this research refers to the introduction of contaminants into the natural environment that could cause adverse change. The ecological footprint used for this criterion parameter was ascertained by measuring the ballast water treatment system operation demand on the Earth's ecosystems, i.e. the amount of natural resource capital used per annum. Efficacy was a testing score parameter, i.e. data collected from the test performance results of ballast water treatment systems recorded via the United States Marine Invasive Species Program (Dobroski et al. 2007). Whether a ballast water treatment system belongs to the G8 or G9 group depends on the test results. Tests were done during the certification process. The Ballast Water Management Convention (BWMC) and the Guidelines for approval of ballast water management systems—G8 must be taken into consideration for the approval of a ballast water management system. If a certain ballast water technology uses an active substance, then, to comply with the Convention, it should be approved by the IMO in accordance with theProcedure for approval of ballast water management systems that make use of Active Substances—G9, adopted by the MEPC (Marine Environment Protection Committee) in session 53 (MEPC 2005; Tsolaki and Diamadopulos 2010). The degree of environmental friendliness of a particular ballast system determines whether it belongs to the G8 or G9 group parameter weighting. The G8 or G9 classification treatment systems and efficacy parameters are the most significant for ports and port ecology. The system that uses peracetic acid will arouse the attention of port staff more as this substance belongs to the G9 group. The number of treatment methods and whether treatment is done in the port or during the voyage is an important parameter from the ship crew members' the point of view. Some systems employ multiple methods to increase effectiveness and some systems use just one method. If a certain ballast water technology uses one treatment method, then, it should be easier to maintain the system. If the system neutralizes the ballast water while the ship is in motion, crewmembers have more time to rest and to go ashore and visit the port during cargo operations. All these criteria parameters were weighted as per the research results collected in Table 1.

Weights simulation:

- Environment pollution 20 %.
- Efficacy 35 %.
- Belonging to the group G8 or G9 20 %.
- Treatment in the port or on the sea—during the voyage 15 %.
- Number of the treatment methods 10 %.

Weighting of criteria is subjective and has a direct influence on the results of prioritizing strategy options. It is therefore critical that criteria weights are determined rationally and truthfully (Mutikanga et al. 2011). Efficacy was the most important criterion. Efficacy is the reason BWTS exist as it shows the amount of quality in terms of treating ballast water. This criterion was given a weighting of 35 % for ranking. The two next most important criteria were ecological: environment pollution and belonging to the group G8 or G9 and were given a ranking weight of 20 % each. The remaining 25 % was given to the two remaining criteria. As the most important criterion to the ship's crew is treatment in port or at sea, it was given 15 %, while the number of methods of a particular BWTS was given the remaining 10 %.

Survey of experiences from ships

The objective of this survey was to ascertain difficulties regarding the operational process of ballast water treatment systems. This was based on ship crew members' operational experience and their personal opinion. The survey raises considerable doubts as to the reliability of quick answers to research questions, whether of a quantitative or qualitative nature (Charmaz 1995). The aim was to collect qualitative information to meet the objectives of the research. A simple exploratory questionnaire was created with four questions. A questionnaire is a research instrument and basic scientific method whose purpose is to gather information from respondents. The discovery of reflexive progression in interviewing is very important for research (Miller and Glassner 1997). This study provides survey findings from a sample of 68 experienced ship captains and bridge officers, 51 of whom attended maritime courses and 18 from the maritime crewing agency. These courses were taught in Split, Sibenik and Rijeka, Croatia, at the Maritime School, and the students were interviewed in the period from 20/12/2014 to 1/3/2015. Interviews were carried out after the survey research data was collected. An on-board experience questionnaire was done for this research and a summary of the survey results are given and explained further in this paper.

The questionnaire

The questionnaire was made up of the following:

Circle one answer only.

How many years have you worked on board ships?

A. ONE YEAR.

B. THREE YEARS.

C. FIVE YEARS.

D. MORE THAN FIVE YEARS.

I work on ships as a:

A. BRIDGE OFFICER.

B. CAPTAIN.

There is a ballast water treatment system installed on the ship that I work on:.

A. YES.

B. NO

There were some functional problems while the ballast water treatment system was in operation:

A. YES.

B. NO.

The Reliability and Availability of the Systems

Malfunction of the ballast water treatment systems on board ships is a criterion that has never been reported or included in any research, but could prove to be the most important factor in ballast water treatment system performance. Performance reliability of certain ballast water treatment systems on board ships have indicators which should be observed, followed and analysed over a longer period of system exploitation. Reliability is the ability of the system for operational work without interruption (Pham 2000). Predictions of potential failures caused by software or hardware errors, as well as potential failures in mechanical part performance, can only be predicted once findings from the experience in handling technologically-similar systems on board ships are found. External factors on ships are one of the most important factors, because the dynamics of movements are constantly under external influences. Likewise, operator reliability and exposure to high temperatures are also very important in assessing the reliability of the proposed system (Siewiorek and Swarz 1982).

The reliability of these types of devices can be shown as an exponential function of the time interval if the time interval is considered to be the useful lifespan of the device. For electronic systems such as sophisticated devices for the on-board ballast water treatment of ships, the failure density function has a form of an exponential distribution, so the failure frequency function is the same by definition (Turban et al. 2003):

$$\lambda(t) = \frac{f(t)}{R(t)} = \frac{\lambda e^{-\lambda t}}{e^{-\lambda t}} = \lambda \qquad (6)$$

In the Eq. (6), we can see that the failure frequency function has a constant value (λ). As such, the equation for the reliability function R(t) (exponential law of reliability) has the form of:

$$R(t) = e^{-\lambda t} \tag{7}$$

The computing subsystem of some of the ballast water treatment systems on board ships can be found in three states: proper function, procedural failure and non-procedural failure (Shooman2002). Computing subsystem reliability depends on the probability of proper function and failures in a specified time. The probability of a computing subsystem of certain ballast water treatment systems on board ships going from a state of proper function to a state of non-procedural failure is:

$$P = (t + \Delta t) = \lambda \Delta t P_c (1 - F) + (1 - \mu \Delta t) P_{i.o}(t) \tag{8}$$

where PPii.ooPPii.oo probability of non-procedural state, PPccPPcc probability of proper functioning state, ***μ***frequency of repairs, ***F*** redundancy failure detection ratio, ***Δt*** time lag/time interval, ***λ***malfunction index.

The Malfunction index is the relation between malfunctioning components and operational components:

$$\lambda = \frac{1}{P_c} \frac{dP_f}{\Delta t} \tag{9}$$

where P_c is the number of components which remained operational after a specified time, and P_f is the number of components which malfunctioned after the specified time of functioning.

Redundancy is a characteristic of the quality of the computing system that ensures failure avoidance when one part of the system fails. This is generally ensured through the use of additional software, through the reliability of the two redundant systems working in parallel and with a known malfunction index for these types of devices in predefined ship conditions.

It can be assumed that the computing part of the subsystem in the system with two parallel subsystems, one of which is redundant, will be regularly maintained by the operator and through the self-diagnosis function for error removal. In this way, the possibility of malfunction of some ballast water treatment systems on board ships is reduced.

RESULTS AND DISCUSSION

Multi-criteria analysis of the systems

The D-Sight computerised visual method or visual projection of PROMETHEE II analysis, as seen in Figure. 1, has projected and showed the most appropriate system in ranking the results.

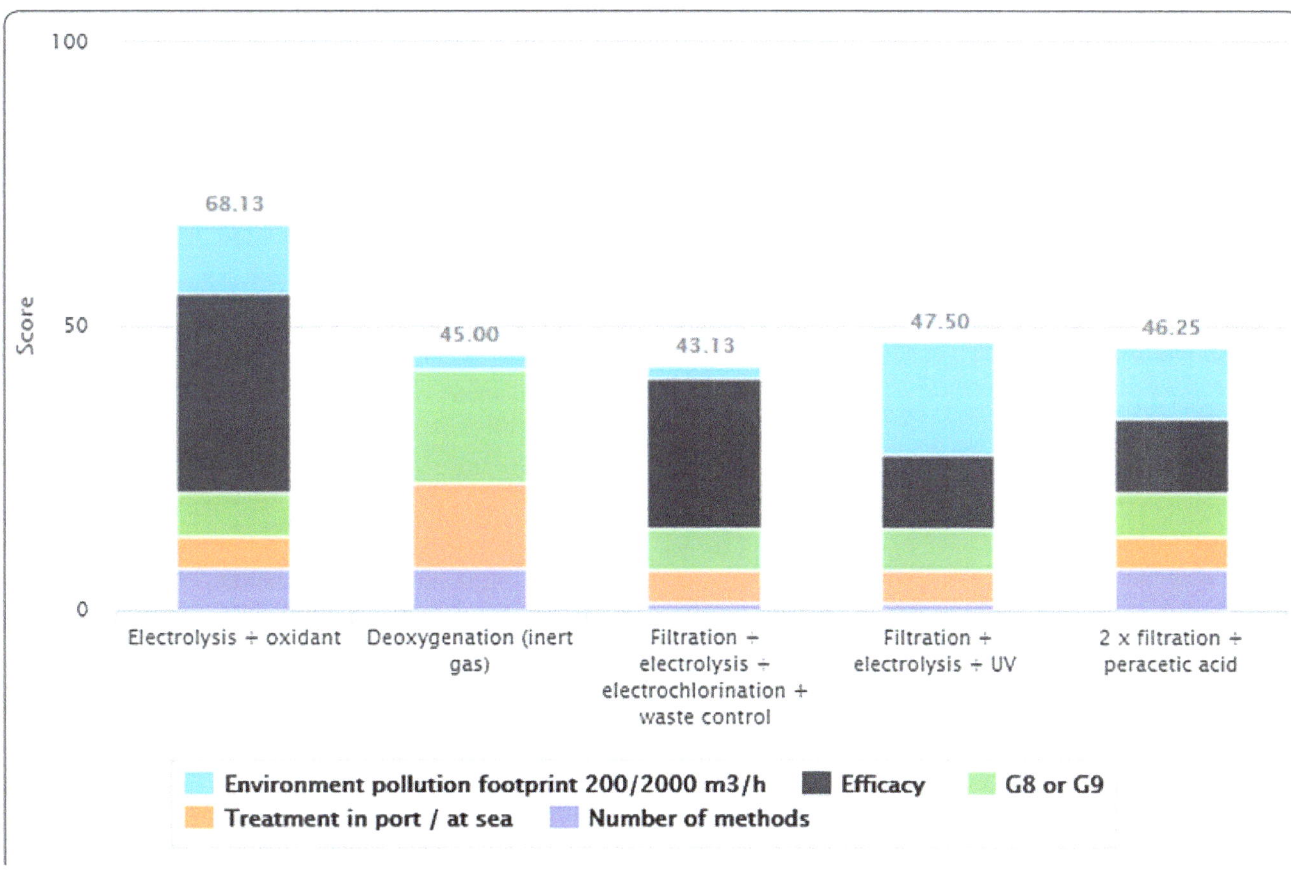

Figure : Multi-criteria analysis PROMETHEE II results.

The criteria parameters contribution to the solutions is shown in five different colours on the Figure. The treatment system of electrolysis + oxidant technology was highly successful in this computational analysis with a percentile of 68.1 %. The value of the criteria parameter environmental pollution for the BWTS filtration with electrolysis and UV, was the most important criterion for ranking second with a percentile of 47.5 %. The system 2 × filtration with peracetic acid technology was ranked third with a percentile of 46.25 %. Crucial to its coming in third place was the high quality of its efficacy value, as well as environmental pollution, in terms of criteria parameters. As one of the analysis criteria was the treatment of ballast water in the port or at sea, the deoxygenation system received a percentile of 45 % as it is the only ballast water treatment system which can be used while at sea. The system which was ranked last was the filtration + electrolysis + electrochlorination + waste control system. It can be confirmed that this system has a high value in terms of the efficacy criteria parameter. However, because of its high environmental pollution footprint, as well as the highest number of treatment methods, received a percentile of 43.1 % and was ranked fifth in this multi criteria analysis.

RESULTS OF THE SURVEY

Three students from the courses (out of the 51 surveyed) responded to questions 3, 4 and 5 of the questionnaire affirmatively. In the maritime crewing agency, two experienced officers responded affirmatively to questions 3, 4 and 5. Of the 5 ship captains who were surveyed and during the interview confirmed that they had problems with the operating of the BWTS, one of the surveyed captains worked on a ship with an installed BWTS using the treatment method of electrolysis and oxidant.

The problem that this chief officer referred to in regards to treatment system that uses electrolysis and oxidants was TRO (total residue oxidant). The measured TRO can provide an indication of treatment performance but not a direct measure of discharge standards. TRO was a problem for the TRO sensor unit in the deballastation process. Sensors would inhibit the operation of the ballast water treatment system due to the TRO value being too high. As the TRO was not at the required level, the residual TRO level of the treated ballast water was over 0.2 mg/L for the entire time, caused by an improperly dosed neutralizer in the neutralization unit of the system. The additional substance was most probably insufficient to automatically neutralize residual oxidants instantly and the system kept shutting down according to the interviewed chief officer. During the interview, two bridge officers that were surveyed said they were on a ship with an installed system that uses filtration, electrolysis and UV light. They said that it can take from three to three and a half hours to replace a UV bulb in a BWTS with a UV generator. Bulbs burn out often and in the case of a bulb burning out a second time, they by-passed the UV ballast water treatment system due to lack of time for cargo operations. According to another bridge officer who was interviewed for the survey, filters in the BWTS that uses filtration with UV light were a constant problem for one particular ship in a Brazilian port. The system was not able to operate due to filter problems. This problem was also referred to by another bridge officer who was on the ship with installed BWTS that uses filtration and UV light. All of officers interviewed stated that they by-passed the treatment system and continued the ballasting or de-ballasting process to avoid shutdowns. No reports were done at the time. The complete survey results can be found in Table 2. Ballast water treatment systems are not always in optimal operational condition and ships' crew members by-pass them, as often occured with automatic oil discharge content monitors in the past (McLaughlin et al. 2014; Bakalar 2014). This is an unacceptable facet of the ballast water treatment systems' implementation. Research regarding this problem should continue in the future so as to observe how survey results vary over time. Trend analysis reports would allow the survey response data to be charted over time and for it to be published in scientific journals.

Table 2: Survey results

	BWTS on board	No BWTS on board	Operational problems	Total sample surveyed	Total % with operational problems (%)
Surveyed Crew members attending courses	5	46	3	51	5.9
Surveyed Crew members in the agency for employment	4	14	2	18	11.1
Total	9	60	5	69	7.3

The Reliability and Availability of the Systems

The malfunction index or maintenance failure which cannot be diagnosed and automatically eliminated for these types of devices ranges from $\lambda = 0.5$ to $\lambda = 0.2$ (Bakalar 2013; Lovric 1989).λ can only be determined from on board ship operational history which is at present unavailable for ballast water treatment systems. The mean value of the malfunction index was the most optional value in this calculation and it was taken for all systems in common. If the mean is taken as being $\lambda = 0.35$ for each of the serially attached subsystems and this value is included in the equation of reliability for 1000 h of work, we can obtain a reliability value for each of the two computing systems individually, where one is redundant:

$$R(t) = e^{-\lambda t} = e^{-0.35} = 0.704688 \tag{10}$$

The total reliability of the whole computing subsystem as part of the ballast water treatment system on board ships which consists of two units, one of which is software redundant, is:

$$R(t) = R_u(1 - \lambda T) = 0.916097 \tag{11}$$

As such, the reliability of this subsystem has been proven according to the analysis of the results of the reliability of the computing system, which is a subsystem of a ballast water treatment system on board ships. The computing subsystem reliability with the possible use of the redundant software is 0.916 or 91.6 %.

CONCLUSIONS

This paper reviews ballast water treatment systems in three parts regarding data and methods: multi-criteria analysis, survey on real experience, reliability and availability study. According to the multi-criteria analysis results, the electrolysis + oxidant technology treatment system had a high success rate of 68.1 %. Many aspects were relevant and important. The electrolysis management system that uses oxidants was the most appropriate system from an ecological standpoint, while determining the quality of the treated ballast water was the most acceptable from an environmental point of view. The multi-criteria analysis parameters and scenarios did not consider the functional and operational problems of the systems. It can be clearly concluded from the survey result of 7.3 % from all participants and interviews that ship officers by-pass ballast water treatment systems due to lack of time for rectifying malfunctions, failures, maintenance or spare parts replacement in ballast water treatment systems. The survey results indicate that ballast water treatment systems can become faulty and shut down when operation sensors indicate improper function of the sub-systems. Even the system which utilises oxidant, which was successful in the multi criteria analysis, was shown to have many problems regarding controlling oxidants. This implies an unclear future for port environment protection management, where the ballast water is pumped into the harbour in order to load cargo. The research in this paper aids in understanding how important it is to learn more about ballast water treatment system experiences from crewmembers, since the survey of operational experiences from ships has proven that more time would be needed for the maintenance of ballast water management systems. The total calculated reliability of the entire computing subsystem as part of a particular ballast water treatment system on board ships which consists of the two units of which one is software redundant, was 0.916 or 91.6 %. This means that for 8.4 % of the operational time, any of the mentioned systems could be in failure or under repair. This is a significant risk for the operation of ballast water treatment systems. The results of this research are a warning to relevant stakeholders in the maritime industry. If the system does not operate well, or not at all, the price of treatment is higher, as is air and water pollution, and the active substances used, e.g. oxidant residue, become unmanageable and out of control. Environmental hazards need to be managed. One possibility would be a mechanism to monitor ballast water systems operation continually on board ships. A check-up system should be close as possible to real time. Sensors should report malfunctions to the port authorities immediately after any questionable or doubtful performance of ballast water treatment system occur.

ACKNOWLEDGEMENTS

I thank the participants of the research survey who provided important data while questioned, surveyed and interviewed by myself. This research did not receive any specific grant from any organization or funding agency in the public, commercial or not-for-profit sector.

COMPETING INTERESTS

The author has no financial interest in any company that makes a product of any method or technology discussed in this article. The author declares no conflict or competing interests, and no financial nor non-financial interests.

REFERENCES

1. Albert RJ, Lishman JM, Saxena JR (2013) Ballast water regulations and the move toward concentration-based numeric discharge limits. Ecol Appl 23(2):289–300
2. Aliotta J, Rogerson A, Campbell CB, Yonge M (2001) Ballast water treatment by electro-ionization. In: 1st International ballast water treatment R&D symposium. Proceedings, pp 61–69. http://globallast.imo.org/wp-content/uploads/2014/11/Mono5.pdf. Accessed 10 Jan 2016
3. Al-Shemmeria T, Al-Klouba B, Pearman A (1997) Model choice in multicriteria decision aid. Eur J Oper Res 97(3):550–560
4. Bakalar G (2013) Automatic control system for ship ballast water treatment by using flow cytometry and satellite communications technologies. Dissertation. Rijeka University Croatia
5. Bakalar G (2014) Review of interdisciplinary devices for detecting the quality of ship ballast water. SpringerPlus 1(3):468. doi:10.1186/2193-1801-3-468
6. Behzadian M, Kazemzadeh R, Albadvi A, Aghdasi M (2009) PROMETHEE: a comprehensive literature review on methodologies and applications. Eur J Oper Res 200(1):198–215
7. Browning WJ Jr, Browning WJ III (2001) Ballast water treatment by de-oxygenation-the AquaHabiStat system. In: 1st International ballast water treatment R&D symposium. Proceedings, pp 51–60. http://globallast.imo.org/wp-content/uploads/2014/11/Mono5.pdf. Accessed 10 Jan 2016
8. Charmaz K (1995) Between positivism and postmodernism: implications for methods. Stud Symb Interact 17:43–72
9. Chelossi E, Faimali M (2006) Comparative assessment of antimicrobial efficacy of new potential biocides for treatment of cooling and ballast waters. Sci Total Environ 356:1–10

10. Dobroski N, Takata L, Scianni C, Falkner M (2007) Assessment of the efficacy, availability and environmental impacts of ballast water treatment systems for use in California waters. California State Lands Commission Marine Facilities Division
11. EPA Environmental Protection Agency (EPA) (2011) Efficacy of ballast water treatment systems: a report by the EPA science advisory board. http://www.EPA-SAB-11-009-unsigned.pdf. Accessed 10 Jan 2016
12. Gavand MR, McClintock BM, Amsler CD, Peters RW, Angus RA (2007) Effects of sonication and advanced chemical oxidants on the unicellular green alga Dunaliella tertiolecta and cysts, larvae and adults of the brine shrimp Artemia salina: a prospective treatment to eradicate invasive organisms from ballast water. Mar Pollut Bull 54(11):1777–1778
13. Gregg M, Rigby G, Hallegraeff GM (2009) Review of two decades of progress in the development of management options for reducing or eradicating phytoplankton, zooplankton and bacteria in ship's ballast water. Aquat Inv 4(3):521–565
14. Hajkowicz S, Collins K (2007) A review of multiple criteria analysis for water resource planning and management. Water Resour Manag 21(9):1553–1566
15. La Carbona S, Viitasalo-Frösen S, Masson D, Sassi J, Pineau S, Lehtiniemi M, Corroler D (2010) Efficacy and environmental acceptability of two ballast water treatment chemicals and an alkylamine based-biocide. Sci Total Environ 409(2):247–255
16. Lacasa E, Tsolaki E, Sbkou Z, Rodrigo MA, Mantzavinos D, Diamandopoulos E (2013) Electrochemical disinfection of simulated ballast water on conductive diamond electrodes. Chem Eng J 223:516–523
17. Lloyd's (2014) Ballast water treatment technology—current status.http://www.lr.org/en/_images/213-35824_Understanding_Ballast_Water_Management_0314_tcm155-248816.pdf, 2014. Accessed 15 Jan 2016
18. Lovric J (1989) Basics of ship's terotechnology. Dubrovnik Maritime Faculty Croatia 44–57.http://www.unidu.hr/datoteke/172izb/osnove_brodske_terotehnologije_-_Josip_Lovric.pdf. Accessed 15 Jan 2016
19. Matheickal JT, Waite TD, Mylvaganam ST (2001) Ballast water treatment by filtration. In: 1st International ballast water treatment R&D symposium. Proceedings, pp 7–20. http://globallast.imo.org/wp-content/uploads/2014/11/Mono5.pdf. Accessed 18 Nov 2015
20. McLaughlin C, Falatko D, Danesi R, Albert R (2014) Characterizing shipboard bilgewater effluent before and after treatment. Environ Sci Pollut Res 21(8):5637–5652
21. MEPC: g/2012/Individual Guidelines for reference/G4.pdf, 2005. Accessed 12 Jan 2016
22. Miller J, Glassner B (1997) The inside and outside funding realities in interviews. In: Silverman D (ed) Qualitative research: theory, method and practise. Sage Publications, Beverley Hills, pp 98–111
23. Mountfort D, Dodgshun T, Taylor M (2001) Ballast water treatment by Heat—New Zealand Laboratory & Shipboard trials. In: 1st International ballast water treatment R&D symposium. Proceedings, pp 45–50. http://globallast.imo.org/wp-content/uploads/2014/11/Mono5.pdf. Accessed 14 Jan 2016
24. Mutikanga HE, Sharma SK, Vairavamoorthy K (2011) Multi-criteria decision analysis: a strategic planning tool for water loss management. Water Resour Manag 25:3947–3969
25. Perrins JC, Cooper WJ, Van Leeuwen JH, Herwing RP (2006) Ozonation of seawater from different locations: formation and decay of total residual oxidant implications for ballast water treatment. Mar Pollut Bull 52:1023–1033
26. Pham H (2000) Software reliability. Springer Google Books, Berlin
27. Rigby G, Taylor AH (2001) Ballast water management and treatment options. Int Marit Technol 113(3):79–99
28. Shooman ML (2002) Reliability of computer systems and networks: fault tolerance, analysis, and design. Wiley, New York
29. Siewiorek DR, Swarz R (1982) The theory and practice of reliable system design. Digital Press Bedford, Maynard
30. Simpson G (2001) Ballast water disinfection with ClO_2. In: Proceedings of the 20th international conference on marine biology, pp 131
31. Smit MGD, Ebbens E, Jak RG, Huijbregts MJA (2008) Time and concentration dependency in the potentially affected fraction of species: the case of hydrogen peroxide treatment of ballast water. Environ Toxicol Chem 27:746–753
32. Sutherland TF, Levings CD, Petersen S, Hesse WW (2013) Mortality of zooplankton and invertebrate larvae exposed to cyclonic pre-treatment and ultraviolet radiation. Mar Technol Soc J 37:3–12
33. Thaillandier P, Stinckwich S (2011) Using the PROMETHEE multi-criteria decision method to define new exploration strategies for rescue robots. In: IEEE international symposium on safety, security, and rescue robotics. Proceedings, pp 321–326. https://hal.archives-ouvertes.fr/hal-00688925v1. Accessed 14 Jan 2016
34. Tsolaki E, Diamadopoulos E (2010) Technologies for ballast water treatment: a review. J Chem Technol Biot 85(1):19–32

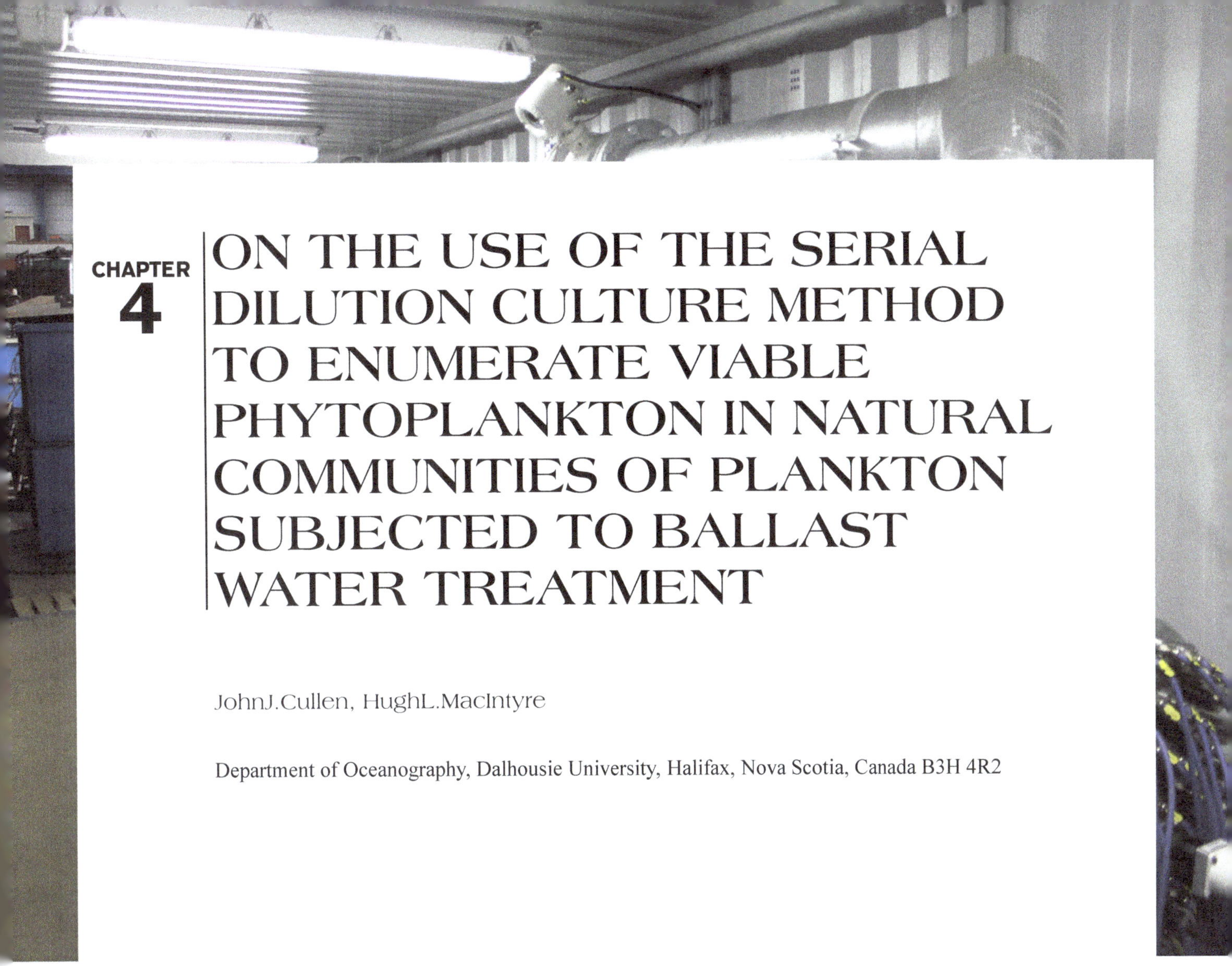

CHAPTER 4

ON THE USE OF THE SERIAL DILUTION CULTURE METHOD TO ENUMERATE VIABLE PHYTOPLANKTON IN NATURAL COMMUNITIES OF PLANKTON SUBJECTED TO BALLAST WATER TREATMENT

JohnJ.Cullen, HughL.MacIntyre

Department of Oceanography, Dalhousie University, Halifax, Nova Scotia, Canada B3H 4R2

ABSTRACT

Discharge standards for ballast water treatment (BWT) systems are based on concentrations of living cells, for example, as determined with vital stains. Ultraviolet radiation (UV) stops the reproduction of microorganisms without killing them outright; they are living, but not viable, and ecologically as good as dead. Consequently, UV-treated discharge can be compliant with the intent of regulation while failing a live/dead test. An alternative evaluation of BWT can be proposed based on the assessment of viable, rather than living, cells in discharge water. In principle, the serial dilution culture-most probable number (SDC-MPN) method provides the appropriate measure for phytoplankton. But, the method has been criticized, particularly because it is thought that many phytoplankton species cannot be cultured. A review of the literature shows that although SDC-MPN has been used for more than 50years—generally to identify and count phytoplankton species that cannot be preserved—its application to enumerate total viable phytoplankton seems to be new, putting past criticisms of the method in a different light. Importantly, viable cells need to grow only enough to be detected, not to be brought into sustained culture, and competition between species in a dilution tube is irrelevant as long as the winner is detectable. Thorough consideration of sources of error leads to recommendations for minimizing and quantifying uncertainties by optimizing growth conditions and conducting systematic comparisons. We conclude that with careful evaluation, SDC-MPN is potentially an effective method for assessing the viability of phytoplankton after BWT.

INTRODUCTION

In response to the threats from continued introductions of aquatic invasive species, the United Nations International Maritime Organization (IMO) adopted the*International Convention for the Control and Management of Ships' Ballast Water and Sediments*(IMO2004). The convention has yet to be ratified, but the US Coast Guard (USCG) has established national regulation of systems to "kill, render harmless, or remove" organisms from ballast water discharge (US Coast Guard2012). Ships will have to treat ballast water to meet regulatory discharge standards. Two size classes of plankton are subject to regulation, classified by size: ≥50, and ≥10 and <50μm (IMO2004; US Coast Guard2012). To meet the

Citation: John J. Cullen, Hugh L. MacIntyre, On the use of the serial dilution culture method to enumerate viable phytoplankton in natural communities of plankton subjected to ballast water treatment, DOI: 10.1007/s10811-015-0601-x standard, any ballast water management system (BWMS) must discharge <10 "living" cellsmL^{-1}(US Coast Guard2012) or <10 "viable" cellsmL^{-1}(IMO2004) in the 10–50μm size range and <10 "living" or "viable" cellsm^{-3}in the >50μm size range. The USCG acknowledges that the two standards are slightly different (US Coast Guard2012) but points out that for the purpose of their approval guidelines, the IMO defines "viable" as "living" (see International Maritime Organization Marine Environment Protection Committee2008).

The distinction between viable, which for our discussion we define as being reproductive, and living, i.e., showing signs of vitality, has important implications for the evaluation of ballast water treatment (BWT) systems. One method, irradiation with ultraviolet radiation (UV, particularly ultraviolet-C), is a proven and widely-applied technology for disinfection of wastewater and drinking water (Hijnen et al.2006) that inactivates microbes by destroying their ability to reproduce but without necessarily killing them outright. As a result, cells that have been effectively treated with UV can be intact and metabolically active—that is, living—but incapable of reproduction and thus nonviable (First and Drake2013a). Consequently, organisms that have been rendered harmless through treatment with UV would be compliant with the intentions of BWT regulations but living and thus noncompliant according to the regulations themselves.

Since a living, but nonreproductive, microbe is ecologically as good as dead (i.e., it is not a viable propagule, a term used by Reavie et al.2010), it can be argued that viability is inherently more accurate than vitality as a measure of invasive potential and that alternative BWT regulations based on viable cells, as compared with living cells, should provide equal protection to the environment while allowing the effectiveness of UV treatment systems at rendering cells harmless to be assessed more accurately. This argument is relevant because the Environmental Technology Verification (ETV) protocol includes consideration and utilization of alternative methods after they have been validated (ETV2010). But, an important question must be considered: Can the concentration of viable cells in natural plankton communities be measured reliably? Addressing the issue in 2012, the USCG opted to use live/dead rather than viable/nonviable as a regulatory criterion, because the determination of viability would require culturing potentially large numbers of different kinds of organisms, many of which, they claimed, scientists are not able to culture (US Coast Guard2012). Consistent with the reasoning that all dead cells are also nonviable and thus noninvasive, the Coast Guard further supported their decision by pointing out that live/dead is more conservative, and thus more protective, than viable/nonviable. But, because the UV doses required to kill microbes greatly exceed those required to inactivate them—past the point of economic practicality--the live/dead criterion could effectively exclude UV technology from being used to treat ballast water.

A recent study by First and Drake (2013a) frames the live/dead/viable issue in the context of assessing the effectiveness of UV for BWT. Arguing that viability is the appropriate measure of invasive potential, they assert that direct measures of growth after treatment ("regrowth assays," e.g., Liebich et al.2012) are definitive. But, they point out that such assays are time-consuming, lasting days to weeks, and that the method applies only to organisms that will grow under laboratory conditions, but that many microorganisms cannot be cultured. The regrowth method employed by Liebich et al. (2012) and others with similar aims (Wright et al.2009) tracked the growth of cells after BWT, but they did not estimate the initial concentration of total viable cells per mL immediately after treatment, the measure most relevant to discharge regulations.

Many assays classify single cells according to signs of life (Zetsche and Meysman2012; Steinberg et al.2012; Reavie et al.2010), including vital stains (Steinberg et al.2011; Reavie et al.2010; Zetsche and Meysman2012), a mortal stain (Steinberg et al.2012; Reavie et al.2010), cellular integrity (Burkholder et al.2007; Wright et al.2009), motility (Gregg and Hallegraeff2007), and indicators of cell division (discussed by First and Drake2013a). Each of these measures is unquestionably related to vitality or viability, but none have been related systematically (i.e., across taxa and subject to varying degrees of debilitation) to quantitative measures of the capability of microbes to reproduce after BWT.

Bulk measurements, such as metrics of variable chlorophyll fluorescence (Drake et al.2014) and changes in chlorophyll concentration (e.g., Wright et al.2009), have been examined as measures of the effects of BWT on phytoplankton. Zetsche and Meysman (2012) argue that the existing regulations based on cell counts preclude the use of such bulk measurements for testing, but Drake et al. (2014) present a validation framework for compliance monitoring that would use bulk measurements when they are related quantitatively to concentrations of cells that satisfy the regulatory criterion—for their examples, living cells. For proxies of the ability to reproduce, such bulk measurements would have to be related to concentrations of viable cells. Even if they do not provide direct estimates of living or viable cells in discharge water, rapid assays based on bulk measurements or single-cell indicators can be important in shipboard testing for compliance with discharge regulations (King and Tamburri2010).

Clearly, there is a need for a method to enumerate viable cells in a sample of plankton. As we will discuss in more detail below, the most direct method for phytoplankton is the serial dilution culture-most probable number assay (SDC-MPN, also called the extinction dilution method, Throndsen1978) (Fig.1). The approach is based on a bacteriological assay developed more than a century ago (McCrady1915; Cochran1950), which was applied to phytoplankton in 1951 (Knight-Jones1951). The SDC-MPN method is considered to be problematic when applied to natural communities of phytoplankton, however, in large part because many planktonic microorganisms are assumed to be unculturable (Steinberg et al.2011; First and Drake2013a; US Coast Guard2012) and also because of concerns about interactions between species[1]in the dilution cultures. Reflecting such concerns, the US Environmental Protection Agency ETV Program (2010) generic protocol for the verification of BWT technology cautioned that the MPN method that they used on cultures of phytoplankton (see also Oemcke and Van Leeuwen2005) "is suitable for pure cultures of heterotrophic protists or phytoplankton, but is not a useful tool for mixed cultures." Here, we revisit prior assessments of the method.

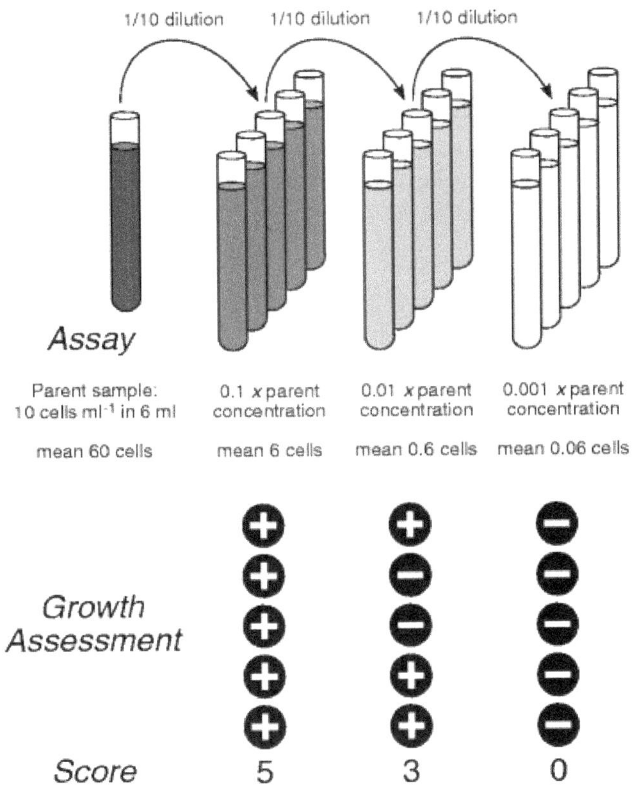

Figure. 1:

Principles and assumptions of the serial dilution culture-most probable number (SDC-MPN) method. The SDC-MPN method (Throndsen1978) estimates the concentration of viable cells in a sample, based on incremental dilution of the sample into a series of replicated liquid subcultures (e.g., test tubes) and statistical determination of the hypothetical dilution that corresponds to one viable cell per subculture tube. Back calculation gives the number of viable cells in the parent sample tube; the concentration of viable cells (cellsmL^{-1}) is determined from that number and the volume of culture in the tube. The test is based on the discrimination of subcultures containing one or more viable cells (assumed for now to result in a positive score for growth) from those with none (a negative score). The proportion of tubes scoring positive at any given dilution is a function of the concentration of viable cells in the parent culture and the dilution factor. It is assumed (Cochran1950; Haas and Heller1988) that (i) organisms are randomly distributed in each tube and evenly distributed between subsamples and (ii) growth will be reliably detected in any tube containing one or more viable phytoplankton cells. • In principle, all replicated tubes expected to have ≫1 viable cell (low dilution) would show growth and score positive. At higher dilutions, negative scores become more likely until at very high dilutions (calculated concentration <<1 per tube), the number of positive scores declines to zero. • Among*n*replicate tubes at any dilution, the likelihood of*s*negative scores follows the binomial distribution set by the probability of a tube being sterile; the most likely number of positive scores thus corresponds to the probability of a tube having one or more viable cells (Cochran1950). • In the illustrated example of a parent tube of 6mL containing 60 cells (10 cellsmL^{-1}), the calculated number of viable cells per tube in each successive tenfold dilution is 6, 0.6, and 0.06 respectively. In a five-replicate tube test, the most likely number of positives at each respective dilution is therefore expected to be 5, 3, and 0, but other combinations of scores are possible due to random chance. • The calculation of MPN, with 95% confidence intervals (CI), is based on statistical comparison of observed scores with calculated probabilities. The MPN corresponding to the test score can be calculated (e.g., Hurley and Roscoe1983; Garthright and Blodgett2003) or found in lookup tables (Blodgett2010); dilutions need not be constrained to tenfold intervals. In the illustrated example for a score of 5, 3, 0, the MPN is 79 cells and expressed as a concentration is 13 cellsmL^{-1}(=79 cells/6mL). The 95% CI, based on the logarithm of the estimate, is 4–37 cellsmL^{-1}. The confidence intervals decrease if the number of replicate tubes at each dilution is increased or if more dilutions, especially with smaller dilution factors (e.g., 5×) are used; larger tube volumes increase the sensitivity of the assay

FRAMEWORK OF THE REVIEW

Motivation

Sharing the widely held misgivings about SDC-MPN for natural communities but considering the demonstrated need for a robust assay of the concentration of viable phytoplankton in treated ballast water, we have reviewed the principles, assumptions, and applications of SDC-MPN. Our investigation has led us to a new appreciation of the method and its application in BWT. The ETV protocol (2010) allows the consideration of alternative methods for measuring the concentration of living organisms in discharge water. Based on our review, we suggest that SDC-MPN may serve as an alternative to the existing method based on vital stains.

Fundamental Postulate

Significantly, the SDC-MPN method enumerates viable phytoplankton but discharge regulations as they stand specify concentrations of living cells. This important distinction would have to be accommodated if the SDC-MPN method were to be adopted for use in regulations. This is not our decision to make. Rather, we present at the outset a fundamental postulate, based on the intent of BWMS regulations:

Because neither a dead organism nor a nonreproductive organism can propagate after discharge from ships' ballast, discharge criteria based on vitality (live/dead) and viability (the ability to reproduce) are equally protective of coastal environments.

This will be referred to as the postulate of equivalent protection.

Approach

Guided by this postulate, we classify potential errors in ballast water testing according to their implications for protection of the environment, not how they conform to live/dead regulatory criteria. In this context, we review the SDC-MPN method for phytoplankton and identify sources of inaccurate results; we discuss their likely influence on the estimation of total viable phytoplankton cells in natural communities of plankton before and after ballast water treatment, and how the resulting errors can be estimated and reduced.

Scope

Given its demands for time, SDC-MPN would be appropriate for land-based verification to gain type approval for ballast water management systems, not rapid assays that are required for shipboard compliance testing (King and Tamburri2010). The method enumerates only photoautotrophs, and its utility is primarily for counting viable cells in the 10–50µm size range that are not readily removed by filtration during BWT. We discuss briefly approaches that can be used to assess other components of the plankton, such as heterotrophs, and the use of SDC-MPN on cultures of phytoplankton to develop proxies of viability for rapid assays that would be suitable for shipboard compliance testing and port state control inspection.

Objective

We intend to support an argument that the SDC-MPN method can potentially produce ecologically valid assessments of the effectiveness of ballast water treatment that are constrained with reasonable estimates of uncertainty, especially if sources of error are specifically assessed.

Background

Well over 100years ago, the dilution method was established as one of several techniques for isolating phytoplankton for growth in unialgal culture (Allen and Nelson1910). The authors dispensed one or two drops of a plankton sample into petri dishes containing, for example, 250mL of growth medium. After a few days, colonies of diatoms appeared, likely originating from individual cells in the dilute culture; they could easily be isolated by pipette and inoculated into fresh medium, ultimately to be maintained in unialgal culture.

Convinced that centrifugation—an established method for concentrating small phytoplankton in natural samples for microscopic enumeration—greatly underestimated the numbers of small cells that could not be retained by meshes, and recognizing that species that grew in diluted cultures represented at least one cell in the inoculum, Allen (1919) estimated the minimum numbers of cells in samples from the Knap Buoy station near Plymouth using a dilution method. He diluted 0.5mL of sea water in 1.5L of growth medium and divided the mixture between 70 flasks, each containing about 21mL, that were "placed in a north light and kept at room temperature without a fire" and examined periodically over the following 6weeks. Altogether, he found 232 different organisms distributed among the 70 flasks. He concluded that this was the minimum number of cells in the original 0.5mL, so the concentration of phytoplankton was at least 464 cellsmL^{-1}. Allen emphasized that the real count would be considerably higher than this because more than one cell of the same species might have been introduced into some flasks and also because not all species could grow under the conditions he provided. Even so, the count was more than 30 times that from a centrifuged sample. Allen concluded that no one approach could provide quantitative estimates of phytoplankton concentrations, so a variety of methods would have to be used to obtain accurate counts.

To support a study on the systematics and abundance of ultraplankton and nanoplankton, many of which are difficult to identify in a counting chamber and do not preserve well for enumeration, Knight-Jones (1951) used a dilution culture method to isolate and enumerate natural phytoplankton. The enumerations were based on procedures used by the UK Ministry of Health for bacteriological assays of water supplies, using statistical tables to estimate counts and errors. Three serial dilutions were prepared with Erd-Schreiber medium, each with five tubes, and these were incubated in windows facing north for 1 to 4months as he waited for cultures to develop. Knight-Jones (1951) tabulated probable numbers for total phytoplankton and also used the same tables to estimate the concentrations of individual species. He found that a 1.5-µm flagellate, now classified as the prasinophyte*Micromonas pusilla*, was the most generally abundant—remarkable because the organism had been previously undescribed. The importance of suitable culture conditions, difficult to provide at the time, was highlighted. Knight-Jones concluded, "If a thermostatically-controlled culture-cabinet were used, quantitative culturing would appear to be a very practicable method of nanoplankton estimation." (p. 154).

The SDC-MPN method was subsequently refined and described in the*Intergovernmental Oceanographic Commission–UNESCO Phytoplankton Manual*by Throndsen (1978) and updated by Andersen and Throndsen (2003), who highlighted that the method is useful for the isolation of phytoplankton while offering "the opportunity to make estimates of the original cell number." Indeed, over the years, SDC-MPN has been applied in studies that required isolation of phytoplankton cultures as well as enumeration of the starting concentrations of taxa from which they came. Modifications of the method have also been applied to enumerate propagules of phytoplankton in sediments (Ishikawa and Furuya2004), to estimate the concentrations of protozoa in sea water (Lighthart1969) and to isolate and enumerate phages (Suttle and Chan1993).

Serial dilution culture has been particularly useful in the enumeration of groups of phytoplankton that do not preserve well, such as flagellates and monads in the subsurface chlorophyll maximum (Furuya and Marumo1983), naked nanoflagellates in the Kiel Bight and Kiel Fjord (Jochem1990), and*Micromonas pusilla*as part of the nanoplankton communities of the Bering Sea (Throndsen and Kristiansen1991).

Throndsen and Kristiansen (1991; see also Backe-Hansen and Throndsen2002) identified sources of error in the SDC-MPN method when they applied it to identify and enumerate co-occurring species in samples: lack of growth of some specimens introduced into the tubes, competitive relations in a single tube, the influence of culture conditions and the medium on growth, and insensitivity of the method for detecting heterotrophs. We return to these issues in following sections.

A study of the diversity and seasonality of cryptomonads in the Gulf of Naples (Cerino and Zingone2006) illustrates strengths and weaknesses of the SDC-MPN method when it is used to describe community composition of the phytoplankton. Although they are identifiable by diagnostic pigments in bulk samples (Gieskes and Kraay1983) and fluorescence and cell-size signatures detectable with flow cytometry (Li and Dickie2001), the diversity of marine cryptomonads is scarcely known because generally they are not reliably identifiable to species under light microscopy and they are usually damaged by common fixatives (Kugrens and Lee1987). But, by growing cryptomonads in culture using SDC-MPN, Cerino and Zingone (2006) were able to identify reliably and describe patterns in abundances of cryptomonad species, in the process bringing about 80 strains into culture for more thorough examination. The authors acknowledged that the SDC-MPN method is subject to error when competition eliminates culturable specimens in some tubes; still, their method yielded information on diversity that could only be obtained by growing and identifying isolates. Flow cytometry with cell sorting offers alternatives to serial dilution, both for isolation of cultures (Sieracki et al.2005; Sinigalliano et al.2009) and genomic quantification of biodiversity without the need for culturing (Kashtan et al.2014; Heywood et al.2010), but it has yet to be examined as a method for quantifying total viable cells in natural samples of phytoplankton.

Inspection of type approvals for BWT systems (IMO2015) reveals that SDC-MPN has been used in regulatory testing for the 10–50μm size class of phytoplankton, but it is difficult to review the method specifically for the assessment of ballast water treatment because much of the information is outside the standard scientific literature (Albert et al.2013). When the serial dilution culture method is mentioned in publications about BWT, it is usually to point out its limitations, described above. However, First and Drake (2013b) conducted SDC-MPN experiments on natural phytoplankton in the ≥10 and <50μm size class, concentrated by screening. The method was used to enumerate viable phytoplankton before and after samples were treated with two doses of UV (200 and 500mJcm^{-2}) in each of three experiments. The MPN estimates of viable phytoplankton cells in the untreated controls were not significantly different from initial cell counts, consistent with most of the cells being viable and accurately detected as such, but measured survival was low in the treated samples: Viable phytoplankton were detected after UV treatment in only one of the three experiments, and only at the lower of the two doses. Detailed discussion of the SDC-MPN method or its results was not warranted in this study, which focused on other approaches for ballast water testing (First and Drake2013b). Others have followed the growth of phytoplankton species in natural samples after treatment (Wright et al.2009; Liebich et al.2012), and SDC-MPN has been used to assess viability after treatment of cultured plankton (Oemcke and Van Leeuwen2005), but to the best of our knowledge, the application of the SDC-MPN method to enumerate total viable phytoplankton after ballast water treatment of natural assemblages has not been examined directly in the scientific literature.

The SDC-MPN Method for total Viable Phytoplankton

The principles of the SDC-MPN method (Fig.1) are straightforward, and calculation of MPN from the scores of a dilution series has been examined thoroughly over the years (e.g., Hurley and Roscoe1983). Agreeing with Cochran's (1950) opinion that it is more important to be clear about the method's assumptions than about the details of the mathematics (which are not controversial), we address here the assumptions of SDC-MPN specifically as they apply to the enumeration of viable phytoplankton in natural communities of plankton subjected to BWT, focusing our discussion on the errors that result when these assumptions are not satisfied. We also discuss briefly the application of the method to enumerate only phytoplankton in the 10–50μm size range and the enumeration of other viable protists.

False Positives and False Negatives

An illustration of the potential outcomes of a live/dead test for BWT compliance (Fig.2) guides our discussion of potential errors in assessing the invasive potential of phytoplankton in ballast water discharge. Embracing the intention of ballast water regulation—to prevent the introduction of propagules of invasive organisms (see, e.g., Reavie et al.2010)—we apply the postulate of equivalent protection and identify the number of viable cellsmL^{-1}in ballast water discharge as the "environmentally relevant concentration" for assessing compliance with the intent of regulations. To reflect existing regulatory criteria and testing methods (e.g., vital stains, seeIntroduction), test results are represented as living cellsmL^{-1}. In this context, two types of error are possible:

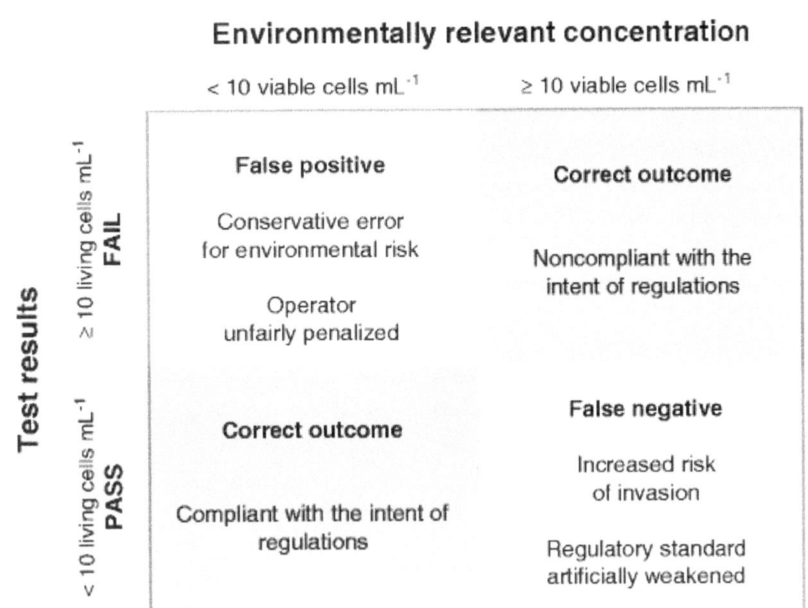

Figure. 2:

Potential results of tests to enumerate the concentration of living plankton in the 10–50μm size range in ballast water discharge—i.e., regulations as they stand—assessed in the context of the intent of regulation as reflected in the postulate of equivalent protection (see text). Viable cellsmL^{-1}in discharge are referred to as environmentally relevant concentrations to reflect the risk of invasion. Test results are in living cellsmL^{-1}, consistent with existing USGS and IMO live/dead testing criteria. Like the US Environmental Protection Agency's Environmental Technology Verification Program (ETV2010), we assume the perspective of a regulator committed to environmental protection and classify a false positive result as a conservative error, even though it will unfairly penalize the BWMS manufacturer or ship operator. Figure is adapted from Frazier et al. (2013)

False positive—Cells that are not viable (i.e., incapable of reproduction) are classified as living and thus potentially invasive: This error (statistically classified as type I, Frazier et al.2013) can lead to regulatory failure of a BWT system that is compliant with the intention of regulations. The error is conservative with respect to environmental protection (ETV2010; US Coast Guard2012).

False negative—Viable (therefore potentially invasive and by definition living) cells are incorrectly classified as dead and incapable of invasion (statistically classified as a type II error, Frazier et al.2013): If this error leads to acceptance of BWT systems that exceed discharge limits, regulatory standards are artificially weakened and the risk of species invasions is increased.

Clearly, the principal concern in BWT testing is the potential for false negatives, because protection of the environment is at stake. False positives are also failures of the test, but when we discuss assumptions or experimental biases that could lead to false positives, we refer to the affected results as "conservative" because the risk of invasion is not increased (Fig.2). But, it should be recognized that consistently false positives could lead to unwarranted exclusion of particular BWT systems or a technology (e.g., UV disinfection, an effective water treatment technology that does not require chemical biocides), and this might indirectly affect environmental protection by compromising the BWT sector's abilities to provide innovative and economical treatment solutions.

The classification of errors and their implications are the same as in Fig.2when results for individual tubes in the SDC-MPN method are considered, but there is no complication from the distinction between viability and vitality. There are only two correct outcomes: A tube with one or more viable cells should register growth and a tube with no viable cells should register no growth; the alternatives are false negatives or false positives, respectively. We now discuss sources of these errors in SDC-MPN and their likelihood of affecting the outcome of tests on natural phytoplankton, particularly in the 10–50μm size range. Interference from, and the enumeration of, viable heterotrophs will be considered peripherally.

Sources of Error

The SDC-MPN method, and other assays based on observing the growth of natural phytoplankton after ballast water treatment, have been identified as being definitive because they assess viability (First and Drake2013a), but such grow-out assays have been discounted for being time-consuming and thus inappropriate for use in rapid assessment of the efficacy of BWT (Steinberg et al.2011), for example, in shipboard compliance testing (King and Tamburri2010) and port state control inspection (Drake et al.2014). Focusing on the use of SDC-MPN for land-based verification testing to gain type approval, we first identify sources of error in the approach, some of which are closely tied to the time allotted for the assay; ways to characterize and minimize the errors will be discussed in following sections.

Failure of viable cells to grow

Potentially, the most potent criticism of SDC-MPN is that because it depends on detecting growth of microbes in culture, it will only account for culturable species (Steinberg et al.2011), and many microorganisms cannot be cultured (First and Drake2013a; US Coast Guard2012). Let us examine the relevance of this criticism to SDC-MPN of phytoplankton.

Consider the term "culturable" when used as a general characterization of a microbe: this is not an inherent property of a species but rather a reflection of the culturist's ability to provide what the organism needs for it to be maintained indefinitely. Studies of microbial diversity reveal many species or ecotypes that are not in culture, for example, the unicellular diazotrophic cyanobacterium UCYN-A (Zehr et al.2001; Krupke et al.2014), marine heterotrophic protists (Heywood et al.2010), and numerous picoeukaryotes (Shi et al.2009). But with skill, insight, and diligence, researchers have brought fastidious species or those with special requirements into culture for further study. For example, Rappé et al. (2002) isolated the key bacterial species*Pelagibacter ubique*12years after its sequence was determined as SAR11 (Giovannoni et al.1990as discussed by Vaulot et al.2008), and it took years to develop effective procedures for culturing a wide range of isolates of*Prochlorococcus*(reviewed by Moore et al.2007), the most abundant unicellular cyanobacterium on Earth (Partensky et al.1999). Notably, reports on microbial biodiversity tend to use the terms "uncultured" (Rodriguez-Martinez et al.2013) or "uncultivated" (Vaulot et al.2008), implying no judgment on the culturability of species that have yet to be isolated and grown in culture. In the abstract of their study of the growth on agar of heterotrophic microbes from marine sediment, Kaeberlein et al. (2002) stated without detailed explanation that, "The majority (>99%) of microorganisms from the environment resist cultivation in the laboratory"; we do not find this generalization to be helpful or well substantiated in assessing the culturability of phytoplankton. While it is true that some plankton in the 10–50μm size range, such as the obligate kleptochloroplastic dinoflagellate*Dinophysis acuminata*, have required extraordinary efforts to bring them into sustained culture (Park et al.2006), we were unable to find a body of evidence to support the belief that a majority of phytoplankton in that size class cannot be cultured. Regardless, it is important to remember that the SDC-MPN method requires only that viable cells in the dilution cultures multiply enough for their growth to be detected; they need not be maintained through successive transfers. Addressing directly the assumption that cells that are viable in the sea will grow under the conditions provided in dilution culture, Throndsen (1978) noted that "Some species with special requirements will regularly grow up in dilution cultures though they will not survive subculturing" (p. 218), and he advised that the number of species that would grow would be increased if the dilution media was based on the same water from which the sample was taken. Even *Dinophysis acuta*, an obligate grazer which can only be cultured when supplied with its preferred prey, the ciliate *Myrionecta rubra*, will continue to divide for three to four generations in the absence of its prey (Nielsen et al. 2013).

Although the vast range of environmental requirements of phytoplankton ensures that no one set of conditions can support the growth of all viable cells, we are aware of no evidence to suggest that a large component of viable natural phytoplankton are inherently unable to grow through enough divisions to be detected in a suitably designed dilution culture. We suggest that if a viable phytoplankton cell fails to reproduce enough to be detected in SDC-MPN, the false negative result is better ascribed to unsuitable growth conditions in the assay leading to slow or no growth, rather than to inherent properties of some, or many species. Fastidious, fragile, or finicky phytoplankton species certainly exist, and each species or strain has its own environmental optima and limits. Below, we discuss how to minimize false negatives due to slow growth of phytoplankton in dilution culture.

Failure to detect growth of the culture

As explained in the Background section, many studies that employed the SDC-MPN method were directed toward the enumeration and ultimate isolation of coexisting species, in part to describe community structure; the dilution-culture tubes were examined microscopically after incubations of weeks to months. If the intention is solely to enumerate viable phytoplankton, as it is with ballast water testing, there is no need for identification, but growth must be detected reliably even when it starts from one cell in a tube.

When SDC-MPN experiments are conducted on unialgal cultures in the laboratory, the likelihood of false negatives for viability can be reduced to near zero through the application of rigorous and sometimes labor-intensive procedures for reliable detection of growth in the dilution tubes. We discuss these in a study of variability among phytoplankton taxa in viability vs. UVC-dose relationships (H.L. MacIntyre et al., submitted for publication). Recognizing that for practical reasons, routine assessment of BWT on natural phytoplankton will need streamlined procedures, provided associated uncertainties can be constrained, we consider the principal influences on accurate detection of growth: (i) the minimum number of cells that can be detected reliably (N_d, cells tube^{-1}); (ii) the time it takes for a dilution culture to reach that threshold (t_d, day); and (iii) when the observations are made (t_{obs}, day). Assuming for now that growth is exponential at rate μ (day^{-1}) from the initiation of a dilution culture with N_0 cells tube^{-1}, the number of detectable cells is reached at time t_d:

$$N_d = N_0 \cdot e^{\mu \cdot t_d} \quad (1)$$

The solution for t_d is

$$t_d = \frac{\ln(N_d/N_0)}{\mu} \quad (2)$$

which reduces to $t_d = \ln(N_d)/\mu$ when a culture starts with one viable cell—a straightforward relationship that illustrates the influences of detectability and growth rate on the time required for detection (Fig. 3, Table 1). If growth is delayed by a lag phase (Wood et al. 2004), as might be associated with photorepair (Liebich et al. 2012; Roy 2000), t_d would be incremented by a lag time, t_{lag}. In either case, if cells are growing in the tube but observations are discontinued prior to t_d, no growth will be detected and a false negative will result. This can be due to slow growth or poor detectability of cells—either because the signal per cell is small (e.g., weakly pigmented phytoplankton) or the instrument is not sufficiently sensitive to detect growth in the time frame of the observations. Remedies include optimizing growth conditions, increasing the sensitivity of detection and extending the period of observations (Fig. 3).

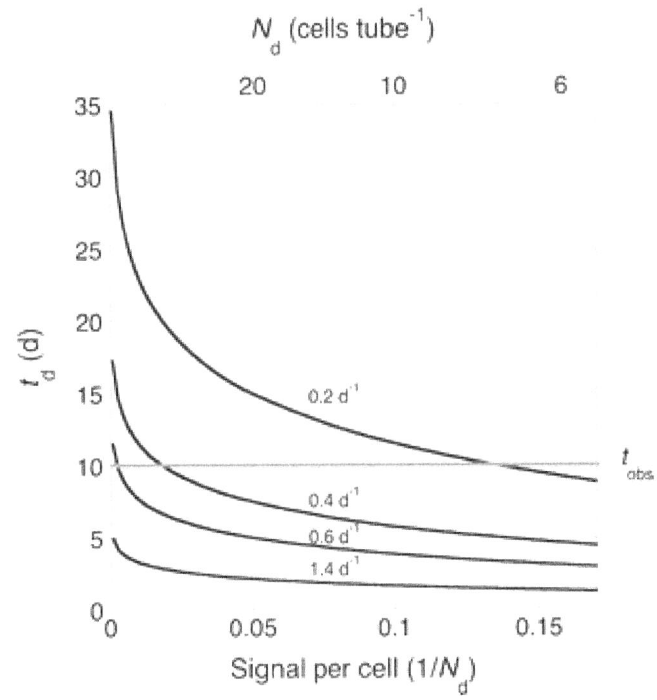

Figure. 3

Influence of the detectability and growth rate of cells on the accurate determination of growth in dilution cultures beginning with one cell per tube. The signal per cell is $1/N_d$, where N_d is the minimum number of cells per tube that can be detected reliably and related to growth (e.g., Table 1). The time required for this detection, t_d (day), is plotted for a range of exponential growth rates. The *gray line* indicates the observation period for the experiment, t_{obs} (day)—in this case, 10 days. *Points below the line* represent combinations of growth rate and cell signal that would lead to reliable detection of growth. Slower growing or less effectively detected cells will register growth if t_{obs} is increased (raising the *gray line*); the sensitivity of the detection method is increased (moving points to the right along the curves for each growth rate); or conditions are optimized to increase growth rates (cells move to a lower line)

Table 1: Time required to detect phytoplankton growth (Eq.2) in dilution cultures beginning with one viable cell in a culture tube containing a volume of 5mL, assuming that the lower limit of detection ($LLD_{0.1}$) corresponds to 0.1mg chlorophyll*am*$^{-3}$, a concentration that is readily resolved with fluorometers deployed in the open ocean (e.g., Letelier et al.2004); the corresponding minimum number of cells per tube isN_d

	Dimension (μm)	Chlorophyll quota (pg cell^{-1})	Growth rate (day^{-1})	N_d(cells tube^{-1})	Time to detection (t_d, day^{-1})
Gymnodinium vitiligo[a]	11	2.31	0.22	216	24.4
Rhodomonas salina[c]	11.6	0.89	0.64	562	9.9
Thalassiosira weissflogii[c]	13.0	3.07	0.83	163	6.1
Heterosigma akashiwo[c]	15.5	2.65	0.56	189	9.4
*Chlamydomonas*sp.[a]	18.6	11.3	0.28	44	13.5
Gymnodinium sanguineum[b]	44	20.4	0.23	9	13.9

The size is minimum dimension (Montagnes et al.1994) or average diameter (Neale et al.1998, H.L. MacIntyre et al. unpublished)

Competition

It has long been recognized that when SDC-MPN is used to enumerate multiple coexisting species in a natural sample, competition between species of phytoplankton is a concern (Andersen and Throndsen2003; Throndsen and Kristiansen1991). For example, Cerino and Zingone (2006) remarked that species that grow in serial dilution cultures often do not survive due to competition. But, competition can only occur when a culture starts with representatives of two or more species in a tube (found at the lower dilutions in a series), and it affects only the estimates of the abundances at the species level; as long as one of the competitors grows to detection, tubes with more than one viable species in the inoculum score positive and the accuracy of SDC-MPN for total viable cells is unaffected. We note that one extreme of competitive exclusion, allelopathic suppression of one species by another through production of growth-inhibiting metabolites, is more likely to occur at higher population densities, when nutrients have been depleted (Granéli and Salomon2010; Hardison et al.2013; Tameishi et al.2009). These conditions are not likely to occur during an SDC-MPN assay.\

Grazing

Acknowledging that viable heterotrophs in the 10–50μm size range would have to be enumerated using alternate methods (see, e.g., Zetsche and Meysman2012; Lighthart1969), we consider the possible effects of grazing (phagotrophy) on the results of SDC-MPN for viable phytoplankton. A false negative will result if a viable grazer co-occurs with one or more viable phytoplankton and prevents them from growing to detection—an outcome that is not simply predicted given the dependence of grazing rates on prey concentration (Strom et al.2000) and the artificially low initial concentrations of prey in a dilution series. It is relevant that oceanographers routinely assess microzooplankton grazing rates in the field by progressively diluting natural plankton up to 1:20 with filtered sea water to reduce the encounter frequency between microzooplankton grazers and phytoplankton prey—and thus grazing losses (Landry and Hassett1982). In addition, the co-occurrence of grazers and phytoplankton becomes increasingly unlikely in the most-diluted tubes. We therefore suggest that grazing is not likely to cause significant numbers of false negatives in SDC-MPN tubes, and when it does, it will not strongly affect the calculated result: The errors will be in the less-diluted series that have a smaller quantitative influence on the estimate of total viable cells. A conservative solution, if practicable, would be to score all tubes containing heterotrophs att_{obs}as positive for phytoplankton growth, thereby ensuring that no viable phytoplankton cell was overlooked due to grazing. This approach would be particularly useful for heterotrophic dinoflagellates, which can be important consumers of diatoms, differ fundamentally from other microzooplankton in their feeding and growth dynamics, and tend to survive well when their food supply is exhausted (Sherr and Sherr2007).

Inclusion of Nonregulated Organisms

If the method of detection, e.g., in vivo fluorescence (Brand and Guillard1981b; Wood et al.2004), is a bulk measurement that does not distinguish cell size, nonregulated phytoplankton <10μm, or more stringently regulated phytoplankton ≥50μm, could grow and produce: (i) false positive scores for the 10–50μm size range if no viable phytoplankton in the range were in the dilution tube or (ii) true positive scores if at least one viable regulated cell was in the tube but was outcompeted. If all tubes with positive scores are scored as such, regardless of cell size, the result is conservative, reducing risk to the environment, but cells <10μm are generally much more abundant than the larger regulated cells (Marañón2015), and significant overestimation of the numbers of viable 10–50μm cells can be expected. The likelihood of error can be reduced greatly by prefiltration to remove larger cells and gentle concentration on a 10-μmmesh to retain cells in the regulated size range (Reavie et al.2010; First and Drake2013b).

Aggregation of Phytoplankton Cells

The SDC-MPN method depends on the assumption that organisms are randomly distributed in each tube and evenly distributed between subsamples (Cochran1950). The assumption will be violated, and numbers of viable cells will be underestimated, if phytoplankton grow in colonies that cannot be broken up without compromising viability or if individual cells or colonies collect in aggregates (Revelante and Gilmartin 1991) that are incompletely disrupted during the sampling and dilution process. Thorough but relatively gentle mixing (e.g., 100 inversions, Andersen and Throndsen 2003) is recommended prior to any subsampling, but harsher methods such as sonication or heating followed by vortex mixing (see Humphries and Widjaja 1979) have been used to break up colonies of cyanobacteria. Even though conditions during ballast water transfers and treatment are anything but placid, harsh disruption of aggregates would likely introduce unacceptable uncertainty into tests of viability after dilution. The importance of aggregation in samples can be assessed with microscopic examination or by testing for uniformity of repeated bulk measurements (e.g., chlorophyll *a*) made on subsamples.

Colonies such as diatom chains would be expected to persist, and they would complicate SDC-MPN because in dilution, they are distributed as entities (after Reavie et al. 2010) that contain more than one cell. The statistics of MPN would estimate the number of viable entities, but this would be less than the number of viable cells. If a colony of x viable cells released into a foreign environment acts more like one propagule rather than x propagules, however, the error would be more numerical than ecological, but the numerical result counts for regulation.

Toward best Practices for Minimizing Error

We suggest that competition and grazing are not important sources of error when SDC-MPN is used to enumerate total viable phytoplankton. The other sources of error might not be totally eliminated, but they could be assessed and minimized wherever practically possible, thereby reducing the uncertainty of SDC-MPN counts of total viable cells for regulation of BWT.

OPTIMIZING GROWTH CONDITIONS

First and foremost, the accuracy of the SDC-MPN method depends on its ability to provide all viable phytoplankton cells with conditions—light, temperature, and chemical milieu—to support growth in highly dilute culture, and the growth must be rapid enough to ensure that cell numbers can increase from a single cell in a tube to the level of reliable detection during the observation period (Fig. 3, Table 1). Given that differentiation between species in growth responses to light, nutrients, and temperature is a foundation of diversity in phytoplankton (Follows et al. 2007; Johnson et al. 2006) [but see Cullen and MacIntyre (1998) for discussions of behavior and physiological plasticity and Verity and Smetacek (1996) for top-down control], it is self-evident that no one set of conditions during SDC-MPN can ensure optimal growth of all viable phytoplankton in a sample. But, decades of research and experience can inform the process of choosing conditions in SDC grow-out so that the exclusion of species due to unsuitable growth conditions is minimized.

A good place to begin is with advice from experts who have used SDC to both enumerate and isolate phytoplankton for culturing. Concluding suggestions from Andersen and Throndsen (2003) are succinct and exemplary: "The success or failure of the method depends on (a) the cleanliness of the equipment; (b) the suitability of the growth medium; and (c) the external culture conditions (temperature and light)" (p. 128).

Growth Media

Cleanliness is a prerequisite for reliable culturing of phytoplankton in dilution culture (discussed by Guillard 2005), and sterile technique is required to avoid contamination (Kawachi and Noël 2005). A variety of growth media can be used, depending on the application (Andersen and Kawachi 2005). Guillard and Morton (2003) provide specific advice on the growth medium, explaining that the purpose of nutrient enrichments is to provide scarce materials to phytoplankton in usable forms at levels that are neither toxic nor limiting for growth rate. For general enrichment cultures, they specify nutrient additions not to exceed concentrations that are lower than for typical culture media (compare their recommendations with recipes from Appendix A in Andersen 2005), and they include nitrate, ammonium, and urea as nitrogen sources, thereby accommodating species that do not utilize nitrate, e.g., many strains of *Prochlorococcus* (Moore et al. 2002). Throndsen (1978) recommends a modified Erd-Schreiber medium (pasteurized) that has higher concentrations of nutrients and includes soil extract; Guillard and Morton (2003) recognize soil extract as often being beneficial, and they provide a recipe. Andersen and Throndsen (2003) mention that dilutions prepared with water from the sampling site yield cultures with higher species diversity than premade growth medium, implying that more species are able to grow in water from their source. In the absence of a comprehensive review that would be beyond the scope of this discussion, it seems reasonable to adopt Andersen and Throndsen's (2003) suggestion that the media for serial dilution culture should be prepared with water sampled together with the inoculum, with nutrient enrichments that are high enough to ensure growth past the point of detection but lower than in conventional growth media (see, e.g., Guillard and Morton 2003; Andersen and Kawachi 2005). Although we expect no one recipe to ensure optimal growth for a maximum number of species in all BWT conditions, we offer provisional guidance in Table 2, based on the expert advice cited above. As with other recommended practices, the way to evaluate them is through experimental comparisons.

Table 2: Provisional recommended practices for SDC-MPN on natural phytoplankton assemblages in land-based testing for BWMS type approval, suitable for being evaluated through systematic comparisons

	Standard	Optimized
Growth medium		
Source water	Filtered water from the original sample	
Nutrient enrichment (options)	Reduced concentrations from typical growth media (include soil extract, multiple nitrogen sources)	
Sterilization	Filter sterilization, pasteurization	
Temperature (°C)	T_{insitu}orT‾insituT‾insitu	T‾insituT‾insitu + 5°C forT‾insituT‾insitu≤20°C decreasing linearly to 0 offset at 28°C
Irradiance (μmol photonsm^{-2}s^{-1})	Light/dark cycle ~10% $_{0cf}$(~60–200)	Longer light period (e.g., 16h) 18% $_0$or82+(4.7·T‾insitu)+ 0.068·((T‾insitu)2)82+(4.7· T‾insitu)+0.068·((T‾insitu)2)

Standard conditions are based on recommendations in the literature, intended to maximize the number of species that will grow. The optimized conditions proposed here are intended to support enhanced growth rates without significant reductions in the number of species that will grow, thereby shortening the time required to detect viable cells in an SDC-MPN assay and minimizing false negatives due to slowly growing isolates. Ambient water temperature isT_{insitu}(°C); the climatological average isT‾insituT‾insitu; climatological cloud-free midday irradiance is $_{0cf}$(μmol photonsm^{-2}s^{-1}PAR); and climatological midday irradiance is $_0$

Options for sterilization of the media are comprehensively reviewed by Kawachi and Noël (2005); they include filter sterilization, which will retain potentially important heat-labile compounds such as vitamins in the source water, but which will allow viruses to pass. Viruses and even proteins can be removed by tangential-flow filtration (van Reis et al.1999), but the benefit of the removal of the viruses might be outweighed by the removal of nutrients essential for growth of auxotrophic species. However, as with grazing, the influence of viruses is expected to be reduced in the most dilute samples (Andersen and Throndsen2003).

Temperature

It is a central tenet of plankton ecology that in the absence of nutrient limitation, light and temperature are the principal influences on the growth rates of phytoplankton (Cullen et al.1993; Yoder1979). Through a combination of competition and the constraints of absolute environmental tolerances, light and temperature regimes select for phytoplankton species and ecotypes, explaining dominant patterns in their distributions (Follows et al.2007; Johnson et al.2006), although these are, of course, also influenced strongly by nutrients, food web interactions, and physical processes such as mixing (Follows and Dutkiewicz2011; Cullen et al.2002).

Generally, discussions of the incubation temperature for dilution culture have focused on controlling it to avoid harmful or lethal variation, rather than choosing a temperature to optimize growth rate (Knight-Jones1951; Guillard and Morton2003; Andersen and Throndsen2003). Considering that the duration of SDC-MPN—that is,t_{obs}in Eq.2—should for practical reasons be no longer than necessary, there are good reasons to optimize growth rate to minimize the time of detection,t_d, for as many species as possible (see Fig.3). Choice of an incubation temperature is an important factor.

The responses of phytoplankton growth rates to temperature (Fig.4) illustrate well-known features that should be considered: Some species grow more quickly than others; all show increasing growth rate with temperature up to a maximum; some have very broad tolerance ranges while others do not; and the decline of growth rate with increasing temperature above the optimum is sharper than the increase with temperature below it, due primarily to reductions in protein and cytochrome functionality (e.g., Fork et al.1979; Gao et al.2000; Nitta et al.2006).

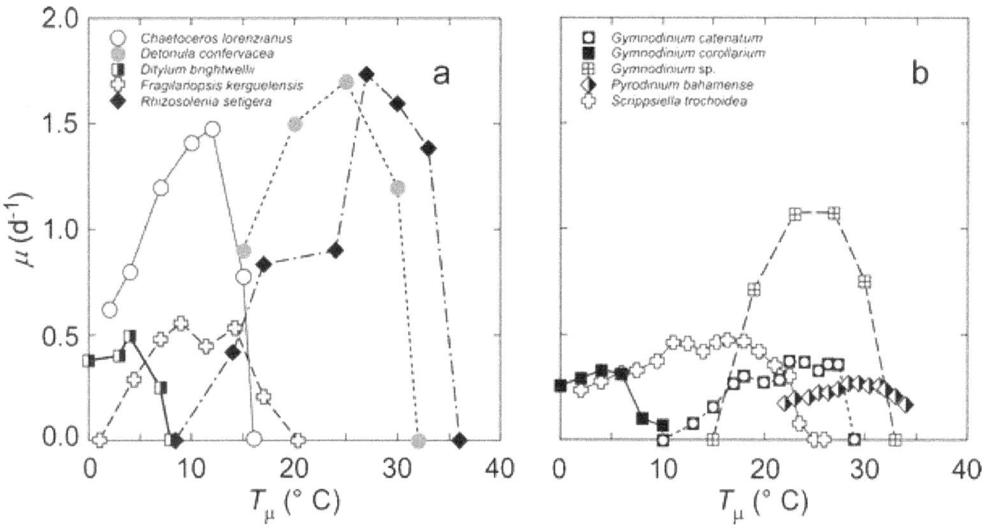

Figure. 4: Specific growth rates, μ(day^{-1}), of cultures of **a** diatoms and **b** dinoflagellates as a function of growth temperature, T_{μ}(°C). Data sources: *Chaetoceros lorenzianus* (Hulburt 1982); *Detonula confervacea* (Smayda 1969); *Ditylum brightwellii* (Paasche 1968); *Fragilariopsis kerguelensis* (Fiala and Oriol 1990); *Gymnodinium catenatum* (Bravo and Anderson 1994); *G. corollarium* (Sundström et al. 2009); *G.* sp. (Thomas 1966); *Pyrodinium bahamense* (Usup et al. 1994); *Scrippsiella trochoidea* (Binder and Anderson 1987)

The response of phytoplankton growth rate to temperature can be generalized by fitting it to the functional form of a model describing the short-term influence of temperature on the photosynthetic capacity of natural benthic diatoms (Blanchard et al. 1996):

$$\mu_{max}(T) = \mu_{max}(T_{opt}) \cdot \left(\frac{T_{fatal}-T}{T_{fatal}-T_{opt}}\right)^{\beta} \cdot \exp\left[-\beta\left(\frac{T_{fatal}-T}{T_{fatal}-T_{opt}}-1\right)\right] \quad (3)$$

where $\mu_{max}(T)$ (day^{-1}) is the maximum growth rate at temperature T(°C), i.e., not limited by light or nutrients, T_{opt}(°C) is the temperature at which growth rate is maximal at the rate $\mu_{max}(T_{opt})$ (day^{-1}), T_{fatal}(°C) is the temperature at which growth rate declines to zero, and is a dimensionless shape factor (Fig. 5a).

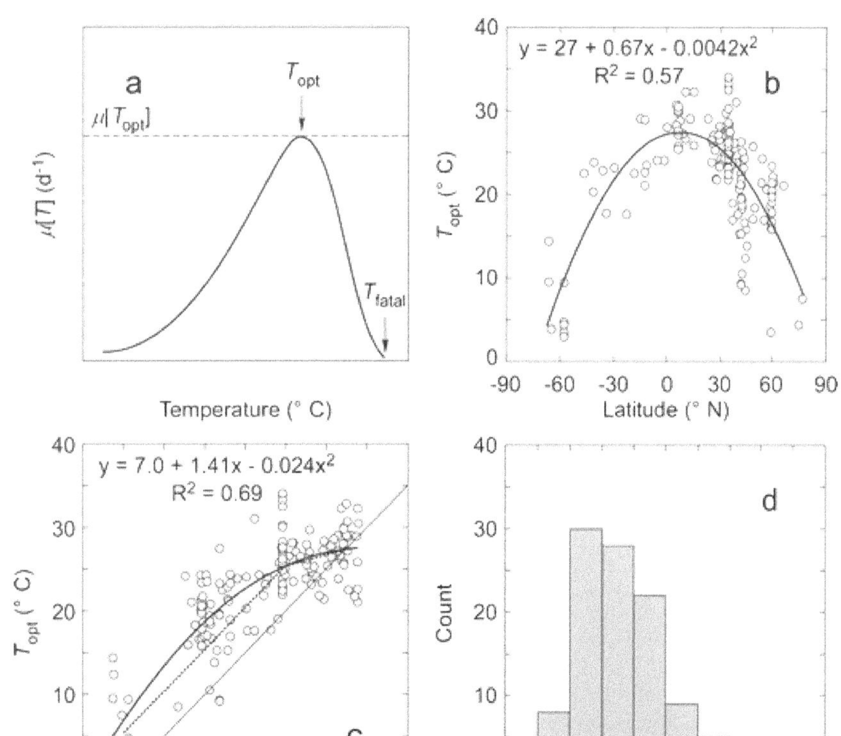

Figure. 5: **a** Generalized form of the relationship between growth rate ($\mu(T)$, day^{-1}) and temperature (Eq.3, based on Blanchard et al. 1996), showing the optimal temperature (T_{opt}) at which growth rate is maximal and the fatal temperature (T_{fatal}) at which growth

is completely abolished. **b** Latitudinal variation in optimal growth temperature in 163 cultures of estuarine and marine phytoplankton obtained by re-fitting data compiled by Thomas et al. (2012) to Eq. 3. The data have been fit to a second-order polynomial (*heavy line and equation*). **c** The same estimates of T_{opt} as a function of the mean annual temperature at their isolation locations (from Thomas et al. 2012). The *straight solid line* shows a 1:1 relationship; the *curved line* is a fit to a second-order polynomial, and the *dotted line* represents a provisional recommendation for incubation temperature during serial dilution culture (Table 2). **d** Frequency distribution of the temperature difference between the fatal and optimal temperatures for the cultures in **b**. Only fits in which the error on T_{fatal} was less than 20% of the estimate ($n = 111$) are included to avoid bias from fits in which the estimate of T_{fatal} is uncertain due to extrapolation. The mean difference is 7.5°C, and the median is 7.1°C

To describe general trends in the temperature responses of cultured phytoplankton with explicit calculation of fatal temperature, we fit to Eq. 3 data on growth rates of cultures of marine and estuarine phytoplankton compiled by Thomas et al. (2012). The equations were fitted using a Levenberg-Marquardt method in Kaleidagraph 4.5 (Synergy Software). Because Eq. 3 contains four parameters, only studies with five or more data points were included. As in the Thomas et al. (2012) analysis that used a different temperature function to estimate T_{opt} (i.e., Norberg 2004), the resulting estimates of optimal growth temperatures for 163 cultures had a significant relationship with the latitude of isolation (Fig. 5b), consistent with annual mean temperature in situ being an important selective factor that is reflected in T_{opt} (Fig. 5c). Note the dispersion of data around the central tendencies: this may reflect a link between seasonal variations in temperature and community composition that cannot be revealed through the analysis. As concluded by Thomas et al. (2012), competition between species with comparable growth rates will tend to favor taxa for which T_{opt} is above, but close to the in situ temperature, possibly subject to an inherent maximum near 30°C. The implication for BWT testing is that if the temperature of an SDC-MPN assay is chosen to optimize growth for the indigenous community, the test is unlikely to select against potentially successful invaders because they are likely to share temperature optima with established members of phytoplankton communities.

We therefore suggest that SDC-MPN assays be conducted at temperatures above ambient, thereby increasing growth rates and minimizing time to detection (Table 2). The relationship between T_{opt} and the annual mean temperature at the location of isolation suggests that growth rates would be higher at temperatures about 5°C above ambient up to about 20°C for the in situ temperature, decreasing to no offset at 28°C, as illustrated with the dotted line in Fig. 5c. For the growth data analyzed in Fig. 5, increasing the temperature to T_{opt} from 5°C below it results in an increase in specific growth rate of 5–660% (median 26%, $n = 163$). The variability arises because of differences in the steepness of the relationship at the growth optimum (see Fig. 4). The increases in growth rates translate to reductions of 5–85% (median 21%) in the time required to detect growth (Eq. 2). However, it is important not to impose thermal stress by approaching closely or exceeding the fatal temperatures for phytoplankton in the samples. The frequency distribution of $T_{fatal} - T_{opt}$ (Fig. 5d)—a measure of the margin for error in overestimating T_{opt}—suggests that the risk of approaching T_{fatal} when targeting T_{opt} for incubation temperature is low until the incubation temperature exceeds T_{opt} of a species by about 5°C. The validity of this inference can be tested through systematic comparison in parallel of SDC-MPN experiments conducted at the in situ temperature and at the purported optimized temperature.

Light

For SDC-MPN of phytoplankton collected from surface waters, recommendations based on experience suggest that incubation light levels, E_{inc}, should be about 10% of full daylight (Andersen and Throndsen 2003). This corresponds roughly to 60–200 μmol photons m^{-2} s^{-1} for the winter vs. summer solstice at 45° latitude (photosynthetically available radiation, clear-sky calculations based on Bird and Riordan 1986), similar to the range suggested by Guillard and Morton (2003), who add that fluorescent bulbs remain the light source of choice and caution that some algae require a dark period to survive (Brand and Guillard 1981a), so light–dark incubation conditions should be used (Table 2).

In an attempt to explore the optimization of light level for SDC-MPN, we consider the relationship between photosynthesis and irradiance (P vs. E), which is not synonymous with growth rate vs. irradiance but which is intimately linked to it (Geider et al. 1998; Cullen 1990). It is described by a maximum rate, P_m (e.g., mg C m^{-3} h^{-1}), and the light saturation parameter, E_k (μmol photons m^{-2} s^{-1}) (Talling 1957). As ambient irradiance exceeds E_k, photosynthetic rate is saturated, absorbed light energy must be otherwise dissipated, and phytoplankton are subject to stress that can lead to the inhibition of photosynthesis and a decrease of growth rate (Fig. 6a; Baroli and Melis 1996; Ritchie and Larkum 2012); in turn, it is expected that incubation of phytoplankton at the E_k for P vs. E should support relatively high growth rates with little stress. Since E_k is an essential reflection of physiological and taxonomic responses of phytoplankton to irradiance—it expresses both phenotypic and genotypic variability—it is rightly called the photoacclimation or photoadaptation parameter and it covaries with the irradiance at which phytoplankton are grown (Bannister and Laws 1980; MacIntyre et al. 2002). A review is well beyond the scope of this discussion, but in the context of SDC-MPN, it is relevant to observe that the ratio of surface irradiance to E_k is a key metric in models of primary productivity for remote sensing (Behrenfeld and Falkowski 1997; Platt and Sathyendranath 1993; Cullen et al. 2012). Platt and Sathyendranath (1993) found that about half of the variability of E_k for the top 40 m of the water column could be explained by variations of surface irradiance: The best estimate of the ratio of E_k to the midday irradiance averaged over 3 days was 0.18 (with individual observations typically in the range of 0.13–0.20), which leads us to recommend $E_{inc} = E_k$ of 18% of the average midday irradiance at the surface (Table 2), corresponding roughly to 100–350 μm photons m^{-2} s^{-1} for the winter- vs. summer solstice at 45° latitude as referenced above.

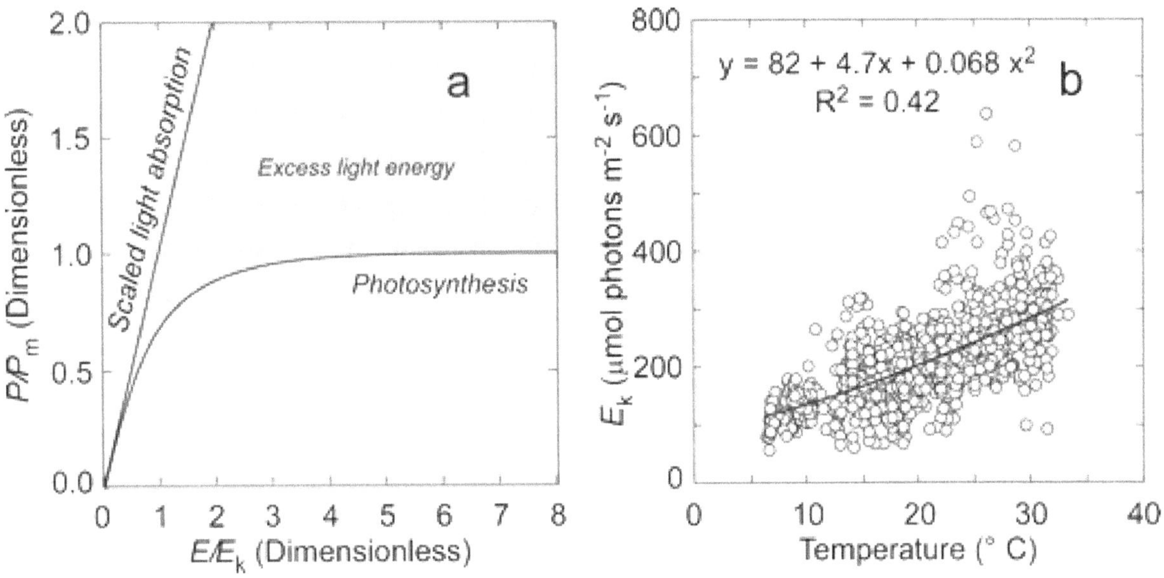

Figure. 6: aA generalized photosynthesis-irradiance (*P*vs.*E*) curve, shown in dimensionless form: Photosynthesis has been scaled to its light-saturated rate,P_m, and irradiance has been scaled to the light-saturation parameter,E_k. The scaled rate of light absorption is also shown: The*shaded area*between the two curves represents light energy that is absorbed by the cell but not used in photosynthesis. This must be dissipated to avoid oxidative stress; incubating cells at irradiances $\gg E_k$is likely to result in reduction in photosynthetic rate (i.e., photoinhibition) and growth rate.**b**The temperature-dependence of$E_k$in 820*P*vs.*E*curves from estuarine, neritic, and oceanic waters off North America (H.L. MacIntyre et al., unpublished data). The data have been fit (*heavy line and equation*) to a second-order polynomial. Note the dispersion of data around the central tendency: 94% are between 0.5 and 2x of the fitted value. Incubating samples at the irradiance predicted by the equation (Table2) should ensure than the overwhelming majority of samples are exposed to irradiances that are between 0.5 and 2xE_k, a range over which photosynthesis achieves 14–61% of the light-saturated rate and over which oxidative stress is minimal. Growth rates are expected to scale directly with the achieved photosynthetic rate

Temperature also influences$E_k$through its effect on enzymatic rates (Hancke et al.2008; Hikosaka et al.2006). Thus, for a given irradiance and nutritional status, models of photosynthesis and growth specify$E_k$and growth rates increasing exponentially with temperature (Lima and Doney2004; Geider et al.1998). Because on a global scale, temperature tends to covary with midday irradiance, relationships between$E_k$and surface irradiance such as that presented by Platt and Sathyendranath (1993) would also include effects of temperature that are more contributory than confounding (cf. Harrison and Platt1980,1986). An extensive set of 820*P*vs.*E*experiments for which temperature (but not average midday irradiance) was measured shows that, consistent with predictions based on light alone,$E_k$varies from about 100–350µm photonsm^{-2}s^{-1}over the range from cold, low-light to warm, high-light waters (Fig.6b).

Our analysis suggests that choosingE_{inc}either on the basis of temperature or a measure of midday irradiance (Table2) will yield estimates with similar ranges, and either would likely support more rapid growth of phytoplankton than incubation at lower irradiance, perhaps faster yet if long day lengths are applied. The effectiveness of optimized conditions would have to be tested.

Assessing the Effects of Culture Conditions

The intention of optimizing growth conditions is to minimize false negatives due to undetected viable cells, thus maximizing the estimate of total viable phytoplankton. Benign conditions could also favor photorepair of DNA damage (Boelen et al.2001; Roy2000) and photosynthetic systems (Tyystjärvi2008; Campbell et al.2013) that might not have occurred in nature, leading to more conservative results.

The most straightforward way to assess the suitability of any experimental regime for SDC-MPN is to compare it to others. Assuming for now that the influence of nonregulated plankton (particularly, rapidly growing phytoplankton <10µm) is minimized or accounted for, the higher counts of viable cells from the same initial sample should be considered more accurate. Two types of comparison can be conducted: parallel and serial.

Parallel experiments to test optimization procedures

Experienced practitioners of dilution culture have suggested growth conditions that are intended to maximize the number of species that will grow, not necessarily how fast they grow. Recognizing that because of expense, time is of the essence in the routine application of SDC-MPN for assessment of ballast water treatment, and that faster growth permits shorter observation periods (Eq.2and Fig.3), we present provisional guidance for optimizing growth conditions, including choices of incubation temperature and irradiance higher than the conventional in situ temperature and ~10% surface irradiance (Table2). The merit of these suggestions can be assessed by conducting experiments in parallel on the same initial samples in which results for optimized temperature and irradiance are compared to those for conventional conditions.

No significant difference between results for the two growth conditions would suggest that growth rates had little influence on results, for example if$t_{obs}$was long enough to detect even the slower growing cells in conventional conditions. Unfortunately, opposing influences of the changes in light and temperature on growth might produce a similar result, so it would be difficult to draw conclusions from this comparison alone. Uncertainties associated with possible light-temperature interactions could be explored using repeated applications of a 2×2 light-temperature design. This would require considerable effort that might be impractical for routine testing programs but would be highly valuable in an effort to characterize rigorously the uncertainties in SDC-MPN and the effectiveness of optimization procedures.

If counts are systematically higher for optimized conditions, it suggests that growth rate influences results over the observation period so that optimized growth conditions lead to more accurate counts. The result does not reveal how many slow-growing species in dilution culture tubes remain undetected under optimal conditions, however. That can be addressed by making repeated observations during the observation period and extending it if necessary, as described below.

Serial sampling to assess false negative results

Repeated observations during the grow-out period, for example using nondestructive measurement of in vivo chlorophyll fluorescence (Brand and Guillard1981b; Wood et al.2004), can obviate uncertainties associated with cultures that grow but then decline below the detection limit prior tot_{obs}(H.L. MacIntyre et al., submitted for publication), and they can provide greater confidence than a single reading above the detection limit that growth of a culture has indeed occurred. But more importantly, perhaps, sequential observations reveal the rate at which new—i.e., slower growing and/or smaller signal-per-cell—cultures are being recruited into the MPN score sheet, thereby increasing the most probable count of total viable cells. Scores that continue to rise day by day (i.e., the consequences of literally raising the$t_d$bar in Fig.3) indicate that the observation period should be extended, whereas little change with time suggests that an asymptote has been reached and the observation period is long enough to ensure stable results. Even one extra observation prior to a terminal assessment att_{obs}, made routinely and analyzed systematically, should provide a strong indication of the influence of the choice of observation period on the accuracy of the results.

What about the cells that do not grow?

If sequential MPN estimates indicate that a steady maximum is consistently reached, we can infer that the observation period is long enough to detect most of the viable cells that will grow in dilution culture. But what of "SDC ungrowables"—viable cells that grow very slowly or not at all: Are they likely to be a significant source of false negatives? As argued above, Throndsen (1978) observed that some species with special requirements will regularly grow in dilution culture although they will not survive subculturing, and we were unable to find strong evidence for inherent unculturability of many species of phytoplankton. Even so, errors in SDC-MPN due to poor or no growth of viable cells are certainly possible and should be addressed. One approach is to compare SDC-MPN estimates of total viable cells with enumerations of total cells using microscopy or flow cytometry, assuming that all the conventionally enumerated cells are viable and keeping in mind that some species are damaged by common fixatives (Kugrens and Lee1987). Because intact but nonviable cells are expected after UV treatment, the comparison is appropriate only for untreated controls. Agreement between counts of total cells and SDC-MPN estimates of viable cells, for example as reported by First and Drake (2013b), would suggest that the most abundant viable cells grew in dilution culture and so-called unculturability of species was not a significant problem in that experiment. But, if the estimates of viable cells are significantly lower than counts of total cells, the discrepancy could be due to (i) false negatives in SDC-MPN, the major concern, or (ii) the enumeration of intact cells that are not viable. The latter possibility is suggested by the observation of intact but metabolically inactive cells in early summer assemblages in Lake Rotorua (Paerl1978) and Blanes Bay (Agustí and Sánchez2002). In the context of regulation and in the absence of other measures of vitality, the false-negative interpretation is conservative and could be used to establish a measure of the type II error associated with viable cells that do not grow.

The comparison of viable cells vs. total cells in untreated controls puts bounds on the uncertainty of the SDC-MPN method for assessing the effectiveness of BWT, but only to the extent that the influence of SDC-ungrowable species on assay results is the same or less in treated samples as in the controls. The requirement would be violated under a hypothetical scenario that there is no reason to expect a priori: SDC-ungrowable species which (i) are rare compared with dominant species and thus do not significantly influence the enumeration of viable cells in untreated controls but (ii) are also very resistant to treatment compared to the majority of cells so that they represent a significant proportion of viable cells after treatment. The comparison of viable vs. total cells in the untreated control would suggest no problem with the SDC-MPN method, but the disproportionate survival of these rare, resistant, and ungrowable cells would lead to an SDC-MPN underestimation of viable cells after treatment proportional to their abundance relative to surviving growable cells. The error would be relevant to regulation only if the concentrations of surviving, originally rare ungrowables were close to or higher than the regulatory limit of 10 cellsmL^{-1}.

Our scenario, presented for completeness, might be highly unlikely, but it should be considered. Hypothetically, the problem could be addressed by identifying species in dilution culture grow-outs from untreated plankton communities to see which consistently fail to grow, and then examining culture-independent single-cell measures of their susceptibility to treatment [e.g., pyrimidine dimers, Buma et al. (1995) or other molecular markers to be developed]. But, it should be remembered that competition will influence the results of SDC (Throndsen and Kristiansen1991; Cerino and Zingone2006) in all but the extreme dilutions (from which rare species are excluded). Although it would be possible to document rare species that do grow, thereby reducing concern about their complicity false-negative errors, there is as yet no straightforward way to document rare species that do not grow well in dilution culture: failure to detect them in mixed-species cultures is not conclusive evidence of their inability to grow to detection on their own.

We conclude that uncertainties associated with poor growth in dilution culture are unlikely to be fully resolved, although agreement between enumerations of viable cells and counts of total cells would indicate that the errors are not large compared to uncertainties in other metrics of phytoplankton or their activities, discussed below. The potential problem of poor growth in dilution culture has another facet: if conditions are chosen to favor local and regional conditions as suggested here, the species that do not grow well in the assay are demonstrating reduced fitness for invasion compared to competitors that do. It follows that false negatives for phytoplankton that are poorly adapted to locally optimized growth conditions represent a reduced environmental risk compared with underestimation of viable cells from more fit species.

Quantifying other Plankton

The SDC-MPN method as described here is designed for plankton that grow well as photoautotrophs, but regulations apply to all plankton. Alternate approaches are required to enumerate resting stages (Gregg and Hallegraeff2007) and heterotrophs (Lighthart1969) in the 10–50μm size range. If methods for enumerating viable heterotrophs as defined here are not practical for application in routine testing, direct counts of living cells could be used (ETV2010), probably leading to false positives for UV treatment of heterotrophs.

Putting Errors in Context

The SDC-MPN method will never be exact, but few if any techniques for quantifying phytoplankton or their activities are. It is well established that the relationship between in vivo fluorescence and chlorophyll*a*is highly variable, as is the ratio of chlorophyll to phytoplankton biomass (Cullen1982). Estimates of phytoplankton biomass from space (e.g., Boyce et al.2010) add yet another level of error that compounds the uncertainty in estimates of primary productivity (Saba et al.2010; Friedrichs et al.2009), models of which are grounded in measurements of photosynthesis that are also subject to significant variability due to methods and interpretations (Richardson1991; Cullen2001). Yet, there is no question that measurements of chlorophyll, in vivo fluorescence, and primary productivity, along with estimates of phytoplankton biomass and productivity from space, have added immeasurably to our knowledge of planktonic processes, despite inaccuracies inherent in each.

Estimates of phytoplankton vitality as assessed by the vital stain fluorescein diacetate—one of the stains recommended by the ETV panel (EPA2010) for verifying treatment efficacy—are subject to significant uncertainties from inter- and intra-specific variability in stain uptake (Selvin et al.1988; Agustí and Sánchez2002; Garvey et al.2007; Peperzak and Brussaard2011). The method has been validated with natural populations by comparison with an enzymic digestion test (Agustí and Sánchez2002) and by comparing staining in control (live) and heat- or cold-treated samples (Steinberg et al.2011), but a systematic and quantitative evaluation of false positive and false negative errors across species and growth conditions has not been published; our results on laboratory cultures suggest that the errors can be large (H.L. MacIntyre et al., in prep.). It follows that estimates of total viable phytoplankton from SDC-MPN, though subject to errors described here, could potentially serve an important role as an alternative technique in BWT verification for type approval (see ETV2010) if uncertainties in the method can be adequately constrained and shown to be acceptable in comparison with other techniques.

Alternate Assays of Viability

Growth is the only direct measure of viability, and for ballast water treatment with UV, viability, as compared to vitality, is the only accurate measure of invasive potential. Further, the viability criterion is suitable for assessing other treatment technologies if it is practicable, because cells that have been killed outright are also nonviable. In principle, the SDC-MPN method is the gold standard for enumerating viable cells, but as we have shown, its application for use on natural samples introduces uncertainties that can be minimized but not entirely eliminated. Regardless, the SDC-MPN method can never be used for rapid assays (e.g., minutes, First and Drake2013b) that are preferred for any BWT testing procedure and required for applications such as routine monitoring of system performance (King and Tamburri2010). Proxy measurements of both vitality and viability are needed (First and Drake2013b; Reavie et al.2010; Steinberg et al.2012), but they must be validated. Given the live/dead/viable issue with UV radiation, we suggest that proxies must ultimately be tested against viability. In our experience, the SDC-MPN method is well suited for quantitative and accurate implementation with cultures of phytoplankton (MacIntyre et al., submitted for publication, Oemcke and Van Leeuwen2005), so with appropriate testing and evaluation methodology, it should be possible to develop rapid assays of viability that are rigorously validated through comparisons with direct measurements of survival and growth after treatment.

SUMMARY AND CONCLUSIONS

This review provides the foundations for the following statements about the use of the SDC-MPN method for enumerating viable phytoplankton in the testing of ballast water management systems:

- *Postulate of equivalent protection*: Because neither a dead organism nor a nonreproductive organism can propagate after discharge from ships' ballast, discharge criteria based on vitality (live/dead) and viability (the ability to reproduce) are equally protective of coastal environments.
- It follows that the concentration of viable phytoplankton cells, as compared to living cells, could serve as an alternative measure for regulating ballast water discharge from any type of management system—if it is established that viable cells in natural phytoplankton samples can be reliably enumerated after BWT.
- Since UV renders phytoplankton nonreproductive and thus harmless without killing them outright, the effectiveness of ballast water treatment with UV for protecting coastal environments can be assessed accurately only with measures of viability, not vital stains.
- The SDC-MPN assay enumerates viable phytoplankton. It has been used for more than 50years in the context of phytoplankton ecology, generally to identify and count phytoplankton species that cannot be preserved for conventional microscopic analysis. To the best of our knowledge, the use of SDC-MPN to enumerate total viable phytoplankton in a natural sample without regard to species composition is a new application emerging from its potential utility in testing BWT systems.

- When the sole objective is to enumerate total viable cells, the accuracy of SDC-MPN depends primarily on its ability to detect growth in culture tubes starting with one or more viable cells. Concerns about SDC-MPN on natural communities can be addressed. Although some species of phytoplankton are difficult to bring into persistent culture, many are expected to grow in a first-round dilution culture for SDC-MPN if conditions are suitable; this expectation can be tested. Competition between phytoplankton species in dilution cultures is irrelevant as long as the winner is detectable. Grazing is expected to have a relatively small influence on the outcome of SDC-MPN, and conservative results can be obtained if it is practicable to detect grazers in tubes and score them as positive for a viable phytoplankton cell. Recovery from damage during treatment, e.g., photorepair, is not a complication: If the cell survives and grows, it is enumerated as viable, and since growth conditions are optimized, repair is more likely than it might be in nature.
- The accuracy of SDC-MPN depends on the ability of a single cell in a diluted sample to reproduce through enough generations to be detected reliably—a function of growth rate, signal per cell, detector sensitivity, and the time period of observation. We propose procedures to maximize growth rate and detectability, and thus accuracy of the method, and to estimate uncertainties, which can be reduced through systematic comparisons that we describe.
- We conclude that SDC-MPN is potentially an effective method for assessing the viability of phytoplankton after BWT. It has been used in phytoplankton studies for many decades but only recently to enumerate total viable phytoplankton in natural samples. In this application, some sources of error are much less important than previously thought and others can be assessed and minimized, though probably not eliminated. When used on cultures in the laboratory, the SDC-MPN method can serve as the standard for validating proxy measurements of viability that could be useful in shipboard compliance testing and port state control inspection. In all applications, experiments must be carefully designed and tested. Evaluation of the method for use in regulation will be facilitated if practitioners share their methods and results.

ACKNOWLEDGMENTS

This work was supported by an NSERC Collaborative Research and Development Grant with Trojan Technologies, an NSERC Discovery Grant (JJC) and Trojan Technologies. Suggestions from B. Petri were very helpful, and comments from two anonymous reviewers led to significant improvements in presentation. We thank R.R.L. Guillard and R.A. Andersen for detailed comments on questions about culturability of phytoplankton, but we take sole responsibility for the interpretations presented here.

REFERENCES

1. Agustí S, Sánchez MC (2002) Cell viability in natural phytoplankton communities quantified by a membrane permeability probe. Limnol Oceanog 47:818–828
2. Albert RJ, Lishman JM, Saxena JR (2013) Ballast water regulations and the move toward concentration-based numeric discharge limits. Ecol Appl 23:289–300
3. Allen E (1919) A contribution to the quantitative study of plankton. J Mar Biol Assoc U K (New Series) 12(01):1–8
4. Allen E, Nelson E (1910) On the artificial culture of marine plankton organisms. J Mar Biol Assoc U K (New Series) 8(05):421–474
5. Andersen RA (ed) (2005) Algal culturing techniques. Academic Press, Amsterdam
6. Andersen RA, Kawachi M (2005) Traditional microalgae isolation techniques. In: Andersen RA (ed) Algal culturing techniques. Academic Press, Amsterdam, pp 83–100
7. Andersen P, Throndsen J (2003) Estimating cell numbers. In: Hallegraeff GM, Anderson DM, Cembella AD (eds) Manual on harmful marine microalgae, vol 11, Monographs on oceanographic methodology. UNESCO, Paris, pp 99–129
8. Backe-Hansen P, Throndsen J (2002) Pico- and nanoplankton from the inner Oslofjord, eastern Norway, including description of two new species of*Luffisphaera*(incerta sedis). Sarsia 87:55–64
9. Bannister TT, Laws EA (1980) Modeling phytoplankton carbon metabolism. In: Falkowski PG (ed) Primary Productivity in the Sea. Plenum, New York, pp 243–248
10. Baroli I, Melis A (1996) Photoinhibition and repair in*Dunaliella salina*acclimated to different growth irradiances. Planta 198:640–646
11. Behrenfeld MJ, Falkowski PG (1997) A consumer›s guide to phytoplankton primary productivity models. Limnol Oceanog 42:1479–1491
12. Binder BJ, Anderson DM (1987) Physiological and environmental-control of germination in*Scrippsiella trochoidea*(Dinophyceae) resting cysts. J Phycol 23:99–107
13. Bird RE, Riordan C (1986) Simple solar spectral model for direct and diffuse irradiance on horizontal and tilted planes at the Earth›s surface for cloudless atmospheres. Climate Appl Meteorol 25:87–97
14. Blanchard GF, Guarini J-M, Richard P, Gros P, Mornet F (1996) Quantifying the short-term temperature effect on light-saturated photosynthesis of intertidal microalgae. Mar Ecol Prog Ser 134:309–313
15. Blodgett R (2010) Bacteriological Analytical Manual Appendix 2. Most Probable Number from Serial Dilutions. http://www.fda.gov/Food/FoodScienceResearch/LaboratoryMethods/ucm109656.htm. Accessed 26 Nov 2014
16. Boelen P, Veldhuis MJ, Buma AG (2001) Accumulation and removal of UVBR-induced DNA damage in marine tropical plankton subjected to mixed and simulated non-mixed conditions. Aquat Microbial Ecol 24:265–274
17. Boyce DG, Lewis MR, Worm B (2010) Global phytoplankton decline over the past century. Nature 466:591–596

18. Brand L, Guillard R (1981a) The effects of continuous light and light intensity on the reproduction rates of twenty-two species of marine phytoplankton. J Exp Mar Biol Ecol 50:119–132

19. Brand LE, Guillard RRL (1981b) A method for the rapid and precise determination of acclimated phytoplankton reproduction rates. J Plankton Res 3:191–201

20. Bravo I, Anderson DM (1994) The effects of temperature, growth-medium and darkness on excystment and growth of the toxic dinoflagellate*Gymnodinium catenatum*from northwest Spain. J Plankton Res 16:513–525

21. Buma AGJ, Van Hannen EJ, Roza L, Veldhuis MJW, Gieskes WWC (1995) Monitoring ultraviolet-B-induced DNA damage in individual diatom cells by immunofluorescent thymine dimer detection. J Phycol 31:314–321

22. Burkholder JM, Hallegraeff GM, Melia G, Cohen A, Bowers HA, Oldach DW, Parrow MW, Sullivan MJ, Zimba PV, Allen EH (2007) Phytoplankton and bacterial assemblages in ballast water of US military ships as a function of port of origin, voyage time, and ocean exchange practices. Harmful Algae 6:486–518

23. Campbell DA, Hossain Z, Cockshutt AM, Zhaxybayeva O, Wu HY, Li G (2013) Photosystem II protein clearance and FtsH function in the diatom*Thalassiosira pseudonana*. Photosynth Res 115:43–54

24. Cerino F, Zingone A (2006) A survey of cryptomonad diversity and seasonality at a coastal Mediterranean site. Eur J Phycol 41:363–378

25. Coast Guard US (2012) Standards for living organisms in ships' ballast water discharged in US waters. Fed Regist 77:17254–17320

26. Cochran WG (1950) Estimation of bacterial densities by means of the "most probable number". Biometrics 6:105–116

27. Cullen JJ (1982) The deep chlorophyll maximum: comparing vertical profiles of chlorophyll a. Can J Fish Aquat Sci 39:791–803

28. Cullen JJ (1990) On models of growth and photosynthesis in phytoplankton. Deep-Sea Res 37:667–683

29. Cullen JJ (2001) Plankton: primary production methods. In: Steele J, Thorpe S, Turekian K (eds) Encyclopedia of Ocean Sciences. Academic Press, pp 2277–2284

30. Cullen JJ, MacIntyre JG (1998) Behavior, physiology and the niche of depth-regulating phytoplankton. In: Anderson DM, Cembella AD, Hallegraeff GM (eds) Physiological Ecology of Harmful Algal Blooms, vol G 41. NATO ASI Series. Springer, Heidelberg, pp 559–580

31. Cullen JJ, Geider RJ, Ishizaka J, Kiefer DA, Marra J, Sakshaug E, Raven JA (1993) Toward a general description of phytoplankton growth for biogeochemical models. In: Evans GT, Fasham MJR (eds) Towards a Model of Ocean Biogeochemical Processes. Springer, Berlin, pp 153–176

32. Cullen JJ, Franks PJS, Karl DM, Longhurst A (2002) Physical influences on marine ecosystem dynamics. In: Robinson AR, McCarthy JJ, Rothschild BJ (eds) The Sea: Biological-Physical Interactions in the Ocean, vol 12. Wiley pp 297–335

33. Cullen JJ, Davis RF, Huot Y (2012) Spectral model of depth-integrated water column photosynthesis and its inhibition by ultraviolet radiation. Glob Biogeochem Cycles 26(1):GB1011. doi:10.1029/2010gb003914

34. Drake LA, Tamburri MN, First MR, Smith GJ, Johengen TH (2014) How many organisms are in ballast water discharge? A framework for validating and selecting compliance monitoring tools. Mar Poll Bull 86:122–128

35. EPA (2010) Generic protocol for the verification of ballast water treatment technology., Version 5.1. EPA/600/R-10/146. US Environmental Protection Agency Environmental Technology Verification Program, Washington, DC

36. ETV (2010) Generic protocol for the verification of ballast water treatment technology. NSF International for USEPA Environmental Technology Verification Program, Ann Arbor

37. Fiala M, Oriol L (1990) Light-temperature interactions on the growth of Antarctic diatoms. Polar Biol 10:629–636

38. First MR, Drake LA (2013a) Approaches for determining the effects of UV radiation on microorganisms in ballast water. Manag Biol Inv 4:87–99

39. First MR, Drake LA (2013b) Life after treatment: detecting living microorganisms following exposure to UV light and chlorine dioxide. J Appl Phycol 26:227–235

40. Follows MJ, Dutkiewicz S (2011) Modeling diverse communities of marine microbes. Annu Rev Mar Sci 3:427–451

41. Follows MJ, Dutkiewicz S, Grant S, Chisholm SW (2007) Emergent biogeography of microbial communities in a model ocean. Science 315:1843–1846

42. Fork DC, Murata N, Sato N (1979) Effect of growth temperature on the lipid and fatty acid composition, and the dependence on temperature of light-induced redox reactions of cytochrome*f*and of light energy redistribution in the thermophilic blue-green alga*Synechococcus lividus*. Plant Physiol 63:524–530Central

43. Frazier M, Miller AW, Lee H, Reusser DA (2013) Counting at low concentrations: the statistical challenges of verifying ballast water discharge standards. Ecol Appl 23:339–351

44. Friedrichs MAM, Carr ME, Barber RT, Scardi M, Antoine D, Armstrong RA, Asanuma I, Behrenfeld MJ, Buitenhuis ET, Chai F, Christian JR, Ciotti AM, Doney SC, Dowell M, Dunne J, Gentili B, Gregg W, Hoepffner N, Ishizaka J, Kameda T, Lima I, Marra J, Melin F, Moore JK, Morel A, O'Malley RT, O'Reilly J, Saba VS, Schmeltz M, Smyth TJ, Tjiputra J, Waters K, Westberry TK, Winguth A (2009) Assessing the uncertainties of model estimates of primary productivity in the tropical Pacific Ocean. J Mar Syst 76:113–133

45. Furuya K, Marumo R (1983) Size distribution of phytoplankton in the Western Pacific Ocean and adjacent waters in summer. Bull Plankton Soc Japan 30:31–32

46. Gao Y, Smith GJ, Alberte RS (2000) Temperature

dependence of nitrate reductase activity in marine phytoplankton: biochemical analysis and ecological implications. J Phycol 36:304–313

47. Garthright W, Blodgett R (2003) FDA's preferred MPN methods for standard, large or unusual tests, with a spreadsheet. Food Microbiol 20:439–445

48. Garvey M, Moriceau B, Passow U (2007) Applicability of the FDA assay to determine the viability of marine phytoplankton under different environmental conditions. Mar Ecol Prog Ser 352:17–26

49. Geider RJ, MacIntyre HL, Kana TM (1998) A dynamic regulatory model of phytoplankton acclimation to light, nutrient, and temperature. Limnol Oceanog 43:679–694

50. Gieskes W, Kraay G (1983) Dominance of Cryptophyceae during the phytoplankton spring bloom in the central North Sea detected by HPLC analysis of pigments. Mar Biol 75:179–185

51. Giovannoni SJ, Britschgi TB, Moyer CL, Field KG (1990) Genetic diversity in Sargasso Sea bacterioplankton. Nature 345:60–63

52. Granéli E, Salomon PS (2010) Factors influencing allelopathy and toxicity in*Prymnesium parvum*. J Am Water Resour Assoc 46:108–120

53. Gregg MD, Hallegraeff GM (2007) Efficacy of three commercially available ballast water biocides against vegetative microalgae, dinoflagellate cysts and bacteria. Harmful Algae 6:567–584

54. Guillard R (2005) Purification methods for microalgae. In: Andersen RA (ed) Algal culturing techniques. Academic Press, Amsterdam, pp 117–132

55. Guillard R, Morton S (2003) Culture methods. Manual on harmful marine microalgae. UNESCO IOC Monographs on Oceanographic Methodology 11:77–97

56. Haas CN, Heller B (1988) Test of the validity of the Poisson assumption for analysis of most-probable-number results. Appl Environ Microbiol 54:2996–3002Central

57. Hancke K, Hancke TB, Olsen LM, Johnsen G, Glud RN (2008) Temperature effects on microalgal photosynthesis-light responses measured by O_2production, pulse-amplitude-modulated fluorescence, and^{14}C assimilation. J Phycol 44:501–514

58. Hardison DR, Sunda WG, Shea D, Litaker RW (2013) Increased toxicity of Karenia brevis during phosphate limited growth: ecological and evolutionary implications. Plos One 8(3). doi:10.1371/journal.pone.0058545

59. Harrison WG, Platt T (1980) Variations in assimilation number of coastal marine phytoplankton: effects of environmental covariates. J Plankton Res 2:249–260

60. Harrison WG, Platt T (1986) Photosynthesis–irradiance relationships in polar and temperate phytoplankton populations. Polar Biol 5:153–164

61. Heywood JL, Sieracki ME, Bellows W, Poulton NJ, Stepanauskas R (2010) Capturing diversity of marine heterotrophic protists: one cell at a time. ISME J 5:674–684Central

62. Hijnen W, Beerendonk E, Medema GJ (2006) Inactivation credit of UV radiation for viruses, bacteria and protozoan (oo)cysts in water: a review. Water Res 40:3–22

63. Hikosaka K, Ishikawa K, Borjigidai A, Muller O, Onoda Y (2006) Temperature acclimation of photosynthesis: mechanisms involved in the changes in temperature dependence of photosynthetic rate. J Exp Bot 57:291–302

64. Hulburt EM (1982) The adaptation of marine phytoplankton species to nutrient and temperature. Ocean Sci Eng 7:187–228

65. Humphries S, Widjaja F (1979) A simple method for separating cells of*Microcystis aeruginosa*for counting. Brit Phycol J 14:313–316

66. Hurley MA, Roscoe M (1983) Automated statistical analysis of microbial enumeration by dilution series. J Appl Bact 55:159–164

67. IMO (2004) Convention BWM/CONF/36 International Convention for the Control and Management of Ships' Ballast Water and Sediments

68. IMO (2015).http://www.imo.org/OurWork/Environment/BallastWaterManagement/Documents/Tableof BA FA TA updated in Oct 2014.pdf Accessed 14 Jan 2015

69. International Maritime Organization Marine Environment Protection Committee (2008) Annex 4 Resolution MEPC. 174 (58). Guidelines for Approval of Ballast Water Management Systems (G8). London, UK

70. Ishikawa A, Furuya K (2004) The role of diatom resting stages in the onset of the spring bloom in the East China Sea. Mar Biol 145:633–639

71. Jochem FJ (1990) On the seasonal occurrence of autotrophic naked nanoflagellates in Kiel Bight, western Baltic. Estuar Coast Shelf Sci 31:189–202

72. Johnson ZI, Zinser ER, Coe A, McNulty NP, Woodward EMS, Chisholm SW (2006) Niche partitioning among*Prochlorococcus*ecotypes along ocean-scale environmental gradients. Science 311:1737–1740

73. Kaeberlein T, Lewis K, Epstein SS (2002) Isolating «uncultivable» microorganisms in pure culture in a simulated natural environment. Science 296:1127–1129

74. Kashtan N, Roggensack SE, Rodrigue S, Thompson JW, Biller SJ, Coe A, Ding H, Marttinen P, Malmstrom RR, Stocker R, Follows MJ, Stepanauskas R, Chisholm SW (2014) Single-cell genomics reveals hundreds of coexisting subpopulations in wild*Prochlorococcus*. Science 344:416–420

75. Kawachi M, Noël M-H (2005) Sterilization and sterile technique. In: Andersen RA (ed) Algal culturing techniques. Academic Press, Amsterdam, pp 65–81

76. King DM, Tamburri MN (2010) Verifying compliance with ballast water discharge regulations. Ocean Dev Int Law 41:152–165

77. Knight-Jones E (1951) Preliminary studies of nanoplankton and ultraplankton systematics and abundance by a quantitative culture method. J Conseil 17:140–155

78. Krupke A, Lavik G, Halm H, Fuchs BM, Amann RI, Kuypers MMM (2014) Distribution of a consortium between unicellular algae and the N_2fixing cyanobacterium UCYN-A in the North Atlantic Ocean.

Env Microbiol 16:3153–3167

79. Kugrens P, Lee RE (1987) An ultrastructural survey of cryptomonad periplasts using quick-freezing freeze fracture techniques. J Phycol 23:365–376
80. Landry MR, Hassett RP (1982) Estimating the grazing impact of marine micro-zooplankton. Mar Biol 67:283–288
81. Letelier RM, Karl DM, Abbott MR, Bidigare RR (2004) Light driven seasonal patterns of chlorophyll and nitrate in the lower euphotic zone of the North Pacific Subtropical Gyre. Limnol Oceanogr 49:508–519
82. Li WKW, Dickie PM (2001) Monitoring phytoplankton, bacterioplankton, and virioplankton in a coastal inlet (Bedford Basin) by flow cytometry. Cytometry 44:236–246
83. Liebich V, Stehouwer PP, Veldhuis M (2012) Re-growth of potential invasive phytoplankton following UV-based ballast water treatment. Aquat Invas 7:29–36
84. Lighthart B (1969) Planktonic and benthic bacteriovorous protozoa at eleven stations in Puget Sound and adjacent Pacific Ocean. J Fish Board Canada 26:299–304
85. Lima ID, Doney SC (2004) A three-dimensional, multi-nutrient, and size-structured ecosystem model for the North Atlantic. Glob Biogeochem Cycles 18:GB3019
86. MacIntyre HL, Kana TM, Anning T, Geider RJ (2002) Photoacclimation of photosynthesis irradiance response curves and photosynthetic pigments in microalgae and cyanobacteria. J Phycol 38:17–38
87. Marañón E (2015) Cell size as a key determinant of phytoplankton metabolism and community structure. Annu Rev Mar Sci 7:241–264
88. McCrady MH (1915) The numerical interpretation of fermentation-tube results. J Infect Dis 17:183–212
89. Montagnes DJ, Berges JA, Harrison PJ, Taylor F (1994) Estimating carbon, nitrogen, protein, and chlorophyll a from volume in marine phytoplankton. Limnol Oceanog 39:1044–1060
90. Moore LR, Post AF, Rocap G, Chisholm SW (2002) Utilization of different nitrogen sources by the marine cyanobacteria*Prochlorococcus*and*Synechococcus*. Limnol Oceanog 47:989–996
91. Moore LR, Coe A, Zinser ER, Saito MA, Sullivan MB, Lindell D, Frois-Moniz K, Waterbury J, Chisholm SW (2007) Culturing the marine cyanobacterium*Prochlorococcus*. Limnol Oceanogr Methods 5:353–362
92. Neale PJ, Banaszak AT, Jarriel CR (1998) Ultraviolet sunscreens in*Gymnodinium sanguineum*(Dinophyceae): mycosporine-like amino acids protect against inhibition of photosynthesis. J Phycol 34:928–938
93. Nielsen LT, Krock B, Hansen PJ (2013) Production and excretion of okadaic acid, pectenotoxin-2 and a novel dinophysistoxin from the DSP-causing marine dinoflagellate*Dinophysis acuta*—effects of light, food availability and growth phase. Harmful Algae 23:34–45
94. Nitta K, Suzuki N, Honman D, Kaneko Y, Nakamoto H (2006) Ultrastructural stability under high temperature or intensive light stress conferred by a small heat shock protein in cyanobacteria. FEBS Lett 579:1235–1242
95. Norberg J (2004) Biodiversity and ecosystem functioning: a complex adaptive systems approach. Limnol Oceanog 49:1269–1277
96. Oemcke D, Van Leeuwen J (2005) Ozonation of the marine dinoflagellate alga*Amphidinium*sp.—implications for ballast water disinfection. Water Res 39:5119–5125
97. Paasche E (1968) Marine plankton algae grown with light–dark cycles. II.*Ditylum brightwellii*and*Nitzschia turgidula*. Physiol Plant 21:66–77
98. Paerl HW (1978) Effectiveness of various counting methods in detecting viable phytoplankton. N Z J Mar Freshw Res 12:67–72
99. Park MG, Kim S, Kim HS, Myung G, Kang YG, Yih W (2006) First successful culture of the marine dinoflagellate*Dinophysis acuminata*. Aquat Microbial Ecol 45:101–106
100. Partensky F, Hess W, Vaulot D (1999)*Prochlorococcus*, a marine photosynthetic prokaryote of global significance. Microbiol Molec Biol Rev 63:106–127
101. Peperzak L, Brussaard CPD (2011) Flow cytometric applicability of fluorescent vitality probes on phytoplankton. J Phycol 47:692–702
102. Platt T, Sathyendranath S (1993) Estimators of primary production for the interpretation of remotely-sensed data on ocean color. J Geophys Res 98:14,561–514,596
103. Rappé MS, Connon SA, Vergin KL, Giovannoni SJ (2002) Cultivation of the ubiquitous SAR11 marine bacterioplankton clade. Nature 418:630–633
104. Reavie ED, Cangelosi AA, Allinger LE (2010) Assessing ballast water treatments: Evaluation of viability methods for ambient freshwater microplankton assemblages. J Great Lakes Res 36:540–547
105. Revelante N, Gilmartin M (1991) The phytoplankton composition and population enrichment in gelatinous macroaggregates in the northern Adriatic during the summer of 1989. J Exp Mar Biol Ecol 146:217–233
106. Richardson K (1991) Comparison of^{14}C primary production determinations made by different laboratories. Mar Ecol Prog Ser 72:189–201
107. Ritchie RJ, Larkum AWD (2012) Modelling photosynthesis in shallow algal production ponds. Photosynthetica 50:481–500
108. Rodriguez-Martinez R, Rocap G, Salazar G, Massana R (2013) Biogeography of the uncultured marine picoeukaryote MAST-4: temperature-driven distribution patterns. Isme J 7:1531–1543Central
109. Roy S (2000) Strategies for the minimisation of UV-induced damage. In: de Mora S, Demers S, Vernet M (eds) The Effects of UV Radiation in the Marine Environment. Cambridge University Press, Cambridge, pp 177–205
110. Saba VS, Friedrichs MAM, Carr M-E, Antoine D, Armstrong RA, Asanuma I, Aumont O, Bates NR, Behrenfeld MJ, Bennington V, Bopp L, Bruggeman J, Buitenhuis ET, Church MJ, Ciotti AM, Doney SC, Dowell M, Dunne J, Dutkiewicz S, Gregg W, Hoepffner N, Hyde

KJW, Ishizaka J, Kameda T, Karl DM, Lima I, Lomas MW, Marra J, McKinley GA, Mélin F, Moore JK, Morel A, O'Reilly J, Salihoglu B, Scardi M, Smyth TJ, Tang S, Tjiputra J, Uitz J, Vichi M, Waters K, Westberry TK, Yool A (2010) Challenges of modeling depth-integrated marine primary productivity over multiple decades: a case study at BATS and HOT. Global Biogeochem Cycl 24:GB3020

111. Selvin R, Requera B, Bravo I, Yentsch CM (1988) Use of fluorescein diacetate (FDA) as a single-cell probe of metabolic activity in dinoflagellate cultures. Biol Oceanogr 6:505–511

112. Sherr EB, Sherr BF (2007) Heterotrophic dinoflagellates: a significant component of microzooplankton biomass and major grazers of diatoms in the sea. Mar Ecol Prog Ser 352:187–197

113. Shi XL, Marie D, Jardillier L, Scanlan DJ, Vaulot D (2009) Groups without cultured representatives dominate eukaryotic picophytoplankton in the oligotrophic South East Pacific Ocean. Plos One 4(10). doi:10.1371/journal.pone.0007657

114. Sieracki M, Poulton N, Crosbie N (2005) Automated isolation techniques for microalgae. In: Andersen RA (ed) Algal culturing techniques. Academic Press, Amsterdam, pp 101–116

115. Sinigalliano CD, Winshell J, Guerrero MA, Scorzetti G, Fell JW, Eaton RW, Brand L, Rein KS (2009) Viable cell sorting of dinoflagellates by multiparametric flow cytometry. Phycologia 48:249–257Central

116. Smayda TJ (1969) Experimental observations on the influence of temperature, light, and salinity on cell division of the marine diatom*Detonula confervacea*(Cleve) Gran. J Phycol 5:150–157

117. Steinberg MK, Lemieux EJ, Drake LA (2011) Determining the viability of marine protists using a combination of vital, fluorescent stains. Mar Biol 158:1431–1437

118. Steinberg MK, First MR, Lemieux EJ, Drake LA, Nelson BN, Kulis DM, Anderson DM, Welschmeyer NA, Herring PR (2012) Comparison of techniques used to count single-celled viable phytoplankton. J Appl Phycol 24:751–758

119. Strom SL, Miller CB, Frost BW (2000) What sets lower limits to phytoplankton stocks in high-nitrate, low-chlorophyll regions of the open ocean? Mar Ecol Prog Ser 193:19–31

120. Sundström AM, Kremp A, Daugbjerg N, Moestrup O, Ellegaard M, Hansen R, Hajdu S (2009)*Gymnodinium corollarium sp nov*(Dinophyceae)—a new cold-water dinoflagellate responsible for cyst sedimentation events in the Baltic Sea. J Phycol 45:938–952

121. Suttle CA, Chan AM (1993) Marine cyanophages infecting oceanic and coastal strains of Synechococcus: abundance, morphology, cross-infectivity and growth characteristics. Mar Ecol Prog Ser 92:99

122. Talling JF (1957) The phytoplankton population as a compound photosynthetic system. New Phytol 56:133–149

123. Tameishi M, Yamasaki Y, Nagasoe S, Shimasaki Y, Oshima Y, Honjo T (2009) Allelopathic effects of the dinophyte Prorocentrum minimum on the growth of the bacillariophyte*Skeletonema costatum*. Harmful Algae 8:421–429

124. Thomas WH (1966) Effects of temperature and illuminance on cell division rates of three species of tropical oceanic phytoplankton. J Phycol 11:17–22

125. Thomas MK, Kremer CT, Klausmeier CA, Litchman E (2012) A global pattern of thermal adaptation in marine phytoplankton. Science 338:1085–1088

126. Throndsen J (1978) The dilution-culture method. In: Sournia A (ed) Phytoplankton manual, vol 6. Monographs on oceanographic methodology UNESCO, Paris, pp 218–224

127. Throndsen J, Kristiansen S (1991)*Micromonas pusilla*(Prasinophyceae) as part of pico- and nanoplankton communities of the Barents Sea. Polar Res 10:201–208

128. Tyystjärvi E (2008) Photoinhibition of Photosystem II and photodamage of the oxygen evolving manganese cluster. Coord Chem Rev 252:361–376

129. U.S. Environmental Protection Agency Environmental Technology Verification Program (2010) Generic protocol for the verification of ballast water treatment technology. Report number EPA/600/R–10/146

130. Usup G, Kulis DM, Anderson DM (1994) Growth and toxin production of the toxic dinoflagellate*Pyrodinium bahamense var. compressum*in laboratory cultures. Nat Toxins 2:254–262

131. van Reis R, Brake JM, Charkoudian J, Burns DB, Zydney AL (1999) High-performance tangential flow filtration using charged membranes. J Membr Sci 159:133–142

132. Vaulot D, Eikrem W, Viprey M, Moreau H (2008) The diversity of small eukaryotic phytoplankton (≤3μm) in marine ecosystems. FEMS Microbiol Rev 32:795–820

133. Verity PG, Smetacek V (1996) Organism life cycles, predation, and the structure of marine pelagic ecosystems. Mar Ecol Prog Ser 130:277–293

134. Wood AM, Everroad RC, Wingard LM (2004) Measuring growth rates in algal cultures. In: Andersen RA (ed) Algal culturing techniques. Academic Press, Amsterdam, pp 269–285

135. Wright D, Dawson R, Caceres V, Orano-Dawson C, Kananen G, Cutler S, Cutler H (2009) Shipboard testing of the efficacy of SeaKleen® as a ballast water treatment to eliminate non-indigenous species aboard a working tanker in Pacific waters. Env Technol 30:893–910

136. Yoder JA (1979) Effect of temperature on light-limited growth and chemical composition of*Skeletonema costatum*(Bacillariophyceae). J Phycol 15:362–370

137. Zehr JP, Waterbury JB, Turner PJ, Montoya JP, Omoregie E, Steward GF, Hansen A, Karl DM (2001) Unicellular cyanobacteria fix N_2in the subtropical North Pacific Ocean. Nature 412:635–638

138. Zetsche E-M, Meysman FJR (2012) Dead or alive? Viability assessment of micro- and mesoplankton. J Plankton Res 34:493–509

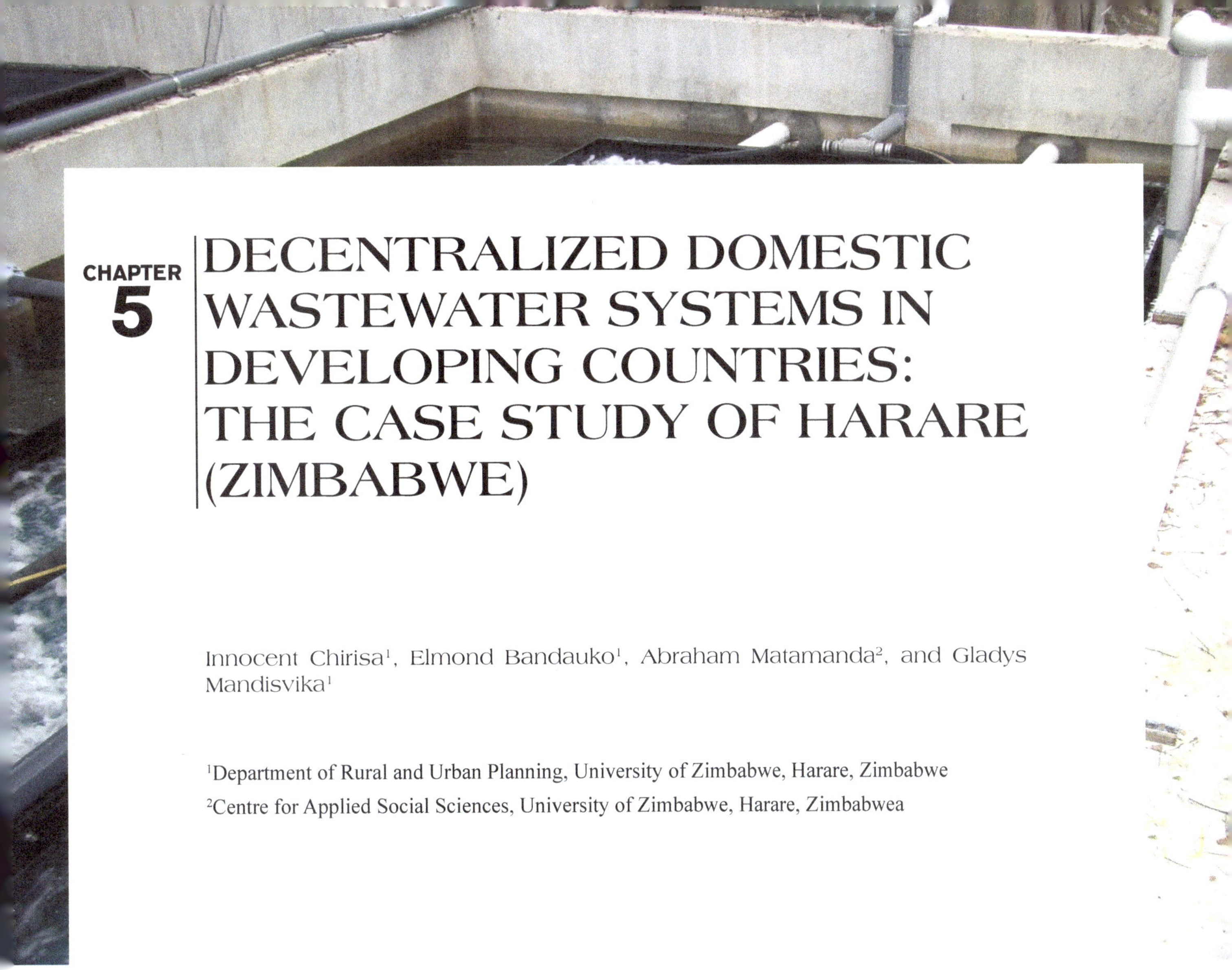

CHAPTER 5

DECENTRALIZED DOMESTIC WASTEWATER SYSTEMS IN DEVELOPING COUNTRIES: THE CASE STUDY OF HARARE (ZIMBABWE)

Innocent Chirisa[1], Elmond Bandauko[1], Abraham Matamanda[2], and Gladys Mandisvika[1]

[1]Department of Rural and Urban Planning, University of Zimbabwe, Harare, Zimbabwe

[2]Centre for Applied Social Sciences, University of Zimbabwe, Harare, Zimbabwea

ABSTRACT

Until recently there has been little, if any, concern over revamping let alone improving wastewater management system in Zimbabwe's urban areas given the dominance and institutionalised water-borne system. Yet, the current constraints in this system and the immensity of urbanisation in the country begs and compels planners, engineers and systems thinkers to rethink what best can work as a sustainable wastewater system. With particular reference to the ever-expanding Harare metropolitan region, this article provides an evaluative analysis on the potentiality, risks and strategies that can be adopted by Harare and its satellites in addressing the problems of the conventional wastewater management system. The suggested framework of operation is a decentralised domestic wastewater collection and treatment system which however has its own multifarious risks. Using systems dynamics conceptualisation of the potentiality, opportunities, risks and strategies, the paper seeks to model the path and outcomes of this decentralised domestic wastewater collection and treatment system and also suggests a number of policy measures and strategies that the city of Harare and its satellites can adopt.

INTRODUCTION

Developing countries are burdened with a multiplicity of problems, and wastewater management is increasingly becoming a priority issue. The management of wastewater system in developing countries is exacerbated by accelerating urbanisation, inadequate management and disposal of wastewater and the implementation of sophisticated treatment technologies that are highly centralised (Libralato et al. 2012). The current wastewater management systems are riddled with a plethora of irregularities that calls for a paradigm shift from the current centralised system to the decentralisation in wastewater treatment management. The principal reason for this has been that the provision of centralised systems is not technically, economically or environmentally feasible as De Gisi et al. (2014) notes that the conventional system regards wastewater as 'waste' yet there is potential to make use of the wastewater. Decentralised systems also offer an alternative approach to providing water, wastewater and storm

Citation: Innocent Chirisa, Elmond Bandauko, Abraham Matamanda, and Gladys Mandisvika, Decentralized domestic wastewater systems in developing countries: the case study of Harare (Zimbabwe), DOI 10.1007/s13201-016-0377-4.

water services to urban areas (Nhapi 2004a, b, c; Libralato et al. 2012). In recent times, the concept of integrating water and wastewater systems through separate collection and treatment of various water and waste streams and recovery of valuable water, nutrients and energy has been proposed. This helps to overcome the limitations of the centralised approach and to move towards more ecologically and economically sound water/wastewater management systems.

Furthermore, innovative decentralised systems are being planned and implemented for new and future urban development either as separate facilities or in combination with a centralised system (Diaper et al. 2007; Brown et al. 2009). Wilsenach (2006: 4) observes that a general critique against classical civil engineering projects is that they have not dealt with transport, drinking water, energy and wastewater in an integrated way. These were all designed as linear systems, without consideration of the cyclic character of most natural systems (De Gisi et al.2014). In urban water management there is a need for a change to improve the sustainability of the systems. This has been a wakeup call from proponents such as Cook et al. (2009) who argue that the conventional centralised systems are not always the most appropriate solution for urban development. Rather, a new approach has to be embraced that should include the integration of social, economic, and environmental aspects with practices such as rainwater management, water conservation, wastewater reuse, rational energy management (incorporating the use of alternative sources), nutrient recovery, and sorting at source (De Gisi et al. 2014; Suriyachan et al. 2012; Brown et al. 2009; Mitchell et al. 2008).

This present paper is an attempted evaluation and analysis of the decentralised wastewater treatment and management systems, putting into picture potentialities, opportunities, risks and strategies. It is a study that emanates from the fact that the conventional centralised wastewater management systems have failed to produce sustainable outcomes in terms of social, economic and environmental concerns. This work reviews some existing technologies applied in decentralised wastewater treatment and disposal in developing countries such as the septic tank, imhoff tank, and soil infiltration device, to name these few (ibid). Technologies that are potentially sound and beneficial for wastewater treatment are also explored. Emphasis is placed on faecal sludge management, which is grossly inadequate, and in some cases totally lacking in rural, peri-urban and some urban areas of developing countries. To demonstrate this, we use documentary and discourse analysis. First, the paper starts by providing an analytical framework, and then proceeds to provide a comparative analysis of the centralised and decentralised systems. The rationale for adoption of decentralised wastewater management system is also explained. Second, the legislative and policy framework for wastewater treatment and management in Zimbabwe is explained, which is then followed by an outline of the current situation in Harare as well as reflections on systems dynamics in the City. Third, the paper ends by a discussion, practical implications, policy options and conclusion.

Decentralised domestic wastewater collection and treatment system: an analytical framework

This section provides a conceptual reflection of the decentralised wastewater treatment and management systems. The aim of this section is to describe and explain the key features or characteristics of the decentralised wastewater management concept.

Decentralised wastewater system is when raw wastewater is simply treated next to the source (point of generation according to Omenka 2010) through a range of simple technologies. Office of Water United States Environmental Protection Agency (2005: 3) defines decentralised wastewater systems as consisting of a wide range of onsite and cluster treatment systems that process household and commercial sewage. The Rocky Mountain Institute (2004) defines decentralised systems as an alternative to conventional, centralised systems. In addition, the decentralised concept can be explained as an organisational paradigm for wastewater management (De Gisi et al. 2014; Sheehan 2011; Nhapi 2004a, b, c) that entails treating and beneficially reusing wastewater as close to where it is generated as practical, using technologies appropriate to the scale of the facilities (Venhuizen 2003). A decentralised system may consist of individual on-site systems and/or cluster systems, either singly or in combination with more highly collectivised facilities as Libralato et al. (2012) advocates for a combination of a centralised and decentralised so as to maximise on the opportunities of each system. The degree of collectivisation at any stage of the treatment and reuse or dispersal processes will be determined by a variety of local circumstances, including topography, site and soil characteristics, development density, type of development, community desires with regard to land use issues, and sites of potential reuse and/or sites where discharge would be allowable. Some systems in arid regions promote evaporation or wastewater uptake by plants.

However, decentralised wastewater treatment is not without its share of challenges resulting from choice of inappropriate technology, improper siting of infrastructure, in adherence to correct design concepts and lack of proper maintenance (Omenka 2010; Libralato et al. 2012). These bring about negative public health and environmental impacts including groundwater nitrate contamination, eutrophication of surface water bodies and contribution to global warming through the emission of green house gases.

Centralised versus decentralised wastewater management systems: an appraisal

In developing countries such as Zimbabwe, onsite systems are generally used in high income and low density areas where the collection of infrastructure discourages sewered systems. The wastewater management systems in Latin America are also centralised systems that are contributing to environmental problems (Noyola 2013). As such, the municipal wastewater treatment systems in Latin America are currently unsustainable because they are characterised by high water consumption, especially for transporting wastes out of cities (Suriyachan et al.2012; Nhapi 2004b). The mixing of industrial waste streams further complicates resource recovery and reuse than if process streams would be kept separate. On this note, De Gisi et al. (2014) castigates centralised system on the basis that they result in high energy and water use as well as increasing the vulnerability of communities to water-borne diseases through leakages of pipes over long distances. Cook et al. (2009) further notes that the centralised collection and treatment of wastewater limits opportunities to harness "waste" as a resource that is otherwise useful if recycled since there is no waste in nature as De Gisi et al. (2014) argues. With, generally, little awareness of environmental consequences, little institutional attention for recovery of resources, and with only degradable organics potentially removed from wastewater effluent or sludge, the remaining resources are either distributed into surface waters or into sludge. As such, wastewater collection and treatment contribute to environmental pollution. Most of the technologies used for centralised treatment of wastewater are expensive (investment and maintenance) and require well-trained staff (Nhapi 2004a, b, c). For these and other reasons, an intermediate or decentralised approach to wastewater management is urgently needed aiming at resource conservation and reducing environmental impacts of current approaches (Cook et al. 2009; Mitchell et al. 2008) (see Table 1).

Table 1: Comparison between centralised and decentralised systems

Parameter	Centralised system	Decentralised system
Collecting system	Large diameters, long distances	Small diameters, short distances
Requirements space	Large area in one place	Small areas in many places
Operation and maintenance	Full time technical staff requirements	Less demanding, can be monitored remotely
Uniformity of water	Many types of water	More uniform water
Dilution grade	Less control over the storm water, more dilution	More control over the storm water, more concentrate
Risk	Risk on a larger scale	Risk distributed
Water transfer	Increase the needs for water transfer	Water is used and reused in the same area
Social control	Social control is lost	More social control
Ease of expansion	High costs, more complexity to implementation	Low cost, less complexity to implementation
Potential to reuse	All water is concentrated in one point	Water can be reused locally

Source CODESAB (2011)

Rationale for Decentralised Wastewater Treatment and Management

Massoud et al. (2009: 652) argue that centralised wastewater collection and treatment systems are expensive to construct and operate, especially in areas with low population densities and dispersed households. Developing countries lack both the funding to construct centralised facilities and the technical expertise to manage and operate them. The decentralised system is therefore not only a long-term solution for small communities but is more reliable and cost effective. Alternatively, the decentralised approach for wastewater treatment which employs a combination of onsite and/or cluster systems is gaining more attention. Such an approach allows for flexibility in management, and simple as well as complex technologies are available. While there are many impediments and challenges towards wastewater management in developing countries, these can be overcome by suitable planning and policy implementation. Understanding the receiving environment is crucial for technology selection and should be accomplished by conducting a comprehensive site evaluation process. Centralised management of the decentralised wastewater treatment systems is essential to ensure they are inspected and maintained regularly (Nhapi 2004a, b, c). In light of this argument by Nhapi (2004a, b, c), Libralato et al. (2012) echoes the same sentiments and argue that a semi-centralised scheme can be a feasible option to introduce decentralisation in an urban area in a developing country, considering that the planning policies and the regulatory framework do not have many components that facilitate a different kind of management other than the traditional "end-of-the pipe" solutions and with use of conventional technologies in centralised systems.

Bernal and Restrepo (2012: 9–12) have pointed out that environmental pollution, water scarcity, population growth, innovation, and technological developments are drivers that encourage rethinking the current approach to urban water management. In this sense, decentralisation encourages us to think of urban water management in a holistic way, integrating all sectors, drinking water, wastewater, and storm water to get the most benefit out of them, thereby reducing costs, improving environmental management, expanding service coverage, and considering social and environmental benefits that are not visible with the current perspective (De Gisi et al. 2014; Werner et al. 2009). Centralised wastewater treatment systems are costly when compared to the decentralised systems. Geisinger and Chartier (2005) notes that sewage collection systems are typically 60–70 % of total project costs for a conventional centralised system hence decentralised systems are economically efficient. According to a study carried out by United Nations (2015) for South-East Asia it emerge that the decentralised system is more economical than the centralised system because treatment facilities for decentralised treatment facilities can be built in an incremental way and requires less initial capital investment than centralised treatment that requires millions of dollars. The above-mentioned should be accompanied by a reform of policies and guidelines that govern urban development plans and water management plans in cities in developing countries. The incorporation of decentralisation as a viable option for wastewater management in urban areas and the regulation of reusing practices such as defining quality criteria are necessary actions to articulate the conceptual framework with the actions that occur in reality. Based on a review of the state of the art and experiences with decentralisation, it can be concluded that the social, financial, and environmental benefits of decentralisation become critical factors when considering this kind of scheme in urban water management plans, mainly in peri-urban areas where wastewater collection and/or treatment is not available. In addition to the benefits, the key issues of each one of the identified economic, social, and environmental categories should be discussed. These include, among others, the cost of collecting and treating wastewater, acceptance and social awareness, and environmental protection, all of which must be considered in implementing decentralisation in urban areas in developing countries (Nhapi2004a, b, c). According to the context of each case, the level of decentralisation may be a critical issue to achieving sustainability of a wastewater management system. In short, wastewater treatment can be decentralised thereby reducing plumbing and pumping costs, possibility of safe reuse of water for gardening open spaces and can be integrated as part of the landscape.

Global Scan on Decentralised Wastewater Treatment Systems

The Decentralised wastewater treatment system was introduced in the Philippines as a way of responding to the problems caused by the conventional system that was in place (UNEP 2012). A new effective system was therefore necessary to prevent further environmental pollution and threats to public health. For the implementation of the decentralized wastewater system in Philippines, the following were required: biogas reactor, settling unit, anaerobic baffled reactor, anaerobic filter and indicator pond (ibid). The same project was also implemented in the Vietnam, where there was training of staff in the Operations and Maintenance of the system. The same impacts of reduced environmental pollution and reduced threats to public health were recorded (UNEP 2012). Gauss (2008) points out that decentralized wastewater treatment systems have been implemented in Central and South America as well as Latin America. The method that was implemented is the constructed wetlands system in Cities such as Masaya (Nicaragua), Lima (Peru), and Pereira (Colombo) and in countries such as Brazil. Recently, the 2009 Melbourne Metropolitan sewerage strategy for Melbourne aims to provide sustainable sewerage services in the city to 2060. The strategy aims to decongest the existing centralised sewerage system for Melbourne through the adoption of decentralised and on-site wastewater systems. The options include secondary and tertiary treatment systems that incorporate re-use of water for non-portable uses, urine separation, black and greywater separation and composting toilets (Brown et al. 2010). The decentralised system has also been used in Georgia, United States where about 40 % of the residents rely on decentralised systems, which prove to be more effective than centralised systems (Sheehan 2011). De Gisi et al. (2014) provides that a low-cost dry sanitation system with or without urine diversion is becoming popular due to a number of factors. The method is based on biological processes that separate urine and faeces resulting in the formation of Terra Preta soils that are eventually used in agriculture. The common technologies currently in use include septic tanks, pit latrines, composting toilets, urine diversion and pour flush. Each strategy and technology comes with its pros and cons and basically fall under two categories namely wet and dry technologies. From the Latin American, Central American and South American Cases, major lessons were drawn about the decentralized wastewater management system. The lessons include the following:

- Low operation and maintenance costs of the system (De Gisi et al. 2014; Bakir 2001).
- Robustness and good contaminant removal (Guo et al. 2014).
- Low environmental impact (Gauss 2008).

Gauss (2008) also argues that decentralized wastewater management systems require an experienced sanitary engineer and community participation is key in the selection of preferred technology.

Regional overview on decentralized wastewater systems

In Southern Africa, countries show different patterns with regards to centralized and decentralized wastewater systems. For example, Kenya uses conventional wastewater treatment systems which are inadequate and non-functional due to higher costs of operation and maintenance (Benard and Omondi 2012). There are also weak institutions mandated with wastewater management. The conventional systems are capital intensive, requiring large sums of shilling which the majority of the local governments do not afford. Windhoek is also one of the Cities in Southern Africa that has shown meaningful progress with regards to the implementation and management of decentralized systems (Moyo 2012). In Malawi and Mozambique De Gisi et al. (2014) maintains that urine diversion technology has been successfully implemented as decentralised wastewater management alternatives.

Local Cases on Decentralized Wastewater Systems

In Zimbabwe, there are cases of decentralised wastewater management system mainly in the rural areas where pit latrines are commonly used. According to Nhapi (2004a, b, c), cases of such a system are in places such as Gweru, Redcliff, Nemanwa and Mupandawana. In Gweru, the decentralized wastewater system started in 1994. In the town of Redcliff, there is evidence of decentralised wastewater treatment and management system as evidenced by the reuse of wastewater for agriculture. Septic tanks are a common wet technology that is used in areas with large plot sizes. The following technologies are used in Redcliff: duckweed pond systems, constructed wetlands and aquaculture. In Mupandawana and Nemanwa, the use and availability of duckweed pond systems provides an opportunity for the implementation of decentralized wastewater treatment and management systems (Nhapi 2004a, b, c). In Zimbabwe, the effective implementation of decentralised wastewater management systems is affected by unavailability of data on basic design of the technologies.

Legislative and Policy Framework for Wastewater Management in Zimbabwe

The framework and national strategy for wastewater management is governed by several pieces of legislation that are the responsibility of different Government Ministries and Agencies. In 2007, Statutory Instrument 6 of 2007 of the Environmental Management Act was gazetted as the Environmental Management (Effluent and Solid Waste Disposal) Regulations, 2007. These regulations generally set for the basic framework for wastewater management in Zimbabwe (Thebe and Mangore 2012). The Public Health Act through Statutory Instrument 638 of 1972 gazetted as the Public Health (Effluent) Regulations sets guidelines for wastewater irrigation concerning public health. These guidelines forbid the irrigation of root crops such as potatoes and sets restrictions that are not greatly enforced presently in Zimbabwe (Thebe and Mangore2012). In terms of Zimbabwean law in the form of Urban Councils Act, Chapter 29.15 and Regional Town and Country Planning Act, Chapter 29.6; all households are compelled to have an acceptable sanitation system before an occupation certificate is issued (Nhapi 2004a, b, c) and this has led to the high sanitation coverage in Zimbabwe urban centres which is estimated at 97 % in the urban areas according to Thebe and Mangore (2012). In a study by Ngwenya (2013) it emerge that there is an average coverage of 83.3 % on average for all councils. The coverage of sewerage network services is at 68.7 %, efficiency of collection of sewerage at 27.4 % indicates that large amount of sewerage does not reach treatment plants. However, the challenges in maintaining wastewater infrastructure mean that water is channelled away from the households and industries but could fail to get adequate treatment before re-entering the water courses.

Harare: the Current Situation

The City of Harare is Zimbabwe's capital city and its administrative, commercial, and communications centre. Harare is situated at an altitude of 1483 m. The topography of the city is hilly in rocky areas, flatter in the south, and undulating in the north. The population of the City is 2 098 199 (Zimstat 2012). Map 1 represents the City of Harare.

The wastewater treatment and management systems in Harare are currently centralised. Harare has a sewerage system that serves approximately 1.8 million people (about 80 % of the total population) in the formal settlements via flush toilets. However, this system is currently dysfunctional and fails to accommodate the informal settlements where pit latrines and open defecation are widespread (Practical Action 2010). The sewage goes to two large activated sludge plants, Crowborough and Firle, which according to Nhapi (2009) are overloaded. The total design capacity of these plants is 208,000 m^3/day compared to total current inflows of about 300,000 m^3/day, resulting in 44 % overloading (Tsiko and Togarepi 2012). The major shortcoming with the existing treatment system in Harare is that it was originally designed for 250,000 people which is a tenth of the current city's population. Efforts to expand the existing system through a US$68 million project by City of Harare are likely not to effectively address the wastewater problems currently bedevilling Harare since over 60 million litres of sewage is generated on a daily basis in the southern part of the city (Chideme 2012). Tsoroti (2015) indicates that even if the plant is completed it will only manage to treat 70 % of the wastewater generated. This is having serious implications for downstream water quality. In Epworth, between 1 and 3 % of households had pour-flush toilets; between 23 and 27 % make use of blair latrines while 2–13 % had no toilet facilities.

Large volumes of inadequately treated wastewater are being discharged to the rivers, Marimba and Mukuvisi, which drain to Lakes Chivero and Manyame, the city's major sources of water. Fifty percent (50 %) of the pollution load discharged into Lake Chivero is attributed to urban wastewater. The city also has two waste stabilisation ponds and an aeration pond; significant quantities of wastewater and sludge are applied to pasture which is a decentralised way of managing waste that just need to be upscaled (Tsoroti 2015). Nhapi et al. (2006) provide very detailed information on wastewater treatment and wastewater impacts on water resources in Harare; they note that a major problem is very high water use (leading to very high wastewater volumes) in relatively wealthy parts of the city (Tsiko and Togarepi 2012). A number of high density areas (Kuwadzana, High field, Dzivaresekwa, Glenview among others) are experiencing unabated sewer bursts for years and remain plagued with pools of raw sewage as council is currently failing to attend to burst sewer pipes citing financial constraints. Such is testimony that the centralised wastewater management and treatment strategies are ineffective and inefficient. The two largest wastewater treatment plants in Harare (Firle and Crowborough) discharge their treated effluents into the Mukuvisi and Marimba rivers, respectively 5 and 8 km upstream of Lake Chivero (Nhapi 2004a, b, c).

Harare: Systems Dynamics Conceptualisation of the Potentiality, Opportunities, Risks and Strategies

The suitability of the decentralised concept was considered for various land uses. A differentiated (different solutions for different areas) and integrated approach was assumed in which first preference was given to onsite treatment and reuse (Nhapi et al. 2002a, 2002b). The second preference was to treat at a decentralised level what cannot be handled onsite. The third preference was to treat the remainder at a centralised level. Only industrial and low-density residential areas have adequate space for the onsite treatment and reuse of wastewater (Nhapi et al. 2002a, 2002b). One of the major constraints to onsite wastewater management is the difficulty of utilising all the nutrients and water for crop irrigation within the plot boundary. As a result medium (500–1500 m^2) and high-density (<500 m^2) residential plots are recommended for decentralised systems because they are normally developed as small housing estates and they produce more wastewater than they can utilise onsite.

Decentralised systems would allow housing developments in isolated areas where sewer connection to existing areas would be difficult and expensive (cf. Nhapi 2004a, b, c, 2009; Thebe and Mangore 2012). In such contexts septic tanks may be an option as well as urine diversion where the faeces may be used to increase agriculture production borrowing from the strategy suggested by De Gisi et al. (2014). Considering the variety of sustainable factors, such as social, cultural, environmental, and technical factors, which must be taken into account to implement a wastewater treatment system, the financial aspect is often the most decisive factor in developing countries like Zimbabwe. This is the major reason why decentralised systems are considered as an alternative option. The United Nations (2015) argues in favour of decentralised systems in that treatment facilities can be built in a piece-meal manner and does not require multi-million dollar investments that is usually difficult to raise considering the financial constraints that burden local authorities in developing countries. However, in the majority of countries, there is a lack of suitable institutional arrangements for managing decentralised systems and a lack of a suitable policy framework that encourages a decentralised approach. Without technical assistance and other capacity-building measures, problems of institutional capacity that existed under a centralised operation are simply passed on to the new structures (Parkinson and Tayler 2003). For Harare, research and scientific work by responsible institutions is all aimed at improving the centralised wastewater systems, yet the developed world is increasingly shunning these methods.

Similarly, there is limited information on how sustainability applies to the general field of wastewater infrastructure, including both centralised and decentralised treatment approaches. The constraints to decentralised wastewater management systems include, among others, the following:

- government policies and regulations founded on centralised infrastructure,
- market failures, with fragmentation and little information,
- distorted rates of water,
- fragmentation of the water and sanitation agencies,
- civil society based on the conventional,
- minimum investment in research,
- lack of local models that combine technology,
- management, financing and customer acceptance,
- segregation of actors (entrepreneurs, professionals, and academics) in three different areas: supply, storm water and wastewater,
- lack of acceptance public,
- lack of economic evaluations procedures,
- institutional constrains and
- Existing practices.

Harare can adopt various strategies in relation to the decentralised wastewater management concept. From the foregoing it emerges that a number of strategies and technologies can be adopted for wastewater management in a decentralised framework, all effectively treating smaller quantities at or near the source. Septic tanks, pit-latrines and urine diversion are some of the strategies that are gaining popularity as decentralised wastewater management systems (De Gisi et al. 2014). There are also less concentrated wastewater and uses natural treatment methods like algae and duckweed based ponds and constructed wetlands, with harvesting of protein biomass. The other strategies include the use of maturation ponds and anaerobic pre-treatment to allow for stabilisation of organic matter. Algae-based systems are also being used for decentralised systems in Harare. This consists of four treatment lines of small algae-based waste stabilisation ponds scattered in the Mabvuku/Tafara high density area (Nhapi 2004a, b,c). The decentralised or onsite handling of wastewater opens opportunities for simpler methods of wastewater treatment (especially natural methods) and encourages wastewater reuse closer to sources of generation. The adoption of decentralised wastewater management strategy as a sustainable wastewater management in Harare has to consider a number of limitations. First, Harare has an existing extensive infrastructure that has been operational for some years and which represents an enormous capital investment. Therefore, this cannot be phased out overnight and a gradual approach is required. Very few residential areas use communal water and sanitation facilities. Second, land use patterns in Harare exhibit vastly different characteristics so much that it would be irrational to adopt a single solution for all areas. Lifestyles and water use characteristics for low, medium and high-density residential areas are quite different. A differentiated approach is therefore required.

DISCUSSION, POLICY IMPLICATIONS AND DIRECTION

The adoption of decentralised wastewater management systems in Harare is still minimal, regardless of the promises of such a concept. The City is used to the conventional centralised systems. A number of factors such as public acceptance, health and cost factors and the availability of alternative water sources rather than technical barriers (Nhapi 2004a, b, c) influence the extent to which the decentralised concept of wastewater treatment and management will be implemented in Harare. Other contributing factors include regulations, convenience and organisations necessary for users if they are to make correct use of wastewater systems. The cost aspect of the decentralised systems is linked to technologies used and the size of population that is to be served by such as system. The technologies are locally available such as duckweed and septic tanks, hence no need for importing such technologies. The availability and cost of alternative water systems (surface, groundwater, rainwater harvesting) is also likely to influence the adoption of decentralised wastewater management in Harare. Water in the Harare metropolitan area is a very serious problem and this may constrain the effective implementation of decentralised systems. Alternative sources of water have to be explored such as rain harvesting. For the effective adoption of decentralised wastewater management strategies, new institutional structures will be required. Cases reviewed from Latin, Central and South America confirms that community engagement is critical factor in the success of the adoption of decentralized wastewater treatment systems. It is therefore important for the City of Harare to engage urban residents in the processes of adopting decentralized wastewater treatment systems. Public awareness is also important so that members of the public are educated on the benefits of implementing such systems in their community.

There is also need to review Effluent Standards (Statutory instrument 274 of 2000) and the Public Health Act and other Health regulations to reflect new environmental health standards that are commensurate with the decentralised wastewater treatment and management concept. The decentralised systems should be complemented by water use reduction, pollution, pollution prevention measures, and sources. The decentralised wastewater management strategy is most applicable in high and medium-density residential areas since they have limited space than the low-density areas. Because treatment plants will be scattered around the city, it is best that these be run by local communities, special interest groups (youth, women, churches, nongovernmental organisations), or by private companies. The possibility of financial recovery from systems such as duckweed-based ponds would make them attractive for community management. The other advantage of community management is that it brings environmental management back to the polluters, making it easier for awareness campaigns. Private developers could also be asked to provide small treatment centres for each new housing subdivision and to manage such treatment centres. Private companies could be allowed to manage treatment plants and charge user fees for their services as in the French Model of privatisation of the water sector. The potentialities, opportunities, risks and strategies of the decentralised wastewater treatment system are represented in tabular format (see Table 2). The issues to be considered under decentralised wastewater treatment systems include finance, public health, space, Operation and maintenance and these are linked to the opportunities, potentialities, risks and strategies in the following table or matrix.

Table 2: Issues, potentialities, opportunities, risks and strategies of decentralised wastewater treatment system

Issue(s)	Potentialities	Opportunities	Risks	Strategy
Financing	Reduction in financial cost	Rateable property in the City	Financial risk in investments	Tapping of low-cost methods, negotiating with government on grants for infrastructure development, levying taxes on households
Design	Unsophisticated design	Availability of engineers	Poorly designed system	Training of engineers on decentralised wastewater systems, capacity building for local authority
Public health	Reduction in public health threats	Availability of city health departments	Public health risks in case of poor designs	Setting standards to ensure public health, review of public health legislation and policies

CONCLUSION

This paper has demonstrated that decentralised wastewater systems are a very important strategy whose adoption can lead to sustainable management of wastewater in the City of Harare. The current socio-economic and environmental challenges currently burdening Harare are drivers that encourage rethinking the current approach to urban water management. In this sense, decentralisation encourages the City of Harare to think of urban water management in a holistic way, integrating all sectors, drinking water, wastewater, and storm water to get the most benefit out of them, thereby reducing costs, improving environmental management, expanding service coverage, and considering social and environmental benefits that are not visible with the current perspective. What is then conspicuous for Harare is a reform of policies and guidelines that govern urban development plans and water management plans. The incorporation of decentralisation as a viable option for wastewater management in urban areas and the regulation of reusing practices such as defining quality criteria are necessary actions to articulate the conceptual framework with the actions that occur in reality.

Based on a review of the state of the art and experiences with decentralisation, it can be concluded that the social, financial, and environmental benefits of decentralisation become critical factors when considering this kind of scheme in developing countries' urban water management plans such as for Harare, mainly in the peri-urban areas, informal settlements and Southern suburbs where wastewater collection and/or treatment is not available. In addition, to the benefits, the key issues of each one of the identified economic, social, and environmental categories should be discussed. These include, among others, the cost of collecting and treating wastewater, acceptance and social awareness, and environmental protection, all of which must be considered in implementing decentralisation in urban areas in developing countries of note in Harare. According to the context of each case, the level of decentralisation may be a critical issue to achieving sustainability of a wastewater management system. Considering the urban form for Harare, a semi-centralised scheme can be a feasible option to introduce decentralisation, considering that the planning policies and the regulatory framework do not have many components that facilitate a different kind of management other than the traditional "end-of-the pipe" solutions and with use of conventional technologies in centralised systems.

REFERENCES

1. Bakir HA (2001) Sustainable wastewater management for small communities in Middle East and North Africa. J Environ Manag 61:319–328
2. Benard O, Omondi G (2012) Wastewater production, treatment and use in Kenya. In: Third regional workshop'safe use of wastewater in agriculture, Johannesburg
3. Bernal DP, Restrepo I (2012) Key issues for decentralisation in municipal wastewater treatment. In: Paper in the 12th edition of the World Wide Workshop for Young Environmental Scientists (WWW-YES-2012), urban waters: resource or risks? Arcueil, France
4. Brown R, Farrelly M, Keath N (2009) Practitioner perceptions of social and institutional barriers to advancing a diverse water source approach in Australia. Water Resour Dev 25(1):15–28CrossRef

5. Brown V, Jackson DW, Khalife M (2010) 2009 Melbourne metropolitan sewerage strategy: a portfolio of decentralised and on-site concept designs. Water Sci Technol 62(3):510–517CrossRef
6. Chideme M (2012) Harare to erect three new sewage treatment plants.http://www.herald.co.zw/harare-to-erect-three-new-sewage-treatment-plants/. Accessed 24 June 2015
7. CODESAB (2011) Los Retos: Tratamiento Descentralisado de Aguas Residuales. ForoMetropolitano de Cochabamba del Agua y Saneamiento. Universidad Mayor de San Simón, Bolivia
8. Cook S, Tjandraatmadja G, Ho A, Sharma A (2009) Definition of decentralised systems in the South East Queensland context. Urban water security research alliance technical report no. 12.http://www.urbanwateralliance.org.au/. Accessed 26 Nov 2014
9. De Gisi S, Petta L, Wendland C (2014) History and technology of Terra Preta sanitation. Sustainability 6:1328–1345CrossRef
10. Diaper C, Sharma A, Tjandraatmadja G (2007) Decentralised water and wastewater systems. In: Newton P (ed) Transitions: pathways towards sustainable urban development in australia, CSIRO Australia Publishing and Springer, The Netherlands
11. Gauss M (2008) Constructed wetlands: a promising wastewater treatment system for small localities: experiences from Latin America. Water and Sanitation Program (WSP)
12. Geisinger D, Chartier G (2005) Managed onsite/decentralised wastewater systems as long-term solutions. Clearwaters 35:6–11
13. Guo X, Liu Z, Chen M, Liu J, Yang M (2014) Decentralized wastewater treatment technologies and management in Chinese villages. Front Environ Sci Eng 8(6):929–936CrossRef
14. Libralato G, Ghirardini AV, Avezzu F (2012) To centralize or to decentralize: an overview of the most recent trends in wastewater Treatment management. J Environ Manage 94(1):61–68CrossRef
15. Massoud MA, Tarhini A, Nasr JA (2009) Decentralised approaches to wastewater treatment and management: applicability in developing countries. J Environ Manage 90:652–659CrossRef
16. Mitchell C, Abeysuriya K, Willetts J (2008) Institutional arrangements for onsite and decentralised systems: needs and opportunities for key players in the field of distributed wastewater management. Paper presented at on site and decentralised sewerage and recycling conference. Australian Water Association in partnership with Environmental Health Australia, Benalla, 12–15 Oct 2008.https://opus.lib.uts.edu.au/bitstream/10453/10708/1/2007004570.pdf. Accessed 23 July 2014
17. Moyo L (2012) Wastewater production, treatment and use in Namibia. In: Third regional workshop'safe use of wastewater in agriculture. Johannesburg
18. Ngwenya ST (2013) Technical constraints to urban wastewater management in Zimbabwe. MSc Dissertation, Department of Civil Engineering, University of Zimbabwe, Harare. Available at:http://ir.uz.ac.zw/jspui/bitstream/10646/1399/1/UZ_WREM%20_NGWENYA_FINAL_THESIS_2014.pdf. Accessed 24 Nov 2015 (Unpublished)
19. Nhapi I (2004a) The potential for the use of duckweed-based pond systems in Zimbabwe. Water SA 30(1):115–120
20. Nhapi I (2004b) A framework for the decentralised management of wasterwater in Zimbabwe. Phys Chem Earth 29:1265–1273CrossRef
21. Nhapi I (2004c) Options for wastewater management in Harare. Taylor and Francis Group, Zimbabwe
22. Nhapi I (2009) The water situation in Harare, Zimbabwe: a policy and management problem. Water Policy 11(2):221–235CrossRef
23. Nhapi I, Hoko Z, Siebel M, Gijzen HJ (2002a) Assessment of major water and nutrient flows in the Chivero catchment area, Zimbabwe. Phys Chem Earth 27(11–12):783–792CrossRef
24. Nhapi I, Siebel M, Gijzen HJ (2002b) Options for the decentralised management of wastewater in Harare Zimbabwe, IWA Water and Environmental Management Series, p 955–964
25. Nhapi I, Siebel MA, Gijzen HJ (2006) A proposal for managing wastewater in Harare, Zimbabwe. Water Environ J 20(2):101–108CrossRef
26. Noyola A (2013) Municipal wastewater treatment in Latin-America: a zoom to Mexico with emphasis in GHG Emissions from municipal wastewater treatment. International Development Research Centre. https://www.globalmethane.org/documents/ww_fs_eng.pdf. Accessed 27 Nov 2014
27. Office of Water US Environmental Protection Agency (2005) Handbook for managing onsite and clustered (decentralised) wastewater treatment systems: an introduction to management tools and information for implementing EPA's management guidelines. Office of Water US Environmental Protection Agency, Cincinnati
28. Omenka E (2010) Improvement of decentralised wastewater treatment in Asaba, Nigeria. Master Thesis, Lund University, Water and Environmental Engineering, Department of Chemical Engineering
29. Parkinson J, Taylor K (2003) Decentralized water management in Periurban areas in low income countries. Environ Urban 15(1):75–89CrossRef
30. Practical Action (2010) Low-cost sanitation technology—Zimbabwe.http://practicalaction.org/practicalanswers/product_info.phd?products_id=408&atrib=2. Accessed 26 July 2015
31. Rocky Mountain Institute (2004) Valuing decentralised wastewater technologies. A catalog of benefits, costs, and economic analysis techniques. Rocky Mountain Institute, New York
32. Sheehan KA (2011). Decentralised wastewater treatment in Georgia: Benefits and management need. In: Proceedings of the 2011 Georgia Water Resources

Conference held April 11–13, 2011 at the University of Georgia

33. Suriyachan C, Nitivattananon ATM, Amin N (2012) Potential of decentralized wastewater management for urban development: case of Bangkok. Habit Int 36(1):85–92
34. Thebe TA, Mangore EN (2012) Wastewater production, treatment and use in Zimbabwe. Department of Civil and Water Engineering, National University of Science and Technology, Bulawayo
35. Tsiko RG, Togarepi S (2012) A situational analysis of waste management in Harare, Zimbabwe. J Am Sci 8(4):692–706
36. Tsoroti S (2015) City aims to double sewage treatment. http://www.hararenews.co.zw/2015/05/city-aims-to-double-sewage-treatment/. Accessed 24 June 2015
37. UNEP (2012) Philippines. Decentralized wastewater treatment systems for the San Fernando City Slaughterhouse. A Borda Dewats Project in the Phillipines
38. United Nations (2015) Market opportunities for decentralized wastewater treatment systems in South-East Asia. In: Regional Policy Workshop on wastewater management and sanitation in South-East Asia, 2 and 3 April 2015, UNCC, Bangkok, Thailand
39. Venhuizen D (2003) The decentralised concept of "waste" water management.http://www.venhuizen-ww.com/html/dcentral.html. Accessed 26 Nov 2014
40. Werner A, Panesar A, Rüd SB, Olt CU (2009) Ecological sanitation: principles, technologies and project examples for sustainable wastewater and excreta management. Desalination 248:392–401

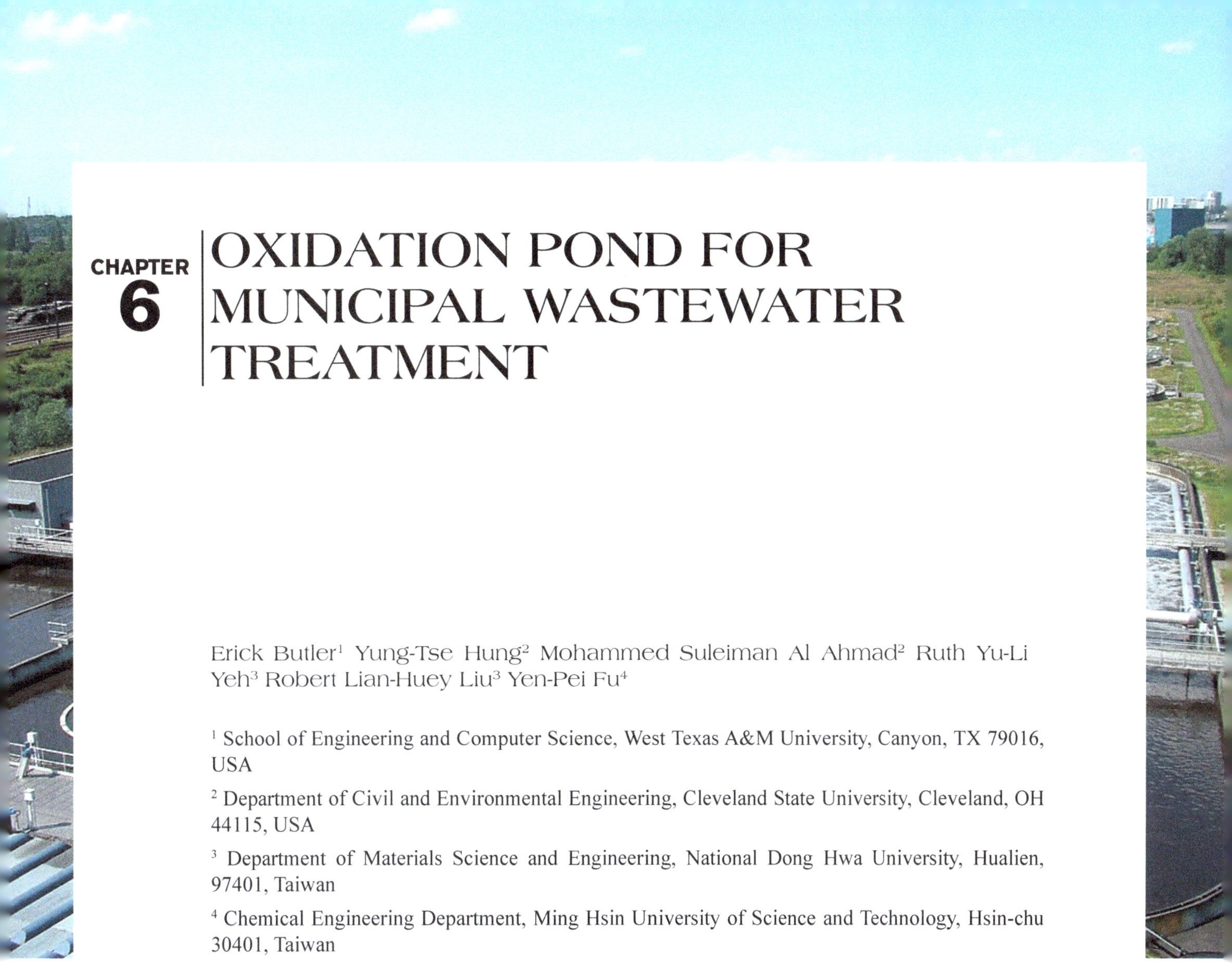

CHAPTER 6

OXIDATION POND FOR MUNICIPAL WASTEWATER TREATMENT

Erick Butler[1] Yung-Tse Hung[2] Mohammed Suleiman Al Ahmad[2] Ruth Yu-Li Yeh[3] Robert Lian-Huey Liu[3] Yen-Pei Fu[4]

[1] School of Engineering and Computer Science, West Texas A&M University, Canyon, TX 79016, USA

[2] Department of Civil and Environmental Engineering, Cleveland State University, Cleveland, OH 44115, USA

[3] Department of Materials Science and Engineering, National Dong Hwa University, Hualien, 97401, Taiwan

[4] Chemical Engineering Department, Ming Hsin University of Science and Technology, Hsin-chu 30401, Taiwan

ABSTRACT

This literature review examines process, design, and cost issues related to using oxidation ponds for wastewater treatment. Many of the topics have applications at either full scale or in isolation for laboratory analysis. Oxidation ponds have many advantages. The oxidation pond treatment process is natural, because it uses microorganisms such as bacteria and algae. This makes the method of treatment cost-effective in terms of its construction, maintenance, and energy requirements. Oxidation ponds are also productive, because it generates effluent that can be used for other applications. Finally, oxidation ponds can be considered a sustainable method for treatment of wastewater.

INTRODUCTION

Oxidation ponds, also known as waste stabilization ponds, provide greater advantages over mechanically based units.

First, ponds can be described as self-sufficient treatment units, because the efficacy of treatment is contingent upon the maintenance of the overall microbial communities of bacteria, viruses, fungi, and protozoa (Hosetti and Frost 1995), and the proper balance of organics, light, dissolved oxygen, nutrients, algal presence (Amengual-Morro et al. 2012), and temperature. Because ponds are self-sufficient, there is a reduction of operator responsibilities to manage treatment, a reduction in labor costs, and an increase in the potential fiscal returns from the tangible products generated by the treatment unit (Hosetti and Frost 1998).

Second, ponds can be used for the purpose of 'polishing,' or providing additional treatment to what has been found within conventional treatment methods (Veeresh et al. 2010). Various authors have conducted studies using oxidation ponds and other treatment methods such as anaerobic digestion (Gumisiriza et al. 2009), upflow anaerobic sludge blanket (Mungraya and Kumar 2008), membranes (Craggs et al. 2004), and reverse osmosis (Lado and Ben-Hur 2010). Third, ponds simplify the treatment process by reducing the need for multiple treatment units. Finally, oxidation ponds are a treatment process that can be used in regions where treating wastewater using

Citation: Erick Butler, Yung-Tse Hung, Mohammed Suleiman Al Ahmad, Ruth Yu-Li Yeh, Robert Lian-Huey Liu, and Yen-Pei Fu, Oxidation pond for municipal wastewater treatment, DOI: 10.1007/s13201-015-0285-z.

conventional treatment methods is very expensive. Indeed, oxidation ponds are commonly used in many regions around the world, specifically in places with year-round mild to warm climates. In many developing countries, effluent from waste stabilization ponds has been reused in aquaculture and irrigation applications. Melbourne, Australia, implements irrigation by using sewage as early as the late-1890s. Latin American countries began projects in the 1960s, while the United Nations Development Programme (UNDP) Resource Recovery Project and the Pan American Health Organization (PAHO) have collaborated in research considering ponds for growing fish (Shuval et al. 1986).

Many European Mediterranean countries also use oxidation ponds to treat municipal wastewaters. Forty-four ponds have served European populations between 500 and 40,000. Specifically in France, 77 % of the ponds have been employed to serve populations under 1000, although the largest population served was a combined 14,000 from the cities of Meze and Louipan, using a facultative pond and a maturation pond system. Ponds are also being used in countries such as Greece, Egypt, Algeria, Morocco, and Tunisia (Mara and Pearson 1998). The additional advantages and disadvantages are summarized in Table 1.

Table 1: Advantages and disadvantages of waste stabilization ponds (Mara et al. 1992)

Advantages	**Disadvantages**
Easy to construct	Requires a large land area
Low cost	High BOD and TSS with algae concentrations
Low maintenance costs	
Effluent does not require disinfection	
Capable of handing a variety of hydraulic loads	
Ideal for small communities and also tropical region	
Completes sludge treatment	
Handles varying wastewater types (industrial or municipal)	
BOD, fecal coliform, and helminth removal is higher than by other treatment methods	
such as activated sludge, biological filters, and rotational biological contactors	

BOD Biochemical oxygen demand, *TSS* total suspended solids

For these reasons, the oxidation pond provides an attractive method of sustainable wastewater treatment with the following caveats: operators should monitor the chemical and biological constituents within the system to ensure proper design parameters are met to correspond with regulatory treatment efficiency standards (Hosetti and Frost 1998), especially if the effluent is to be reused.

While the authors' research supports the use of oxidation ponds, it makes no claim concerning the closely related lagoon. Waste stabilization and oxidation ponds are synonymous, but oxidation ponds are different from lagoons. Lagoons are single or multi-celled designed natural wastewater treatment reservoirs that house diluted manure for the purpose of removing organics and other constituents by means of microorganisms or other biological processes. Similar to waste stabilization ponds, lagoons can operate under various conditions of dissolved oxygen depending on the type of microorganisms that are present. However, lagoons are more commonly affiliated with agricultural wastewater (Hamilton et al. 2006) and should not be confused with oxidation ponds.

The purpose of this text is to review the significance of oxidation ponds within wastewater treatment by discussing the types, design considerations, and treatments that have been completed by this particular method.

TYPES OF OXIDATION PONDS

There are four major types of oxidation ponds: aerobic (high-rate), anaerobic, facultative, and maturation ponds.

Aerobic (high-rate) Ponds

Aerobic ponds, also known as high-rate algal ponds, can maintain dissolved oxygen throughout the 30–45 cm-deep pond because of algal photosynthetic activity (USEPA 2011). Photosynthetic activity supplies oxygen during the day, while at night the wind creates aeration due to the shallow depth of the pond (Davis and Cornwell 2008). Aerobic ponds are well known for having high biochemical oxygen demand (BOD) removal potential and are ideal for areas where the cost of land is not expensive. Other characteristics of these ponds include a detention time of 2–6 days, a BOD loading rate between 112 and 225 kg/1000 m^3 day, and a BOD removal efficiency of 95 % (USEPA 2011).

Anaerobic Ponds

Anaerobic ponds operate without the presence of dissolved oxygen. Under methanogenic conditions, the major products are carbon dioxide and methane (Quiroga 2011). Typically, these ponds are designed to have a depth of 2–5 m, with a detention time between 1 and 1.5 days, an optimum pH less than 6.2, temperatures greater than 15 °C (Kayombo et al. 2010), and an organic loading rate of 3000 kg ha/day (Quiroga 2011; Kayombo et al. 2010). Anaerobic ponds can remove 60 % BOD. However, this efficiency is climate dependent (USEPA 2011). The driving force behind treatment is sedimentation. Helminths settle to the bottom of the pond, and bacteria and viruses are removed by attaching to settling solids within the pond or die with the loss of available food or by the presence of predators. In practice, anaerobic ponds are usually incorporated alongside facultative ponds (Martinez et al. 2014).

Facultative Ponds

A facultative pond is a treatment unit with anaerobic and aerobic conditions. A typical pond is divided into an aerobic surface region consisting of bacteria and algae, an anaerobic bottom region, consisting of anaerobic bacteria, and a region in between anaerobic and aerobic conditions where bacteria can thrive in both conditions (Kayombo et al. 2010; Joint Departments of the Army, the Navy, and the Air Force 1988). If used in series, effluent from a previously treated source enters the pond (Quiroga 2011; Kayombo et al. 2010). Facultative ponds treat BOD, typically within a range of 100–400 kg BOD/ha/day, by removing BOD by 95 %. Because facultative ponds employ algae as decomposers, the treatment time can range between 2 and 3 weeks, which is attributed to the photosynthetic processes that occur within the unit. A facultative pond on average has a depth of 1–2 m (Kayombo et al. 2010).

Maturation Ponds

Similar to facultative ponds, maturation ponds use algae as a primary driving force in the treatment. Nevertheless, while facultative ponds typically treat BOD, maturation ponds remove fecal coliform, pathogens, and nutrients (Cinara 2004). In comparison with the other pond types, the characteristics of the maturation pond include a depth range between 1 and 1.15 m (Kayombo et al. 2010), which makes it shallower than all of the ponds besides the aerobic. Generally, maturation ponds maintain anaerobic conditions (Martinez et al. 2014).

Arrangement of Ponds

Treatment by waste stabilization ponds occurs by using a single pond to handle treatment or by a multiple pond system. Figures 1, 2, 3, 4, and 5 depict a multiple pond system in Wellington, Texas, a small town in the Texas Panhandle. The system treats the town's wastewater which is used for irrigation (Fig. 6). There are two arrangements for a multiple pond system—series and parallel. In the series arrangement, wastewater is treated in the initial and subsequent ponds and then polished in the final pond. On the contrary, wastewater flow is evenly divided in the parallel pond arrangement. The USEPA states that wastewater flow division normally occurs in the first two waste stabilization ponds (USEPA 2011).

Figure. 1: A multiple pond treatment system in Wellington, TX, USA. The pond system consists of a facultative lagoon (*1*) and three oxidation ponds (*2–4*) (Google Map 2014).

Figure. 2: Facultative lagoon in the treatment system in Wellington, TX, USA. Photograph taken by the author.

Figure. 3: Oxidation pond 1 in the treatment system in Wellington, TX, USA. Photograph taken by the author.

Figure. 4: Oxidation pond 2 in the treatment system in Wellington, TX, USA. Photograph taken by the author.

Figure. 5: Oxidation pond 3 in the treatment system in Wellington, TX, USA. Photograph taken by the author.

Figure. 6: Central pivot irrigation system directly applies effluent from the pond system in Wellington, TX, USA. Photograph taken by the author.

Each multiple pond arrangement has its benefits and therefore an operator can change the pond arrangement depending on the situation. For example, ponds operating in parallel prevent interruption of treatment during the cooler months of the year. This is when a pond can experience low biological activity. Low biological activity can create anaerobic conditions within a pond. In addition, the application of ponds in parallel can reduce problems related to periodic low dissolved oxygen concentrations, particularly in the morning hours (USEPA 2011). Also, Mara and Pearson (1998) recommend this arrangement when the population of the city reaches 10,000. On the other hand, ponds in series are ideal during the summer months and also during periods of low biological loading (USEPA 2011). Nevertheless, the choice of applying multiple ponds can be beneficial for treatment as compared to a single pond arrangement.

Design of Oxidation Ponds

Oxidation ponds are designed to function as either completely stirred or plug flow reactors, but mass transport mechanisms have a greater impact than the type of reactor model chosen. There are four major mass transport mechanisms acting in oxidation ponds—diffusion, advection, gravity, and interception. The mechanism(s) observed are contingent on the type of wastewater treated (Peña and Mara 2003).

There are many methods available for the design of waste stabilization ponds. For example, facultative pond design can use the areal loading rate, Gloyna equation, plug flow model, Marais and Shaw model, and the Thirumurthi application. The design equations for each procedure together with design limitations are shown below.

Areal Loading Rate

The areal loading rate design procedure optimizes an organic loading rate into a waste stabilization pond by examining various factors such as the volumetric loading, organic constituents within the wastewater, the ability of algae to use sunlight to grow and supply oxygen, and BOD loading per unit area (USEPA 1983). Climate impacts the BOD loading rate, as the loading rate is directly related to the ability of a pond to avoid becoming anaerobic (Gloyna 1971). The areal loading rate procedure limits BOD_5 loading rates between 11 and 22 kg/ha/day when the temperature is below 0 °C and a hydraulic detention time between 120 and 180 days (Gloyna 1971; USEPA 2011), as opposed to tropical climates which can handle higher BOD_5 loading rates. Therefore, the use of areal loading rate is climate dependent and may not be as predictable as other methods.

Gloyna Equation

Authors Hermann and Gloyna attempted to address the limitations of the areal rate procedure with a method of determining the volume of a waste stabilization pond that would maintain a high BOD removal (80–90 %) despite the change in temperatures (Marais 1966). The authors state that if the average coldest water temperature of the year is known along with a loading factor based on the product of the total population and waste per capita, the volume of the pond can be determined graphically. The procedure is as follows. The design engineer begins by computing the loading factor. Because the graph consists of lines representing water temperatures, the engineer would find the computed loading factor on the abscissa, move vertically toward the average coldest temperature line, and read the volume on the ordinate. After selecting the volume, the depth was found using a table of recommended depths (Gloyna1971).

Gloyna further expanded on this to create an empirical equation that summarizes the relationship between pond volume, BOD concentration, and flow rate (Finney and Middlebrooks 1980):

$$V = 3.5 \times 10^{-5} Q L_a \left(\theta^{(35-T)} \right) f f' \quad (1)$$

where V is the pond volume, m^3; Q the influent flow rate, l/day; L_a the ultimate BOD or COD, mg/L; the temperature coefficient; T the pond temperature, Celsius; f the algal toxicity factor (1.0 for domestic and industrial wastes); f the sulfide oxygen demand (1.0 for SO_4 concentration less than 500 mg/L).

In addition, Gloyna attempted to adjust for sunlight by multiplying the computed volume from Eq. (1) and the quotient of solar radiation in the area and the solar radiation found in the southwest.

However, there are several limitations within the Gloyna equation. Finney and Middlebrooks argued that the use of Gloyna's equation cannot be applied to all ponds, because it does not always correctly predict pond depths or BOD removal efficiencies. The authors also mention that the sunlight correction can have more of an impact on the volume than the other variables in the equation (Finney and

Middlebrooks 1980). In addition, Marais (1966) concluded that the sizing of ponds by Gloyna's equation results in the design of smaller ponds in series, rather than using one larger pond to meet the same treatment objective.

Plug Flow Model

The plug flow model, derived from first-order kinetics, considers not only the BOD_5concentration, but also the rate reaction rate (k_p). The k_p is chosen based on the BOD_5 loading rate and the pond temperature. Table 2 provides the k_o based on BOD_5 loading rate at 20 °C (USEAP 1983). The plug flow model is presented in Eq. (2) at a temperature of 20 °C (USEPA1983):

Table 2: BOD_5 loading rate based on k_p at 20 °C (USEPA 1983)

BOD_5 loading rate (kg/ha/d)	kp20kp20 (d^{-1})
22	0.045
45	0.071
67	0.083
90	0.096
112	0.129

$$C_e/C_o = e^{-k_p t}, \quad (2)$$

where C_e is the effluent BOD_5 concentration, mg/L; C_o the influent BOD_5 concentration, mg/L;e the base of natural logarithm, 2.7183; k_p the plug flow first-order reaction rate, day^{-1}; t the hydraulic resident time in each pond, days.

To use the plug flow model at other temperatures, a subsequent equation is required. Equation (3) converts the temperature from 20° to the desired temperature (USEPA 1983):

$$K = K_8 C_{Te} C_o C_{T0x}, \quad (3)$$

where kpTkpT is the reaction rate at minimum operating water temperature, day^{-1}, kp20kp20 the reaction rate at 20 °C; T the minimum operating water temperature, °C.

Marais and Shaw

Marais and Shaw developed a pond equation that combines both first-order kinetics and completely mixed conditions (Crites et al. 2006). This can also be used by designers for aerobic pond design (USEPA 2011; Crites et al. 2006):

$$C_n/C_o = 1/(1 + k_c t_n)^n, \quad (4)$$

where C_n is the effluent BOD_5 concentration, mg/L; C_o the influent BOD_5 concentration, mg/L; k_c the complete-mix flow first-order reaction rate, day^{-1}; t_n the hydraulic resident time in each cell, days; n the number of equal-sized pond cells in series.

- The developed equation is based on the following assumptions (Marais 1966; Finney and Middlebrooks 1980):
- Assume that the pond is completely mixed, with no changes in pond state.
- Treat the pond effluent as equal to the influent under equilibrium conditions, with slight variations of pollutant concentration within the influent.
- Pollution removal increases within a series of ponds until reaching a desired removal rate.
- BOD does not settle as sludge (Finney and Middlebrooks 1980).

Thirumurthi application

Thirumurthi argued that Marais and Shaw's assumption of completely mixed conditions is not ideal for designing a waste stabilization pond, but suggests using a chemical reaction (Eq. 5) (Finney and Middlebrooks 1980; Thirumurthi 1974):

$$C_e/C_i = 4ae^{1/2d}/(1+a)^2 e^{a/2d} - (1-a)^2 e^{-a/2d} \quad (5)$$

where C_e is the effluent biochemical oxygen demand (BOD), mg/L; C_i is the influent BOD, mg/L; K is the first-order BOD removal coefficient, day^{-1}; t is the mean detention time, days;$a = \sqrt{(1 + 4Ktd)}$; $d = Dt/L^2$; d is the dimensionless dispersion number; D is the axial dispersion coefficient, ft^2/h.

The thesis of this application is to focus on the first-order BOD removal coefficient (K), making corrections due to temperature, organic load, toxic contaminants, and solar radiation (Thirumurthi 1974). Equation (6) summarizes the factors that are associated with K (Finney and Middlebrooks 1980; Thirumurthi 1974):

$$K = K_8 C_{Te} C_o C_{T0x}, \quad (6)$$

where K is the first-order BOD removal coefficient, day^{-1}; K_8 the BOD_5 removal coefficient, day^{-1}; C_{Te} the correction factor for temperature; C_o the correction factor for organic load; C_{T0x} the correction factor for toxic chemicals.

Anaerobic pond design

Anaerobic pond design is contingent on the volumetric loading rate into the pond. This is based on the water temperature and hydraulic retention time. Table 3 provides treatment efficiency based on major operating parameters.

Table 3: Operating parameters for anaerobic ponds (USEPA 2011)

ALR BOD_5(kg/ha/d)		Est. VLR (kg/1000 m^3)		Removal (%)		Depth (m)	HRT (day)	References
Summer	Winter	Summer	Winter	Summer	Winter			
66.2		37.6		75		0.9–1.2		Parker (1970)
51.5		29.5		65		0.9–1.2		Parker (1970)
18.4		10.6		86		0.9–1.2		Parker (1970)
31.3		17.8		52		0.9–1.2		(Parker1970)

103.0	73.6	58.9	42.1	89	60	0.9–1.2		(Parker1970)
73.6	18.4			70				Oswald (1968)
165.6–220.8	124.2	83.0–110.6	62.3	60–70		0.9–1.5	2–5	Parker et al. (1959)
						2.4–3.1	30–50	Eckenfelder et al. (1961)
36.8–110.4			8.2–22.2				15–160	Cooper (1968)
92.0			18.5	70		2.4–3.1	5	Oswald et al. (1967)
						2.4–3.7	2 (s) 5 (w)	Malina and Rios (1976)

ALR BOD$_5$ areal loading rate BOD_5 (kg/ha/d), *Est. VLR* estimated volumetric loading rate (kg/1000 m^3), *HRT* hydraulic retention time (day). This has been converted from English to SI units

Overall, there are different methods that can be utilized to design a waste stabilization pond. Choosing the appropriate design technique requires careful consideration of assumptions and limitations, pond type, and the wastewater quality indicators that are desired to be analyzed and removed.

OPERATION AND DESIGN PARAMETERS

Operation Parameters

The major operation parameters for oxidation ponds include light penetration, temperature, wind, pond geometry, and oxygen concentration.

Light Penetration

Light penetration affects the process of photosynthetic organisms utilizing sunlight to produce oxygen. Therefore, recognizing such factors as an organism's ability to absorb light at given wavelengths (optimal wavelength range for photosynthesis is between 400 and 700 nm), also known as photosynthetic active radiation (PAR) (Curtis et al. 1994), the presence of suspended material, and the climate and geographic location of the pond (USEPA 2011) will assist in providing a framework for evaluation during operation. In addition, the natural scattering of light particles due to water characteristics and the presence of microorganisms can further impact the usefulness of light that enters into the pond (Curtis et al. 1994).

Temperature

Since this is a natural treatment system, it is also important to understand that the temperature cannot be regulated. Rather, temperature analysis plays a significant role in understanding the efficiency of the oxidation pond as a treatment system, the control of nitrification (Kirby et al.2009), methane production during anaerobic digestion (Sukias and Craggs 2011), COD removal (Daviescolley et al. 1995), the handling of heavy metals (Mona et al. 2011), bacteria growth (Halpern et al. 2009; Olaniran et al. 2001), and the influence of bacteria and algal presence on pond productivity (Quiroga 2011).

One of the major ways that temperature impacts pond productivity is during the summer and winter months, when the surface water heated by the sunlight remains at the top of the pond and cooler denser water is at the bottom. The noticeable change in temperature with increasing depth creates layers or 'strata', a phenomenon known as stratification. Stratification is intensified in the winter months when sheets of ice can form within the pond layers, reducing light penetration and further water layer formation (USEPA 2011). Access to sunlight is limited during both seasons producing anaerobic conditions (Marais 1966). Stratification limits the metabolism of algae and aerobic bacteria in the pond.

Conversely, during the spring months, temperature differences are lessened, and surface water and water at lower depths will combine by mixing (USEPA 2011). Mixing also causes turnover of sediments from the bottom of the pond increasing the available suspended solids concentration to be removed by microorganisms (Finney and Middlebrooks 1980). Seasonal changes impact the water depth through rainfall and evaporation (Hamilton et al. 2006).

Wind

Wind is important in preventing anaerobic conditions by mixing warmer and cooler layers of water (USEPA 2011; Marais 1966). This reduces the potential for stratification, odors, and short-circuiting (USEPA 2011). However, the effectiveness of wind energy is contingent on the temperature of the pond, as the required energy to reduce 1° of pond temperature increases with increasing temperature of the pond (Marais 1966).

Wind patterns and wind speed control the overall removal of bacteria. Various studies compare wind speeds, direction, and also the implementation of baffles within the simulated waste stabilization ponds. The factors related to bacteria removal include the direction, speed of wind, and the presence of baffles, particularly L-shaped baffles, where winds blowing parallel to the inlet–outlet of the pond produce poorer results as compared to the orthogonal orientation of the pond. Nevertheless, the results may vary with increasing wind speeds (Badrot-Nico et al. 2010). Wind coupled with pond geometry can likewise have an impact on treatment performance. For example, in the aerobic layer of facultative ponds the presence of wind leads to vertical mixing and the distribution of dissolved oxygen, bacteria, algae, and BOD. This creates a quality effluent. However, without the wind, algae create a 20 cm-thick layer moving through 50 cm of the pond. This leads to a variable quality effluent (Mara and Pearson 1998).

Pond geometry

Having analyzed two different pond types with varying dimensions and depths, Pearson et al. (1995) confirmed that pond geometry alone does not indicate a strong relationship with treatment performance. Consequently, optimum treatment can be achieved at low dimension ratios, along with shallower pool depths. This can eliminate high construction costs. While the pond geometry may not have indicated an effect on a pond's performance, Hamdan and Mara's comparison between horizontal and vertical pond orientation for the treatment of total Kjeldahl nitrogen (TKN) showed that the vertical-flow orientation has a better treatment performance as compared to the horizontal flow (Hamdan and Mara 2011).

In addition, Abbas et al. (2006) concluded that the dimensions of the waste stabilization pond affect the removal of BOD_5. Their model, initiated from the conservation of moment and mass equations solved by using Newton–Raphson non-linear iterations, find increasing BOD_5removal when the number of baffles was 2. High performance is found regardless of the dimensions at four baffles (peak performance recorded at a dimension ratio of 4, $BOD_5 = 95.8$ %), while low performance is seen at no baffles (ranging from 16.1 to 21.6 %, from 1 to 4 dimension ratio). Therefore, pond geometry can monitor the efficiency of pollutant removal in a pond.

Design Factors

Understanding design factors is important in controlling pollutants such as BOD_5. There are many factors that affect the efficiency of BOD_5 removal in waste stabilization ponds. These factors include raw wastewater strength, food-to-microorganism ratio, organic loading rates, pH, and hydraulic detention time (HRT).

Hydraulic retention time is important because it determines not only how long the wastewater remains within the pond, but also the treatment efficiency. In general, the HRT was found to be as low in literature as 18 h (Gumisiriza et al. 2009) and as high as 300 days (Shpiner et al.2007). Various authors propose different HRT values because of the type of wastewater treated and the type of pond treatment applied. For example, 18 h is sufficient to treat fish-processing wastewater by combining the oxidation pond with anaerobic digestion (Gumisiriza et al. 2009), while Quiroga (2011) recommends that the retention time for the design of an anaerobic pond in Canada range between 2 and 5 days.

The significance of the pH range becomes important when taking into account the presence of microorganisms and other biological species that are needed for the purpose of enhancing treatment. The pH has been regulated for optimum growth of algal species (Bhatnagar et al.2010), sulfate-reducing bacteria (Burns et al. 2012), macrophytes (Kirby et al. 2009), and*Salvinia rotundifolia* (Banerjee and Sarker 1997). For example, photosynthesis by algae increases pH through the presence of hydroxide ions that are formed by the use of carbon dioxide (USEPA 2011). Various authors set the pH range as low as 3–5 (Burns et al. 2012), while the highest pH range is 4–11 (Bhatnagar et al. 2010).

While this text studies the aforementioned as the most frequent parameters, other parameters may have an effect on an oxidation pond's performance, such as flow rate (Kirby et al. 2009; Faleschini et al. 2012), volume (Fyfe et al. 2007), and area (Craggs et al. 2003). However, there are still other parameters that can be considered, but the previously mentioned have been surveyed as having the most effect on the overall treatment of wastewater within oxidation ponds.

Wastewater Characteristics

The characteristics of wastewater are very important when determining the most efficient treatment system. Oxidation ponds treat various wastewaters consisting of nitrogen and phosphorus, heavy metals, organics, and pharmaceuticals. The following includes the mechanisms involved in treating for these specific constituents, along with results that have been outlined in the literature.

Nutrients

In the case of nitrogen, volatilization, specifically ammonia volatilization, is the most common method applied (USEPA 2011; Kayombo et al. 2010). Other methods for reducing nitrogen include deposition, adsorption, nitrification, and denitrification (USEPA 2011). On the other hand, phosphorus treatment uses various processes such as precipitation, sedimentation, and uptake by algal biomass (Kayombo et al. 2010). Nevertheless, oxidation ponds have been effective in removing nutrients.

Nutrient removal is contingent upon the type of oxidation pond (Kayombo et al. 2010). For example, Ke et al. (2012) combined submerged macrophyte oxidation ponds (SMOPs) and subsurface vertical-flow (SVFW) ponds to treat particulate phosphorus (PP) and total dissolved phosphorus (TDP). Garcia et al. (2002) combine two high-rate oxidation ponds (HROPs) for phosphorus removal. Other treatment processes include a hybrid system (oxidation pond, a two-stage surface wetland system, and one subsurface-flow wetland) for the treatment of total Kjeldahl nitrogen (TKN) and ammonium (Yeh et al. 2010), four 3.09-ha high-rate algal ponds (HRAPs) treating ammonia-N and dissolved reactive phosphorus (Craggs et al. 2012), high-rate pending (HRP) treatment system for nitrogen and phosphorus (Nurdogan and Oswald 1995), and shallow waste ponds for removing ammonia in landfill leachate (Leite et al. 2011). Tsalkatidou et al. (2013) combined eight vertical construction wetlands, three facultative ponds, and aerobic and maturation ponds to treat nutrient constituents such as nitrate-N, ammonia-N, and total phosphorus. Table 4 summarizes the values for the treatment efficiencies for each nutrient.

Table 4: Treatment efficiencies of nutrients in oxidation ponds

Nutrient	Efficiency (%)	HRT (days)	References
Particulate phosphorus (PP) Total dissolved phosphorus (TDP)	82.9	3.5	Ke et al. (2012)
Phosphorus	43	10	Garcia et al. (2002)
Phosphorus	32	3	Garcia et al. (2002)
Total Kjeldahl nitrogen	41		Yeh et al. (2010)
Ammonium	48		Yeh et al. (2010)
Ammonia-N	99.5	39.5	Leite et al. (2011)
Ammonia-N	65		Craggs et al. (2012)
Dissolved reactive phosphorus	19		Craggs et al. (2012)

Heavy metals

Without proper treatment, heavy metals at high concentrations can significantly impact the environment and human health, causing damage to the colon, kidneys, skin, and nervous, reproductive, and urinary systems, yet the impact of a heavy metal is contingent on the type and concentration (Ogunfowokan et al. 2008). Nevertheless, oxidation ponds impact human health, causing damage to the colon, kidneys, skin, and nervous, reproductive, and urinary systems, yet the impact of a heavy metal is contingent on the type and concentration (Ogunfowokan et al.2008). Nevertheless, oxidation ponds can be utilized to treat wastewater with heavy metals by reducing its concentration prior to discharge.

Several studies have been done to examine the result of oxidation ponds in treating heavy metals. Ali et al. (2011) found that the effluent from treating heavy metals enhanced the quality of soil and increased plant growth potential. Other authors have considered the effects of heavy metal removal in oxidation ponds. Table 5 summarizes the treatment efficiency of heavy metals.

Table 5: Treatment efficiencies of heavy metals in oxidation ponds

Heavy metal	Wastewater	Treatment efficiency (%)	References
Iron(II)	Alkaline anthracite mine waste	99.7	Kirby et al. (2009)
Cobalt (Co II)	Textile mill	58–60	Mona et al. (2011)
Chromium (Cr(VI))			
Iron	Acid mine drainage	95	Burns et al. (2012)
Lead(II)	Industrial wastewater	85–95	Banerjee and Sarker (1997)
Nickel (Ni)	Acid mine drainage	37–87	Kalin and Chaves (2003)
Aluminum (Al)		77–98	
Zinc (Zn)		74–82	
Iron (Fe)		75–98	
		98	Batty et. al. (2008)
Chromium	Kraft pulp and paper mill	41	Achoka (2002)
Copper		39	Achoka (2002)
Nickel		16	Achoka (2002)

Organics

Organic removal can be accomplished using either anaerobic or aerobic conditions. In aerobic conditions, organic degradation is a two-stage process. During the first stage, bacteria degrade organic matter into carbon dioxide, water, phosphorus, and ammonia in the presence of oxygen. Next, algae in the presence of light use carbon dioxide and water from bacteria to produce oxygen, water, and new algae. The cycle is repeated—oxygen from the second stage is used by the bacteria to restart the first stage of organic degradation. Under anaerobic conditions (commonly known as anaerobic digestion), the process consists of hydrolysis, acidogeneis, acetogenesis, and methanogenesis. In anaerobic digestion, organic matter is degraded into methane, carbon dioxide, and water (Arthur 1983). During these processes, bacteria and algal biomass increase with decreasing organic material.

Various authors report removal of various forms of organics by oxidation ponds—54 % removal of fluorescent organic matter (Musikavong et al. 2007), 64.77 and 97.75 % of polycyclic biphenyls (PCBs) (Badawy et al. 2010), and 54 % of dissolved organic matter (Musikavong, and Wattanachira 2007). The BOD removal efficiency is dependent on numerous factors and can vary from approximately 81 % (Yeh et al. 2010), 50 % (Craggs et al. 2003), 50.3 % (Meneses et al. 2005), and others (Banerjee and Sarker 1997; Faleschini et al. 2012; Mtethiwa et al. 2008; Sukias et al. 2001; Tanner and Sukias 2003; Abbasi and Abbasi 2010). The BOD removal is not only dependent on the type of oxidation pond, but on other factors such as the environment (Mara and Pearson 1998).

Other applications

Oxidation pond treatment analysis has been made in other applications. Spongberg et al. (2011) analyzed oxidation pond treatment performance on pharmaceuticals such as doxycyclines, salicylic acid, triclosan, and caffeine. Gomez et al. (2007) observed 90–95 % removal of estrogenicity. Ahmad et al. (2004) measured low concentrations of pesticides such as heptachlor, dieldrin, and *pp*-dichlorodiphenyltrichloroethane from oxidation pond sludge samples. Khan et al. (2010) analyzed the growth of sorghum using treated wastewater from oxidation ponds.

Pathogens

From a public health perspective, the overall effects of the presence of pathogens must be taken into account. There are several problematic pathogens such as bacteria, protozoa, helminths, and viruses, which can cause various diseases such as cholera, gastroenteritis, typhoid fever, dysentery, and hepatitis (Parker 1970). The World Health Organization (WHO) provides stringent requirements for effluent based on the amount of pathogens present within the system. For example, nematode discharge is limited to 1 egg/L into the system (Mara and Pearson 1998). Human health impacts of microorganisms such as chironomid eggs (Senderovich et al. 2008) have also been discussed for their potential effects on public health and the environment. WHO discusses additional concerns for bacteriophage species, F-RNA, and somatic coliphages 10^3–10^4 plaque-forming units/mL (Gino et al. 2007), and somatic coliphage (phiX-174) and F-specific RNA phage (MS2) (Benyahya et al. 1998).

Oxidation ponds are capable of treating pathogens as well. There are several parameters that are necessary for treating bacteria—time and temperature, pH, light intensity, and dissolved concentration (Mara and Pearson 1998). Maiga et al. (2009) found *E. coli* inactivation to be contingent on sunlight, pH, and dissolved oxygen concentration. In addition, Reinoso and Becares (2008) observe that the prevalence of sunlight reduces *Cryptosporidium parvum* by 40 %. Oxidation ponds can also reduce other pathogens such as *Campylobacter jejuni*,*Salmonella enterica*, and *E. coli* (Sinton et al. 2007), reducing salmonella in particular by 96.4 % (Gopo et al. 1997). Oxidation ponds can also remove viruses (NoV), 47 % geogroup I (GI) and 67 % geogroup II (GII) (Da Silva et al. 2008).

Oxidation ponds can eliminate fecal coliform, fecal enterococci, F^+ coliphage, somatic coliphage, and *Ascaris* eggs by using the four ponds (Nelson et al. 2004). A treatment system consisting of activated sludge, extended aeration, physical, chemical, and biological treatment (BIOFORE), and oxidation ponds successfully removes pathogens (Jamwal et al. 2009), while a pond and wetland system minimizes bacteria by at least one log (Tanner and Sukias 2003). Overall, oxidation ponds can remove pathogens given that the detention time is between a few days to several weeks (Gloyna 1971).

Algae

Algae are one of the more perennial driving forces with respect to proper treatment within oxidation ponds and are important because they are capable of taking up phosphates, carbon dioxide (CO_2), and nitrogen compounds such as ammonia and nitrates, incorporating these constituents into algal biomass production. At the same time, algae supply the oxygen necessary for heterotrophic bacteria to degrade organic material (USEPA 1983). Literature describes that a strong algae presence is determined by nutrients, temperatures, and sunlight. Sunlight is also linked to dissolved oxygen concentration (mg/L) and can predict photosynthetic activity.

There are various algal species utilized such as *Chlamydomonas* and *Euglena*. Nevertheless, algae such as *Scenedesmus acutus* within high-rate oxidation ponds (HROPs)-sedimentation systems reduce TSS (Garcia et al. 2000), while algae species *Chlorella vulgaris* and *Oscillatoria brevis* are also effective in treatment (Tharavathi and Hosetti 2003). In addition, microalgae treat polycyclic aromatic hydrocarbons (PAHs), phenolics, and organic solvents (Munoz and Guieysse 2006). Advanced oxidation systems also use microalgae (Ahmad et al. 2004). Other examples of algae species can include *Phacus* from the phylum *Euglenophyta*,*Chlamydomonas*, and *Eudorina*, (*Chlorophyta*); *Navicula* and *Cyclotella* (*Chrysophyta*); and*Anabaena* (*Cyanophyta*) (Mara and Pearson 1998).

Yet, there are limitations with algae. One of the biggest known problems is finding algae following discharge from the pond (USEPA 1983). Kaya et al. (2007) found that the algae problem is related to high suspended solids (SS) concentrations. Having observed these issues, the authors find that using a laboratory-scale step feed dual treatment (SFDT) and trickling filter (TF) as an addendum to the oxidation ponds improves chlorophyll-*a* (Chl-*a*) from 89.4 to 97 %, as long as the hydraulic loading rate is 2 m^3/m^2 day. In this study, the TF alone removes 88 and 95 % Chl-*a*. Besides the SS concentrations, additional problems may include the clogging of the system when improperly discharged (Arthur 1983).

Despite these limitations, algae grown within the pond can be harvested (converted into biomass) for the purpose of creating an alternative fuel source, provided that the influent does not contain heavy metals. As a fuel source, algae have many advantages—high productivity, short growth time, low land requirements, and the formation of lipids capable of being converted into fuels (Pittman et al. 2011). Studies show that lipids have a dry weight range between 10 and 30 %, where the values can be contingent on the type of wastewater and its associated concentrations (Pittman et al. 2011; Rawat et al. 2011). Algal harvesting creates biofuels such as gas or oil-based fuels, hydrogen gas (Pittman et al. 2011), ethanol, biogas from methane for electricity (Rawat et al. 2011), and biodiesel, a fuel that is biodegradable, renewable, and low in CO_2 emissions (Brennan and Owende 2010). Depending on the species, the amount of gas production will vary. For example, *Chlorella vulgaris* has a potential methane production between 0.6 and 0.8 L/g volatile suspended solids (VSS), while *Chlorella pyrenoidosa* produces 0.8 L/g VSS. Other algae species such as *Scenedesmus obliquus* ranged from 0.6 to 0.7 L/g VSS (Singh and Olsen 2011).

Biofuel formation from algae consists of two major processes—conversion of algae into biomass, followed by the formation of gas based on biochemical and thermochemical processes (Pittman et al. 2011; Rawat et al. 2011). Figure 7 provides a process flow diagram of the algal harvesting process. Conventional algal harvesting begins with flocculation, or the neutralization of the negatively charged algal constituents for the purpose of aggregation (Pittman et al. 2011; Rawat et al. 2011). Flocculation occurs by the use of alum (Pittman et al. 2011) or polymers (Rawat et al. 2011). Following flocculation, biomass is then separated from algal cells and recovered by the use of either centrifugation or gravity sedimentation. An additional biomass recovery procedure consists of biomass attached onto constituents such as alginates in a process known as immobilization (Pittman et al. 2011; Rawat et al. 2011).

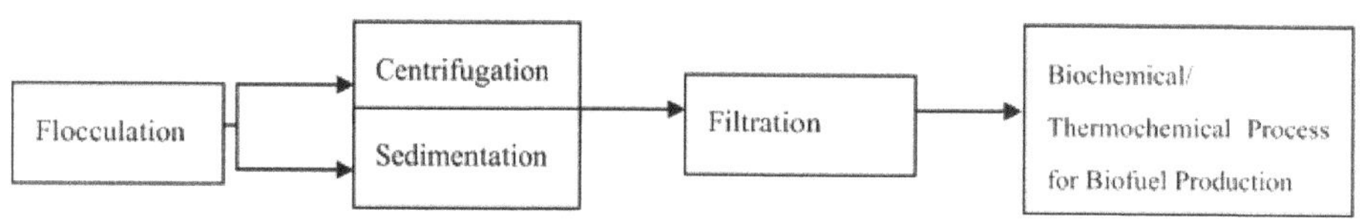

Figure.7: Conventional process flow diagram of algal harvesting process for biofuel production (adapted from Pittman et al. 2011; Rawat et al. 2011)

Following algal harvesting, the biomass undergoes thermo or biochemical processes, converting biomass into biofuels, where thermochemical processes can include gasification, pyrolysis, liquefaction, and combustion, while anaerobic digestion and fermentation encompass the major biochemical processes (Pittman et al. 2011; Rawat et al. 2011). Table 6 describes some of the major thermochemical processes that have been employed for biomass conversion.

Nonetheless, Algal Harvesting has Several Limitations.

Table 6: Thermochemical processes used in biomass conversion

Process	Description	Requirements
Gasification	Formation of syngas or combustible gas	CO_2, CH_4, H_2
Pyrolysis	Biomass from charcoal, based on temperature	O_2/air (not required)
Liquefaction	Breakdown of biomass by low temperature and high pressure	Catalyst
Transesterification	Conversion of lipids into biodiesel	Alcohol, catalyst
Combustion	Converts biomass into biofuel at high temperatures	Air

Based on the descriptions recorded from Rawat et al. (2011)

First, the majority of the algae are unicellular and low in density, making harvesting very expensive (Rawat et al. 2011; Brennan and Owende 2010). Second, several life cycle assessments (LCAs) indicate that various setbacks need to be addressed. An LCA completed by Campbell et al. (2011) found that while biodiesel reduces greenhouse gas emissions when compared with fossil fuel and canola-based biodiesel production of 15 g/m^2 day and 30 g/m^2 day, biodiesel is more expensive than fossil fuel and canola-based fuels (Campbell et al.2011). Singh and Olsen (2011) concluded that biodiesel production has high electricity costs and water requirements—for every 1 kg of biodiesel produced, 3726 kg of freshwater is required. Algal harvesting also requires a supply of nitrogen, usually from chemical fertilizers (Clarens et al. 2010), phosphorus, potassium, and sulfur (Singh and Olsen 2011). Another LCA finds that when comparing algae, corn, canola, and switchgrass as a potential biofuel, algae have the highest energy production, lowest land requirement, and eutrophication potential, but corn and switchgrass reduce greenhouse gases more efficiently (Clarens et al. 2010).

On the other hand, there are ways to resolve these difficulties and continue using algae for biofuel production. For example, if algae are grown within wastewater or saltwater, water requirements are reduced to approximately 400 kg/1 kg of biodiesel, along with nutrients due to their availability within wastewater (Singh and Olsen 2011; Clarens et al. 2010). Still, future studies need to be conducted to address many of the concerns that have been mentioned in the various life cycle analyses, specifically the costs associated with using algae biomass for the purpose of forming biofuels.

A more recent application is the uptake of phosphorus by algae within waste stabilization ponds. Phosphorus uptake by microalgae is necessary, but only in limited quantities. Nevertheless, there are some instances where the uptake of phosphorus by microalgae is greater than what is required for viability. This process is known as luxury uptake. In luxury uptake, microalgae take available phosphorus from the environment and store unused phosphorus as polyphosphate. The forms of polyphosphate stored are either acid-soluble polyphosphate (ASP) or acid-insoluble polyphosphate (AISP). Stored polyphosphates are used in situations when phosphate concentrations are low.

Luxury uptake has been observed in natural environments. In literature, the authors evaluate luxury uptake on three factors—phosphorus concentration, light intensity, and temperature. Results from the study indicate that the use of phosphorus concentration is contingent on the difference between batch and continuous reactors. Light intensity and temperature affect the rate of polyphosphate accumulation by algae. Full-scale waste stabilization ponds validate the experiments. However, luxury uptake research is still in infant stages. Questions such as determining the effects of mixing and detention time on phosphorus uptake and the optimum ideal phosphorus, bacteria, and algal concentrations still need to be answered (Brown and Shilton 2014).

Greenhouse Gas Emissions

Despite many advantages, waste stabilization ponds have been linked with emitting greenhouse gases. There are three recognized greenhouse gases—methane (CH_4), carbon dioxide (CO_2), and nitrous oxide (NO_x) (Hernandez-Paniagua et al. 2014; Khatiwada and Silveira 2011). In general, wastewater treatment processes contribute to 5 % of the greenhouse gas production in the world, where anaerobic treatment methods are the most responsible (Cakir and Stenstrom2005). According to Hernandez-Paniagua et al. (2014), stabilization ponds typically emit 85 g/m^2 day of CO_2 and 86 g/m^2 day of CH_4, with nitrous oxide levels significantly lower. In fact, methane emissions in a waste stabilization pond were reported to be twice as high compared to activated sludge systems (Hernandez-Paniagua et al. 2014).

At the aerobic layer near the surface of the pond, methane will oxidize to CO_2. However, it has been observed that methane emissions are much higher in ponds as compared to carbon dioxide (Silva et al. 2012). This has been attributed to the fact that during photosynthesis algae, use atmospheric carbon dioxide, raising the pH. At higher pH levels, dissolved carbon dioxide is converted to carbonic acid and bicarbonates (Silva et al. 2012). Algae are also responsible for biologically fixing carbon dioxide and converting atmospheric carbon dioxide into a solid, thereby reducing the emission of CO_2 (Shilton et al. 2008). Nevertheless, the release of CO_2 into the atmosphere is still dependent upon the pH value of the system.

While waste stabilization ponds are responsible for a small portion of the world's greenhouse gas emissions (particularly methane), there is an opportunity for the gas, commonly known as biogas, to be recovered and used as a source of energy (Konate et al. 2013). While biogas primarily consists of methane, it is also made up of carbon dioxide, oxygen, and other gases (NIWA 2008). Nevertheless, methane is the main gas targeted when considering biogas recovery. Biogas is formed either during anaerobic digestion of organic compounds or the reduction of carbon dioxide and hydrogen (El-Fadel and Massoud 2001). As a viable energy option, it has been seen as being ideal for electricity generation in a wastewater treatment plant as natural gas, provided that it has undergone several treatment processes (Craggs et al. 2014), or for heat and power (NIWA 2008). If biogas is not recovered for energy, it can be flared off or transitioned into an aerobic pond if a multiple pond system exists (NIWA 2008).

The recovery of biogas is not a novel idea. As a matter of fact, recovery of biogas in pond systems began appearing in peer-reviewed journals within animal waste operations no later than the 1970s (Safley and Westerman 1988). Recently, authors have studied biogas recovery in stabilization ponds from various types of wastewater. For example, Park and Craggs (2007) found a mean areal biogas production of 0.78 m^3/m^2 day from piggery wastewater and 0.03 m^3/m^2 day from dairy wastewater (Park and Craggs 2007). Konate et al. (2013) recorded a mean areal biogas production of 0.121 m/m^2 day from domestic wastewater in an anaerobic pond. McGrath and Mason (2004) observed biogas production between 0.002 and 0.039 m^3/m^2 day in dairy wastewater. Heubeck and Craggs (2010) found an average annual methane production rate of 0.263 m^3 CH_4/kg VSS_{added} from a piggery pond.

For a pond to be efficient in producing a substantial amount of biogas to be recovered, several parameters must be considered. Two of the most important parameters are the ambient air and pond temperatures. It has been determined that there is a power law relationship between biogas production and ambient temperature. This is expressed in the following equation (McGrath and Mason 2004):

$$R = 1.241(T_a)^{1.127} \tag{7}$$

where R is the areal biogas production (L/m^2 day) and T_a the ambient air temperature (°C).

With regard to pond temperature, typical biogas production in stabilization ponds occurs in the temperature range between 10° and 30° where a linear relationship exists between 10° and 20° (McGrath and Mason 2004). Other parameters can include pond geometry (Konate et al. 2013), pH, retention time, treatment method, and microorganism presence (El-Fadel and Massoud2001). BOD and COD loading rate is also important as the production of methane increases with increase in wastewater strength (El-Fadel and Massoud 2001; Cakir and Stenstrom 2005).

The most efficient way found is to fully or partially cover the pond surface with a material made usually of high-density polyethylene (HDPE) or polypropylene (PP). The pond cover serves as a way to capture the gas generated from anaerobic digestion. Covers are usually designed to extend through the pond's perimeter where it is either placed in a trench or held in place by means of a concrete slab. After the biogas is captured, it is collected by perforated pipes located below the cover. From the pipes, the biogas is removed from the cover by a vacuum (e.g., centrifugal fan) (NIWA 2008). Several authors have recorded the application of covers to capture biogas in municipal wastewater treatment systems. DeGarie et al. (2000) discussed the implementation of a three-layer cover consisting of geomembrane, polyform, and HDPE in Melbourne (Australia). The biogas cover was placed on the anaerobic lagoons in the wastewater treatment plant. Shelef and Azov (2000) constructed a waste stabilization pond system in the Negev desert of Israel. The pond treatment system consisted of anaerobic ponds, stabilization ponds, rock filters, and stabilization reservoirs. The authors included a cover on the anaerobic pond.

There are some points that need to be considered when recovering biogas from a pond. One of the most important is cost. The associated costs for biogas recovery include construction costs of the pond, capital and operation and maintenance costs for gas purification such as gas scrubbing, any capital and operation and maintenance costs to apply biogas for other uses, and the cost for cover design (Craggs 2004). Other potential design considerations include sludge management and the maintenance of a proper organic removal with biogas recovery (DeGarie et al. 2000). Optimizing the pH, nutrients, temperature, organic loading, and retention time are also important because it will assist in producing an effective biogas system (Kaewmai et al.2013). These and many other considerations must be addressed when attempting to recover biogas from a pond.

In summary, waste stabilization ponds are capable of producing greenhouse gases, particularly methane. However, applying techniques such as pond covers can mitigate the emission of greenhouse gases and provide a viable source of energy.

Costs

The cost of an oxidation pond is contingent on the year and the particular country where it is to be built. This is because labor and material costs are not uniform around the world. Nevertheless, there are various analyses available to determine the feasibility of installing oxidation ponds. As with any treatment project, the major costs are capital costs (construction costs) and operation and maintenance (O&M) costs. Examples of capital costs include construction, administration/legal, land, structures, and architecture/engineering (A/E) fees, while energy is the primary operation and maintenance (O&M) cost. Cost data to build an oxidation pond facility in the USA is presented below.

There are two heavily cited sources for construction cost data in the USA—the Environmental Protection Agency (EPA) and the World Bank. In 1978, the EPA compiled construction costs based on data collected from oxidation pond use in Kansas City/St. Joseph, Missouri. The initial costs of an oxidation pond include 15 different parameters. Results from the EPA study conclude that oxidation ponds have a lower cost per capita as compared to primary and secondary treatment when considering populations between 100 and 100,000. The study also finds that estimated capital costs ranged between $0.20 and $1.00/population equivalent/year (Gloyna 1971). Recently, the EPA published updated construction cost data for the year 2006. In this publication, the EPA states that this data can be calculated for any US city. This is accomplished by multiplying the cost and the ratio of the Engineering News Record Construction Cost (ENR CC) index for a given city by the primary operation and maintenance (O&M) cost. Cost data to build an oxidation pond facility in the USA is presented below. The ENR CC index is for Kansas City in 2006. The ENR CC indexes are accessible athttp://www.enr.com (USEPA 1983). In 1983, the World Bank published data comparing the capital and operation and maintenance costs of oxidation ponds with other natural treatment methods such as aerated lagoons, oxidation ditches, and biological filters. The study calculates capital costs by totaling the land, earthworks, structure, and equipment costs (Arthur 1983). Results from this report estimate that the capital costs of an oxidation pond totaled 7.3 million US$, while annual operation costs were $50,000/year. Furthermore, the report discovers that oxidation ponds have the lowest capital and operational costs as compared to other natural treatment methods (Arthur 1983; Varon and Mara 2004). Table 7 provides an example of costs between the three natural treatment methods.

Table 7: Capital and operation costs for natural wastewater treatment methods (Arthur 1983; Varon and Mara 2004)

Treatment systems	Capital costs (US$, millions)	Operational costs (US$/capita)
Waste stabilization pond	5.7	0.2
Aerated lagoon system	7.0	1.2
Oxidation ditch system	4.8	1.4
Biological filter	7.8	0.8

On the other hand, the Water Environmental Federation (WEF) published data to approximate the cost of an oxidation pond for a residential area. The WEF conclude that the size of the population appropriates the value of the oxidation pond. For example, the $7.3 million dollar estimate would be at the higher end of treatment for a population of about 250,000. Considering that many oxidation ponds would be used for smaller communities, the total costs, including capital and operation and maintenance, can range between $316,400 for 20 homes and $4.7 million for 200 homes (WEF 2010) or approximately $2600–$7600 per home (USEPA 2002).

Modeling

When evaluating the performance of a waste stabilization pond, it is imperative to consider additional factors that impact the performance of the pond beyond the mere parameters that have been previously discussed. One of the best ways to make the determinations is by using modeling software. Although its popularity began in the mid-1990s, computational fluid dynamics (CFD) has evolved parallel to the advancement of computer technology. CFD models the performance of the pond in removing nutrients, BOD, sedimentation, and pathogens. CFD can vary with meteorological conditions, pond orientation, and design adjustments (Sah et al.2012). CFD benefits the researcher insomuch as he or she is able to determine how varying the effects of conditions within waste stabilization ponds would affect the performance of removing various constituents. This is beneficial whenever the conditions are unfavorable for sampling (Karteris et al. 2005).

Various authors discuss the significance of the use of CFD in its ability to analyze treatment performance. Wood et al. (1995) initiated the use of CFD by using two-dimensional computational fluid dynamics (2-D CFD) models for the purpose of designing an optimum treatment performance.

The attempt was to derive conclusions by trying to transfer a three-dimensional entity into two-dimensional space, without the consideration of the depth changes, turbulence-caused mechanical aeration, and the non-isothermal conditions that can potentially exist within a pond under given circumstances. An additional study by Wood et al. (1998) determined that the inlet geometry had a major impact on the water flow, concluding that models are capable of determining water flow effects by varying different parameters. Also Olukanni and Ducoste (2011) determined bacteria removal by using both 2-D CFD and optimization. The authors analyzed baffle adjusted by CFD, while optimization was used to determine the minimal construction costs of waste stabilization ponds.

CFD can incorporate various parameters such as sludge accumulation, temperature, and sunlight. Alvarado et al. (2012) determined that sludge accumulation and settling of sludge within waste stabilization ponds increases as water moves away from the inlet and the velocity decreases. Karteris et al. (2005) created a temperature model within a covered anaerobic pond under unsteady state conditions. Sweeney et al. (2007) analyzed the effects of the presence of sunlight when considering the water quality between the summer and winter months.

Recent investigations into CFD modeling have been conducted to further improve the performance of waste stabilization ponds. Alvarado et al. (2013) deduced the appropriate number of aerators to achieve proper mixing within a waste stabilization pond. The results from the study are further supported by the use of tracer studies completed on a full-scale waste stabilization pond. Hadiyanto et al. (2013) analyzed the length-to-width (L:W) ratio using dimensions from a currently operating pond in the Netherlands to investigate power consumption, dead zones, and large eddy formation. Also, the authors not only incorporate a discussion on the relationship with velocity and shear stress, but include the effects of shear stress and turbulence on the growth of microorganisms and algae.

Martinez et al. (2014) compared traditional methods of pond design with mathematical software (Matlab) to find a cost-effective facultative pond that can reduce fecal coliform and BOD. The authors examine each method based on the number of baffles and the hydraulic efficiency. Results from the study conclude that the mathematical model can reduce cost by 11 %, the area requirement by 18 %, and the hydraulic retention time by 19 days.

Lee and Cheong (2014) used numerical modeling to vary the L:W ratio, pond shape, flow rate, and depth under constant pressure and a steady state flow to understand its effects on pond flow characteristics to treat acid mine drainage (AMD). The authors found that the retention time in rectangular ponds decreased with increasing elongation. Pond depth also has an impact on the retention time. An optimum depth for a rectangular pond is 2 m given a pond area of 500 m^2.

As seen within this section, modeling for waste stabilization ponds has been conducted and improved to not only consider specific individual constituents, but also model various phenomena that can occur within the system. Yet, various parameters are highlighted individually within computation models. Currently, no published work has been able to provide a complete profile of wastewater entering the system or through multiple pond systems. With the constant evolution of software, modeling waste stabilization ponds could experience further advancement. Sah et al. (2012) agreed with this sentiment. In addition, the authors recommended that upto-date full-scale data should be available to verify the accuracy of the models (Sah et al. 2012).

Oxidation ponds or waste stabilization ponds are a very effective treatment option not only as tertiary treatment, but also in areas that are not capable of having conventional domestic wastewater treatment or other technologically advanced methods, as seen in Pena et al. (2002) in Columbia. Yet, the design of an oxidation pond is congruent to the location of the pond, as loading rates will vary based on temperature and climate (Mara and Pearson 1998; Gloyna1971).

To ensure proper treatment, the operator must avoid the possibility of effluent discharge that can contaminate the environment or jeopardize public health. There are many ways that these dangers can be avoided. First, maintain all piping and other treatment unit equipment properly, along with monitoring the conditions within the pond to provide enclosure of treated waste without exposure to the external environment or the formation of residuals. Second, evaluate pathogen presence by taking samples. This will determine if the pond system is having issues. Finally, design the pond system to produce the best efficiency based on the given standards. Computer modeling provides an effective way of confirming the best treatment necessary for the given condition. If the efficiency cannot be met by one pond, another alternative is to create a multiple-staged treatment system of ponds, or include another method of treatment for maximum results.

Computer modeling provides an effective way of confirming the best treatment necessary for the given condition. If the efficiency cannot be met by one pond, another alternative is to create a multiple-staged treatment system of ponds, or include another method of treatment for maximum results.

CONCLUSIONS

Waste stabilization ponds have demonstrated the capabilities of being a viable treatment technology, specifically in small communities and developing countries because of their inexpensive maintenance and simple design. As the worldwide water crisis increases, wastewater effluent reuse, specifically from waste stabilization ponds in some fashion, will continue to increase (Mara 2012).

Two recommendations will need to be considered for the purpose of expanding this treatment technology. First, reevaluate how the treatment efficiency relates to the corresponding treatment standards for the purpose of effluent reuse. WHO has published guidelines for wastewater reuse in a four-volume publication titled: Guidelines for the safe use of wastewater, excreta and greywater in agriculture and aquaculture (WHO 2006). This document outlines standards that are necessary to ensure that the effluent from the pond is suitable for use in these applications, a very important factor in the current attempt to advocate pond wastewater as a viable alternative. The efforts to implement WHO's guidelines will assure both citizens and governmental agencies that human health and safety is a top priority. Modeling software can be used as a tool that will provide the engineer an opportunity to design a pond that will comply with these standards. Second, retrofit current infrastructure and add stations for fuel production can provide a return on the investment when transitioning waste stabilization ponds for use in larger communities. However, biofuel production will depend on the economic viability of recovery. Despite many technological advances, the main challenge is overcoming the preference of using conventional treatment methods such as activated sludge and trickling filters to comply with treatment standards. Nevertheless, future research addressing the concerns with this section, along with those made throughout the course of this text, may persuade designers and treatment operators to reconsider their stance of minimizing the use of oxidation ponds.

ACKNOWLEDGMENTS

The authors would like to thank Mr. Jon Sessions, City Manager, and the City of Wellington, TX, for granting permission to take pictures of the city's pond treatment system.

REFERENCES

1. Abbas H, Nasr R, Seif H (2006) Study of waste stabilization pond geometry for the wastewater treatment efficiency. Ecol Eng 28:25–34
2. Abbasi T, Abbasi SA (2010) Enhancement in the efficiency of existing oxidation ponds by using aquatic weeds at little or no extra cost—the macrophyte-upgraded oxidation pond (MUOP). Bioremediation J 14:67–80
3. Achoka JD (2002) The efficiency of oxidation ponds at the kraft pulp and paper mill at Webuye in Kenya. Water Res 36:1203–1212
4. Ahmad UK, Ujang Z, Woon CH, Indran S, Mian MN (2004) Development of extraction procedures for the analysis of polycyclic aromatic hydrocarbons and organochlorine pesticides in municipal sewage sludge. Water Sci Technol 50:137–144
5. Ali HM, EL- EM, Hassan FA, EL-Tarawy MA (2011) Usage of sewage effluent in irrigation of some woody tree seedlings. Part 3: Swietenia mahagoni (L.) Jacq. Saudi J Biol Sci 18:201–207
6. Alvarado A, Sanchez E, Durazno G, Vesvikar M, Nopens I (2012) CFD analysis of sludge accumulation and hydraulic performance of a waste stabilization pond. Water Sci Technol 66:2370–2377
7. Alvarado A, Vesvikar M, Cisneros JF, Maere T, Goethals P, Nopens I (2013) CFD study to determine the optimal configuration of aerators in a full-scale waste stabilization pond Water Res 47:4528–4537
8. Amengual-Morro C, Moya Niell G, Martinez-Taberner A (2012) Phytoplankton as bioindicator for waste stabilization ponds. J Environ Manag 95:S71–S76
9. Arthur JP (1983) Notes on the design and operation of waste stabilization ponds in warm climates of developing countries. [Online] The World Bank, Washington. http://www-wds.worldbank.org/external/default/WDSContentServer/WDSP/IB/2002/08/27/000178830_9810190416457/Rendered/PDF/multi0page.pdf. Accessed 18 Oct 2013
10. Badawy MI, El-Wahaab RA, Moawad A, Ali MEM (2010) Assessment of the performance of aerated oxidation ponds in the removal of persistent organic pollutants (POPs): a case study. Desalination 251:29–33
11. Badrot-Nico F, Guinot V, Brissaud F (2010) Taking wind into account in the design of waste stabilisation ponds. Water Sci Technol 61:937–944
12. Banerjee G, Sarker S (1997) The role of salvinia rotundifolia in scavenging aquatic Pb(II) pollution: a case study. Bioprocess Eng 17:295–300
13. Benyahya M, Bohatier J, Laveran H, Senaud J, Ettayebi M (1998) Comparison of elimination of bacteriophages MS2 and phi X-174 during sewage treatment by natural lagooning or activated sludges. A study on laboratory-scale pilot plants. Environ Technol 19:513–519
14. Bhatnagar A, Bhatnagar M, Chinnasamy S, Das KC (2010) Chlorella minutissima-A promising fuel alga for cultivation in municipal wastewaters. Appl Biochem Biotechnol 161:523–536
15. Brennan L, Owende P (2010) Biofuels from microalgae—a review of technologies for production, processing, and extractions of biofuels and co-products. Renew. Sust Energ Rev 14:557–577.https://wiki.umn.edu/pub/Biodiesel/WebHome/An_intensive_continuousnext_term_culture_system_using_tubular_photobioreactors_for_producing_previous_termmicroalgae.pdf. Accessed 20 Aug 2013
16. Brown N, Shilton A (2014) Luxury uptake of phosphate by microalgae in waste stabilization ponds: current understanding and future direction. Rev Environ Sci Biotechnol 13:321–328
17. Burns AS, Pugh CW, Segid YT, Behum PT, Lefticariu L, Bender KS (2012) Performance and microbial community dynamics of a sulfate-reducing bioreactor treating coal generated acid mine drainage. Biodegradation 23:415–429

18. Cakir FY, Stenstrom MK (2005) Greenhouse gas production: a comparison between aerobic and anaerobic wastewater treatment technology. Water Res 39:4197–4203
19. Campbell PK, Beer T, Batten D (2011) Life cycle assessment of biodiesel production from microalgae in ponds. Bioresour Technol 102:50–56
20. Cinara Columbia (2004). Waste stabilization ponds for wastewater treatment: FAQ sheet on waste stabilization ponds. http://www.irc.nl/page/8237. Accessed 10 May 2013
21. Clarens AF, Resurreccion EP, White MA, Colosi LM (2010) Environmental life cycle comparison of algae to other bioenergy feedstocks. Environ Sci Technol 44:1813–1819
22. Cooper RC (1968) Industrial waste oxidation ponds. SW Water Works J 5:21–26
23. Craggs R (2004) (n.d.) Potential for biogas recovery from anaerobic ponds in New Zealand.http://www.bioenergy.org.nz/documents/Christchurch%20Workshop/7_Potential%20for%20biogas%20recovery%20from%20anaerobic%20ponds%20in%20NZ.pdf. Accessed 6 Jan 2015
24. Craggs RJ, Davies-Colley RJ, Tanner CC, Suklas JP (2003) Advanced pond system: performance with high rate ponds of different depths and areas. Water Sci Technol 48:259–267
25. Craggs R, Sukias J, Tanner C, Davies-Colley R (2004) Advanced pond system for dairy-farm effluent treatment. NZ J Agric Res 47:449–460
26. Craggs R, Sutherland D, Campbell H (2012) Hectare-scale demonstration of high rate algal ponds for enhanced wastewater treatment and biofuel production. J Appl Phycol 24:329–337
27. Craggs R, Park J, Heubeck S, Sutherland D (2014) High rate algal pond systems for low-energy wastewater treatment, nutrient recovery and energy production. NZ J Bot 52:60–73
28. Crites RW, Middlebrooks EJ, Reed SC (2006) Natural wastewater treatment systems CRC Press, Boca Raton. http://books.google.com/books?id=X5uuikcVyfEC&pg=PA193&dq=Natural+Wastewater+Treatment+Completely+Mixed&hl=en&sa=X&ei=Nv1fUtv7KoPZ2AWk1ICQDA&ved=0CEkQ6AEwAQ#v=onepage&q=Natural%20Wastewater%20Treatment%20Completely%20Mixed&f=false. Accessed 17 Oct 2013
29. Curtis T, Mara DD, Dixo N, Silva SA (1994) Light penetration in waste stabilization ponds. Water Res 28:1031–1038
30. Da Silva AK, Le Guyader FS, Le Saux J, Pommepuy M, Montgomery MA, Elimelech M (2008) Norovirus removal and particle association in a waste stabilization pond. Environ Sci Technol 42:9151–9157
31. Daviescolley RJ, Hickey CW, Quinn JM (1995) Organic-matter, nutrients, and optical characteristics of sewage lagoon effluents. NZ J Mar Freshw Res 29:235–250
32. Davis ML, Cornwell DA (2008) Introduction to environmental engineering, 4th edn. McGraw-Hill, New York
33. DeGarie C, Crapper T, Howe B, Burke B, McCarthy P (2000) Floating geomembrane covers for odour control and biogas collection and utilization in municipal lagoons. Water Sci Technol 42:291–298
34. Eckenfelder WW (1961) Biological waste treatment. Pergamon Press, London
35. El-Fadel M, Massoud M (2001) Methane emissions from wastewater management. Environ Poll 114:117–185
36. Faleschini M, Esteves JL, Valero MAC (2012) The effects of hydraulic and organic loadings on the performance of a full-scale facultative pond in a temperate climate region (Argentine Patagonia). Water Air Soil Poll 223:2483–2493
37. Finney BA, Middlebrooks EJ (1980) Facultative waste stabilization design. J Water Poll Control Fed 134–147. http://www.jstor.org/stable/25040557. Accessed 15 Oct 2013
38. Fyfe J, Smalley J, Hagare D, Sivakumar M (2007) Physical and hydrodynamic characteristics of a dairy shed waste stabilisation pond system. Water Sci Technol 55:11–20
39. Garcia J, Hernandez-Marine M, Mujeriego R (2000) Influence of phytoplankton composition on biomass removal from high-rate oxidation lagoons by means of sedimentation and spontaneous flocculation. Water Environ Res 72:230–237
40. Garcia J, Hernandez-Marine M, Mujeriego R (2002) Analysis of key variables controlling phosphorus removal in high rate oxidation ponds provided with clarifiers. Water SA 28:55–62
41. Gino E, Starosvetsky J, Armon R (2007) Bacteriophage ecology in a small community sewer system related to their indicative role in sewage pollution of drinking water. Environ Microbiol 9:2407–2416
42. Gloyna EF (1971) Waste stabilization ponds. World Health Organization, Geneva.http://apps.who.int/iris/bitstream/10665/41786/1/WHO_MONO_60_%28part1%29.pdf. Accessed 18 Oct 2013
43. Gomez E, Wang X, Dagnino S, Leclercq M, Escande A, Casellas C, Picot B, Fenet H (2007) Fate of endocrine disrupters in waste stabilization pond systems. Water Sci Technol 55:157–163
44. Google Maps (2014) [City of Wellington, Wellington, Texas] [Google Earth].https://www.google.com/maps/@34.8454567,-100.2300974,681m/data=!3m1!1e3
45. Gopo J, Setoaba M, Lesufi W, Sibara M (1997) Incidence of Salmonella spp. in sewage and semi-urban waste water treated by pond oxidation method at the University of the North. Water SA 23:333–337
46. Gumisiriza R, Mshandete AM, Rubindamayugi MST, Kansiime F, Kivaisi AK (2009) Enhancement of anaerobic digestion of Nile perch fish processing wastewater. Afr J Biotechnol 8:328–333
47. GvR Marais (1966) New factors in the design, operation and performance of waste-stabilization ponds. Bull World Health Organ 34:737
48. Hadiyanto H, Elmore S, Van Gerven T, Stankiewicz A

(2013) Hydrodynamic evaluations in high rate algae pond (HRAP) design. Chem Eng J 217:231–239

49. Halpern M, Shaked T, Schumann P (2009) Brachymonas chironomi sp nov., isolated from a chironomid egg mass, and emended description of the genus *Brachymonas*. Int J Syst Evol Microbiol 59:3025–3029

50. Hamdan R, Mara DD (2011) The effect of aerated rock filter geometry on the rate of nitrogen removal from facultative pond effluents. Water Sci Technol 63:841–844

51. Hamilton DW, Fathepure B, Fulhage CD, Clarkson W, Lalman J (2006) Treatment lagoons for animal agriculture. Animal agriculture and the environment: National Center for Manure and Animal Waste Management White Papers 547–574.http://osuwastemanage.bae.okstate.edu/articles-1/lagoonwp.pdf. Accessed 17 Oct 2013

52. Hernandez-Paniagua IY, Ramirez-Vargas R, Ramos-Gomez MS, Dendooven L, Avelar-Gonzalez FJ, Thalasso F (2014) Greenhouse gas emissions from stabilization ponds in subtropical climates. Environ Technol 35:727–734

53. Heubeck S, Craggs R (2010) Biogas recovery from a temperate climate covered anaerobic pond. Water Sci Technol 61:1019–1026

54. Hosetti B, Frost S (1995) A review of the sustainable value of effluents and sludges from wastewater stabilization ponds. Ecol Eng 5:421–431

55. Hosetti B, Frost S (1998) A review of the control of biological waste treatment in stabilization ponds. Crit Rev Environ Sci Technol 28:193–218

56. Jamwal P, Mittal AK, Mouchel J (2009) Efficiency evaluation of sewage treatment plants with different technologies in Delhi (India). Environ Monit Assess 153:293–305

57. Joint Departments of the Army, the Navy, and the Air Force (1988) Chapter 14: Wastewater treatment ponds. In: Domestic wastewater treatment. Army TM 5-814-3 and Air Force AFM 88-11, vol. 3. TM 5-814-3/AFM 88-11, Volume III, Washington DChttp://www.discountpdh.com/course/domestic/Domestic%20Wastewater%20Treatment%20chap14.pdf. Accessed 18 Oct 2013

58. Kaewmai R, H-Kittikun A, Suksaroj C, Musikavong C (2013) Alternative technologies for the reduction of greenhouse gas emissions from palm oil mills in Thailand. J Environ Sci Technol 47:12417–12425

59. Kalin M, Chaves W (2003) Acid reduction using microbiology: treating AMD effluent emerging from an abandoned mine portal. Hydrometallurgy 71:217–225

60. Karteris A, Papadopoulos A, Balafoutas G (2005) Modeling the temperature pattern of a covered anaerobic pond with computational fluid dynamics. Water Air Soil Poll 162:107–125

61. Kaya D, Dilek FB, Goekcay CF (2007) Reuse of lagoon effluents in agriculture by post-treatment in a step feed dual treatment process. Desalination 215:29–36

62. Kayombo S, Mbwette TSA, Katima JHY, Ladegaard N, Jorgensen SE (2010) Waste stabilization ponds and constructed wetland design manual. UNEP International Environmental Technology Center.http://www.unep.or.jp/Ietc/Publications/Water_Sanitation/ponds_and_wetlands/Design_Manual.pdf. Accessed 18 Oct 2013

63. Ke F, Li WC, Li HY, Xiong F, Zhao AN (2012) Advanced phosphorus removal for secondary effluent using a natural treatment system. Water Sci Technol 65:1412–1419

64. Khan MA, Shaukat SS, Hany O, Jabeen S (2010) Irrigation of sorghum crop with waste stabilization pond effluent: growth and yield responses. Pak J Bot 42:1665–1674

65. Khatiwada D, Silveira S (2011) Greenhouse gas balances of molasses based ethanol in Nepal. J Clean Prod 19:1471–1485

66. Kirby CS, Dennis A, Kahler A (2009) Aeration to degas CO_2, increase pH, and increase iron oxidation rates for efficient treatment of net alkaline mine drainage. Appl Geochem 24:1175–1184

67. Konate Y, Maiga AH, Casellas C, Picot B (2013) Biogas production from an anaerobic pond treating domestic wastewater in Burkina Faso. Desalin Water Treat 51:2445–2452

68. Lado M, Ben-Hur M (2010) Effects of irrigation with different effluents on saturated hydraulic conductivity of arid and semiarid soils. Soil Sci Soc Am J 74:23–32

69. Lee DK, Cheong YW (2014) A numerical flow analysis using the concept of inflow age for oxidation pond design. J Environ Manag 133:388–396

70. Leite VD, Pearson HW, de Sousa JT, Lopes WS, de Luna MLD (2011) The removal of ammonia from sanitary landfill leachate using a series of shallow waste stabilization ponds. Water Sci Technol 63:666–670

71. Batty L, Hooley L, Younger P (2008) Iron and manganese removal in wetland treatment systems: rates, processes and implications for management. Sci Total Environ 394:1–8

72. Maiga Y, Wethe J, Denyigba K, Ouattara AS (2009) The impact of pond depth and environmental conditions on sunlight inactivation of *Escherichia coli* and enterococci in wastewater in a warm climate. Can J Microbiol 55:1364–1374

73. Malina JF Jr, Rios RA (1976) Anaerobic ponds. In: Gloyna EF, Malina JF Jr, Davis EM (eds) Ponds as a wastewater alternative, water resources symposium, No. 9, University of Texas Press, Austin, TX

74. Mara D (2012) Waste stabilization ponds: past, present and future. Desalination Water Treat 4:85–88.http://www.tandfonline.com.databases.wtamu.edu:2048/doi/pdf/10.5004/dwt.2009.359. Accessed 3 Nov 2013

75. Mara D, Pearson H (1998) Design manual for waste stabilization ponds in Mediterranean countries [Online] Lagoon Technology International Ltd, Leeds.http://www.efm.leeds.ac.uk/CIVE/Sewerage/articles/medall/medall.pdf. Accessed 18 Oct 2013

76. Mara DD, Mills SW, Pearson HW, Alabaster GP (1992) Waste stabilization ponds—a viable alternative for small community treatment systems. J Inst Water Environ Manag 6:72–78

77. Martinez FC, Cansino AT, Garcia MAA, Kalashnikov V, Rojas RL (2014) Mathematical analysis for the optimization of a design in a facultative pond: indicator organism and organic matter. Math Probl Eng 1–12
78. McGrath RJ, Mason IG (2004) An observational method for the assessment of biogas production from an anaerobic waste stabilisation pond treating farm dairy wastewater. Biosyst Eng 87:471–478
79. Meneses CGR, Saraiva LB, Mello HND, de Melo JLS, Pearson HW (2005) Variations in BOD, algal biomass and organic matter biodegradation constants in a wind-mixed tropical facultative waste stabilization pond. Water Sci Technol 51:183–190
80. Mona S, Kaushik A, Kaushik CP (2011) Hydrogen production and metal-dye bioremoval by a Nostoc linckia strain isolated from textile mill oxidation pond. Bioresour Technol 102:3200–3205
81. Mtethiwa AH, Munyenyembe A, Jere W, Nyali E (2008) Efficiency of oxidation ponds in wastewater treatment. Int J Environ Res 2:149–152
82. Mungraya AK, Kumar P (2008) Occurrence of anionic surfactants in treated sewage: risk assessment to aquatic environment. J Hazard Mater 160:362–370
83. Munoz R, Guieysse B (2006) Algal-bacterial processes for the treatment of hazardous contaminants: a review. Water Res 40:2799–2815
84. Musikavong C, Wattanachira S (2007) Reduction of dissolved organic matter in terms of DOC, UV-254, SUVA and THMFP in industrial estate wastewater treated by stabilization ponds. Environ Monit Assess 134:489–497
85. Musikavong C, Wattanachira S, Nakajima F, Furumai H (2007) Three dimensional fluorescent spectroscopy analysis for the evaluation of organic matter removal from industrial estate wastewater by stabilization ponds. Water Sci Technol 55:201–210
86. National Institute of Water and Atmospheric Research Ltd (2008) Covered anaerobic ponds for anaerobic digestion and biogas capture piggeries [Online] NIWA Information Series No. 32. Available at http://www.biogas.org.nz/Publications/WhosWho/biogas-pond-booklet.pdf. Accessed 30 Dec 2014
87. Nelson KL, Cisneros BJ, Tchobanoglous G, Darby JL (2004) Sludge accumulation, characteristics and pathogen inactivation in four primary waste stabilization ponds in central Mexico. Water Res 38:111–127
88. Nurdogan Y, Oswald W (1995) Enhanced nutrient removal in high-rate ponds. Water Sci Technol 31:33–43
89. Ogunfowokan AO, Adenuga AA, Torto N, Okoh EK (2008) Heavy metals pollution in a sewage treatment oxidation pond and the receiving stream of the Obafemi Awolowo University, Ile Ife, Nigeria. Enivron Monit Assess 143:25–41
90. Olaniran AO, Babaloa GO, Okoh AI (2001) Aerobic dehalogenation potentials of four bacterial species isolated from soil and sewage sludge. Chemosphere 45:45–50
91. Olukanni DO, Ducoste JJ (2011) Optimization of waste stabilization pond design for developing nations using computational fluid dynamics. Ecol Eng 37:1878–1888
92. Oswald WJ (1968) Advances in anaerobic pond systems design. In: Gloyna EF, Eckenfelder WW Jr (eds) Advances in water quality improvement. University of Texas Press, Austin
93. Oswald WJ, Golueke CG, Tyler RW (1967) Integrated pond systems for subdivisions. J Water Pollut Control Fed 39:1289–1304
94. Park J, Craggs R (2007) Biogas production from anaerobic waste stabilisation ponds treating dairy and piggery wastewater in New Zealand. Water Sci Technol 55:257–264
95. Parker CD (1970) Experiences with anaerobic lagoons in Australia. In: Proceedings of the second international symposium for waste treatment lagoons, Kansas City
96. Parker CD, Jones HL, Greene NC (1959) Performance of large sewage lagoons at Melbourne, Australia. Sewage Ind Wastes 31:133
97. Pearson H, Mara D, Arridge H (1995) The influence of pond geometry and configuration on facultative and maturation waste stabilization pond performance and efficiency. Water Sci Technol 31:129–139
98. Peña MR, Mara DD (2003) High-rate anaerobic pond concept for domestic wastewater treatment: results from pilot scale experience. In: Water and environmental management series (WEMS). IWA Publishing, London. http://www.bvsde.paho.org/bvsacd/agua2003/pilot.pdf. Accessed 18 Oct 2013
99. Pena M, Madera C, Mara D (2002) Feasibility of waste stabilization pond technology for small municipalities in Colombia. Water Sci Technol 45:1–8
100. Pittman JK, Dean AP, Osundeko O (2011) The potential of sustainable algal biofuel production using wastewater resources. Bioresour Technol 102:17–25
101. Quiroga FJ (2011) Waste stabilization ponds for waste water treatment, anaerobic pond.http://home.eng.iastate.edu/~tge/ce421-521/Fernando%20J.%20Trevino%20Quiroga.pdf. Accessed 9 May 2013
102. Rawat I, Kumar RR, Mutanda T, Bux F (2011) Dual role of microalgae: phycoremediation of domestic wastewater and biomass production for sustainable biofuels production. Appl Energy 88:3411–3424
103. Reinoso R, Becares E (2008) Environmental inactivation of *Cryptosporidium parvum* oocysts in waste stabilization ponds. Microb Ecol 56:585–592
104. Safley LM Jr, Westerman PW (1988) Biogas production from anaerobic lagoons. Biol Waste 23:181–193
105. Sah L, Rousseau DPL, Hooijmans CM (2012) Numerical modelling of waste stabilization ponds: where do we stand? Water Air Soil Poll 223:3155–3171
106. Senderovich Y, Gershtein Y, Halewa E, Halpern M (2008) Vibrio cholerae and aeromonas: do they share a mutual host? ISME J 2:276–283
107. Shelef G, Azov Y (2000) Meeting stringent environmental and reuse requirements with an integrated pond system for the twenty-first century. Water Sci Technol 42:299–305

108. Shilton AN, Mara DD, Craggs R, Powell N (2008) Solar-powered aeration and disinfection, anaerobic co-digestion, biological CO_2 scrubbing and biofuel production: the energy and carbon management opportunities of waste stabilisation ponds. Water Sci Technol 58:253–258

109. Shpiner R, Vathi S, Stuckey DC (2007) Treatment of oilfield produced water by waste stabilization ponds. Water Sci Technol 55:265–271

110. Shuval H, Adin A, Fattal B, Rawitz E, Yekutiel P (1986) Integrated resource recovery—wastewater irrigation in developing countries—health effects and technical solutions [Online] World Bank Technical Paper Number 51. The World Bank, Washington DC. http://www-wds.worldbank.org/external/default/WDSContentServer/WDSP/IB/1999/09/17/00178830_98101904164938/Rendered/PDF/multi_page.pdf. Accessed 4 Nov 2013

111. Silva JP, Ruiz JL, Pena MR, Lubberding H, Gijzen H (2012) Influence of photoperiod on carbon dioxide and methane emissions from two pilot-scale stabilization ponds. Water Sci Technol 66:1930–1940

112. Singh A, Olsen SI (2011) A critical review of biochemical conversion, sustainability and life cycle assessment of algal biofuels. Appl Energy 88:3548–3555.http://www.researchgate.net/publication/230238121_A_viable_technology_to_generate_thirdgeneration_biofuel/file/79e415033319adeb75.pdf. Accessed 20 Aug 2013

113. Sinton L, Hall C, Braithwaite R (2007) Sunlight inactivation of *Campylobacter jejuni* and*Salmonella enterica*, compared with *Escherichia coli*, in seawater and river water. J Water Health 5:357–365

114. Spongberg AL, Witter JD, Acuna J, Vargas J, Murillo M, Umana G, Gomez E, Perez G (2011) Reconnaissance of selected PPCP compounds in Costa Rican surface waters. Water Res 45:6709–6717

115. Sukias JPS, Craggs RJ (2011) Digestion of wastewater pond microalgae and potential inhibition by alum and ammoniacal-N. Water Sci Technol 63:835–840

116. Sukias JPS, Tanner CC, Davies-Colley RJ, Nagels JW, Wolters R (2001) Algal abundance, organic matter, and physico-chemical characteristics of dairy farm facultative ponds: implications for treatment performance. NZ J Agric Res 44:279–296

117. Sweeney DG, Nixon JB, Cromar NJ, Fallowfield HJ (2007) Temporal and spatial variation of physical, biological, and chemical parameters in a large waste stabilisation pond, and the implications for WSP modelling. Water Sci Technol 55:1–9

118. Tanner CC, Sukias JPS (2003) Linking pond and wetland treatment: performance of domestic and farm systems in New Zealand. Water Sci Technol 48:331–339

119. Tharavathi N, Hosetti B (2003) Biodiversity of algae and protozoa in a natural waste stabilization pond: a field study. J Environ Biol 24:193–199

120. Thirumurthi D (1974) Design criteria for waste stabilization ponds. J Water Poll Control Fed 2094–2106. http://www.jstor.org/stable/25038241. Accessed 15 Oct 2013

121. Tsalkatidou M, Gratziou M, Kotsovinos N (2013) Septage treatment using a combined waste stabilization ponds-vertical flow constructed wetland system. Desalin Water Treat 51:3011–3017

122. United States Environmental Protection Agency (1983) Design manual: municipal waste stabilization ponds. EPA-625-1-83-0115. Office of Water, Cincinnati. http://nepis.epa.gov/Exe/ZyNET.exe/300044QA.TXT?ZyActionD=ZyDocument&Client=EPA&Index=1981+Thru+1985&Docs=&Query=&Time=&EndTime=&SearchMethod=1&TocRestrict=n&Toc=&TocEntry=&QField=&QFieldYear=&QFieldMonth=&QFieldDay=&IntQFieldOp=0&ExtQFieldOp=0&XmlQuery=&File=D%3A\zyfiles\Index%20Data\81thru85\Txt\00000001\300044QA.txt&User=ANONYMOUS&Password=anonymous&SortMethod=h|-&MaximumDocuments=1&FuzzyDegree=0&ImageQuality=r75g8/r75g8/x150y150g16/i425&Display=p|f&DefSeekPage=x&SearchBack=ZyActionL&Back=ZyActionS&BackDesc=Results%20page&MaximumPages=1&ZyEntry=1&SeekPage=x&ZyPURL. Accessed 18 Oct 2013

123. United States Environmental Protection Agency (2002) Onsite wastewater treatment systems Manual. EPA- 625-R-008. Office of Water, Washington DC.http://water.epa.gov/aboutow/owm/upload/2004_07_07_septics_septic_2002_osdm_all.pdf. Accessed 16 Apr 2015

124. United States Environmental Protection Agency (2011) Principles of design and operations of wastewater treatment pond systems for plant operators, engineers, and managers. EPA 600-R-11-088. Office of Research and Development, Cincinnati.http://www.epa.gov/ordntrnt/ORD/NRMRL/lrpcd/projects/ponds.htm. Accessed 18 Oct 2013

125. Veeresh M, Veeresh AV, Huddar BD, Hosetti BB (2010) Dynamics of industrial waste stabilization pond treatment process. Environ Monit Assess 169:55–65

126. Varon MP, Mara D (2004) Waste stablization ponds. IRC International Water and Sanitation Centre, Leeds, UK. http://www.bvsde.paho.org/bvsacd/cd27/ponds.pdf. Accessed 16 Apr 2015

127. Water Environmental Federation (2010) "Fact sheet T5: performance and cost of decentralized unit processes treatment series: lagoons". Water Environmental Federation, Alexandria.http://www.werf.org/c/DecentralizedCost/T5_Lagoons.aspx. Accessed 18 Oct 2013

128. Wood M, Greenfield P, Howes T, Johns M, Keller J (1995) Computational fluid dynamic modeling of waste-water ponds to improve design. Water Sci Technol 31:111–118

129. Wood M, Howes T, Keller J, Johns M (1998) Two dimensional computational fluid dynamic models for waste stabilisation ponds. Water Res 32:958–963

130. World Health Organization (2006) Guidelines for the safe use of wastewater, excreta and greywater. http://www.who.int/water_sanitation_health/wastewater/gsuww/en/index.html. Accessed 3 Nov 2013

131. Yeh TY, Pan CT, Ke TY, Kuo TW (2010) Organic matter and nitrogen removal within field-scale constructed wetlands: reduction performance and microbial identification studies. Water Environ Res 82:27–33

CHAPTER 7

RESEARCH TRENDS IN ELECTROCHEMICAL TECHNOLOGY FOR WATER AND WASTEWATER TREATMENT

Tianlong Zheng[1] Juan Wang[1] Qunhui Wang[1] Huimin Meng[2] Lihong Wang[1,3]

[1] Department of Environmental Engineering, University of Science and Technology Beijing, 30 Xueyuan Road, Haidian District, Beijing 100083, China

[2] Laboratory for Corrosion-Erosion and Surface Technology, Corrosion and Protection Center, University of Science and Technology Beijing, 30 Xueyuan Road, Haidian District, Beijing 100083, China

[3] Department of Architectural Engineering, Handan Polytechnic College, 141 Zhuhe Road, Handan 056001, Hebei, China

ABSTRACT

It is difficult to completely degrade wastewater containing refractory pollutants without secondary pollution by biological treatment, as well as physical–chemical process. Therefore, electrochemical technology has attracted much attention for its environmental compatibility, high removal efficiency, and potential cost effectiveness, especially on the industrial wastewater treatment. An effective bibliometric analysis based on the Science Citation Index Core Collection database was conducted to evaluate electrochemical technology for water and wastewater treatment related research from 1994 to 2013. The amount of publications significantly increased in the last two decades. Journal of the Electrochemical Society published the most articles in this field with a top h-index of 90, taking 5.8 % of all, followed by Electrochimica Acta and Journal of Electroanalytical Chemistry. The researchers focused on categories of chemistry, electrochemistry, and materials science. China and Chinese Academy of Sciences were the most productive country and institution, respectively, while the USA, with the most international collaborative articles and highest h-index of 130, was the major collaborator with 15 other countries in top 20 most productive countries. Moreover, based on the analysis of author keywords, title, abstract, and 'KeyWords Plus', a new method named "word cluster analysis" was successfully applied to trace the research hotspot. Nowadays, researchers mainly focused on novel anodic electrode, especially on its physiochemical and electrochemical properties.

INTRODUCTION

With the world's rapid population growth, the water pollution has become a serious global problem for the indiscriminate discharge of wastewater. Therefore, more efficient and cost-saving wastewater treatment technologies are vital to satisfy the ever-increasing requirements of human beings. Generally, wastewater treatment is carried out by the conventional wastewater treatment process (Grady Jr et al. 2011; Afzal et al. 2011), such as, physical–chemical methods and biological methods. However, it is difficult to degrade the complex refractory organic pollutant in the wastewater by biological methods. In addition, physical–chemical methods are not always effective due to the unreacted chemicals and sludge treatment (Feng et al. 2003; Afzal et al. 2011). Nowadays, electrochemical technology, which was applied to water and wastewater treatment, has attracted great attention for its environmental

Citation: Tianlong Zheng Juan Wang Qunhui Wang Huimin Meng Lihong Wang, Research trends in electrochemical technology for water and wastewater treatment, DOI 10.1007/s13201-015-0280-4

compatibility, high removal efficiency, and potential cost effectiveness. Electrochemical technology including electrocoagulation, electrodeposition, electrooxidation, electrodisinfection, electrofenton, electroflotation, and electrosorption, may be one of the most effective technologies. Among them, electrodisinfection process (Ghernaout and Ghernaout 2010), which is extremely popular in recent years, has great potential to eliminate and disinfect drinking water, swimming pool water and industrial cooling water. Compared with electrodisinfection, electrofenton was widely applied in the treatment of wastewater containing toxic and non-biodegradable materials (Jiang and Zhang 2007). In addition, electroflotation technology was used in the elimination of flocs and suspended sludge of wastewater (Chen et al. 2002; Ge et al. 2004), while electrosorption technology effectively removed polar molecules or ions from wastewater (Huang and Su 2010). However, electrocoagulation, electrodeposition and electrooxidation still were the uppermost electrochemical technologies during water and wastewater treatment in the history. Electrooxidation, which was extensively investigated since the late 1970s (Mayeda et al. 1972; Nilsson et al. 1973), had been used worldwide in various refractory wastewater treatments, such as textile wastewater (Andrade et al. 2007), olive mill wastewater (Un et al. 2008), distillery industry wastewater (Piya-Areetham et al. 2006), dye-containing wastewater (Valero et al.2010), brown-colored molasses wastewater (Canizares et al. 2008), and coking wastewater (Zhu et al. 2009). Recently, researchers have paid more concern on anodic material of electrodes and its electrochemical performance and degradation mechanism (Marselli et al. 2003; Chen 2004; Chang et al. 2009). The common electrodes (Gao et al. 1994; Chen 2004; Martinez-Huitle and Ferro 2006; Anglada et al. 2009; Zhu et al. 2009), such as, graphite, Pt, PbO_2, SnO_2, Ti/Pt, Ti/SnO_2, Ti/PbO_2, Ti/RuO_2–TiO_2, Ti/IrO_2–TiO_2, Ti/IrO_2–RuO_2–TiO_2, boron-doped diamond electrode (BDD), Titanium-based boron-doped diamond electrode (BDD/Ti) are usually found in the wastewater treatment field. Moreover, novel anodic material will be the future frontier of electrochemical technology for water and wastewater treatment related research.

Bibliometrics, which gained insight into the development trends of various research fields through a quantitative analysis and statistics, firstly introduced by Pritchard in 1969 (Pritchard1969). Science Citation Index Expanded™, accessed via Web of Science™ Core Collection, was the most important and frequently used source database for an overview of scientific production. Bibliometric methods have been widely applied to assess the scientific outputs of various fields, especially on environmental concerns and energy utilization frontier, such as drinking water (Hu et al. 2010), nitrate removal (Huang et al. 2012), sediment (Niu et al. 2014), solid waste (Fu et al. 2010), volatile organic compounds (Zhang et al. 2010a), and materials of lithium-based batteries (Wagner et al. 2013). In addition, research patterns including journals, categories, countries, and institutes, as well as the international cooperation had been broadly applied (Fu et al. 2010, 2014; Niu et al. 2014; Tan et al. 2014). Moreover, h-index, namely the number h of publications with at least h citations, which was introduced by Hirsch in 2005 (2005), had become a valid indicator on scientific achievement (Hirsch 2007; Ciriminna and Pagliaro 2013; Tan et al. 2014). In addition, the analysis of author keywords, title, abstract, and 'KeyWords Plus' in different periods also had been used to evaluate the research trend in recent years (Xie et al. 2008; Fu et al. 2010, 2014; Wang et al. 2013).

In this study, a comprehensive bibliometric analysis was used to trace and obtain the global trend on electrochemical technology for water and wastewater treatment research from 1994 to 2013. Document types, languages, categories, journals, countries/territories, institutions, and h-index were analyzed to dissect the publication pattern. Further, research tendency and hotspot including the distribution of author keywords, title, abstract, and 'KeyWords Plus' in different periods, and hot issues were also applied. Besides, an effective method named "word cluster analysis" (Mao et al. 2010) was applied to trace the hotspot of related research field in previous work. The results could provide a basis for comprehensive understanding of global development and serve as a potential guide for future direction of electrochemical technology for water and wastewater treatment related research.

Data Sources and Methodology

One common method of bibliometric research used to trace publications was using the SCI of the Institute for Scientific Information (ISI). The original information sources were based on the database of the Web of Science Core Collection and the Journal Citation Report (JCR) of 2012 by Thomson Reuters, where we extracted information about scientific output on 7 April, 2014. According to Journal Citation Reports (JCR) of 2012, it indexes 8471 journals with citation references across 176 scientific disciplines. Since 1991, abstract information has been included in the SCI. The topic search can trace the related information in the title, abstract, and keywords at one time. (electrochemi* or electro-chemi*) and [wastewater or (waste water) or waste-water or water] were used as a search phrase to search topics in SCI for the period from 1994 to 2013. Aspects referring to document type, language, output, subject category, journal, country, institute, source title, abstract keyword, and h-index were all analyzed by Microsoft Excel 2010. The contribution of different countries and institutes were estimated by the location of the affiliation of at least one author of the published papers. Articles that originated from Hong Kong were included under the China heading for analysis. Articles that originated from England, Scotland, Northern Ireland, and Wales were grouped under the UK heading. "Internationally collaborative publication" was designated to those articles that were coauthored by researchers from more than one country, where "independent type" was assigned if the researchers' addresses were from the same country. The term "single institute publication" was assigned if the researchers' addresses were from the same institute. The term "inter-institutionally collaborative publication" was assigned if the authors were from different institutes. h-index was defined by the h of Np papers having at least h citations each and the other (Np-h) papers have $\leq$h citations each. The reported IF of each journal was obtained from the 2012 JCR Science Edition.

RESULTS AND DISCUSSION.

Publication pattern

Document Type and Language of Publication

The 29,144 publications related to the electrochemical technology for water and wastewater treatment research identified by the ISI between 1994 and 2013 included 8 document types. Articles were the dominant document type comprising 90 % or 26,330 of the total production. The remaining publications were proceedings papers (1557), review (1048), meeting abstract (77), letter (57), editorial material (25), notes (23), book chapter (15), correction (9), and reprint (3). On average, there were 17 articles per proceedings papers and 25 articles per review.

As articles (26330) were the dominant type of publications, these were analyzed in the following study. For language analysis, 25,551 articles (97 %) were published in English, followed by Chinese (439), Japanese (88), German (43), French (39), Portuguese (36), Spanish (32), Rumanian (31), Russian (21), Polish (21), Serbo-Croatian (9), Hungarian (6), Czech (4), Korean (4), Italian (3), and only one article each in Turkish, Ukrainian, and Malay, respectively. Obviously, English was by far the dominant language in the journals listed in SCI.

Characteristics of Publication Outputs

As there are no abstracts in the publications of the SCI before 1991, (electrochemi* or electro-chemi*) and (wastewater or (waste water) or waste-water or water) were used as a search phrase in the title only to obtain a general long-term trend of electrochemical technology for water and wastewater treatment. The earliest research Web of Science™ Core Collection on electrochemical technology for water and wastewater treatment related was published in 1911. Figure 1 shows that the amount of electrochemical technology for water and wastewater treatment research has slowly grown in the long period from 1911 to 1990. The amount of publications goes up significantly in 1991 and rockets during the period since 2013. The electrochemical technology for water and wastewater treatment research has become a hot topic worldwide, because of high rate increase of related literatures and human beings urgent necessary for energy and environment.

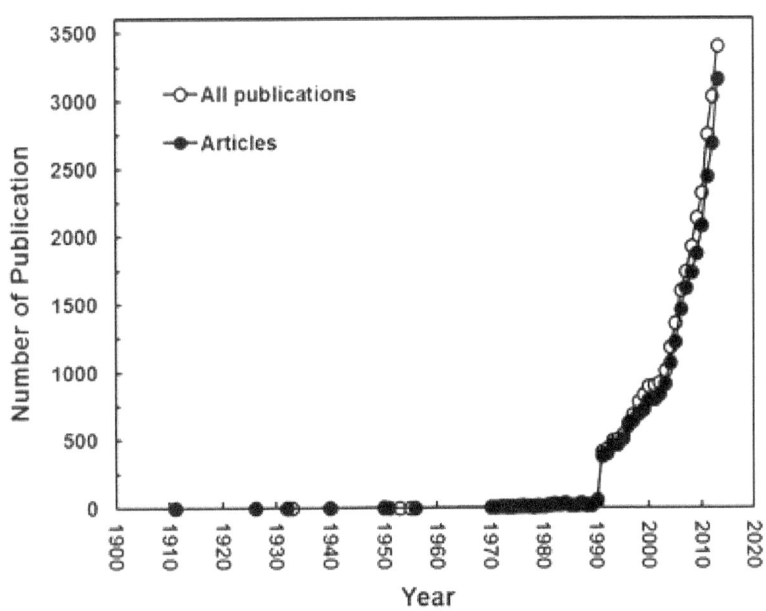

Figure. 1: Number of SCI publications on electrochemical technology for water and wastewater treatment research during the last 110 years.

During the study period, the cumulative number of articles grew from 461 in 1994 to 26,330 in 2013. A power model and an exponential model described the relationship between the annual cumulative number of articles and the year published for the two periods from 1994 to 2003 and 2003 to 2013, respectively (Fig. 2). A significant correlation between the number of articles and the year was observed with a high coefficient of determination ($R^2 > 0.998$). Both the power and the exponential curve fitting indicated that there was a high growth rate for annual articles. The power and exponential fitting curves for research of electrochemical technology for water and wastewater treatment were found to be: $C_1 = 438.17Y^{1.19}$ ($R^2 = 0.999$) and $C_2 = \exp(0.13Y + 7.55)$ ($R^2 = 0.9989$), respectively, where C is the cumulative number of articles and Y is the number of years since 1994. Based on the exponential model during 2003–2013, the number of articles in 2018 (6406) is predicted to be twice more than that in 2013 (3160).

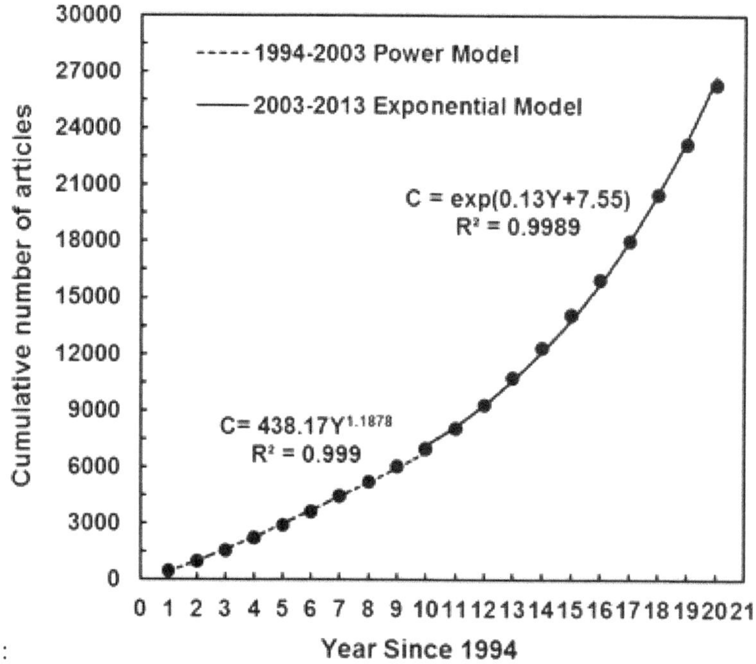

Figure. 2: The relationship between cumulative number of publications and published year.

The number of authors per article increased from 3.3 in 1994 to 4.8 in 2013 and the increasement was 47 %, with a similar increase in the number of references cited per article and the number of institutions per article, and the corresponding increasement was 43 and 33.5 %. The average article length fluctuated slightly, with an overall average length of 7.9 pages. The number of countries per article was stable at 1.2 countries. Of the 26,330 articles, 183 (0.7 %) articles did not include abstracts. Of the 183 articles without abstracts, 71 (38.8 %) articles were published during 2010–2013. Similar to the analysis of articles abstract, 24 (0.1 %) of 26,330 articles did not cite references, and the most common number of references per article was 27 and 30 which were found in 751 articles (2.9 %). Of the 24 articles without references, 9 (38 %) articles were published in 1994 and 1995.

Articles (26,330) were analyzed, including 985 (3.7 %) articles without reprint author information. The most frequent number of authors was three, accounting for 5829 articles (22.1 %). Two to five authors accounted for 19811 articles (75.2 %) and 848 articles (3.2 %) had one author, and the maximum number of authors per article was 31 [which was published on corrosion science named "effect of the marine environment on reinforced concrete durability in Iberoamerican countries: DURACON project/CYTED" (Trocónis De Rincón et al. 2007)]. The 26,330 articles contained a total of 208,848 pages, and the maximum number of pages per article was 98 [which was a review of the research literature published in 2010 dealing with physico-chemical processes used for water and wastewater treatment (Afzal et al. 2011)], and the minimum number was one. Six to eight pages (12,212; 46.4 %) were the most frequent number of pages (Table 1).

Table 1: Characteristics by year of publication outputs from 1994 to 2013

Year	TP	AU	AU/TP	NR	NR/TP	PG	PG/TP	NI	NI/TP	NC	NC/TP
1994	461	1509	3.3	12,931	28.0	3641	7.9	637	1.4	506	1.1
1995	510	1681	3.3	14,613	28.7	4127	8.1	721	1.4	585	1.1
1996	601	2064	3.4	16,493	27.4	4664	7.8	846	1.4	689	1.1
1997	644	2225	3.5	17,973	27.9	5187	8.1	939	1.5	741	1.2
1998	690	2504	3.6	19,412	28.1	5701	8.3	1062	1.5	802	1.2
1999	726	2681	3.7	19,551	26.9	5865	8.1	1057	1.5	830	1.1
2000	796	2938	3.7	22,583	28.4	6337	8.0	1246	1.6	946	1.2
2001	801	3042	3.8	22,052	27.5	6308	7.9	1256	1.6	966	1.2
2002	838	3250	3.9	24,562	29.3	6803	8.1	1349	1.6	999	1.2
2003	917	3450	3.8	25,854	28.2	7264	7.9	1457	1.6	1098	1.2
2004	1070	4231	4.0	32,765	30.6	8614	8.1	1711	1.6	1268	1.2
2005	1221	4944	4.0	37,608	30.8	9684	7.9	1986	1.6	1475	1.2

2006	1462	5950	4.1	44,592	30.5	11,633	8.0	2426	1.7	1765	1.2
2007	1620	6759	4.2	51,964	32.1	12,587	7.8	2646	1.6	1927	1.2
2008	1733	7349	4.2	57,299	33.1	13,201	7.6	2964	1.7	2091	1.2
2009	1875	8062	4.3	63,351	33.8	14,305	7.6	3230	1.7	2248	1.2
2010	2080	9203	4.4	76,224	36.6	15,802	7.6	3588	1.7	2522	1.2
2011	2440	11,123	4.6	89,474	36.7	19,180	7.9	4307	1.8	2995	1.2
2012	2685	12,574	4.7	104,263	38.8	21,901	8.2	4758	1.8	3295	1.2
2013	3160	15,213	4.8	126,336	40.0	26,044	8.2	5831	1.8	3973	1.3
Total	26,330	110,752		879,900		208,848		44,017		31,721	
Average			4.0		31.2		7.9		1.6		1.2

TP number of articles, *AU* number of authors, *NR* cited reference count, *PG* page count, *NI*institution count, *NC* country count, *AU/TP*, *NR/P*, *PG/TP*, *NI/TP*, *NC/TP* average number of authors, pages, references, institutions, countries, per articles

Output in Subject Categories and Journals

Articles (26,330) were published in 1790 journals in 106 subject categories in SCI. Among these journals, 1432 (80 %) journals contained less than 10 articles. Top 20 most productive journals (TP > 240) and their ISI category of journals, and the position of the journal in its category, IF in 2012, h-index and Journal Country are shown in Table 2.

Table 2: Top 20 most productive journals (1994–2013) with the number of papers, ISI category of journals, and the position of the journal in its category, IF, h-index, and Journal Country

Journal	TP (%)	ISI category and position	IF (*R*)	h-index (*R*)	Journal Country
Journal of the Electrochemical Society	1540 (5.8)	Electrochemistry (11/26) Materials Science, Coatings and Films (1/17)	2.588 (18)	90 (1)	USA
Electrochimica Acta	1334 (5.1)	Electrochemistry (6/26)	3.777 (10)	66 (2)	UK (England)
Journal of Electroanalytical Chemistry	801 (3.0)	Analytical Chemistry (23/75) Electrochemistry (10/26)	2.672 (17)	53 (7)	Switzerland
Journal of Power Sources	554 (2.1)	Electrochemistry (3/26) Energy and Fuels (9/81)	4.675 (4)	55 (6)	Netherlands
International Journal of Hydrogen Energy	472 (1.8)	Physical Chemistry (38/135) Electrochemistry (7/26) Energy and Fuels (16/81)	3.548 (13)	37 (17)	UK (England)
Electroanalysis	458 (1.7)	Electrochemistry (7/26)	2.817 (16)	33 (19)	Germany
Langmuir	391 (1.5)	Multidisciplinary Chemistry (31/152) Physical Chemistry (29/135) Multidisciplinary Materials Science (31/241)	4.187 (8)	49 (8)	USA
Journal of Applied Electrochemistry	377 (1.4)	Electrochemistry (16/26)	1.836 (20)	39 (15)	Netherlands
Journal of Physical Chemistry C	376 (1.4)	Physical Chemistry (28/135) Multidisciplinary Materials Science (27/241) Nanoscience and Nanotechnology (19/69)	4.814 (3)	45 (11)	USA
Journal of Hazardous Materials	359 (1.4)	Civil Engineering (2/122) Environmental Engineering (5/42) Environmental Sciences (16/210)	3.925 (9)	45 (12)	Netherlands
Analytica Chimica Acta	342 (1.3)	Analytical Chemistry (7/75)	4.387 (7)	48 (10)	Netherlands

Electrochemistry Communications	339 (1.3)	Electrochemistry (4/26)	4.425 (6)	44 (13)	USA
Journal of Physical Chemistry B	330 (1.3)	Physical Chemistry (35/135)	3.607 (12)	56 (5)	USA
Analytical Chemistry	327 (1.2)	Analytical Chemistry (3/75)	5.695 (2)	58 (4)	USA
Corrosion Science	316 (1.2)	Multidisciplinary Materials Science (37/241) Metallurgy and Metallurgical Engineering (2/76)	3.615 (11)	38 (16)	USA
International Journal of Electrochemical Science	307 (1.2)	Physical, Chemical and Earth Sciences (/)	–	13 (21)	Serbia
Talanta	276 (1.0)	Analytical Chemistry (12/75)	3.498 (15)	40 (14)	UK (England)
Journal of the American Chemical Society	254 (1.0)	Multidisciplinary Chemistry (11/152)	10.677 (1)	66 (3)	USA
Inorganic Chemistry	251 (1.0)	Inorganic and Nuclear Chemistry (5/44)	4.593 (5)	49 (9)	USA
Sensors and Actuators B-Chemical	249 (0.9)	Analytical Chemistry (11/75) Electrochemistry (8/26) Instruments and Instrumentation (2/57)	3.535 (14)	35 (18)	Switzerland
Electrochemical and Solid State Letters	249 (0.9)	Electrochemistry (14/26) Multidisciplinary Materials Science (64/241)	2.01 (19)	31 (20)	USA

TP number of articles, *IF* impact factor, *R* the rank during top 20 most productive journals

Both IF and h-index were used to evaluate the articles in a journal. There was a high concentration of electrochemical technology for water and wastewater treatment research publications in these top journals. Table 2 shows the top 20 most productive journals, accounting for appropriately 38 % of the total articles. *Journal of the Electrochemical Society*with the highest h-index of 90, published the most articles with 1540 articles (5.8 %), followed by *Electrochimica Acta* (1334; 5.1 %), *Journal of Electroanalytical Chemistry* (801; 3.0 %), and *Journal of Power Sources* (554; 2.1 %).The percentage of the top productive journal was not high, which indicates the breadth of article distribution as well as the broad interest in electrochemical technology for water and wastewater treatment research from various research angles. This phenomenon also appears in other environment related research areas, such as Water research (4.2 %) in drinking water (Fu et al. 2013), Water research (2.2 %) in nitrate removal research (Huang et al. 2012), and Marine Pollution Bulletin (8.4 %) in estuary pollution research (Sun et al. 2012a). *Journal of the Electrochemical Society*, which remained one of the most-highly cited journals in electrochemistry since 1902, ranked on top in the category of Materials Science, Coatings and Films (17) with an impact factor of 2.588. It is worth mentioning that *International Journal of Electrochemical Science* from Serbia with a high IF (3.729) in 2011 and the least h-index (13) in 2012 according to JCR, which published tenth most articles with 116 (1.5 %), was suppressed by JCR 2012 because of anomalous citation. The relationship between h-index, IF and the rank order of the 20 most productive journals revealed extremely different trend. For example, *Journal of Applied Electrochemistry* ranked 8th in the top 20 of most productive journals, while IF (1.836) and h-index (39) ranked 20th and 15th, respectively; on the contrary, *Journal of the American Chemical Society* ranked 18th in top 20 of most productive journals, when IF (10.677) and h-index (66) ranked 1st and 3rd. This result might attribute to the large number of subject categories in our current research. Among the top 20 journals, ten journals are from USA, four from Netherlands, three from the UK (England), two from Switzerland, one from Germany and Serbia, which revealed the dominated status of developed countries in journal country.

Due to the assignment of journals to multiple subject categories, electrochemical technology for water and wastewater treatment research covered 106 Web of Science categories. As illustrated in Fig. 3, based on the continuous increase in the number of articles of each category, electrochemical technology for water and wastewater treatment research in the six subject categories had steadily increased especially at the beginning of the 21st century. The most common subject category was chemistry, followed distantly by other categories. The other two relatively popular subject categories belonged to the second echelon were electrochemistry and materials science, which had kept a sound momentum. Moreover, the research trend in these two fields was the hotspots in the current study.

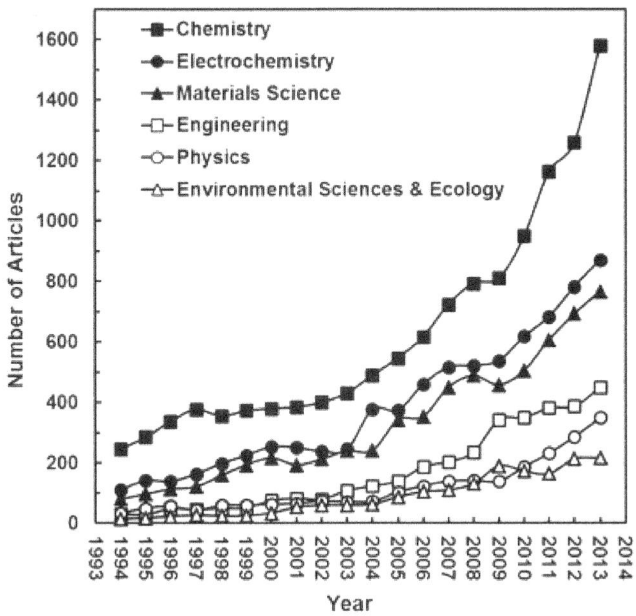

Figure. 3: The growth trends of the top 6 subject categories.

Of the 1790 journals publishing the 26,330 articles, 705 (39.4 %) journals contained 1 article, 272 (15.2 %) journals contained two articles, 145 (8.1 %) journals contained 3 articles and 82 (4.6 %) journals contained four articles. Approximately, 5159 (19.6 %) articles were published in 6 core journals, and for comparison, the trends in the six journals with the greatest number of articles are shown in Fig. 4. The number of articles in *Journal of the Electrochemical Society*and *Electrochimica Acta* had slowly increased from 1993 to 1998, and significantly went up in general since 1998, and then fluctuated in recent years. The other four journals concluding*Journal of Electroanalytical Chemistry, Journal of Power Sources, International Journal of Hydrogen Energy, and Electroanalysis,* which revealed the same growth trend in the research period, rose undulate from 1993 to 2011 and showed a promising development since 2011.

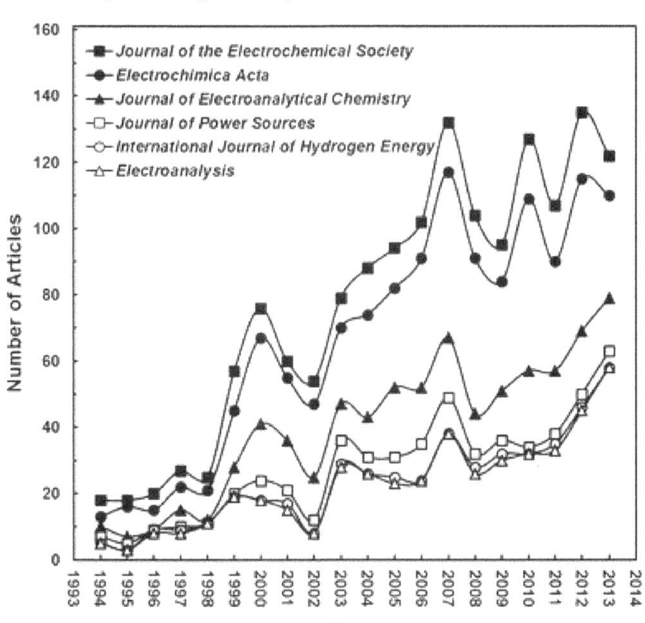

Figure. 4: The growth trends of the top 6 journals.

Publication Performances: Institutes and Countries

The contribution of different countries and territories were estimated by the affiliation of at least one author to the publications. Moreover, the publication, number of the countries and territories could reveal the research performance in the related field. Excluding 34 articles without any author address information on the ISI Web of Science, of the 26,296 articles with author address information, published from 1994 to 2013 were analyses. This covered 118 countries/territories, of which 21,556 (82.0 %) were independent articles covering 100 countries/territories and 4740 (18.0 %) were international collaborative articles covering 114 countries/territories. Table 3 shows the top 20 most productive countries for total articles. Number of total, single-country, internationally collaborated articles, the percentage of single-country articles in total articles, the percentage of international collaborative articles in total articles, and the h-index of countries/territories were exhibited in Table 3. The number of countries involved in electrochemical technology for water and wastewater treatment research increased rapidly. Nine European countries, six Asian countries/territories, four American countries, and one Oceania country were ranked in the top 20 productive countries of articles. The top 20 most productive countries included 80.7 % of the total number of articles. The seven major industrialized countries (G7: Canada, France, Germany, Italy, Japan, the UK, and USA), ranked in the top 11, had 40.7 % over the investigation period. The result indicated that the developed countries hold a dominant position in electrochemical technology for water and wastewater treatment research as this pattern has occurred in most scientific fields (Zhang et al.2010b; Fu et al. 2010). Moreover, some developing countries, such as BRIC (China, India, Brazil, and Russia) were also listed as productive countries with 7441 or 23.2 % of total articles.

China dominated, ranking first in total articles (4939), independent articles (4231), first authored articles (4640) and corresponding authored articles (4561), and had 2nd h-index of 83. However, China was ranked 19th of the 20 countries on percentage of collaboration with outside authors, while the UK ranked 9th in the number of articles, that had the highest proportion of collaborative articles to total articles. Simultaneously, the UK was ranked 1st with nearly 64 % of international collaborative articles in its total articles. The proportions of internationally collaborated articles to the total outputs of these 20 countries showed significant disparity. Yet, USA with the highest h-index of 130 not only predominantly completed the most collaborative articles (1286), but also was the major collaborator of 15 other countries (Table 3), which indicated that USA still had unparalleled strength and leadership in the scientific research. In addition, although single-country articles still dominated in the electrochemical technology for water and wastewater treatment research, based on intensifying academic exchanges in the electrochemical technology for water and wastewater treatment study area, international collaboration was popular worldwide in recent years, which had also been a general trend in other fields (Fu et al. 2010; Niu et al. 2014). In the current study, the relationship between h-index and the rank order of the 20 most productive countries of articles revealed almost similar trends.

Table 3: Top 20 most productive countries of articles during 1994–2013

Country/territories	TP	*R*(%)	Single-Country		Internationally-collaborated			h-index [*R*]
			SP	%	CA	% [*R*]	MC [*P*]	
China	4939	1 (18.8)	4231	85.67	708	14.33 (19)	USA (204)	83 (2)
USA	4481	2 (17.0)	3195	71.30	1286	28.70 (12)	China (204)	130 (1)
Japan	2180	3 (8.3)	1745	80.05	435	19.95 (16)	China (115)	73 (3)
France	1663	4 (6.3)	927	55.74	736	44.26 (4)	USA (77)	71 (5)
Germany	1392	5 (5.3)	715	51.36	677	48.64 (2)	USA (104)	73 (3)
Spain	1309	6 (5.0)	831	63.48	478	36.52 (7)	USA (58)	71 (5)
India	1267	7 (4.8)	1004	79.24	263	20.76 (15)	USA (38)	51 (11)
UK	1141	8 (4.3)	621	54.43	520	45.57 (3)	USA (74)	64 (8)
South Korea	1082	9 (4.1)	787	72.74	295	27.26 (13)	USA (132)	50 (12)
Italy	937	10 (3.6)	587	62.65	350	37.35 (6)	USA (64)	70 (7)
Canada	879	11 (3.3)	610	69.40	269	30.6 (11)	USA (79)	63 (9)
Brazil	700	12 (2.7)	537	76.71	163	23.29 (14)	USA (27)	42 (14)
Taiwan	659	13 (2.5)	562	85.28	97	14.72 (18)	USA (33)	41 (15)
Iran	564	14 (2.1)	505	89.54	59	10.46 (20)	USA (13)	38 (17)
Russia	535	15 (2.0)	352	65.79	183	34.21 (10)	Germany (34)	32 (19)
Australia	518	16 (2.0)	296	57.14	222	42.86 (5)	China (61)	46 (13)
Poland	487	17 (1.9)	319	65.50	168	34.5 (9)	USA (32)	34 (18)
Turkey	437	18 (1.7)	350	80.09	87	19.91 (17)	USA (26)	40 (16)
Switzerland	393	19 (1.5)	185	47.07	208	52.93 (1)	Germany (42)	62 (10)
Mexico	361	20 (1.4)	235	65.10	126	34.9 (8)	USA (42)	30 (20)

TP articles in the study period, *R* (%) the rank of total articles, *SP* single-country articles, *CP*internationally collaborated articles, *MC* [*P*] major collaborator (the number of collaborated articles between two countries)

USA had the highest h-index of 130 and six countries had an h-index of 70–85. It is notable that Switzerland was ranked 19th in the number of articles but unexpectedly 10th in h-index and India was ranked 7th in the number of articles but only 11th in h-index. Figure 5 showed that China continuously ranked 1st in the number of articles per year since 1994. Moreover, China still had the highest growth rate and energy in recent years. The number of electrochemical technology for water and wastewater treatment articles in China slowly increased from 1994 to 2004 and rocketed during the period from 2005 to 2013. With a similar develop trends, the numbers of ariticles in USA was also close to China. The other countries followed distantly and showed a similar increasing curve during the research period.

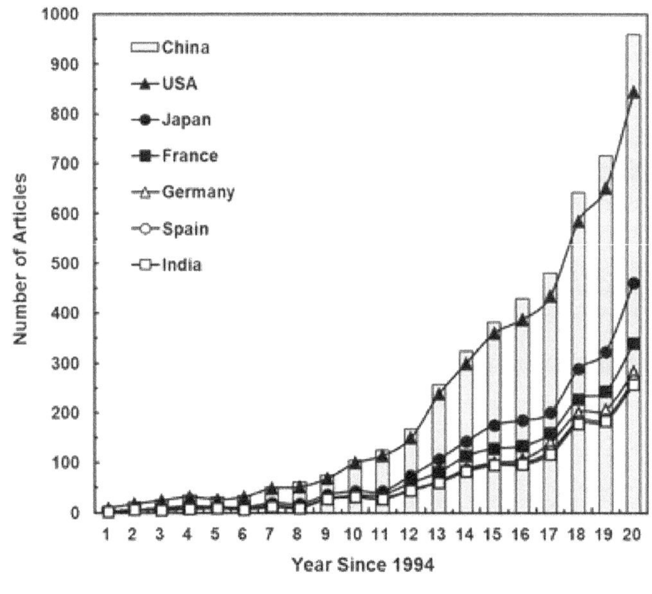

Figure. 5: he growth trends of the 7 most productive countries (TP > 1200)

At the institution level, 26,296 articles from 8752 institutes in 118 countries, 12,327 (46.9 %) were inter-institutionally collaborative publications, and 13,969 (53.1 %) were independent publications. The percentage of collaboration between institutes was much higher than that between countries (28.9 %). Six institutes in China, four in Japan, two each in USA and France, and one each in Switzerland, Brazil, UK, India, Spain, and Russia, all of which were ranked in the top 20 most productive institutes as shown in Table 4.

Table 4: Top 20 most productive institutions of articles during 1994–2013

Institution	TP	TP*R*(%)	Single-institution		Inner-institution collaborated		
			SP	%	CP	%	MC [*P*]
Chinese Acad Sci, China	796	1 (3.0)	342	42.96	454	57.04 (1)	Zhejiang Univ, China (22)
Univ Sao Paulo, Brazil	248	2 (0.9)	112	45.16	136	17.09 (3)	Univ Fed ABC, Brazil (22)
Russian Acad Sci, Russia	248	2 (0.9)	112	45.16	136	17.09 (4)	Moscow Mv Lomonosov State Univ, Russia (19)
Penn State Univ, USA	203	4 (0.8)	107	52.71	96	12.06 (10)	Sandia Natl Labs, USA (11)
Nanjing Univ, China	198	5 (0.8)	102	51.52	96	12.06 (11)	City Univ Hong Kong, China (8)
Univ Paris 06, France	175	6 (0.7)	43	24.57	132	16.58 (5)	CNRS, France (19)
CNRS, France	175	6 (0.7)	27	15.43	148	18.59 (2)	Univ Paris 06, France (19)
Zhejiang Univ, China	171	8 (0.7)	81	47.37	90	11.31 (14)	Chinese Acad Sci, China (22)
Wuhan Univ, China	169	9 (0.6)	72	42.60	97	12.19 (9)	Chinese Acad Sci, China (17)
Univ Oxford, UK	164	10 (0.6)	55	33.54	109	13.69 (7)	Queens Univ Belfast, UK (20)
Univ Barcelona, Spain	162	11 (0.6)	59	36.42	103	12.94 (8)	Univ Autonoma Barcelona, Spain (8)
Tokyo Inst Technol, Japan	160	12 (0.6)	86	53.75	74	9.3 (18)	Tokyo Inst Technol, Japan (9)
Univ Tokyo, Japan	156	13 (0.6)	42	26.92	114	14.32 (6)	Univ Tokyo, Japan (9)
Indian Inst Technol, India	148	14 (0.6)	87	58.78	61	7.66 (19)	Bhabha Atom Res Ctr, India (7)
Ecole Polytech Fed Lausanne, Switzerland	141	15 (0.5)	48	34.04	93	11.68 (12)	Univ Lausanne, Switzerland (10)
Univ Calif Berkeley, USA	136	16 (0.5)	59	43.38	77	9.67 (16)	Caltech, USA (5)
Dalian Univ Technol, China	135	17 (0.5)	58	42.96	77	9.67 (17)	Hong Kong Univ Sci and Technol, China (9)
Tohoku Univ, Japan	132	18 (0.5)	41	31.06	91	11.43 (13)	Hitachi Ltd, Japan (11)
Kyoto Univ, Japan	122	19 (0.5)	67	54.92	55	6.91 (20)	Ryukoku Univ, Japan (5)
Tsinghua Univ, China	121	20 (0.5)	41	33.88	80	10.05 (15)	Chinese Acad Sci, China (9)

TP articles in the study period, *R* (%) the rank of total articles, *SP* Single-institution articles, *CP* Inner-institution collaborated articles, *MC* [*P*] major collaborator (the number of collaborated articles between two institutions)

It is worth noting that the Chinese Academy of Sciences with a highest h-index of 55 was the most productive institute for the total articles (796), as well as single institute articles (342), first authored articles (542), and corresponding authored articles (528), while the inter-institutionally collaborative articles (454) ranked 1st. There is a bias for the Chinese Academy of Sciences, as it has many branches in many cities. In this study, the articles of these institutes were regarded as one heading, dividing the articles into branches that would make difference in ranking (Fu et al. 2014). With the exception of Chinese Academy of Sciences, University of São Paulo, with 248 articles, Russian Academy of Sciences, with 248 articles, and Pennsylvania State University, with 203 articles, ranked 2nd, 2nd, and 4th, with which the outputs of these three institutes were all above 200. The CNRS, France ranked 6th in the most productive institutes, whereas, ranked 2nd in the number of inter-institutionally collaborated articles. In addition, 6 of the top 20 most productive institutions were from China, the other 14 institutions were scattered in nine countries. 58.7 % of articles produced by the 20 most productive institutions in electrochemical technology for water and wastewater treatment articles were inter-institutionally collaborated. It is obvious that institutions from the same country tended to have higher rates of collaboration, and the major collaborative institutions of these 20 institutions were all from the same country.

It is apparent that collaboration plays an ever-growing important role in electrochemical technology for water and wastewater treatment articles. Among the 26,296 articles with address information, 13,969 (53.1 %) were independent articles published by single institution, and the other articles were inter-institutional collaborative works, including both national (28.9 %) and international (18.0 %) collaborations. As seen in Fig. 6, the number of collaborative articles exceeded the number of independent articles in 2011 and ultimately accounted for 55.7 % of the total articles in 2013. In general, the increasing trend of collaborative articles relative to the total can be partly explained by the fact that the number of institutes and countries engaged in the electrochemical technology for water and wastewater treatment research increased. In collaborative articles, national articles consistently dominated, with an overall average of 61.7 % over the 20 years. The proportion of international articles has recently gradually increased, from 32 % in 1994 to 55.7 % in 2013. We can anticipate that international collaboration will become the major form of collaborative articles in the future.

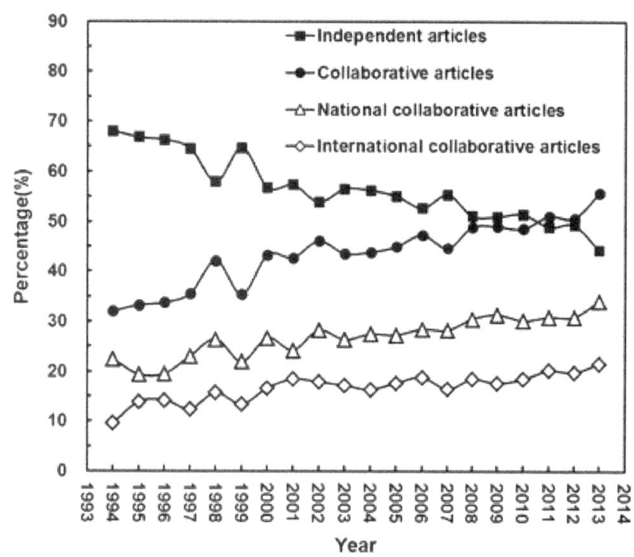

Figure. 6: Independent and collaborative share of articles per year during the research period Research tendency and hotspot

Author Keywords

The technique of statistical analysis of author keywords, articles title, article abstract, and keyword plus could be used to identify directions in science, and has proved to be efficient in comprehending and grasping the progress of new frontiers of science. Nowadays, bibliometric analysis of author keywords in different periods was further applied to analyze research trends (Chiu and Ho 2007; Xie et al. 2008), whereas using author keywords was much more infrequent (Imbelloni 2012). Excluding 8661 articles without author keywords information on the ISI Web of Science, the other 17,669 articles with author keywords information were analyzed. There were 5764 keywords listed by authors, however, 24,417 (74.18 %) keywords were used only once, 3690 (11.30 %) keywords were used twice, and the large number of once-only author keywords most likely indicated a lack of continuity in research and a wide disparity in research focuses (Chuang et al. 2007). Only 1078 (3.3 %) keywords were used more than ten times, which showed that the mainstream research in electrochemical technology for water and wastewater treatment was considered to focus on a small field. Keyword ranking changes in the four 5-year intervals are indicative of changes in the hot fields. The top 30 most frequently used keywords for the study period are listed in Table 5.

Table 5: Top 30 frequently used substantives in author keywords during 1994–2013 and four 5-year periods

Author keywords	**94-13 TP**	**94-13*R* (%)**	**94-98*R* (%)**	**99-03*R* (%)**	**04-08*R* (%)**	**09-13*R* (%)**
Electrochemistry	955	1 (5.4)	1 (6.3)	1 (5.5)	1 (5.3)	1 (5.3)
Electrochemical Impedance Spectroscopy (EIS)	718	2 (4.1)	4 (3.0)	3 (3.5)	2 (3.7)	2 (4.6)
Cyclic voltammetry	599	3 (3.4)	2 (4.1)	2 (4.3)	3 (3.6)	4 (2.9)
Corrosion	543	4 (3.1)	3 (3.1)	4 (3.2)	4 (3.0)	3 (3.1)
Electrochemical oxidation	422	5 (2.4)	24 (1.0)	7 (2.1)	5 (2.4)	5 (2.7)
Electrocatalysis	360	6 (2.0)	6 (2.5)	5 (2.4)	7 (2.2)	9 (1.8)
Proton exchange membrane fuel cells (PEMFC)	331	7 (1.9)	421 (0.1)	50 (0.6)	6 (2.3)	7 (2.0)
Voltammetry	325	8 (1.8)	6 (2.5)	5 (2.4)	8 (1.8)	12 (1.6)
Copper	260	9 (1.5)	5 (2.9)	11 (1.3)	13 (1.3)	15 (1.4)
Fuel cell(s)	255	10 (1.4)	80 (0.5)	29 (0.8)	9 (1.8)	13 (1.6)
Electrocoagulation	247	11 (1.4)	186 (0.3)	62 (0.5)	11 (1.4)	9 (1.8)
Electrodeposition	247	11 (1.4)	80 (0.5)	12 (1.2)	13 (1.3)	11 (1.6)
Ionic liquid(s)	239	13 (1.4)	NA	184 (0.3)	10 (1.5)	6 (2.1)
Carbon nanotube(s)	232	14 (1.3)	NA	242 (0.2)	23 (1.1)	8 (2.0)

Adsorption	224	15 (1.3)	8 (2.3)	12 (1.2)	19 (1.2)	23 (1.2)
Wastewater treatment	220	16 (1.2)	142 (0.3)	8 (1.5)	16 (1.3)	16 (1.3)
Electrolysis	219	17 (1.2)	67 (0.6)	9 (1.4)	11 (1.4)	19 (1.2)
Biosensor(s)	218	18 (1.2)	20 (1.0)	20 (1.0)	20 (1.1)	14 (1.4)
Phenol(s)	217	19 (1.2)	13 (1.2)	9 (1.4)	15 (1.3)	23 (1.2)
Titanium dioxide (TiO_2)	188	20 (1.1)	13 (1.2)	38 (0.7)	17 (1.2)	26 (1.0)
Electrooxidation	187	21 (1.1)	54 (0.6)	34 (0.8)	20 (1.1)	18 (1.3)
Oxidation	186	22 (1.1)	24 (1.0)	19 (1)	33 (0.9)	21 (1.2)
Electrochemical	183	23 (1.0)	80 (0.5)	29 (0.8)	20 (1.1)	25 (1.1)
Hydrogen peroxide	167	24 (0.9)	24 (1.0)	16 (1.2)	17 (1.2)	51 (0.7)
Hydrogen	165	25 (0.9)	80 (0.5)	24 (0.9)	31 (0.9)	28 (1.0)
Hydroxyl radical(s)	159	26 (0.9)	80 (0.5)	25 (0.9)	26 (1.0)	32 (0.9)
Wastewater	159	26 (0.9)	186 (0.3)	17 (1.2)	25 (1.0)	35 (0.9)
Electrochemical properties	152	28 (0.9)	30 (0.9)	20 (1)	36 (0.8)	38 (0.8)
Stainless steel	151	29 (0.9)	36 (0.8)	29 (0.8)	44 (0.7)	32 (0.9)
Polypyrrole	148	30 (0.8)	13 (1.2)	29 (0.8)	26 (1.0)	55 (0.7)

TP articles in the study period, *R* (%) the rank and percentage of the author keywords

With the exception of most frequently used author keyword "electrochemistry", which were search words in this study, the three most frequently used keywords were "Electrochemical Impedance Spectroscopy (EIS)" (718; 4.1 %), "Cyclic Voltammetry" (599; 3.4 %), "Corrosion"(543; 3.1 %). "Proton Exchange Membrane Fuel Cells (PEMFC)" received increased attention with a rise in rankings from 421 to 7 over the 20 years. In addition, "ionic liquid(s)" and "carbon nanotube(s)" might be new hot topics in the future since both of them did not appear in 1994–1998, but ranked 6th and 8th in 2009–2013, respectively. Ionic liquid(s), which could supply a suitable and stable environment, which draw a great attention on electrode modification (Wadhawan et al. 2000; Rozniecka et al. 2005), electrolyte solutions (Markevich et al. 2006; de Souza et al. 2007), and extraction of solvent and metal ions (Wei et al. 2003; Khachatryan et al. 2005). In previous researches (Munoz-Sandoval 2014), carbon nanotube(s) related literatures had increased enormously and represented a very important material in nanotechnology and nanoscience area.

Most of the top 30 author keywords were related to the following methods: electrochemical oxidation (electrochemical oxidation, electrolysis, electrooxidation, and oxidation) (1014; 5.8 %), oxidants [hydrogen peroxide and hydroxyl radical(s)] (326; 1.8 %), wastewater treatment (wastewater treatment and wastewater) (379; 2.1 %), which are also the most important topics in electrochemical technology for water and wastewater treatment related research. Especially in the contemporary, electrochemical technology, which had promising potential to be developed as a robust, cost-effective and environmental friendly alternative, was universally applied to water and wastewater treatment. Electrocoagulation, electrodeposition, electrooxidation, electrodisinfection, electrofenton, electroflotation, and electrosorption were frequently found in electrochemical technologies. However, during the research, trends on electrochemical technology for water and wastewater treatment, electrocoagulation, electrodeposition and electrooxidation, which were ranked in the top 11 frequently used substantives in author keywords, were the three main electrochemical technologies. Both "electrocoagulation" (247; 1.4 %) and "electrodeposition" (247; 1.4 %) had an apparent upward movement in their ranks, from 186 and 80 in 1994–1998 to 9 and 11 in 2009–2013. Electrocoagulation of wastewater using aluminum (Mouedhen et al. 2008; Heidmann and Calmano 2008a), iron (Heidmann and Calmano 2008b) or the hybrid Al/Fe electrodes (Kobya et al. 2003; Gomes et al. 2007) has been reported by a number of authors. Electrodeposition was not only applied to the removal (Al-Shammari et al. 2004) and recycle (Szpyrkowicz et al. 2000) of heavy metal ions from wastewater, but also improved the performance of electrode by cathodic deposition (Habazaki et al. 2002). Electrooxidation, which was extensively investigated since the late 1970s (Mayeda et al. 1972; Nilsson et al. 1973), was quickly put into application of industrial wastewater (Piya-Areetham et al. 2006; Andrade et al. 2007; Zhu et al. 2009) and refractory organic matters (Rodrigo et al. 2001; Jiang et al. 2010; Choi et al. 2010) treatment.

Articles Title and Articles Abstract

The titles containing the information of viewpoint and thought in articles were clearly expressed in front of readers (Baskerville 1904). In addition, the title, which provided a reasonably detailed picture of an article's theme, could draw great attention and called research enthusiasm of the readers. Recently, distribution of words in article titles of different periods was applied in sediment-related trend research in earth science (Niu et al. 2014) and methane trend research in anaerobic digestion (Wang et al. 2013). In this study, all single words within the title of articles were statistically analyzed. However, the meaning of the single words in titles which sometimes does not make sense were discarded, such as, "a/an", "for", "of", "the", "study", and "research". Articles title ranking changes in the four 5-year intervals are indicative of changes in the hot fields. The top 30 most frequency substantives in articles title used for the study period are listed in Table 6.

Table 6: Top 30 frequency substantives in articles title during 1994–2013 and four 5-year periods

Title keywords	94-13 TP	94-13 *R*(%)	94-98 *R*(%)	99-03 *R*(%)	04-08 *R*(%)	09-13 *R*(%)
Electrochemical	6927	1 (26.3)	1 (29.2)	1 (28.8)	1 (26.1)	1 (24.9)
Electrode(s)	3638	2 (13.8)	3 (12.1)	3 (12.6)	2 (13.4)	2 (14.9)
Water	3462	3 (13.1)	2 (12.7)	2 (12.9)	4 (12.3)	3 (13.8)
Carbon	2075	4 (7.9)	21 (3.3)	13 (4.7)	6 (7.0)	4 (10.5)
Cell(s)	1846	5 (7.0)	40 (2.1)	14 (4.5)	5 (7.8)	5 (8.6)
Film(s)	1834	6 (7.0)	7 (6.5)	4 (7.8)	3 (12.7)	9 (6.2)
Determination	1650	7 (6.3)	4 (7.9)	7 (6.3)	11 (5.9)	10 (6.1)
Corrosion	1643	8 (6.2)	6 (7.2)	5 (6.8)	7 (6.5)	11 (5.7)
Oxidation	1595	9 (6.1)	9 (5.0)	8 (5.8)	14 (5.5)	7 (6.7)
Synthesis	1521	10 (5.8)	11 (4.4)	20 (3.9)	12 (5.7)	6 (6.8)
Solution(s)	1485	11 (5.6)	5 (7.7)	6 (6.7)	13 (5.5)	16 (4.9)
Properties	1473	12 (5.6)	8 (5.2)	9 (5.2)	8 (6.2)	12 (5.5)
Fuel	1340	13 (5.1)	127 (1.0)	33 (2.8)	10 (5.9)	8 (6.3)
Membrane(s)	1333	14 (5.1)	31 (2.6)	18 (4.3)	9 (6.1)	13 (5.3)
Surface(s)	1248	15 (4.7)	13 (4.3)	10 (5.1)	16 (5.1)	18 (4.5)
Liquid(s)	1203	16 (4.6)	13 (4.3)	11 (4.9)	15 (5.2)	23 (4.2)
Application(s)	1161	17 (4.4)	26 (3.1)	23 (3.6)	18 (4.1)	14 (5.1)
Characterization	1137	18 (4.3)	24 (3.2)	14 (4.5)	17 (4.7)	20 (4.3)
Detection	1089	19 (4.1)	16 (4.1)	20 (3.9)	20 (3.9)	19 (4.3)
Acid	1046	20 (4.0)	17 (4.1)	16 (4.4)	19 (3.9)	26 (3.8)
Hydrogen	1001	21 (3.8)	24 (3.2)	26 (3.3)	22 (3.6)	21 (4.2)
Ion(s)	998	22 (3.8)	11 (4.4)	17 (4.4)	23 (3.5)	28 (3.6)
Aqueous	997	23 (3.8)	18 (3.7)	12 (4.8)	25 (3.5)	27 (3.6)
Oxide	973	24 (3.7)	42 (2.1)	29 (3.0)	28 (3.2)	17 (4.6)
Reduction	932	25 (3.5)	10 (4.9)	22 (3.9)	21 (3.6)	35 (3.1)
Electrolyte(s)	898	26 (3.4)	22 (3.3)	24 (3.5)	24 (3.5)	30 (3.4)
Treatment	870	27 (3.3)	49 (2.0)	34 (2.8)	27 (3.2)	25 (3.8)
Steel	868	28 (3.3)	27 (3.0)	26 (3.3)	26 (3.3)	29 (3.4)
Modified	840	29 (3.2)	68 (1.5)	39 (2.4)	38 (2.6)	22 (4.2)
Nanotube(s)	792	30 (3.0)	NA	509 (0.3)	50 (2.2)	15 (5.1)

TP the number of articles in the study period, % the percentage of the source title

The most frequency substantives used in article titles were "electrochemistry" and "water", however, which should be ascribed to these words included in search topics of this study. Excluding the above factors, "Electrode(s)" which was found in 3638 article titles still might be the hottest topics during the last twenty years. Both "carbon" (2075; 7.9 %) and "cell(s)" (1846; 7.0 %) related to electrode(s) had an apparent upward progress in their ranks, from 21st and 40th in 1994–1998 to 4th and 5th in 2009–2013. "Fuel", "application(s)", "oxide", "treatment", "modified", and "nanotube(s)" showed a continuous increasing trend in titles, while the terms "solution(s)" showed a decreasing trend. It is worth mentioning that "Nanotube(s)" had a notable promotion, the same with "carbon nanotube(s)" in author keywords, which did not appear in 1994–1998, but ranked 15th in 2009–2013.

Distribution of words in article abstracts of different periods could also be the information of research trend analysis (Zhang et al. 2010a; Fu et al. 2014). In the analysis of article abstracts in current study, the words in article abstracts showed that "electrochemical", "electrode(s)", "water", "film(s)", "oxidation", "solution", "current", "potential", "temperature", "surface", "corrosion", "carbon", "reaction", "reduction", "cell", "detection", "process", "hydrogen", "acid", "oxygen", "cyclic", "properties", "aqueous", and "pH" appeared at least 4,000 times. In all, "electrode(s)" still might be one of most popular issue during the electrochemical technology for the wastewater treatment. This result was similar to the preceding analysis in articles title (Zhang et al. 2010a).

'KeyWords Plus'

'KeyWords Plus', which were different from the author keywords, articles title and articles abstract, were automatically generated by the SCI from an algorithm that extracts key words from all the titles referenced or cited in the documents

(Garfield 1990). In recent years, 'KeyWords Plus' in different periods have been presented as information of research trends during previous literature (Xie et al. 2008; Fu et al. 2010; Wang et al. 2013; Fu et al. 2014). 'KeyWords Plus' ranking changes in the four 5-year intervals are indicative of changes in the hot fields. The top 30 most frequently used 'KeyWords Plus' for the study period are listed in Table 7.

Table 7: Top 30 frequently used substantives in author "KeyWords Plus" during 1994–2013 and four 5-year periods

Author Keyword Plus	94-13 TP	94-13 *R*(%)	94-98 *R*(%)	99-03 *R*(%)	04-08 *R*(%)	09-13 *R*(%)
Water	4436	1 (17.6)	1 (16.8)	1 (17.1)	1 (17.3)	1 (18.2)
Electrode(s)	2339	2 (9.3)	2 (9.1)	2 (8.0)	2 (8.8)	2 (10.0)
Oxidation	1923	3 (7.6)	5 (5.1)	3 (6.5)	3 (7.9)	3 (8.4)
Film(s)	1593	4 (6.3)	4 (5.4)	4 (5.5)	4 (6.3)	5 (6.8)
Performance	1230	5 (4.9)	67 (1.0)	25 (2.4)	7 (4.3)	4 (6.8)
Behavior	1211	6 (4.8)	10 (4.0)	5 (4.9)	5 (5.2)	11 (4.7)
Surface(s)	1188	7 (4.7)	7 (5.0)	6 (4.6)	8 (4.3)	8 (5.0)
Aqueous-solution(s)	1092	8 (4.3)	16 (2.6)	15 (2.9)	9 (4.3)	7 (5.2)
Degradation	1090	9 (4.3)	52 (1.1)	17 (2.6)	6 (4.4)	6 (5.5)
System(s)	1006	10 (4)	9 (4.5)	11 (3.8)	11 (4.0)	14 (3.9)
Waste-water treatment	948	11 (3.8)	72 (0.9)	12 (3.4)	10 (4.1)	13 (4.3)
Adsorption	945	12 (3.8)	8 (4.8)	7 (4.2)	13 (3.4)	16 (3.6)
Reduction	923	13 (3.7)	11 (3.9)	10 (3.8)	15 (3.2)	15 (3.8)
Waste-water	839	14 (3.3)	347 (0.3)	55 (1.3)	21 (2.9)	9 (4.9)
Electrochemistry	830	15 (3.3)	3 (6.1)	8 (3.9)	17 (3.1)	22 (2.6)
Removal	815	16 (3.2)	125 (0.6)	44 (1.5)	19 (2.9)	12 (4.5)
Acid	804	17 (3.2)	24 (2.0)	23 (2.4)	12 (3.5)	17 (3.5)
Electrochemical detection	802	18 (3.2)	5 (5.1)	9 (3.8)	14 (3.4)	26 (2.5)
Kinetics	751	19 (3.0)	12 (3.8)	12 (3.4)	15 (3.2)	23 (2.6)
Mechanism	713	20 (2.8)	13 (3.7)	14 (3.0)	19 (2.9)	24 (2.5)
Nanoparticles	705	21 (2.8)	1990 (0.04)	543 (0.2)	35 (1.8)	10 (4.7)
Thin-Films	645	22 (2.6)	44 (1.4)	30 (1.9)	27 (2.3)	18 (3.2)
Transport	625	23 (2.5)	27 (1.8)	16 (2.6)	18 (3.0)	31 (2.3)
Electrochemical oxidation	613	24 (2.4)	59 (1.1)	43 (1.6)	23 (2.6)	19 (2.9)
Spectroscopy	577	25 (2.3)	14 (3.4)	20 (2.5)	25 (2.3)	35 (2.0)
Oxide(s)	573	26 (2.3)	29 (1.8)	31 (1.8)	35 (1.8)	20 (2.8)
Anodic-oxidation	571	27 (2.3)	75 (0.9)	22 (2.5)	22 (2.7)	33 (2.2)
Membrane(s)	548	28 (2.2)	39 (1.4)	39 (1.6)	24 (2.5)	29 (2.3)
Sensor(s)	548	28 (2.2)	84 (0.9)	37 (1.7)	32 (2.0)	21 (2.7)
Ion(s)	537	30 (2.1)	17 (2.5)	24 (2.4)	31 (2.0)	34 (2.0)

TP articles in the study period, *R* (%): the rank and percentage of the author "KeyWords Plus"

The analysis of 'KeyWords Plus' shows that 25,173 'KeyWords Plus' were used in 26,330 articles. The top ten most frequently used 'KeyWords Plus' were "water (4436; 17.6 %)", "electrode(s) (2339; 9.3 %)", "oxidation (1923; 7.6 %)", "film(s) (1593; 6.3 %)", "performance (1230; 4.9 %)", "behavior (1211; 4.8 %)", "surface(s) (1188; 4.7 %)", "aqueous-solution(s) (1092; 4.3 %)", "degradation (1090; 4.3 %)", "system(s) (1006; 4 %)", respectively. Some new research areas could also be found. "Nanoparticles" was ranked 1990 in 1994–1998, while ranked 10th in 2009–2013. It is worth mentioning that the notable promotion of "Nanoparticles" was similar to "carbon nanotube(s)" in author keywords, and "Nanotube(s)" in articles title. China was the leading country in nanoscience and nanotechnology research and published more papers per year than USA since 2010 (The word used in the search was "nano*" under the "title" from 1997 to 2012) (Munoz-Sandoval 2014). Both "Waste-Water" (839; 3.3 %) and "Removal" (815; 3.2 %) had an apparent upward movement in their ranks, from 349 h and 125th in 1994–1998 to 9th and 12th in 2009–2013.

However, "water and wastewater treatment", "electrooxidation or electrochemical oxidation", and "electrode(s) or titanium dioxide(TIO_2)" were present in words in the author keywords, articles title, and 'KeyWords Plus' list, which still might be identified as current electrochemical technology for water and wastewater treatment research hotspots. Some words "anodic-oxidation", "aqueous-solution(s)", "behavior", "degradation", "kinetics", "removal", "sensor(s)", "spectroscopy", "thin-films", only appeared in the rank of the 'KeyWords Plus' list describing the article's contents with greater variety. The analysis of 'KeyWords Plus' was particularly useful for the analysis and delimitation of a scientific domain, which expanded and refined the perspective of scientific research field.

Hot Issues

Based on comprehensive analysis of the distribution of author keywords, title, abstract, and 'KeyWords Plus', researchers could obtain few valuable clues of the potential research hotspots. Moreover, a new method named "word cluster analysis" (Mao et al. 2010), which combined the author keywords, article title, and 'KeyWords Plus', had been successfully applied to trace the hotspot of related research field in previous work (Tanaka and Ho 2011; Huang et al. 2012; Fu et al. 2013; Tan et al. 2014). Therefore, it could comprehensively and accurately identify the overview of the research hotspots through the refinement of synonymic single words and congeneric phrases. Research tendency on electrochemical technology for water and wastewater treatment were extracted and separated into two categories research methods including electrochemical impedance spectroscopy (EIS), Cyclic Voltammetry (CV), and Mass Spectrometry (MS), and research items including electrode(s) and corrosion. In the topic of research methods category, the EIS included the words "electrochemical impedance spectroscopy (EIS)", "electrochemical impedance spectroscopy" and "EIS", CV included the words "Cyclic Voltammetric", "Cyclic Voltammetric", and "Cyclic Voltammetric Technique", and the MS included the words "Electrochemical Mass-Spectrometry", "Mass-Spectrometry", "Mass Spectrometry", "Mass Spectrometry (MS)", "MS", and "the other MS-related Method". Accordingly, the topic electrode(s) included the words "electrode" and "electrode", and the topic corrosion included only the word "corrosion" on research items.

Growth trend of hotpot-related articles during the last 20 years is shown in Fig. 7. All research related had revealed an escalating trend along with the increase of the total publications. The investigation by word cluster analysis showed that EIS, CV, and MS could be the most commonly studied technique on electrochemical technology for water and wastewater treatment related research. EIS had attained tremendous increase in popularity and dominated the research methods since 2002, when EIS surpassed the research methods of CV and MS. Further, the EIS, which was a mature and well-known technique, had been widely used to analyze the characterization of electrode processes and corrosion performance. During the related researches on electrochemical technology for water and wastewater treatment, EIS mainly applied in the field of corrosion (Liu et al. 2003; Rios et al. 2013), coatings (Deflorian et al. 1994; Deflorian and Fedrizzi 1999), and fuel cells (Yi and Van Nguyen 1999; Lin and Van Nguyen 2005) since 1990s. In addition, EIS, which had been shown to be a powerful and useful tool in different scientific areas, was adopted to investigate the hotspot research on microbial fuel cells (MFCs) (He and Mansfeld 2009; Hosseini and Ahadzadeh 2012). Then, CV, which offered a first view on electrode/electrolyte systems in the potential window of interest, was the most commonly used method in electroanalysis and always combined with other analysis method (e.g., EIS, chronoamperometry, differential pulse voltammetry, square wave voltammetry, scanning electron microscope (SEM), MS and X-ray diffraction); Finally, MS measurements, which were carried out in the gas phase on ionized analytes, were generally applied to analyze the chemical and element. Nowadays, integrated with other techniques, MS had obtained technical and conceptual advances by liquid chromatography mass-spectrometry (LC–MS), gas chromatography mass-spectrometry (GC–MS), electrospray ionization mass-spectrometry (ESI–MS), desorption/ionization mass-spectrometry, capillary electrophoresis-mass spectrometry, Fourier transform ion cyclotron resonance mass spectrometry, time-of-flight mass spectrometry, plasma-mass-spectrometry, trap mass-spectrometry, and differential electrochemical mass spectrometry. These techniques had their own strength and weakness because of the difference of design and performance. Moreover, these analysers could separately analyze specific analysis or put together in tandem to overcome the weakness of each. In addition, it is worth mentioning that the role of MS-based proteomics had been successfully illustrated as an indispensable tool for molecular and cellular biology (Aebersold and Mann2003). The related result indicated that the ability of mass spectrometry to precisely identify and quantify complex molecular chemicals could be expected to impact broadly on different scientific areas.

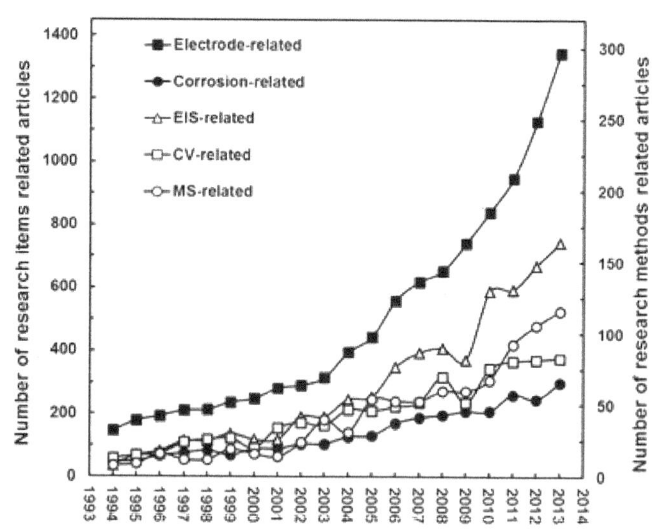

Figure7: Growth trend of hotpot-related articles during the last 20 years (note: research items related articles including electrode-related and corrosion-related articles, research methods related articles including EIS-related, CV-related and MS-related articles, *solid symbols* in main coordinates, and *blank symbols* in sub-coordinates)

As shown in Fig. 7, the amount of the research items on corrosion continuously increased from 1994 to 2013, but decreased a little bit in 2012. "Corrosion" was one of the most important research directions on electrochemical technology for water and wastewater treatment, which were mainly focused on pitting corrosion, localized corrosion, atmospheric corrosion, corrosion protection, corrosion resistance, and corrosion inhibition. Pitting corrosion (Frankel 1998) which belonged to localized corrosion accelerated the dissolution of metal when the otherwise protective passive film on metal surface was broken down. Moreover, researchers had paid more concern on the effects of alloy composition (Williams et al. 2010), environment (Jafarian et al.2008), potential (Lee et al. 2011), and temperature (Lee et al. 2010). Atmospheric corrosion which was a new scientific focus on the corrosion process had been studied by engineers for nearly a century. Leygraf and Graedel (2000) firstly comprehensively described the corrosion on structures and materials, and properly used latest scientific tools to prevent or minimize the corrosion damage. Therefore, corrosion protection, corrosion resistance, and corrosion inhibition had appeared in anticorrosion field for the prevention of different metals and alloys, which was achieved by specific coatings or composite films on their surface. Besides, nano-structured surface treatment coatings was widely used in the manufacture of the specific coatings or composite films, such as, tantalum oxide nano-particle coatings (Shen et al. 2005; Díaz et al.2013), nano-siloxane epoxy coating (Bagherzadeh et al. 2012), and nano-structured polymeric coatings (Aglan et al. 2007). In addition, novel area of corrosion mainly focused on copper corrosion, microbiological corrosion, and CO_2 corrosion in recent years.

Compared with the well growth trend of corrosion, the amount of the research items on electrode(s) fluctuated from 1994 to 2003 and significantly rocketed since 2008, which revealed that the electrode-related research was the hotspot on electrochemical technology for water and wastewater treatment. Recently, researchers (Chen 2004; Anglada et al. 2009; Wang et al. 2010; Sun et al. 2012b) had paid great attention on the removal efficiency in degrading various pollutants on all kinds of electrodes, enhancement of electrochemical stability of anodic material of electrodes, investigation of factors affecting the electrodes performance, and explanation of the mechanisms and kinetics of pollutant degradation. During the anodic materials of common electrode, such as, graphite, carbon fibers, platinum, diamond, gold (Au), aluminum, iron, PbO_2, SnO_2, TiO_2, Ti/Pt, Ti/SnO_2, Ti/PbO_2, Ti/RuO_2–TiO_2, Ti/IrO_2–TiO_2, Ti/IrO_2–RuO_2–TiO_2, boron-doped diamond (BDD) electrode, Titanium-based boron-doped diamond electrode (BDD/Ti) were widely used in the wastewater treatment. Among them, graphite, platinum, and the dimensionally stable anodes (DSA®) was identified as the most common insoluble anodes, whereas, revealed a lower oxygen evolution overpotential and durability; Aluminum and iron was usually applied in electrocoagulation, which were effective on the treatment of water and wastewater, respectively. However, PbO_2, SnO_2, TiO_2 and diamond electrode seemed to be the potential anode materials with high oxygen evolution overpotentials. Moreover, PbO_2 electrode are relatively cheap and widely used in oxidizing pollutants, but there was some concern for its toxic Pb^{2+} ions; SnO_2 was always not effective because of its poor electrochemical stability; TiO_2, combined with ultraviolet technique, was usually used as a photocatalyst in wastewater treatment; Diamond electrode was well known for its high strength, strong hardness, and long durability and usually served as anodic substrates with other metal materials. Nowadays, Titanium or other noble metal-based electrodes including Ti/RuO_2–TiO_2, Ti/IrO_2–TiO_2, Ti/IrO_2–RuO_2–TiO_2 showed high oxygen evolution overpotential and well electrochemical performance, which would be the promising anode materials in industrial application. In addition, BDD/Ti electrode with high oxygen evolution overpotential (Chen et al. 2003), which was a new frontier of electrode application, was found to be the most active anodic material for degradation of refractory pollutants. Therefore, the hotspots on electrochemical technology for water and wastewater treatment related research was novel electrode, which owned high oxygen evolution overpotential and good physiochemical and electrochemical properties.

CONCLUSIONS

Based on 26,330 electrochemical technologies for water and wastewater treatment related publications dealing with SCI, this bibliometric study provided an overview of research in electrochemical technology for water and wastewater treatment and identified some significant points in the research throughout the investigation period. Electrochemical technology for water and wastewater treatment related researches have significantly increased in the last 20 years. A power model and an exponential model were applied to illustrate the relations between cumulative number of articles and the year. Furthermore, based on the exponential model, it can be calculated that the number of articles in 2018 will be twice than that in 2013. Articles were published in 1790 journals in 106 subject categories. Journal of the Electrochemical Society with a top h-index of 90 published the most articles in this field, taking 5.8 % of all, followed by Electrochimica Acta and Journal of Electroanalytical Chemistry. The researchers focused on categories of chemistry, electrochemistry, and materials science. China and Chinese Academy of Sciences were the most productive country and institution, respectively, while USA, with the most international collaborative articles and highest h-index of 130, was the major collaborator with other 15 countries in top 20 most productive countries. The G7 and BRIC accounted for 40.7 and 23.2 % of the total world production. Word cluster analysis was successfully used to trace the hotspot on electrochemical technology for water and wastewater treatment related research, which showed that the novel anodic electrode with high oxygen evolution overpotential and good physiochemical and electrochemical properties was the hotspot.

ACKNOWLEDGMENTS

The authors gratefully acknowledge the financial support of the Major Science and Technology Program for Water Pollution Control and Treatment (2012ZX07201002-6).

REFERENCES

1. Aebersold R, Mann M (2003) Mass spectrometry-based proteomics. Nature 6928:198–207
2. Afzal A, Pourrezaei P, Ding N et al (2011) Physico-Chemical Processes. Water Environ Res 10:994–1091
3. Aglan A, Allie A, Ludwick A et al (2007) Formulation and evaluation of nano-structured polymeric coatings for corrosion protection. Surf Coat Tech 2:370–378

4. Al-Shammari AA, Rahman SU, Chin D (2004) An oblique rotating barrel electrochemical reactor for removal of copper ions from wastewater. J Appl Electrochem 4:447–453
5. Andrade LS, Ruotolo LAM, Rocha-Filho RC et al (2007) On the performance of Fe and Fe, F doped Ti-Pt/PbO_2 electrodes in the electrooxidation of the Blue Reactive 19 dye in simulated textile wastewater. Chemosphere 11:2035–2043
6. Anglada A, Urtiaga A, Ortiz I (2009) Contributions of electrochemical oxidation to waste-water treatment: fundamentals and review of applications. J Chem Technol Biot 12:1747–1755
7. Bagherzadeh MR, Daneshvar A, Shariatpanahi H (2012) Novel water-based nanosiloxane epoxy coating for corrosion protection of carbon steel. Surf Coat Tech 8:2057–2063
8. Baskerville C (1904) The Titles of Papers. Science (New York, NY) 487:702–703
9. Canizares P, Hernandez M, Rodrigo MA et al (2008) Electrooxidation of brown-colored molasses wastewater. Effect of the electrolyte salt on the process efficiency. Ind Eng Chem Res 3:1298–1301
10. Chang M, Gao C, Jiang J (2009) Electrochemical oxidation of organic compounds using boron-doped diamond electrode. J Electrochem Soc 2:E50–E54
11. Chen G (2004) Electrochemical technologies in wastewater treatment. Sep Purif Technol 1:11–41
12. Chen X, Chen G, Yue PL (2002) Novel electrode system for electroflotation of wastewater. Environ Sci Technol 4:778–783
13. Chen X, Chen G, Yue PL (2003) Anodic oxidation of dyes at novel Ti/B-diamond electrodes. Chem Eng Sci 3:995–1001
14. Chiu W, Ho Y (2007) Bibliometric analysis of tsunami research. Scientometrics 1:3–17
15. Choi JY, Lee Y, Shin J et al (2010) Anodic oxidation of 1,4-dioxane on boron-doped diamond electrodes for wastewater treatment. J Hazard Mater 1:762–768
16. Chuang K, Huang Y, Ho Y (2007) A bibliometric and citation analysis of stroke-related research in Taiwan. Scientometrics 2:201–212
17. Ciriminna R, Pagliaro M (2013) On the use of the h-index in evaluating chemical research. Chem Cent J 1:132
18. de Souza RF, Padilha JC, Gonçalves RS et al (2007) Electrochemical hydrogen production from water electrolysis using ionic liquid as electrolytes: towards the best device. J Power Sources 2:792–798
19. Deflorian F, Fedrizzi L (1999) Adhesion characterization of protective organic coatings by electrochemical impedance spectroscopy. J Adhes Sci Technol 5:629–645
20. Deflorian F, Miskovic-Stankovic VB, Bonora PL et al (1994) Degradation of epoxy coatings on phosphatized zinc-electroplated steel. Corrosion 6:438–446
21. Díaz B, Światowska J, Maurice V et al (2013) Tantalum oxide nanocoatings prepared by atomic layer and filtered cathodic arc deposition for corrosion protection of steel: Comparative surface and electrochemical analysis. Electrochim Acta 90:232–245
22. Feng C, Sugiura N, Shimada S et al (2003) Development of a high performance electrochemical wastewater treatment system. J Hazard Mater 1:65–78
23. Frankel GS (1998) Pitting corrosion of metals a review of the critical factors. J Electrochem Soc 6:2186–2198
24. Fu HZ, Ho YS, Sui YM et al (2010) A bibliometric analysis of solid waste research during the period 1993–2008. Waste Manage 12:2410–2417
25. Fu H, Wang M, Ho Y (2013) Mapping of drinking water research: a bibliometric analysis of research output during 1992–2011. Sci Total Environ 443:757–765
26. Fu HZ, Long X, Ho YS (2014) China's research in chemical engineering journals in Science Citation Index Expanded: a bibliometric analysis. Scientometrics 1:119–136
27. Gao Y, Tsuji H, Hattori H et al (1994) New on-line mass spectrometer system designed for platinum-single crystal electrode and electroreduction of acetylene. J Electroanal Chem 1:195–200
28. Garfield E (1990) Key-words-plus takes you beyond title words. 2. Expanded Journal Coverage for Current-Contents-on-Diskette Includes Social and Behavioral-Sciences. Current Contents 33:5–9
29. Ge J, Qu J, Lei P et al (2004) New bipolar electrocoagulation–electroflotation process for the treatment of laundry wastewater. Sep Purif Technol 1:33–39
30. Ghernaout D, Ghernaout B (2010) From chemical disinfection to electrodisinfection: the obligatory itinerary? Desalin Water Treat 1–3:156–175
31. Gomes JA, Daida P, Kesmez M et al (2007) Arsenic removal by electrocoagulation using combined Al–Fe electrode system and characterization of products. J Hazard Mater 2:220–231
32. Grady CL Jr, Daigger GT, Love NG et al (2011) Biological wastewater treatment. IWA Publishing, UK
33. Habazaki H, Hayashi Y, Konno H (2002) Characterization of electrodeposited WO_3 films and its application to electrochemical wastewater treatment. Electrochim Acta 26:4181–4188
34. He Z, Mansfeld F (2009) Exploring the use of electrochemical impedance spectroscopy (EIS) in microbial fuel cell studies. Energy Environ Sci 2:215–219
35. Heidmann I, Calmano W (2008a) Removal of Zn (II), Cu (II), Ni (II), Ag (I) and Cr(VI) present in aqueous solutions by aluminium electrocoagulation. J Hazard Mater 3:934–941
36. Heidmann I, Calmano W (2008b) Removal of Cr(VI) from model wastewaters by electrocoagulation with Fe electrodes. Sep Purif Technol 1:15–21
37. Hirsch JE (2005) An index to quantify an individual's scientific research output. P Natl Acad Sci USA 46:16569–16572
38. Hirsch JE (2007) Does the h index have predictive

power? Proc Natl Acad Sci 49:19193–19198

39. Hosseini MG, Ahadzadeh I (2012) A dual-chambered microbial fuel cell with Ti/nano-TiO< sub> $_2$/Pd nano-structure cathode. J Power Sources 292-297

40. Hu J, Ma YW, Zhang L et al (2010) A historical review and bibliometric analysis of research on lead in drinking water field from 1991 to 2007. Sci Total Environ 7:1738–1744

41. Huang C, Su Y (2010) Removal of copper ions from wastewater by adsorption/electrosorption on modified activated carbon cloths. J Hazard Mater 1:477–483

42. Huang W, Zhang B, Feng C et al (2012) Research trends on nitrate removal: a bibliometric analysis. Desalin Water Treat 1–3:67–77

43. Imbelloni LE (2012) Scientific articles' titles: thanks for the information contained in your title. Revista brasileira de anestesiologia 2:140

44. Jafarian M, Gobal F, Danaee I et al (2008) Electrochemical studies of the pitting corrosion of tin in citric acid solution containing Cl^-. Electrochim Acta 13:4528–4536

45. Jiang C, Zhang J (2007) Progress and prospect in electro-Fenton process for wastewater treatment. J Zhejiang Univ Sci A 7:1118–1125

46. Jiang Y, Zhu X, Li H et al (2010) Effect of nitro substituent on electrochemical oxidation of phenols at boron-doped diamond anodes. Chemosphere 9:1093–1099

47. Khachatryan KS, Smirnova SV, Torocheshnikova II et al (2005) Solvent extraction and extraction–voltammetric determination of phenols using room temperature ionic liquid. Anal Bioanal Chem 2:464–470

48. Kobya M, Can OT, Bayramoglu M (2003) Treatment of textile wastewaters by electrocoagulation using iron and aluminum electrodes. J Hazard Mater 1:163–178

49. Lee S, Kim J, Koo J (2010) Investigation of pitting corrosion of a copper tube in a heating system. Eng Fail Anal 6:1424–1435

50. Lee C, Batchelor B, Park SH et al (2011) Perchlorate reduction during electrochemically induced pitting corrosion of zero-valent titanium (ZVT). J Hazard Mater 197:183–189

51. Leygraf C, Graedel T (2000) Atmospheric corrosion. Wiley-Interscience, New york

52. Lin G, Van Nguyen T (2005) Effect of thickness and hydrophobic polymer content of the gas diffusion layer on electrode flooding level in a PEMFC. J Electrochem Soc 10:A1942–A1948

53. Liu C, Bi Q, Leyland A et al (2003) An electrochemical impedance spectroscopy study of the corrosion behaviour of PVD coated steels in 0.5 N NaCl aqueous solution: part II: EIS interpretation of corrosion behaviour. Corros Sci 6:1257–1273

54. Mao N, Wang M, Ho Y (2010) A bibliometric study of the trend in articles related to risk assessment published in Science Citation Index. Hum Ecol Risk Assess 4:801–824

55. Markevich E, Baranchugov V, Aurbach D (2006) On the possibility of using ionic liquids as electrolyte solutions for rechargeable 5V Li ion batteries. Electrochem Commun 8:1331–1334

56. Marselli B, Garcia-Gomez J, Michaud P et al (2003) Electrogeneration of hydroxyl radicals on boron-doped diamond electrodes. J Electrochem Soc 3:D79–D83

57. Martinez-Huitle CA, Ferro S (2006) Electrochemical oxidation of organic pollutants for the wastewater treatment: direct and indirect processes. Chem Soc Rev 12:1324–1340

58. Mayeda EA, Miller LL, Wolf JF (1972) Electrooxidation of benzylic ethers, esters, alcohols, and phenyl epoxides. J Am Chem Soc 19:6812–6816

59. Mouedhen G, Feki M, Wery M et al (2008) Behavior of aluminum electrodes in electrocoagulation process. J Hazard Mater 1:124–135

60. Munoz-Sandoval E (2014) Trends in nanoscience, nanotechnology, and carbon nanotubes: a bibliometric approach. J Nanopart Res 1:1–22

61. Nilsson A, Ronlán A, Parker VD (1973) Anodic oxidation of phenolic compounds. Part III. Anodic hydroxylation of phenols. A simple general synthesis of 4-alkyl-4-hydroxycyclo-hexa-2, 5-dienones from 4-alkylphenols. J. Chem. Soc. Perkin Trans 1:2337–2345

62. Niu BB, Hong S, Yuan JF et al (2014) Global trends in sediment-related research in earth science during 1992–2011: a bibliometric analysis. Scientometrics 1:511–529

63. Piya-Areetham P, Shenchunthichai K, Hunsom M (2006) Application of electrooxidation process for treating concentrated wastewater from distillery industry with a voluminous electrode. Water Res 15:2857–2864

64. Pritchard A (1969) Statistical bibliography or bibliometrics. J Doc 4:348–349

65. Rios JF, Calderón JA, Nogueira RP (2013) Electrochemical behavior of metals used in drinking water distribution systems: a rotating cylinder electrode's study. Corrosion 9:875–885

66. Rodrigo MA, Michaud PA, Duo I et al (2001) Oxidation of 4-chlorophenol at boron-doped diamond electrode for wastewater treatment. J Electrochem Soc 5:D60–D64

67. Rozniecka E, Shul G, Sirieix-Plenet J et al (2005) Electroactive ceramic carbon electrode modified with ionic liquid. Electrochem Commun 3:299–304

68. Shen GX, Chen YC, Lin CJ (2005) Corrosion protection of 316 L stainless steel by a TiO_2nanoparticle coating prepared by sol–gel method. Thin Solid Films 1:130–136

69. Sun J, Wang M, Ho Y (2012a) A historical review and bibliometric analysis of research on estuary pollution. Mar Pollut Bull 1:13–21

70. Sun Z, Wei X, Hu X et al (2012b) Electrocatalytic dechlorination of 2,4-dichlorophenol in aqueous solution on palladium loaded meshed titanium electrode modified with polymeric pyrrole and surfactant. Colloid Surface A 414:314–319

71. Szpyrkowicz L, Kaul SN, Molga E et al (2000) Comparison of the performance of a reactor equipped with a Ti/Pt and an SS anode for simultaneous cyanide removal and copper recovery. Electrochim Acta 2:381–387

72. Tan J, Fu H, Ho Y (2014) A bibliometric analysis of research on proteomics in science citation index expanded. Scientometrics 2:1473–1490

73. Tanaka H, Ho Y (2011) Global trends and performances of desalination research. Desalin Water Treat 1–3:1–12

74. Trocónis De Rincón O, Sánchez M, Millano V et al (2007) Effect of the marine environment on reinforced concrete durability in Iberoamerican countries: DURACON project/CYTED. Corros Sci 7:2832–2843

75. Un UT, Altay U, Koparal AS et al (2008) Complete treatment of olive mill wastewaters by electrooxidation. Chem Eng J 3:445–452

76. Valero D, Ortiz JM, Expósito E et al (2010) Electrochemical wastewater treatment directly powered by photovoltaic panels: electrooxidation of a dye-containing wastewater. Environ Sci Technol 13:5182–5187

77. Wadhawan JD, Schröder U, Neudeck A et al (2000) Ionic liquid modified electrodes. Unusual partitioning and diffusion effects of $Fe(CN)_6^{4-/3-}$ in droplet and thin layer deposits of 1-methyl-3-(2, 6-(<i>S)-dimethylocten-2-yl)-imidazolium tetrafluoroborate. J Electroanal Chem 1:75–83

78. Wagner R, Preschitschek N, Passerini S et al (2013) Current research trends and prospects among the various materials and designs used in lithium-based batteries. J Appl Electrochem 5:481–496

79. Wang G, Gao Y, Wang Z et al (2010) Investigation of PtNi/C anode electrocatalysts for direct borohydride fuel cell. J Power Sour 1:185–189

80. Wang L, Wang Q, Zhang X et al (2013) A bibliometric analysis of anaerobic digestion for methane research during the period 1994–2011. J Mater Cycles Waste Manage 1:1–8

81. Wei G, Yang Z, Chen C (2003) Room temperature ionic liquid as a novel medium for liquid/liquid extraction of metal ions. Anal Chim Acta 2:183–192

82. Williams DE, Kilburn MR, Cliff J et al (2010) Composition changes around sulphide inclusions in stainless steels, and implications for the initiation of pitting corrosion. Corros Sci 11:3702–3716

83. Xie SD, Zhang J, Ho YS (2008) Assessment of world aerosol research trends by bibliometric analysis. Scientometrics 1:113–130

84. Yi JS, Van Nguyen T (1999) Multicomponent transport in porous electrodes of proton exchange membrane fuel cells using the interdigitated gas distributors. J Electrochem Soc 1:38–45

85. Zhang GF, Xie SD, Ho YS (2010a) A bibliometric analysis of world volatile organic compounds research trends. Scientometrics 2:477–492

86. Zhang LA, Wang MH, Hu J et al (2010b) A review of published wetland research, 1991–2008: ecological engineering and ecosystem restoration. Ecol Eng 8:973–980

87. Zhu X, Ni J, Lai P (2009) Advanced treatment of biologically pretreated coking wastewater by electrochemical oxidation using boron-doped diamond electrodes. Water Res 17:4347–4355

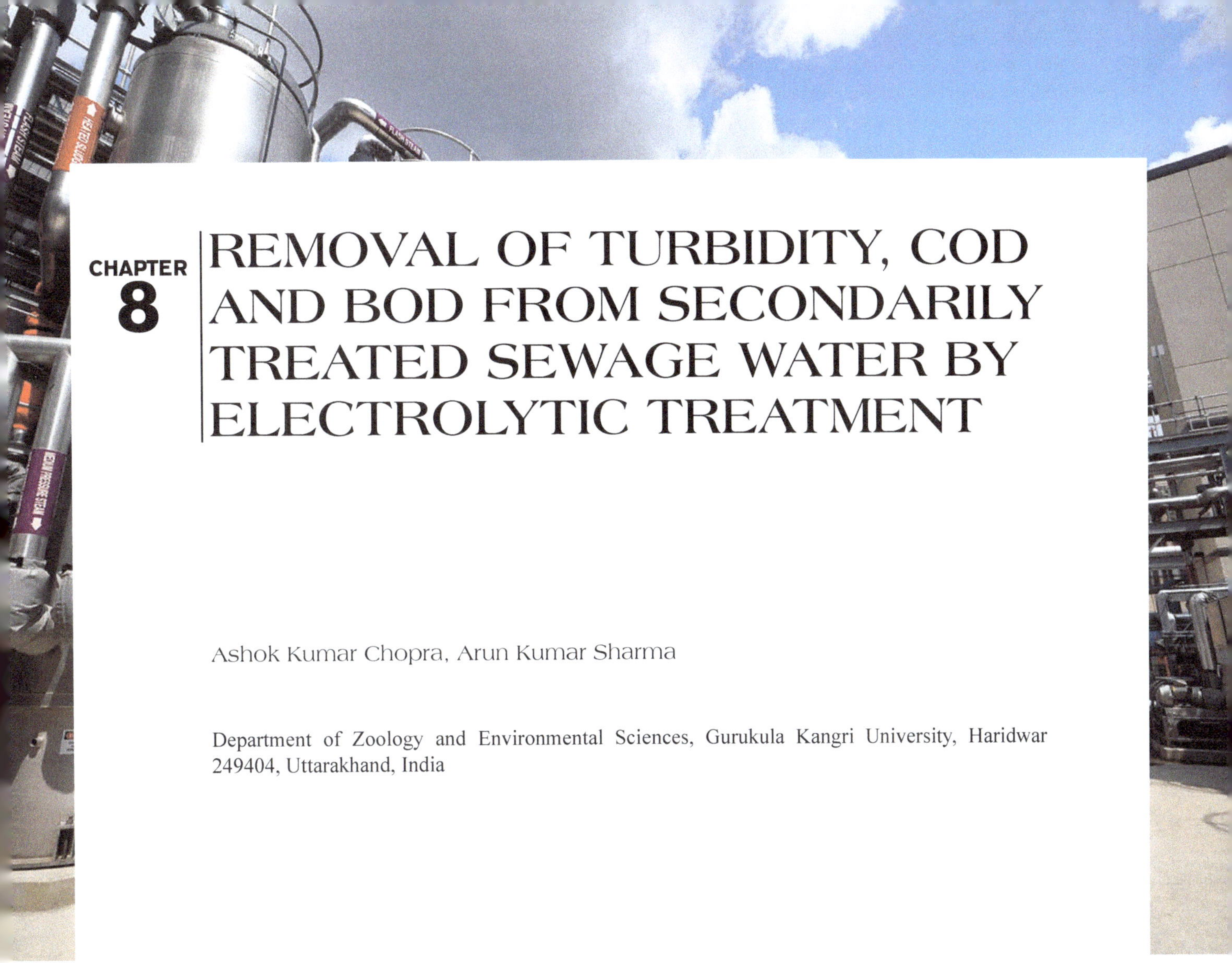

CHAPTER 8

REMOVAL OF TURBIDITY, COD AND BOD FROM SECONDARILY TREATED SEWAGE WATER BY ELECTROLYTIC TREATMENT

Ashok Kumar Chopra, Arun Kumar Sharma

Department of Zoology and Environmental Sciences, Gurukula Kangri University, Haridwar 249404, Uttarakhand, India

ABSTRACT

A preliminary study was conducted for the removal of turbidity (TD), chemical oxygen demand (COD) and biochemical oxygen demand (BOD) from secondarily treated sewage (STS) water through the electrolytic batch mode experiments with DC power supply (12 V) up to 30 min and using a novel concept of electrode combinations of different metals. The different surface areas (40, 80, 120 and 160 cm^2) of the electrodes as a function of cross-sectional area of the reactor and the effect of inter-electrode distances (2.5–10 cm) on the electrolysis of STS water were studied. This study revealed that the effluent can be effectively treated with the aluminum (Al) and iron (Fe) electrode combinations (Al–Fe and Fe–Al). The maximum removal of TD (81.51 %), COD (74.36 %) and BOD (70.86 %) was recorded with Al–Fe electrode system, while the removal of these parameters was found to be 71.11, 64.95 and 61.87 %, respectively, with Fe–Al electrode combination. The Al–Fe electrode combination had lower electrical energy consumption (2.29 kWh/m^3) as compared to Fe–Al electrode combination (2.50 kWh/m^3). The economic evaluation of electrodes showed that Al–Fe electrode combination was better than Fe–Al electrode combination. This revealed the superiority of aluminum as a sacrificial electrode over that of iron which can probably be attributed to better flocculation capabilities of aluminum than that of iron.

INTRODUCTION

Water is an essential substance for living system as it allows the transport of nutrients as well as waste products in the living systems. However, sustainable water supply is becoming more challenging by the day due to ever increasing demand of growing population as well as increasing contamination of water resources. At the same time, huge quantities of wastewater generated by industries of every hue and kind and also by exponential growth in the number of households are becoming a serious concern for society.

The role of electrochemistry in water and effluent treatment is relatively small, since conventional electrode materials achieve only low current efficiencies due to the water electrolysis side reactions (Comninellis 1994; Simonsson 1997). However, the use of sacrificial electrodes of metals which can give rise to multiple charged ions and their corresponding salts in the electrolytic systems results in coagulation and

Citation: Ashok Kumar Chopra, Arun Kumar Sharma, Removal of turbidity, COD and BOD from secondarily treated sewage water by electrolytic treatment, DOI: 10.1007/s13201-012-0066-x flocculation of dissolved and undissolved water impurities. This helps in the removal of contaminants from wastewater. Matteson et al. (1995) described a device, referred to as an "electronic coagulator" which electrochemically dissolved aluminum (from the anode) into the solution, reacting this with the hydroxyl ion (from the cathode) to form aluminum hydroxide. The aluminum hydroxide, thus formed, flocculates and coagulates the suspended solids and thereby purifies waste water. Carmona et al. (2006) reported that Al or Fe was usually used as electrode material and their actions were generated by the dissolution of sacrificial anodes upon the application of a direct current. This electrolytic process of generating metallic hydroxide flocks in situ via electro-dissolution of the sacrificial anode immersed in the waste water is referred to as electrocoagulation (EC). The generation rate of flocks can be controlled by applying varying amount of current. The electrochemically generated metallic ions can be hydrolyzed next to the anode and generate a series of metal hydroxides that are able to destabilize the dispersed particles present in the wastewater. The destabilized particles are believed to be responsible for the aggregation and precipitation of the suspended particles and for the adsorption of the dissolved and/or colloidal pollutants which are subsequently removed by sedimentation and/or flotation (Bayramoglu et al. 2004; Lung Chou 2010). Thus, the EC process offers the possibility of anodic oxidation which leads to in situ generation of adsorbents such as hydrous ferric oxides, hydroxides of aluminum, etc. (Kumar et al. 2004). The electrode material has a significant effect on the treatment efficiency in terms of cost of the treatment and removal of pollutants, and iron and aluminum electrodes are reasonably inexpensive and are easily available. These electrodes are anodically soluble leading to high wear and tear and thus generate sludge (Mollah et al. 2001; Holt et al. 2002; Kobaya and Can 2003; Daneshvar et al. 2003; Bayramoglu et al. 2004 and Chen 2004).

Electrolytic Mechanism with Al and Fe electrodes

The electrolytic process involves the generation of coagulants in situ by electrolytic oxidation of the sacrificial electrode material. Aluminum or iron is usually used as electrodes and their cations are generated by the dissolution of sacrificial anodes upon the application of direct current. The metal ions generated are hydrolyzed in the electrochemical cell to produce metal hydroxide ions according to the reactions (1)–(7) and the solubility of the metal hydroxide complexes formed depends on pH and ionic strength. Insoluble flocs are generated in a pH range between 6.0 and 7.0 as seen from the solubility diagrams of aluminum hydroxide at various pH values (Bensadok et al. 2008). The Al plates are also finding applications in wastewater treatment either alone or in combination with Fe plates due to the high coagulation efficiency of Al^{3+} (Chen 2004). Mollah et al. (2001) had reported that the electrolytic dissolution of the Al anode produces the cationic monomeric species such as Al^{3+} and $Al(OH)_2^+$ under acidic conditions. At appropriate pH values, they are transformed initially into $Al(OH)_3$ and finally polymerized to $Al_n(OH)_{3n}$ according to the following reactions:

$$Al \rightarrow Al_{(aq)}^{3+} + 3e^-. \tag{1}$$

$$Al_{(aq)}^{3+} + 3H_2O \rightarrow Al(OH)_3 + 3H_{(aq)}^+ \tag{2}$$

$$nAl(OH)_3 \rightarrow Al_n(OH)_{3n}. \tag{3}$$

However, depending on the pH of the aqueous medium, other ionic species, such as $Al(OH)_2^+$, $Al_2(OH)_2^{4+}$ and $Al(OH)_4^-$ may also be present in the system.

In addition, various forms of charged multimeric hydroxo Al^{3+} species may be formed under appropriate conditions. These gelatinous charged hydroxo cationic complexes can effectively remove pollutants by adsorption (Yetilmezsoy et al. 2009).

When a DC electric field is applied, the following electrolysis reactions are expected in the vicinity of the iron electrodes (Ofir et al. 2007).

At anode:

$$Fe_{(s)} \rightarrow Fe_{(aq)}^{2+} + 2e^- \tag{4}$$

$$Fe_{(aq)}^{2+} + 2OH_{(aq)}^- \rightarrow Fe(OH)_{2(s)} \tag{5}$$

At the cathode:

$$2H_2O_{(l)} + 2e^- \rightarrow 2OH_{(aq)}^- + H_{2(g)} \tag{6}$$

Overall

$$Fe_{(s)} + 2H_2O(l) \rightarrow Fe(OH)_{2(s)} + H_{2(g)} \tag{7}$$

Generation of iron hydroxide is followed by an electrophoretic concentration of colloids (usually negatively charged), which are then swept by the electric field in the region close to the anode. The particles subsequently interact with the iron hydroxide and can be removed by the electrostatic attraction. In the region close to the anode, the high concentration of local iron hydroxide increases the probability of coagulation of colloids.

The present investigation was focused on the electrolytic treatment of secondarily treated sewage (STS) water and to find out the removal efficiency of Al–Fe and Fe–Al electrode combinations with different electrode surface areas and inter-electrode distances.

MATERIALS AND METHODS

Collection of Wastewater Samples

The samples of STS water were collected from the outlet of activated sludge process (ASP) of the sewage treatment plant (STP), Jagjeetpur, Haridwar (Uttarakhand), India, brought to the laboratory and then used for electrolytic treatment using Al–Fe and Fe–Al electrode combinations.

Electrolytic Experimental Set Up

The schematic arrangement of the experimental setup is shown in Fig. 1. The experiments were carried out in a cylindrical reactor having a capacity of 5 L. Al and Fe electrode plates in two combinations (Al–Fe and Fe–Al) having different surface areas (40, 80, 120 and 160 cm^2) were connected to the respective anode and cathode leading to the DC rectifier and energized for a required duration of time at a fixed voltage. The inter-electrode distances between the two neighboring electrode plates varied between 2.5 and 10 cm (Table 1). All the experiments were performed at room temperature (30 ± 2 °C) and at a constant stirring speed (100 rpm) to maintain the uniform mixing of effluent during the electrolytic procedure. Before conducting an experiment, the electrodes were washed with water, dipped in dilute hydrochloric acid (HCl, 5 % v/v) for 5 min, thoroughly washed with water and finally rinsed twice with distilled water. After electrolytic treatment, the effluent was allowed to stand for 2 h and then sampled for analysis.

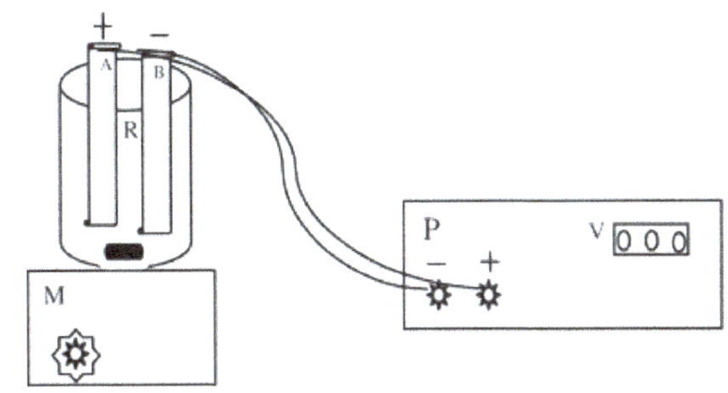

Figure. 1: Systematic design of experimental set-up. *A* Anode, *C* cathode, *R* reactor, *M* magnetic stirrer, *P* DC power supply.

Table 1: Operating conditions for electrolytic treatment of STS water

Parameters	Conditions
Electrode material (anode/cathode)	Al/Fe and Fe/Al
Applied voltage	12 V
Shape of electrode	Rectangular
Electrode area (cm^2)	40, 80, 120 and 160
Inter-electrode space (cm)	0.5, 1.0, 1.5 and 2.0
Time of operation (min)	10, 20 and 30
Volume of sample	5 L

Analytical Methods

The TD, COD and BOD of wastewater were analyzed before and after the electrolytic treatment following the standard methods for examination of water and wastewater (APHA 2005). The calculation of TD, COD and BOD removal efficiencies after electrolytic treatment was carried out using the formula:

$$\text{CR } \% = \frac{C_0 - C}{C_0} \times 100,$$

where C_0 and C are TD, COD or BOD of wastewater before and after electrolysis.

RESULTS AND DISCUSSION

Removal of chemical and biological impurities from the contaminated water system by the process of electrolysis is governed by several factors including electrode material and distance between them, time of electrolysis, electrical parameters such as voltage and current densities, pH of the system and last but not the least the presence of other coagulants in the system, This preliminary study was devoted to figure out effects of electrode systems consisting of different materials on STS water treatment. Changes in surface area and distance between the electrodes were studied in detail. The characteristics of different parameters of the STS water are given in Table 2. The removal of TD, COD and BOD of STS water with electrode combinations (Al–Fe and Fe–Al) using different surface areas (40–160 cm^2) and different inter-electrode distances (2.5–10 cm) are shown in Figs. 2, 3, 4, 5 and 6.

Table 2: Characteristics of STS water

Parameter	Mean ± SD
pH	7.40 ± 0.17
Conductivity (μS)	727.8 ± 23.05
TDS (mg/L)	472.5 ± 24.66
TD (NTU)	16.34 ± 1.28
BOD (mg/L)	53.31 ± 2.37
COD (mg/L)	106.69 ± 8.11

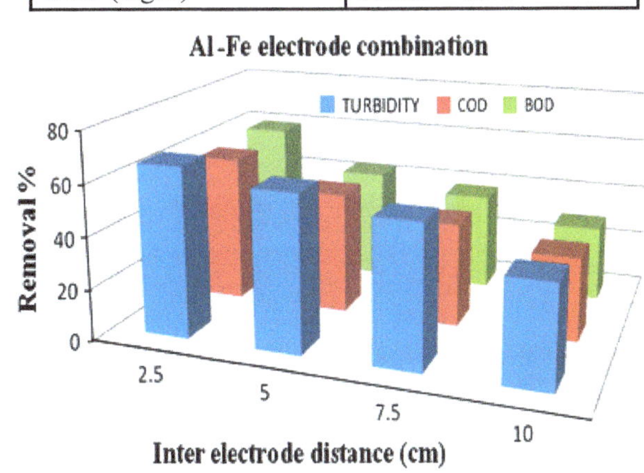

Figure. 2: Percentage removal of TD, COD and BOD of STS water using Al–Fe electrode combination with different inter-electrode distances at constant voltage (12 V), time (30 min) and electrode area (80 cm^2).

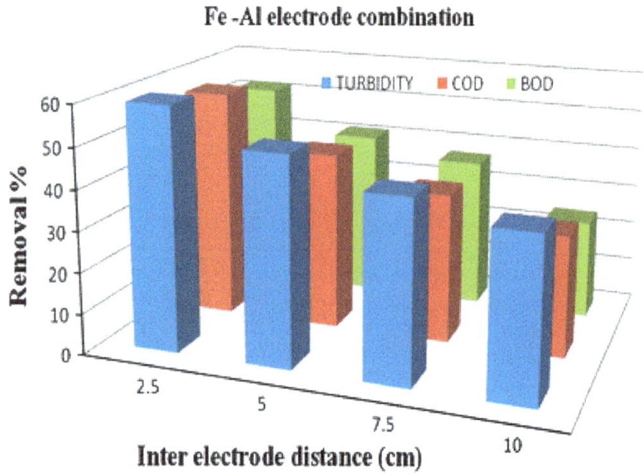

Figure. 3: Percentage removal of TD, COD and BOD of STS water using Fe–Al electrode combination with different inter-electrode distances at constant voltage (12 V), time (30 min) and electrode area (80 cm²).

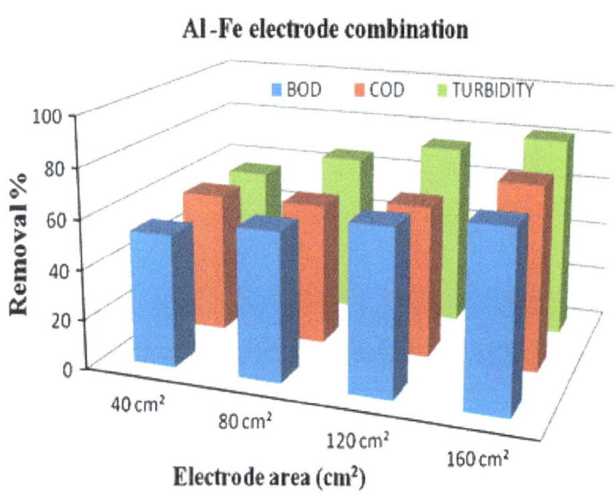

Figure. 4: Percentage removal of TD, COD and BOD of STS water using Al–Fe electrode combination with different electrode areas at constant voltage (12 V), time (30 min) and inter-electrode distance (2.5 cm).

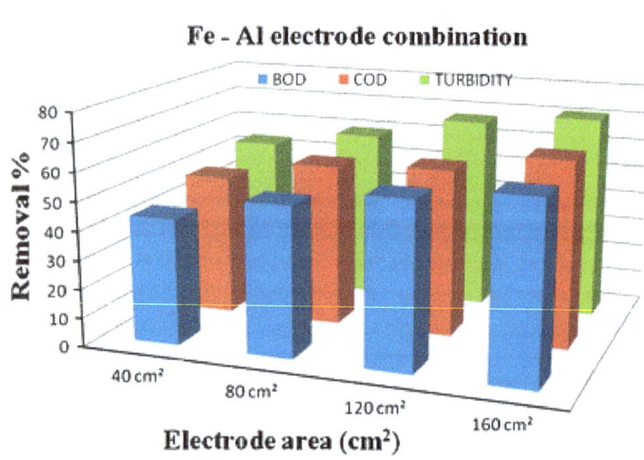

Figure. 5: Percentage removal of TD, COD and BOD of STS water using Fe–Al electrode combination with different electrode areas at constant voltage (12 V), time (30 min) and inter-electrode distance (2.5 cm).

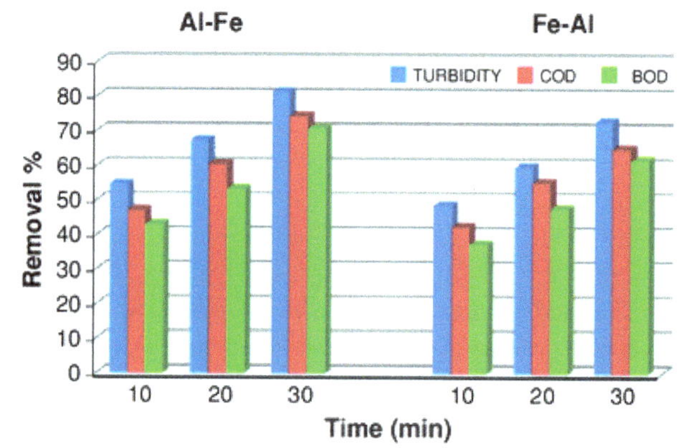

Figure. 6: Percentage removal of TD, COD and BOD of STS water using Al–Fe and Fe–Al electrode combination at constant voltage (12 V), inter-electrode distance (2.5 cm) and electrode area (160 cm²) with different time,

Inter-electrode distance was observed to be an effective factor in the electrolytic treatment of STS water. The removal percentage of TD, COD and BOD increased progressively with decrease in inter-electrode distance from 10.0 to 2.5 cm, whereby it exhibited the maximum removal of TD (65.9 %), COD (57.41 %) and BOD (59.56 %) at the shortest distance (2.5 cm) between the electrodes (Al and Fe) with each electrode area of 80 cm², whereas the Fe and Al electrode combination showed the removal of TD (59.66 %), COD (56.46 %) and BOD (51.99 %) (Figs. 2,3). Similar observations have also been reported by Li et al. (2008) that COD decreases with the decrease in distance between electrodes of the same composition. This is because the shorter distance speeds up the anion discharge on the anode and improves the oxidation. It also reduces resistance, the electricity consumption and the cost of the wastewater treatment. Ghosh et al. (2008) have also observed that with the increase of inter-electrode distance, the percentage removal of dye products from waste water decreases. At a lower inter-electrode distance, the resistance encountered by current flowing in the solution medium decreases thereby facilitating the electrolytic process and resulting in enhanced dye removal. The above results also indicated the superiority of aluminum as sacrificial electrode when compared to that of iron as sacrificial electrode. This can probably be attributed to better coagulating properties of Al^{3+} to those of oxidized products of Fe. It may be due to the fact that the majority of Al^{3+} ions subsequently precipitates in the form of hydroxides. The adsorption of Al^{3+} ion with colloidal pollutants results in coagulation, and resulting coagulants can be more efficiently removed by settling, surface complexation and electrostatic attraction in comparison to Fe^{2+} ions.

With a fourfold increase in the electrode area of Al–Fe from 40 to 160 cm², the current increased from 0.24 to 0.58 A; this resulted in an increase in the removal percentage of TD, COD and BOD. Highest removal efficiencies of 81.51 % (TD), 74.36 % (COD) and 70.86 % (BOD) were achieved at electrode area of 160 and at 2.5 cm inter-electrode distance. The removal efficiency can be attributed to a greater electrode area that produced larger amounts of anions and cations from the anode and cathode. The greater electrode area increased the rate of flock's formation, which in turn influenced the removal efficiency (Figs. 4, 5). Escobara et al. (2006) have also observed logistical relationship between electrode geometric area (A_G) and copper removal efficiency and concluded that the increase in copper removal was related to an increase in A_G, reaching an optimal value of 35 cm^2, with an asymptotic value near 80 %. In the case of Fe–Al electrode combination the removal efficiency of TD, COD and BOD was 71.11, 64.95 and 61.87 %, respectively, which was somewhat lower than the values obtained with Al–Fe electrode.

Also in the present study, the electrolytic reactor equipped with a higher electrode area of Al and Fe electrode combinations was able to produce significant quantities of coagulants, thereby indicating the enhancement in removal efficiency of TD, COD and BOD from STS water. The increase in the electrode area during electrolytic treatment was predicted to an increase in number of hydroxide ions (OH^-) in solution resulting from water reduction at the cathode.

$$2H_2O(l) + 2e^- \Leftrightarrow H_2(g) + 2OH^-$$

In the electrolytic treatment, the selection of suitable electrode material is important and so is the time required to effect an acceptable removal of dissolved and undissolved impurities. Therefore, we studied the electrolytic treatment using both electrode combinations under the same conditions but as a function of time. The comparative results of TD, COD and BOD removal, obtained with the same voltage (12 V), same inter-electrode spacing (2.5 cm) and same area of electrode (160 cm^2) but a varying time of up to 30 min again demonstrated the superiority of Al–Fe electrode combination over that of Fe–Al electrode combination (Fig. 6). During their study of the electrolytic treatment of latex wastewater, Vijayaraghavan et al. (2008) also observed that the increase in the electrolysis period resulted in a decrease in residual COD and BOD concentrations irrespective of the current densities. An increase in the operating time from 10 to 60 min in the treatment of the baker's yeast wastewater by electrocoagulation resulted in an increase in the removal efficiencies of COD, TOC and turbidity as reported by Kobya and Delipinar (2008).

Metallic hydroxides are produced up to a sufficient concentration of coagulant inducing the formation of white and slightly greenish precipitate using Al–Fe and Fe–Al electrode combination as the OH^- of Al and Fe, respectively. This indicates that the STS water can be efficiently treated with Al–Fe combination that ensures better adsorption of soluble and colloidal species which settles down in the form of $Al(OH)_3$ from STS water. Zongo et al. (2009) in their investigation of electro-coagulation for the treatment of textile wastewater with Al or Fe electrode elucidated that the Fe electrode is easily dissolved in water in comparison with Al. However, the use of iron electrodes often results in the formation of very fine brown particles which are less prone to settling than the gel flows formed with aluminum. For further re-use of the treated water, the post-treatment downstream of the electro-coagulation–electro-flotation system might represent a penalty to the use of iron over aluminum.

The present finding is in support of Lai and Lin (2003) who observed that the Al–Fe electrode pair is deemed to be a better choice out of the five electrode pair combinations tested. They also observed that Al–Fe electrode pair offers good overall COD and copper removal, low final wastewater NTU and reasonably low sludge production. Adhoum and Monser (2004) using Al electrodes achieved a COD removal efficiency of 76 % in the treatment of olive mill effluents. Ilhan et al. (2008) indicated the maximum removal of COD 56 and 35 % on the EC of leachate using Al and Fe electrode, respectively, in 30-min contact time. According to Lung Chou (2010), removal efficiency of polyvinyl alcohol (PVA) from aqueous solutions for Fe–Al and Fe–Fe pairs using Fe as the anode was greater than those of Al–Al and Al–Fe pairs using Al as the anode. This has been explained by the chemical reactions that take place at the aluminum anode and the iron anode. Katal and Pahlavanzadeh (2011) observed that the Fe–Al electrode combination has higher COD removal efficiency in comparison to Al–Fe electrode combination, while in present study, Al–Fe electrode combination was more efficient in comparison to Fe–Al electrode combination. In our opinion, this difference can probably be attributed to the different types of contaminants present in the waste water being studied. Iron is a *3d* block transition metal and has better complexing properties with organic/inorganic impurities present in water which can act as complexing ligands than aluminum which is a *p* block metal and lacks empty *d*-orbitals necessary for making coordination compounds. Therefore, Fe–Al electrode system may be a better choice if electron-rich nucleophilic organic compounds such as dyes, their intermediates and degradation products are present in waste water. However, in the present study, Al–Fe system was found to be more efficient. Therefore, chemical behavior of the contaminants present in waste water may be a deciding factor in the selection of anodic sacrificial electrode.

Energy consumption and operating cost

Electrical energy and electrode consumption are important economical parameters in EC process. In EC process, the operating cost includes material, mainly electrodes and electrical energy costs, as well as labor, maintenance, sludge dewatering and its disposal. In the present study, energy and electrode material costs have been taken into account as major cost items in the calculation of the operating cost (US \$/m^3) (Ghosh et al. 2008) as follows:

$$\text{Operating cost} = a\,C_{\text{energy}} + b\,C_{\text{electrode}},$$

where C_{energy} (kWh/m^3) and $C_{\text{electrode}}$ (kg Al/m^3) are the consumption quantities for the turbidity, COD and BOD removal, "*a*" is the electrical energy price 0.1 US\$/kWh, "*b*" is the electrode material price 3.4 US\$/kg for Al electrode and 1.3 US\$/kg for Fe electrode. Cost due to electrical energy (kWh/m^3) is calculated as:

$$C_{\text{energy}} = \frac{U \times I \times t\,EC}{v}.$$

Cost for electrode (kg Al/m^3) was calculated as follows using the equation:

$$C_{\text{electrode}} = \frac{I \times t \times M_w}{z \times F \times v},$$

where *U* is the cell voltage (*V*), *I* current (*A*), t_{EC} time of electrolysis (*s*), *v* volume (m^3) of STS water, MW molecular mass of aluminum (26.98 g/mol) and iron (55.84 g/mol), *z* no. of electrons transferred ($z = 3$ for Al and 2 for Fe) and *F* is the Faraday's constant (96487C/mol) .

For both electrode combinations (Al–Fe and Fe–Al), the energy consumption increased from 1.04 to 2.5 kWh/m^3 with an increase in current from 0.24 to 0.58 A that resulted in increasing the electrode consumption (0.85×10^{-5} to $6.04 \times$

10^{-5} kg/m^3). The cost due to electrical energy consumption as well as an electrode assembly was calculated for both electrode combinations at optimum operating condition. The operating cost of Fe–Al (0.25006 US\$/m^3) electrode combination was found to be slightly higher than Al–Fe (0.22906 US\$/m^3) electrode combination (Table 3).

Table 3: Economic evaluation of Al–Fe and Fe–Al electrode combination at optimum operating condition (current 0.53 and 0.58 A; voltage 12 V; time 30 min)

Electrode combination	Electrode consumption (kg/m^3)	Energy consumption (kWh/m^3)	Operating cost US\$/m^3
Al–Fe	1.77×10^{-5}	2.29	0.22906
Fe–Al	6.04×10^{-5}	2.5	0.25006

Kinetic Study of Turbidity, COD and BOD

The rate of removal of TD, COD and BOD is represented by the following first-order mechanism (El-Ashtoukhy and Amin 2010):

ln(C0Ct)=kt,ln(C0Ct)=kt,

where C_0 is the initial concentration (mg/L), C_t final concentration with respect to time, t the time (min) and k is the rate constant (min^{-1}) for TD, COD and BOD for electrolytic treatment using Al–Fe and Fe–Al electrode combination. Rate constants for electrolytic treatment of TD, COD and BOD from STS water using two types of electrode combination are given in Table 4.

Table 4: Rate constant (k) (min^{-1}) values at variable distances between electrode and their correlation coefficients (r^2)

Distance between electrodes	Al–Fe electrode combination						Fe–Al electrode combination					
	TD k (min^{-1})	r^2	BOD k (min^{-1})	r^2	COD k (min^{-1})	r^2	TD k (min^{-1})	r^2	BOD k (min^{-1})	r^2	COD k (min^{-1})	r^2
2.5	0.026	0.991	0.018	0.999	0.020	0.998	0.019	0.990	0.014	0.998	0.015	0.994
5.0	0.015	0.997	0.013	0.994	0.015	0.967	0.015	0.992	0.010	0.984	0.012	0.994
7.5	0.013	0.999	0.009	0.996	0.010	0.987	0.012	0.995	0.009	0.988	0.008	0.978
10.0	0.010	0.990	0.006	0.994	0.007	0.982	0.008	0.959	0.005	0.966	0.006	0.998

The kinetic study on the distances between electrodes for electrolytic treatment has not been given due consideration so far. There appears to be no work with regard to the kinetic study on the distance between electrodes during electrolytic treatment. In the present study, it was revealed that there is a strong positive correlation between inter-electrode space, and TD, COD as well as BOD abatement rates and rate of coefficients. The pseudo-first-order abatement kinetic was relatively fitted. The decrease in the distance between electrodes from 10.0 to 2.5 cm increased the rate constant from 0.01 to 0.026 min^{-1} for TD, 0.007 to 0.020 min^{-1} for COD and 0.006 to 0.018 min^{-1} for BOD using Al–Fe electrode combination and from 0.008 to 0.019 min^{-1}for TD, 0.006 to 0.015 min^{-1} for COD and 0.005 to 0.014 min^{-1} for BOD using Fe–Al electrode combination. The increase in the rate constant of both Al–Fe and Fe–Al electrode combinations may be ascribed to the decrease of TD, COD and BOD of the STS water. The use of this kinetic study showed high correlation coefficients ($r^2 \geq 0.959$). Thus, the kinetic study is more suitable for explaining the efficiency of distance between electrodes for electrolytic treatment.

CONCLUSION

The use of electrode systems using different metals for anodes and cathodes was studied in an attempt to improve upon the existing system and to further understand the process of electrolysis. The removal of TD, COD and BOD was found to be dependent on the inter-electrode distances, electrode areas and the electrode combinations (Al–Fe and Fe–Al) in the treatment of STS water. An increase in the surface area of the electrodes and a decrease in the distance between them resulted in better removal of contaminants from the waste water; the optimal removal has been obtained with the use of an electrode area of 160 cm^2 and a short distance of 2.5 cm between electrodes in a 5-L reactor. Acquired results of the present study could be specified and evaluated by employing pseudo-first-order kinetics. The electrical energy consumption was calculated as 0.229 kWh/m^3 for Al–Fe and 0.25 kWh/m^3 for Fe–Al electrode combination. Al–Fe electrode combination proved more effective in comparison to Fe–Al electrode combination for the treatment of STS water. Due to economical constraint, EC with Al–Fe electrode combination should be preferred in comparison to Fe–Al electrode combination. Al anode was more efficient in comparison to Fe anode establishing the superiority of aluminum as the preferred material for sacrificial electrode for the treatment of sewage water obtained from STP, Jagjeetpur, Haridwar (Uttarakhand), India.

ACKNOWLEDGMENTS

The University Grant Commission, New Delhi, India is acknowledged for providing the financial support in the form of UGC research fellowship (F.4-1/2006 (BSR) 7-70/2007 BSR) to Mr. Arun Kumar Sharma.

REFERENCES

1. Adhoum N, Monser L (2004) Decolourization and removal of phenolic compounds from olive mill wastewater by electro coagulation. Chem Eng Process 43:1281–1287
2. APHA (2005) Standard methods for the examination of water and wastewater, American Public Health Association, 21st edn. Washington, DC
3. Bayramoglu M, Kobya M, Can OT, Sozbir M (2004) Operating costs analysis of electrocoagulation of textile dye wastewater. Sep Purif Technol 37:117–125
4. Bensadok K, Benammar S, Lapicque F, Nezzal G (2008) Electrocoagulation of cutting oil emulsions using aluminum plate electrodes. J Hazard Mater 152(1):423–430
5. Carmona M, Khemis M, Leclerc JP, Lapicque F (2006) A simple model to predict the removal of oil suspensions from water using the electrocoagulation technique. Chem Eng Sci 61:1237–1246
6. Chen G (2004) Electrochemical technologies in wastewater treatment. Sep Purif Technol 38:11–41
7. Comninellis Ch (1994) Electrocatalysis in the electrochemical conversion/combustion of organic pollutants for wastewater treatment. Electrochim Acta 39(11):1857–1862
8. Daneshvar N, Ashassi-Sorkhabi H, Tizpar A (2003) Decolorization of orange II by electrocoagulation meted. Sep Purif Technol 31:153–162
9. El-Ashtoukhy ESZ, Amin NK (2010) Removal of acid green dye 50 from wastewater by anodic oxidation and electrocoagulation—a comparative study. J Hazard Mater 179:113–119
10. Escobara C, Cesar SS, Toral M (2006) Optimization of the electrocoagulation process for the removal of copper, lead and cadmium in natural waters and simulated wastewater. J Environ Manag 81(4):384–391
11. Ghosh D, Medhi CR, Solanki H, Purkait MK (2008) Decolorization of crystal violet solution by electrocoagulation. J Environ Prot Sci 2:25–35
12. Holt PK, Barton GW, Wark M, Mitchell CA (2002) A quantitative comparison between chemical dosing and electrocoagulation. Colloids Surf A Physicochem Eng Aspects 211:233–248
13. Ilhan F, Kurt U, Apaydin O, Gonullu MT (2008) Treatment of leachate by electrocoagulation using aluminum and iron electrodes. J Hazard Mater 154:381–389
14. Katal R, Pahlavanzadeh H (2011) Influence of different combinations of aluminum and iron electrode on electrocoagulation efficiency: Application to the treatment of paper mill wastewater. Desalination 265:199–205
15. Kobaya MOT, Can Bayramoglu M (2003) Treatment of textile wastewaters by electrocoagulation using iron and aluminum electrodes. J Hazard Mater B 100:163–178
16. Kobya M, Delipinar S (2008) Treatment of the baker's yeast wastewater by electrocoagulation. J Hazard Mater 154:1133–1140
17. Kumar PR, Chaudhari S, Khilar KC, Mahajan SP (2004) Removal arsenic from water by electrocoagulation. Chemosphere 55:1245–1252
18. Lai CL, Lin SH (2003) Electrocoagulation of chemical mechanical polishing (CMP) wastewater from semiconductor fabrication. Chem Eng J 95:205–211
19. Li Xu, Wang Wei, Wang Mingyu, Cai Yongyi (2008) Electrochemical degradation of tridecane dicarboxylic acid wastewater with tantalum-based diamond film electrode. Desalination 222:388–393
20. Lung Chou W (2010) Removal and adsorption characteristics of polyvinyl alcohol from aqueous solutions using electrocoagulation. J Hazard Mater 177:842–850
21. Matteson MJ, Dobson RL, Glenn RW Jr, Kukunoor NS, Waits WH III, Clayfield EJ (1995) Electrocoagulation and separation of aqueous suspensions of ultrafine particles. Colloids Surf A 104(1):101–109
22. Mollah MYA, Schennach R, Parga JR, Cocke DL (2001) Electrocoagulation (EC)—science and applications. J Hazard Mater B 84:29–41
23. Ofir E, Oren Y, Adin A (2007) Comparing pretreatment by iron of electro flocculation and chemical flocculation. Desalination 204:87–93
24. Simonsson D (1997) Electrochemistry for a cleaner environment. Chem Soc Rev 26:181–189
25. Vijayaraghavan K, Ahmad D, Yuzri A, Yazid A (2008) Electrolytic treatment of latex wastewater. Desalination 219:214–221
26. Yetilmezsoy K, Ilhan F, Zengin ZS, Sakar S, Gonullu MT (2009) Decolorization and COD reduction of UASB pretreated poultry manure wastewater by electrocoagulation process: A post-treatment study. J Hazard Mater 162:120–132
27. Zongo I, Maiga AH, Wéthé J, Valentina G, Leclerca JP, Paternottea G, Lapicquea F (2009) Electrocoagulation for the treatment of textile wastewaters with Al or Fe electrodes: compared variations in COD levels, turbidity and absorbance. J Hazard Mater 169:70

CHAPTER 9

BIOLOGICAL REMOVAL OF PHENOL FROM WASTEWATERS: A MINI REVIEW

N. V. Pradeep [1], S. Anupama[2], K. Navya[3] H. N. Shalini[3], M. Idris[4], U. S. Hampannavar[5]

[1]Department of Civil Engineering, Angadi Institute of Technology and Management, Belgaum, India

[2]Department of Civil Engineering, M. V. J. College of Engineering, Bangalore, India

[3]Department of Biotechnology, Ballari Institute of Technology and Management, Bellary, India

[4]Department of Civil Engineering, University Visvesvaraya, College of Engineering, Bangalore University, Bangalore, India

[5]Angadi Institute of Technology and Management, Belgaum,India

ABSTRACT

Phenol and its derivatives are common water pollutants and include wide variety of organic chemicals. Phenol poisoning can occur by skin absorption, inhalation, ingestion and various other methods which can result in health effects. High exposures to phenol may be fatal to human beings. Accumulation of phenol creates toxicity both for flora and fauna. Therefore, removal of phenol is crucial to perpetuate the environment and individual. Among various treatment methods available for removal of phenols, biodegradation is environmental friendly. Biological methods are gaining importance as they convert the wastes into harmless end products. The present work focuses on assessment of biological removal (biodegradation) of phenol. Various factors influence the efficiency of biodegradation of phenol such as ability of the microorganism, enzymes involved, the mechanism of degradation and influencing factors. This study describes about the sources of phenol, adverse effects on the environment, microorganisms involved in the biodegradation (aerobic and anaerobic) and enzymes that polymerize phenol.

INTRODUCTION

One of the most important environmental problems faced by the world is management of wastes (Anupama et al. 2013). Industrial processes generate a variety of molecules that may pollute air and waters due to negative impacts such as toxicity, carcinogenic and mutagenic properties (Busca et al. 2008). Phenols are the major organic constituents found in effluents of coal conversion processes, coke ovens, petroleum refineries, phenolic resin manufacturing, herbicide manufacturing, fiberglass manufacturing and petrochemicals (El-Ashtoukhy et al.2013; Veeresh et al. 2004; Jadhav and Vanjara 2004). Phenol and its derivatives are a major source of environmental pollutants (Said et al. 2013; Varma and Gaikwad 2008). Phenol, a waste product of industrial processes that is introduced into aquatic ecosystems, adversely affects the indigenous biota, including algae, protozoa, invertebrates, and vertebrates (Babich and Davis 1981). The concentrations of these compounds can range from one to several hundred mg/L (Moussavi et al. 2008). Industrial wastewaters associated with the manufacture of halogenated organics characteristically have concentrations as high as hundreds of mg/L (Annachatre and Gheewala 1996). Water pollution by organic and inorganic compounds is of great public concern

Citation: N. V. Pradeep, S. Anupama, K. Navya, H. N. Shalini, M. Idris, U. S. Hampannavar, Biological removal of phenol from wastewaters: a mini review, doi:10.1007/s13201-014-0176-8

(Pradeep et al. 2014). Their fate in the environment is of great importance as they are toxic, recalcitrant and bioaccumulating in organisms (Annachatre and Gheewala1996). As it adversely affects the aquatic biota, phenol is one of the 129 specific priority chemicals that is considered toxic under the 1977 Amendments to the Clean Water Act and for which the US Environmental Protection Agency (EPA) has issued water quality criteria (Singh et al. 2013; Babich and Davis 1981). Phenol at concentration as low as 5 mg/L imparts typical smell upon chlorination and the World Health Organization (WHO) prescribed 1 mg/L as the maximum permissible concentration of phenol in drinking water (Saravanan et al. 2008).

Effects of Phenol

Phenol and its derivatives penetrate ecosystems as the result of drainage of the municipal or industrial sewage to surface water. The pollution of the aquatic environment by phenols could modify the biota of this environment because most of these compounds exhibit a high degree of toxicity (Lika and Papadakis 2009). Acute exposure to phenol is known to cause skin irritation, gastrointestinal discomfort, and headaches. Phenol is toxic to the nervous system, the heart, the kidneys, and the liver and is readily absorbed through skin and mucosa (Wang et al. 2011). Toxicity of phenol towards plants has been proved. The willow trees exposed to 1,000 mg/L phenol wilted and eventually died (Wang et al. 2011). Phenol can also inhibit synthesis and replication of DNA in cells. A study revealed that phenol stopped preparation of DNA in diploid human fibroblasts (Michalowicz and Duda 2007).

Biological Removal of Phenol

Biological treatment transforms the wastes into simple end products (Pradeep et al. 2011). For this reason the interest in the use of biological methods is increasing (Lika and Papadakis2009). Although the rate and extent of biodegradation of a chemical compound largely depends upon its structure and the environment in its vicinity, halogens containing organic compounds in particular are found to be biochemically resistant (Annachatre and Gheewala 1996). Biological removal can be classified into microbial and enzymatic methods.

Microbial Removal of Phenol

Microbial degradation is a useful strategy to eliminate organic compounds and detoxify wastewaters and polluted environments (Gallego et al. 2003). Phenol is degraded by diverse microorganisms including yeasts, fungi and bacteria (Table 1). Because of widespread occurrence of phenol in the environment, many microorganisms utilize phenol as the sole carbon and energy source which includes both aerobic and anaerobic microorganisms (Basha et al. 2010).

Table 1: Various microorganisms capable of degrading phenol

Microorganism	References
Pseudomonas putida	Seker et al. (1997)
Corynebacterium sp., *S. aureus*, *Proteus* sp., *B. subtilis* and *Staphylococcus* sp.	Ajaz et al. (2004)
Kelibsiella, *Citrobacter* and *Shigella*	Kafilzadeh et al. (2010)
Pseudomonas aeruginosa and *Pseudomonas fluorescence*	Sgountzos et al. (2006), Agarry et al. (2008), Wang et al. (2011)
Acinetobacter baumannii	Prasad et al. (2010)
Alcaligenes faecalis	Thomas et al. (2002)
Candida species	Ehrhardt and Rehm (1985), Varma and Gaikwad (2008), Jiang et al. (2010)
Aspergillus sp.	Passos et al. (2010)
Rhodococcus erythropolis	Zidkova et al. (2012)
Pseudomonas stutzeri	Viggiani et al. (2006)
Bacillus brevis	Arutchelvan et al. (2006)

Aerobic Biodegradation of Phenol

Aerobic biodegradation of phenol, resulting in its complete mineralization, is generally preferred in wastewater clean-up. Phenol is recognized as an inhibitory substrate at relatively low concentrations (100 mg/L) and is a convenient model for studying the kinetics of aromatic molecule degradation (Christen et al. 2012). Since microbial growth is inhibited by higher concentrations of phenol, various approaches were used to overcome this substrate inhibition. Microorganisms can be adapted to higher phenol concentration by stepwise increasing the phenol concentration (Banerjee 1997), cell immobilization (biofilm formation) or procedures of genetic manipulations. Industrial strain improvement has currently moved from simple adaptation and selection to a targeted metabolic engineering (Zidkova et al. 2012).

Chakraborty et al. (2010) studied the phenol degradation using native microorganisms isolated from coke-oven wastewater. Initial phenol concentration of 200 mg/L was given as sole carbon source. Phenol removal was 76.76 % at optimal conditions. Glucose addition up to a specific low concentration could improve the degradation rate, but impeded the degradation process at higher concentrations. Paraskevi and Euripides (2005) used enriched mixed culture from samples of petroleum-contaminated soil in Denmark for the isolation of microorganisms, which were capable of growing on phenol as a sole source of carbon and energy. Above 300 mg/L of initial phenol concentration no considerable depletion was recorded. Maximum degradation rates for phenol were recorded at 30 °C. Seker et al. (1997) carried out phenol degradation studies using pure culture of *Pseudomonas putida* at 30 °C. The adaptation of species to phenol was done using mineral salt medium (except glucose and yeast extract). Continuos phenol degradation was carried out in a fermentor. Dual-substrate growth kinetics, Haldane kinetics for phenol and Monod kinetics for oxygen were derived with high correlation coefficients. Baek et al. (2003) isolated and characterized the bacteria capable of degrading phenol and reducing nitrate under low-oxygen conditions. The bacteria were isolated from natural and contaminated environments under low-oxygen conditions and were grown using phenol as sole carbon source. Isolated bacteria were classified according to the best similarity with the sequences of the 16S rRNA in GenBank database. The phylogenetic relationships of the phenol-degrading isolates were determined by comparison of the sequences of the 16S rRNA genes. The order of*Rhizobiales* has not been well-known for its ability to degrade aromatic compounds. This study showed that *Rhizobiales* can degrade phenol. Ajaz et al. (2004) carried out a study on phenol degradation using microorganisms isolated from garden soil. The identification, characterization and genetical studies of these strains were carried out. The microorganisms were identified as *Corynebacterium* sp., *S. aureus*, *Proteus* sp., *B. subtilis* and *Staphylococcus*sp. Microorganisms were able to degrade phenol without any adaptation period. This is due to the fact that the microorganisms were able to sustain and use phenol as carbon which was present in the rhizosphere environment. Kafilzadeh et al. (2010) isolated and identified phenol-degrading bacteria from Lake Parishan and they assayed their kinetic growth. The ability of bacteria to degrade different concentrations of phenol was measured using culturing bacteria in different concentrations of phenol from 0.2 to 0.9 g/L. Most of the isolated bacteria showed a good ability of degradation of phenol, where *Pseudomonas* and *Acinetobacter* showed 0.8–0.9 g/L, and *Kelibsiella*, *Citrobacter* and *Shigella* showed 0.6–0.7 g/L and the rest showed 0.2–0.3 g/L of phenol degradation. Agarry et al. (2008) studied the bioremediation potential of an indigenous binary mixed culture of *Pseudomonas aeruginosa* and *Pseudomonas fluorescence*in batch culture using synthetic phenol in water in the concentration range of 100–500 mg/L as a model limiting substrate. The effect of initial phenol concentration on the degradation process was investigated. Prasad et al. (2010) isolated *Acinetobacter baumannii* and tested for phenol degradation. This bacterium had high phenol degradation activity and high tolerance to phenol. The biodegradation assays were performed in liquid medium with phenol as single substrate. Different concentrations of phenol samples ranging from 125 to 1,000 mg/L were taken for observations. Phenol was completely degraded at different cultivation times with different initial phenol concentrations. Sameer et al. (2002) studied the biodegradation of synthetically prepared phenol wastewater in a single stage, bench-scale rotating biological contactor (RBC). The effect of process variables, namely rotational speed (40–175 rpm), input phenol loading and temperature of wastewater (20–30 °C) on the amount of phenol removed in the system was investigated. It was observed that an increase in the speed of rotation significantly improved the performance. An increase in the hydraulic loading rates caused a reduction in the phenol removal rate, while an increase in the organic loading rate resulted in an improvement in performance. An increase in temperature caused an increase in the microbial activity and, therefore, gave better performance.

Anaerobic Biodegradation of Phenol

During anaerobic digestion of organic waste, both energy-rich biogas and a nutrient-rich digestate are produced. The digestate can be used as a fertilizer in agricultural soils if the levels of hazardous compounds and pathogens are low. Chemical analysis of digestate from bioreactors operating at thermophilic temperature has detected higher content of phenols compared to mesophilic bioreactors, verifying the degradation results. Digestate with the highest phenol content has the greatest negative impact on soil microbial activity (Leven et al.2012).

Veeresh et al. (2004) used upflow anaerobic sludge blanket (UASB) for treatment of phenolic wastewater. Phenol concentrations greater than 500 mg/L was effectively treated with acclimatization of inocula. Bench-scale studies showed that phenol and cresols were degraded with and without co-substrate in a UASB process with a suitable operational strategy. Maintenance of substrate concentration within their inhibitory range in the reactor is the utmost control strategy in the treatment of phenolic wastes. Azbar et al. (2009) studied phenol removal using anaerobic hybrid reactor. Phenol removal efficiency ranged from 39 to 80 % at different conditions. Biogas was obtained during this process. Shibata et al. (2006) examined microbial degradation of various phenol and its derivatives under both aerobic and anaerobic conditions in seven Japanese paddy soils. Under flooded and anaerobic conditions, 4-*n*-propylphenol would be degraded as well as phenol and *p*-cresol while alkylphenols with long and branched alkyl chains were hardly degraded.

Phenol can also be degraded in the absence of oxygen and it is less advanced than the aerobic process. It is based on the analogy with the anaerobic benzoate pathway proposed for*Paracoccus denitrificans* in 1970. In this pathway, phenol is carboxylated in the para position to 4-hydroxybenzoate which is the first step in the anaerobic pathway. Here, the enzyme involved is the 4-hydroxybenzoate carboxylase. The anaerobic degradation of several other aromatic compounds has been shown to include a carboxylation reaction. The organisms capable of degrading phenol under anaerobic conditions were *Thauera aromatica* and *Desulphobacterium phenolicum* (Basha et al. 2010).

Enzymatic Polymerization of Phenol

Enzymes are proteins and are found in all types of living cells. A given cell will characteristically contain hundreds of enzymes. Enzymes are biological catalysts that increase the rate of chemical reactions taking place within living cells. Although an enzyme increases the rate of reaction, it does not affect the reaction equilibrium. The acceleration achieved by enzymatic catalysis is often tremendous, with some reaction rates propelled to one million times faster than the rate in the absence of enzymes (Bailey and Ollis 1986).

Enzymes catalyze specific reactions and mostly act under moderate conditions (temperature, pH, solvents and ionic strength). Hence enzymes represent a promising tool for the selective removal of pollutants from waste streams (Demarche et al. 2011). Enzyme-based processes act with high specificity and are very efficient in removing targeted compounds. The use of enzyme-based techniques to remove organic compounds from aqueous solution was first proposed by Klibanov and colleagues and has been continuously improved since then. Peroxidases are the most widely reported enzymes for phenol polymerization when compared with other enzymes. Authors have reported the use of purified horseradish peroxidase (HRP) to remove 30 different phenols and aromatics amines (Wilberg et al. 2000; Cooper and Nicell 1996).

Phenol conversion is activated by H_2O_2; the enzyme catalyzes the oxidation of aromatic compounds, forming free radicals which undergo spontaneous polymerisation (Wilberg et al.2002). Soybean seed hulls have been identified as a rich source of peroxidase, the soybean peroxidase (SBP), and being a by-product of soybean food industry, they provide a cheap and abundant source of peroxidase (Wilberg et al. 2002; Hejri and Saboora 2009). Radish roots contain peroxidase enzyme and can be used for the removal of phenol from wastewaters (Naghibi et al. 2003).

Enzyme (peroxidases) mechanism for degradation of aromatic compounds is as follows:

One-electron oxidation of aromatic substrates (AH_2) catalyzed by peroxidase is depicted by the Chance-George mechanism as the following:

$$E + H_2O_2 \rightarrow E_i + H_2O \quad (1)$$

$$E_i + AH_2 \rightarrow E_{ii} + AH \quad (2)$$

$$E_{ii} + AH_2 \rightarrow E + AH\cdot + H_2O \quad (3)$$

$$E_{ii} + H_2O_2 \rightarrow E_{iii} + H_2O. \quad (4)$$

The native enzyme (E) is oxidized by hydrogen peroxide (H_2O_2) to an active intermediate enzymatic form called compound I (E_i). Compound I accepts an aromatic compound (AH_2) into its active site and carries out its oxidation. A free radical (AH•) is produced and released into solution leaving the enzyme in the compound II (E_{ii}) state. Compound II oxidizes a second aromatic molecule, releasing another free radical product and returning the enzyme to its native state, thereby completing the cycle. The overall peroxidase reaction consists of the reactions described by Eqs. (1)–(3).

In the presence of excess hydrogen peroxide, the reaction of Eq. (4) becomes important because compound III (E_{iii}) is a reversibly inactivated form of the enzyme. This implies that enzyme is inhibited by H_2O_2 in excess. On the other hand, lack of hydrogen peroxide during the reaction step limits the rate of reaction. The semi-batch addition of H_2O_2 to maintain an optimized ratio between hydrogen peroxide and enzyme concentrations was found to suppress this inhibition (Kulkarni and Kaware 2013; Wilberg et al. 2000).

In the process of reducing the enzyme, phenol and its derivatives are transformed to phenolic-free radicals. Phenolic-free radicals catalytically generated during the peroxidase cycle will either combine with each other to form phenolic polymers or undergo radical transfer with phenolic monomers or polymers present in solution converting them to radical species (Zou and Taylor 1994). Phenolic polymers may take part in the peroxidase cycle and be converted to polymer radicals. Polymer radicals may react with other radical species to form larger polymers or undergo radical transfer reactions. The end result is a solution containing polymers of various sizes (Dec and Bollag 1990). Since the phenolic polymers are less soluble than their precursors, they precipitate out of solution and can be physically removed by either gravity settling or filtration, aided in some cases by the use of coagulants (Wright 1995). Various enzymes used for phenol polymerization and their sources are given in Table 2.

Table 2: Various microbial and plant enzymes and their sources involved in biodegradation of phenol

Enzyme	Source	References
Phenol hydroxylase	*Bacillus stearothermophilus*	Gurujeyalakshmi and Oriel (1989)
Polyphenol oxidase	Mushroom	Burton et al. (1993)
Polyphenol oxidase	*Trametes trogii*	Garzillo et al. (1998)
Polyphenol oxidase	Transgenic tomato	Steffens and Li (2002)
Polyphenol oxidase	Potato	Loncar et al. (2011)
Phenol oxidase	*Lentinula edodes LE2*	Okeke et al. (1997)
Catechol 2,3 dioxygenase	*Bacillus* sp.	Ali et al. (1998)

Laccase	*Rhizoctonia praticola*	Bollag et al. (1988)
Laccase	*Aspergillus oryzae*	Schneider et al. (1999)
Laccase	*Pleurotus ostreatus*	Hublik and Schinner (2000)
Laccase	*Chalara paradoxa*	Robles et al. (2000)
Phenol oxidase	*Termitomyces albuminosus*	Johjima et al. (2003)
Horseradish peroxidase	Horseradish	Cooper and Nicell (1996), Wu et al. (1998), Zahida et al. (1998), Ghioureliotis and Nicell (1999); Wilberg et al. (2000); Stanisavljevic and Nedic (2004), Lai and Lin (2005), Mossallam et al. (2009), Iran and Siamak (2009)
Soybean peroxidase	Soybean	Caza et al. (1999), Wright and Nicell (1999), Ghioureliotis and Nicell (1999), Hejri and Saboora (2009)
Radish peroxidase	*Raphanus sativus*	Naghibi et al. (2003)
Peroxidase	*Momordica charantia*	Akhtar and Husain (2006)
Turnip peroxidase	*Brassica napus*	Guerrero et al. (2008)

Potential advantages of enzyme-based treatment (Wright 1995; Duran and Esposito 2000)

- Selective treatment, only phenolic and similar reducing substrates removed.
- Faster reaction velocity, reduced residence time.
- Action on substances which are toxic to microbes.
- Operation over a wide range of substrate concentrations.
- Operation over a wide range of pH, temperature, and salinity.
- No shock loading effects, no acclimatization required.
- Process control is simpler, reliable and predictable.

Wilberg et al. (2000) used an enzymatic process for phenol removal comprising a reaction step in which phenol was polymerized in the presence of an HRP enzyme. Experimental results showed the potential of the proposed technique having a conversion higher than 99 %. Cooper and Nicell (1996) used HRP for the oxidation of phenols by hydrogen peroxide which resulted in the formation of water-insoluble polymers. The feasibility of the enzyme process to treat a foundry wastewater containing 3.5 mM of total phenols (330 mg/L as phenol) was examined. Two enzyme stocks of different purities were used but total phenols removal was independent of enzyme purity. For both stocks, 97–99 % of the phenolic contaminants were removed, despite the presence of other contaminants such as organic compounds and iron in the waste matrix. Lai and Lin (2005) reported the development and application of an immobilized HRP system with porous aminopropyl glass (APG) beads. After 3 h of reaction with the addition of hydrogen peroxide a maximal removal efficiency of 25 % was observed at pH 7.5. The addition of polyethylene glycol (PEG) significantly enhanced the removal efficiency presumably via the formation of a protective shield in the vicinity of the active site of HRP from the free radicals formed during polymerization. The reusability of the HRP immobilized APG was also demonstrated. Mossallam et al. (2009) conducted phenol degradation studies using horseradish peroxidase (HRP) enzyme. Wastewater from petroleum oil reservoirs was used for this study which contained 4.5 mg/L phenol, 300 mg/L petroleum oil, Cl-62 mg/L and pH 6.0. It was found that petroleum oil has strong depressing effect on the activity of HRP at concentrations up to 1 g/L of petroleum oil. The most effective concentrations for treating wastewater were peroxidase at 2 U/ml and 0.5 mM hydrogen peroxide (H_2O_2) at neutral pH. Stanisavljevic and Nedic (2004) reported a process for phenol removal comprising a reaction step in which phenol was polymerized in the presence of an enzyme HRP. Experimental results showed the potential of the proposed technique. A phenol conversion higher than 90 % was observed using this polymerization process. Wright and Nicell (1999) studied phenol removal using SBP enzyme, activity was optimal at pH 6.4, with significant activity observed between pH 3 and 9. SBP was very stable at 25 °C at neutral and alkaline conditions but experienced rapid inactivation below pH 3. SBP was most effective when used to treat phenolic solutions between pH 6 and 9. In comparison with HRP, the activity of SBP was only slightly more sensitive to pH, more stable at elevated temperatures, and less susceptible to permanent inactivation by hydrogen peroxide.

Iran and Siamak (2009) reported the use of encapsulated HRP in calcium alginate for the purpose of phenol removal. The ratio of hydrogen peroxide/phenol at which highest phenol removal obtained was found to be dependent on initial phenol concentration and in the solution of 2–8 mM phenol it was 1.15 and 0.94, respectively. Loncar et al. (2011) reported an inexpensive immobilized enzyme for the removal of phenols. Partially purified potato polyphenol oxidase (PPO) was immobilized onto different commercial and laboratory produced carriers. 45 % removal of 4-bromophenol was achieved, while the 4-chlorophenol and phenol were 35 and 20 %, respectively. After eight repeated tests, the efficiency of 4-chlorophenol removal had decreased to 55 %.

Caza et al. (1999) conducted experiments to investigate the efficiency of using SBP to remove phenol and its derivatives from unbuffered synthetic wastewater. A removal efficiency of

95 % was observed during this study. Hejri and Saboora (2009) treated synthetic wastewater containing phenol, *o*-cresol and *m*-cresol, with SBP. The results showed that, an increase in hydrogen peroxide up to the optimum amount leads to an elevated removal of phenol and its derivatives. Higher concentrations of H_2O_2 inhibited the reaction. Treatment in the presence of PEG as an additive increased the effect of enzymatic removal. Wilberg et al. (2002) studied the application of low-purity soybean peroxidase (LP-SBP), obtained from wasted seed hulls, as a catalyst for phenol polymerizations in an aqueous solution in the presence of hydrogen peroxide. At all phenol concentrations tested, a retention time of about 100 min was sufficient to achieve good results.

Naghibi et al. (2003) conducted phenol removal using synthetic wastewater buffered at pH 7.4 containing 0.9 mM phenol which was treated with cut *Raphanus sativus* root and its juice. More than 90 % of phenol was removed in both cases.

Plant materials have been found useful in decontamination of water polluted with phenol and its derivatives. The detoxification effect is due to peroxidases contained in plant tissue. Enzyme-mediated oxidative coupling of phenol is followed by precipitation of the formed polymer and its removal from the aqueous phase (Naghibi et al. 2003).

CONCLUSIVE COMMENTS

The data summarized in this mini review paper provide research efforts carried out for the treatment of phenolic wastewaters. Degradation of phenol and its derivatives using various microorganisms has been the topic of scientific interest for a number of decades. Practical applications of microorganisms for the degradation of phenol are presently used exclusively for treatment of industrial wastewaters, both pure and mixed cultures of microorganisms and enzymes can serve the purpose. The findings of this review suggest that microbial and enzymatic methods can be robust and promising for effectively treating wastewaters containing phenol and its derivatives. Aerobic and anaerobic microbial treatment methods can be used for phenol biodegradation. Pure culture and mixed cultures can be used in different reactors for phenol degradation. Enzymatic methods have been widely reported the polymerization of phenol. These enzymes can be used in free and immobilized forms. Immobilized enzymes can be reused which will reduce the operating costs.

REFERENCES

1. Agarry SE, Solomon BO, Layokun SK (2008) Kinetics of batch microbial degradation of phenols by indigenous binary mixed culture of *Pseudomonas aeruginosa* and *Pseudomonas fluorescence*. Afr J Biotechnol 7(14):2417–2423
2. Ajaz M, Noor N, Rasool SA, Khan SA (2004) Phenol resistant bacteria from soil: identification, characterization and genetical studies. Pak J Bot 36(2):415–424
3. Akhtar S, Husain Q (2006) Potential applications of immobilized bitter gourd (*Momordica charantia*) peroxidase in the removal of phenols from polluted water. Chemosphere 65:1228–1235
4. Ali S, Lafuente RL, Cowan DA (1998) Meta-pathway degradation of phenolics by thermophilic*Bacilli*. Enzyme Microb Technol 23(7–8):462–468
5. Annachatre AP, Gheewala SH (1996) Biodegradation of chlorinated phenolic compounds. Biotechnol Adv 14(1):35–56
6. Anupama S, Pradeep NV, Hampannavar US (2013) Anaerobic followed by aerobic treatment approaches for Spentwash using MFC and RBC. Sugar Tech 15(2):192–202
7. Arutchelvan V, Kanakasabai V, Elangovan R, Nagarajan S, Muralikrishnan V (2006) Kinetics of high strength phenol degradation using *Bacillus brevis*. J Hazard Mater B 129:216–222
8. Azbar N, Tutuk F, Keskin T (2009) Biodegradation performance of an anaerobic hybrid reactor treating olive mill effluent under various organic loading rates. Int Biodeterior Biodegrad 63:690–698
9. Babich H, Davis DL (1981) Phenol: a review of environmental and health risks. Regul Toxicol Pharmacol 1:90–109
10. Baek SH, Kim KH, Yin CR, Jeon CO, Im WT, Kim KK, Lee ST (2003) Isolation and characterization of bacteria capable of degrading phenol and reducing nitrate under low-oxygen conditions. Curr Microbiol 47:462–466
11. Bailey JE, Ollis DF (1986) Biochemical engineering fundamentals, 2nd edn. McGraw-Hill, Maidenheach
12. Banerjee G (1997) Treatment of phenolic wastewater in RBC reactor. Water Res 31(4):705–714
13. Basha KM, Rajendran A, Thangavelu V (2010) Recent advances in the biodegradation of phenol: a review. Asian J Exp Biol Sci 1(2):219–234
14. Bollag JM, Shuttleworth KL, Anderson DH (1988) Laccase-mediated detoxification of phenolic compounds. Appl Environ Microbiol 54(12):3086–3091
15. Burton SG, Duncan JR, Perry T (1993) Activity of mushroom polyphenol oxidase in organic medium. Biotechnol Bioeng 42(8):938–944
16. Busca G, Berardinelli S, Resini C, Arrighi L (2008) Technologies for the removal of phenol from fluid streams: a short review of recent developments. J Hazard Mater 160:265–288
17. Caza N, Bewtra JK, Biswas N, Taylor KE (1999) Removal of phenolic compounds from synthetic wastewater using soybean peroxidase. Water Res 33(13):3012–3018
18. Chakraborty S, Bhattacharya T, Patel TN, Tiwari KK (2010) Biodegradation of phenol by native microorganisms isolated from coke processing wastewater. J Environ Biol 31:293–296
19. Christen P, Vega A, Casalot L, Simon G, Auria R (2012) Kinetics of aerobic phenol biodegradation by the acidophilic and hyper thermophilic archaeon *Sulfolobus solfataricus*98/2. Biochem Eng J 62:56–61
20. Cooper VA, Nicell JA (1996) Removal of phenols from a foundry wastewater using horseradish peroxidase. Water Res 30(4):954–964
21. Dec J, Bollag JM (1990) Detoxification of substituted phenols by oxidoreductive enzymes through

polymerization reactions. Archiv Environ Contam Toxicol 19:543–550

22. Demarche P, Junghanns C, Nair RR, Agathos SN (2011) Harnessing the power of enzymes for environmental stewardship. Biotechnol Adv 30:933–953
23. Duran N, Esposito E (2000) Potential applications of oxidative enzymes and phenoloxidase-like compounds in wastewater and soil treatment: a review. Appl Catal B 28:83–99
24. Ehrhardt HM, Rehm HJ (1985) Phenol degradation by microorganisms adsorbed on activated carbon. Appl Microbiol Biotechnol 21:32–36
25. El-Ashtoukhy SZ, El-Taweel YA, Abdelwahab O, Nassef EM (2013) Treatment of petrochemical wastewater containing phenolic compounds by electrocoagulation using a fixed bed electrochemical reactor. Int J Electrochem Sci 8:1534–1550
26. Gallego A, Fortunato MS, Foglia J et al (2003) Biodegradation and detoxification of phenolic compounds by pure and mixed indigenous cultures in aerobic reactors. Int Biodeterior Biodegrad 52:261–267
27. Garzillo AMV, Colao MC, Caruso C, Caporale C, Celletti D, Buonocore V (1998) Laccase from the white-rot fungus *Trametes trogii*. Appl Environ Microbiol 49:545–551
28. Ghioureliotis M, Nicell JA (1999) Assessment of soluble products of peroxidase-catalyzed polymerization of aqueous phenol. Enzyme Microb Technol 25(3–5):185–193
29. Guerrero QF, Duarte-Vazquez MA, Garcia-Almendarez BE, Tinoco R, Vazquez-Duhalt R, Regalado C (2008) Polyethylene glycol improves phenol removal by immobilized turnip peroxidase. Bioresour Technol 99:8605–8611
30. Gurujeyalakshmi G, Oriel P (1989) Isolation of phenol degrading *Bacillus stearothermophilus*and partial characterization of the phenol hydroxylase. Appl Environ Microbiol 55(2):500–502
31. Hejri S, Saboora A (2009) Removal of phenolic compounds from synthetic wastewaters by enzymatic treatments. J Sci 35(1):13–19
32. Hublik G, Schinner F (2000) Characterization and immobilization of the laccase from *Pleurotus ostreatus* and its use for the continuous elimination of phenolic pollutants. Enzyme Microb Technol 27(3–5):330–336
33. Iran A, Siamak N (2009) Removal of phenols with encapsulated horseradish peroxidase in calcium alginate. Iran J Chem Eng 28(2):43–49
34. Jadhav DN, Vanjara AK (2004) Removal of phenol from wastewater using sawdust, polymerized sawdust and sawdust carbon. Indian J Chem Technol 11:35–41
35. Jiang Y, Cai X, Wu D, Ren N (2010) Biodegradation of phenol and *m*-cresol by mutated*Candida tropicalis*. J Environ Sci 22(4):621–626
36. Johjima T, Ohkuma M, Kudo T (2003) Isolation and cDNA cloning of novel hydrogen peroxide-dependent phenol oxidase from the basidiomycete *Termitomyces albuminosus*. Appl Environ Microbiol 61:220–225
37. Kafilzadeh F, Farhangdoost MS, Tahery Y (2010) Isolation and identification of phenol degrading bacteria from Lake Parishan and their growth kinetic assay. Afr J Biotechnol 9(40):6721–6726
38. Kulkarni S, Kaware JP (2013) Review on research for removal of phenol from wastewater. Int J Sci Res Publ 3(4):1–5
39. Lai YC, Lin SC (2005) Application of immobilized horseradish peroxidase for the removal of*p*-chlorophenol from aqueous solution. Process Biochem 40:1167–1174
40. Leven L, Nyberg K, Schnurer A (2012) Conversion of phenols during anaerobic digestion of organic solid waste—a review of important microorganisms and impact of temperature. J Environ Manag 95:S99–S103
41. Lika K, Papadakis IA (2009) Modeling the biodegradation of phenolic compounds by microalgae. J Sea Res 62:135–146
42. Loncar N, Bozic N, Andelkovic I, Milovanovic A et al (2011) Removal of aqueous phenol and phenol derivatives by immobilized potato polyphenol oxidase. J Serbian Chem Soc 76:1–10
43. Michalowicz J, Duda W (2007) Phenols–sources and toxicity. Pol J Environ Stud 16(3):347–362
44. Mossallam KF, Sultanova FM, Salimova NA (2009) Enzymatic removal of phenol from produced water and the effect of petroleum oil content. 13th international water technology conference, Hurghada, Egypt, pp 1009–1020
45. Moussavi G, Mahmoudi M, Barikbin B (2008) Biological removal of phenol from strong wastewaters using a novel MSBR. Water Res 43:1295–1302
46. Naghibi F, Pourmorad F, Honary S, Shamsi M (2003) Decontamination of water polluted with phenol using *Raphanus sativus* root. Iran J Pharm Res 2(1):29–32
47. Okeke BC, Paterson A, Smith JE, Watson IA (1997) Comparative biotransformation of pentachlorophenol in soils by solid substrate cultures of *Lentinula edodes*. Appl Environ Microbiol 48:563–569
48. Paraskevi NP, Euripides GS (2005) Effect of temperature and additional carbon sources on phenol degradation by an indigenous soil *Pseudomonad*. Biodegradation 16:403–413
49. Passos CT, Michelon M, Burkert JFM, Kalil SJ, Burkert CAV (2010) Biodegradation of phenol by free and encapsulated cells of a new *Aspergillus* sp. isolated from a contaminated site in southern Brazil. Afr J Biotechnol 9(40):6716–6720
50. Pradeep NV, Anupama, Hampannavar US (2011) Biodegradation of phenol using rotating biological contactor. Int J Environ Sci 2(1):105–113
51. Pradeep NV, Anupama S, Arunkumar JM, Vidyashree KG, Ankitha K, Lakshmi P, Pooja J (2014) Treatment of sugar industry wastewater in anaerobic downflow stationary fixed film (DSFF) reactor. Sugar Tech 16(1):9–14

52. Prasad SBC, Babu RS, Chakrapani R, Ramachandra Rao CSV (2010) Kinetics of high concentrated phenol biodegradation by *Acinetobacter baumannii*. Int J Biotechnol Biochem 6(4):609–615
53. Robles A, Lucas R, Cienfuegos GA, Galvez A (2000) Phenol-oxidase (laccase) activity in strains of the hyphomycete *Chalara paradoxa* isolated from olive mill wastewater disposal ponds. Enzyme Microb Technol 26(7):484–490
54. Said M, Ahmad A, Abdul Wahab M (2013) Removal of phenol during ultrafiltration of Palm oil mill effluent (POME): effect of pH, ionic strength, pressure and temperature. Der Pharma Chemica 5(3):190–196
55. Sameer HI, Shrikant SK, Ashwin WP, Melo JS, Dsouza SF (2002) Phenol degradation in rotating biological contactors. J Chem Technol Biotechnol 77:1050–1057
56. Saravanan P, Pakshirajan K, Saha P (2008) Growth kinetics of an indigenous mixed microbial consortium during phenol degradation in a batch reactor. Bioresour Technol 99:205–209
57. Schneider P, Caspersen MB, Mondorf K et al (1999) Characterization of a *Coprinus cinereus*laccase. Enzyme Microb Technol 25(6):502–506
58. Seker S, Beyenal H, Salih B, Tanyolac A (1997) Multi-substrate growth kinetics of *Pseudomonas putida* for phenol removal. Appl Microbiol Biotechnol 47:610–614
59. Sgountzos IN, Pavlou S, Paraskeva CA, Payatakes AC (2006) Growth kinetics of *Pseudomonas fluorescens* in sand beds during biodegradation of phenol. Biochem Eng J 30:164–173
60. Shibata A, Inoue Y, Katayama A (2006) Aerobic and anaerobic biodegradation of phenol derivatives in various paddy soils. Sci Total Environ 367:979–987
61. Singh A, Kumar V, Srivastava JN (2013) Assessment of bioremediation of oil and phenol contents in refinery waste water via bacterial consortium. Pet Environ Biotechnol 4(3):1–4
62. Stanisavljevic M, Nedic L (2004) Removal of phenol from industrial wastewaters by horseradish (*Cochlearia armoracia* L) peroxidase. Work Living Environ Prot 2(4):345–349
63. Steffens JC, Li L (2002) Overexpression of polyphenol oxidase in transgenic tomato plants results in enhanced bacterial disease resistance. Planta 215:239–247
64. Thomas S, Sarfaraz S, Mishraz LC, Iyengar L (2002) Degradation of phenol and phenolic compounds by a defined denitrifying bacterial culture. World J Microbiol Biotechnol 18:57–63
65. Varma RJ, Gaikwad BG (2008) Rapid and high biodegradation of phenols catalyzed by *Candida tropicalis* NCIM 3556 cells. Enzyme Microb Technol 43:431–435
66. Veeresh GS, Kumar P, Mehrotra I (2004) Treatment of phenol and cresols in upflow anaerobic sludge blanket (UASB) process: a review. Water Res 39:154–170
67. Viggiani A, Olivieri G, Siani L, Di Donato A, Marzocchella A, Salatino P, Barbieri P, Galli E (2006) An airlift biofilm reactor for the biodegradation of phenol by *Pseudomonas stutzeri*OX1. J Biotechnol 123:464–477
68. Wang Y, Song J, Zhao W et al (2011) In situ degradation of phenol and promotion of plant growth in contaminated environments by a single *Pseudomonas aeruginosa* strain. J Hazard Mater 192:354–360
69. Wilberg KQ, Nunes DG, Rubio J (2000) Removal of phenol by enzymatic oxidation and flotation. Braz J Chem Eng 77(4):1–13
70. Wilberg K, Assenhaimer C, Rubio J (2002) Removal of aqueous phenol catalysed by a low purity soybean peroxidase. J Chem Technol Biotechnol 77:851–857
71. Wright H (1995) Characterization of soybean peroxidase for the treatment of phenolic wastewaters. M. E thesis, Department of Civil Engineering and Applied Mechanics, McGill University, Montreal
72. Wright H, Nicell JA (1999) Characterization of soybean peroxidase for the treatment of aqueous phenols. Bioresour Technol 70:69–79
73. Wu Y, Taylor KE, Biswas N, Bewtra JK (1998) A model for the protective effect of additives on the activity of horseradish peroxidase in the removal of phenol. Enzyme Microb Technol 22(5):315–322
74. Zahida MS, Deva W, Peerzada GM, Behere DV (1998) Oxidation of phenols by horseradish peroxidase and lactoperoxidase compound II—kinetic considerations. Indian J Biochem Biophys 35(6):353–357
75. Zidkova L, Szokol J, Rucka L, Patek M, Nesvera J (2012) Biodegradation of phenol using recombinant plasmid-carrying *Rhodococcus erythropolis* strains. Int Biodeterior Biodegrad. doi:10.1016/j.ibiod.2012.05.017
76. Zou H, Taylor KE (1994) Products of oxidative coupling of phenol by horseradish peroxidase. Chemosphere 28(10):1807–1817

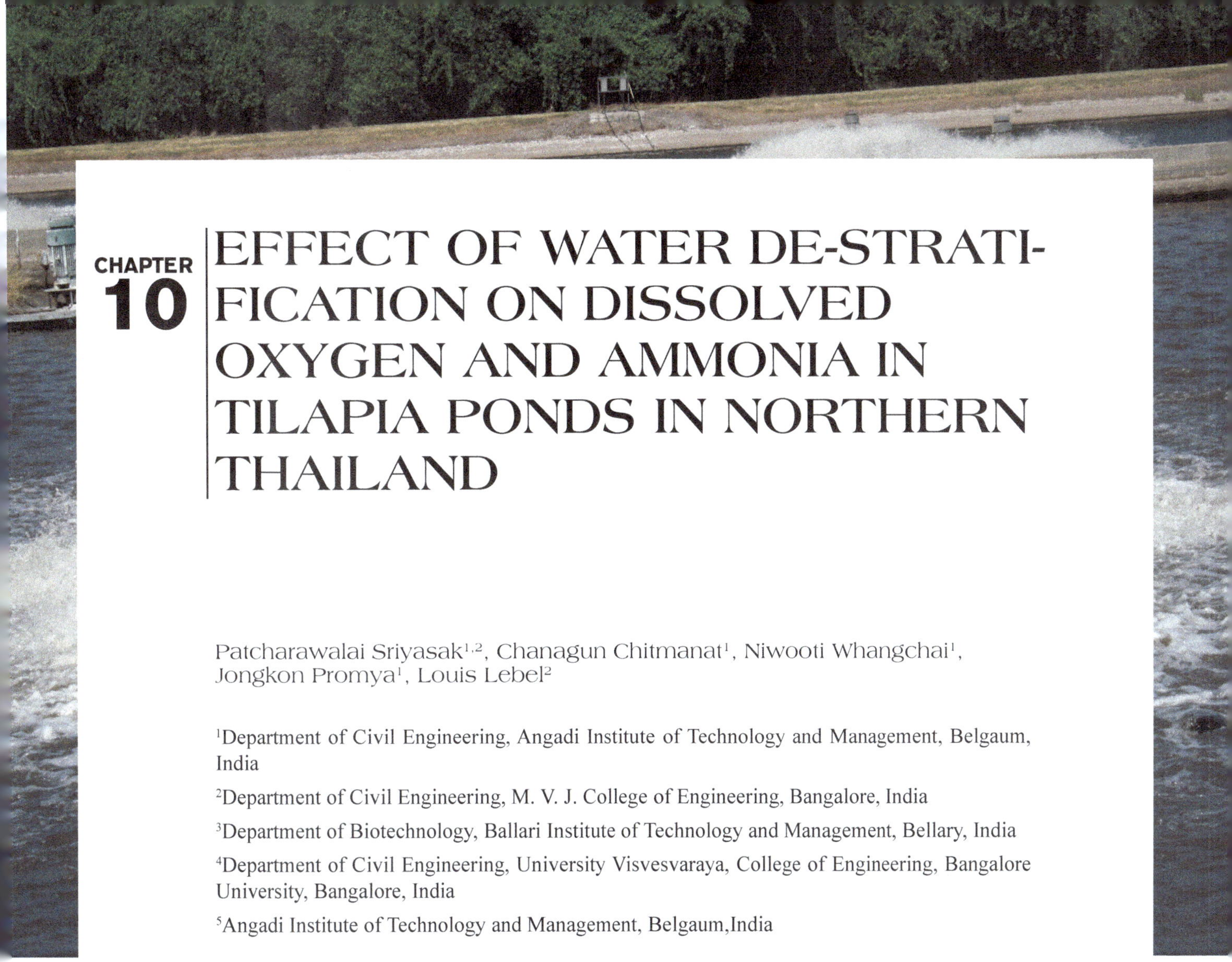

CHAPTER 10

EFFECT OF WATER DE-STRATIFICATION ON DISSOLVED OXYGEN AND AMMONIA IN TILAPIA PONDS IN NORTHERN THAILAND

Patcharawalai Sriyasak[1,2], Chanagun Chitmanat[1], Niwooti Whangchai[1], Jongkon Promya[1], Louis Lebel[2]

[1]Department of Civil Engineering, Angadi Institute of Technology and Management, Belgaum, India

[2]Department of Civil Engineering, M. V. J. College of Engineering, Bangalore, India

[3]Department of Biotechnology, Ballari Institute of Technology and Management, Bellary, India

[4]Department of Civil Engineering, University Visvesvaraya, College of Engineering, Bangalore University, Bangalore, India

[5]Angadi Institute of Technology and Management, Belgaum,India

ABSTRACT

Episodes of low concentrations of dissolved oxygen and high concentrations of ammonia are major causes of fish stress, which in turn, reduces growth and increases mortality rates in aquaculture ponds. This study measured the effects of water de-stratification on dissolved oxygen and ammonia concentrations in tilapia ponds in Northern Thailand. Fifteen ponds in five provinces in Northern Thailand were sampled on multiple dates in the hot, wet, and dry seasons. Thermal water stratification peaked around 14:00–16:00 h each day; wherein the differences between surface and lower water temperatures in the 0.8–2.0 m deep ponds reached 1.3–4.0 °C. Thermal de-stratification in the hot season and dry season usually occurred late at night; in the wet season however, it occurred early in the evening due to the cooling effects of rain. The mixing of surface and bottom waters decreased dissolved oxygen levels in water near the surface, and, increased concentration levels in the bottom water layer. Mean DO concentrations of integrated and commercial ponds were lower than 1 mg/L from 02:00 to 06:00 h, in 68 % of the observations. Repeated measures ANOVA analysis showed that water depth and fish culture systems have significant effects on total ammonia nitrogen (TAN). TAN concentrations near the surface were lower than at the bottom, and increased after water de-stratification. TAN in the integrated culture system was significantly higher than in commercial and subsistence systems. Toxic, un-ionized ammonia (NH_3) concentrations in all culture systems, however, were still lower than standard toxicity level for tilapia. The improved understanding of the seasonal and culture-system specific effects of de-stratification on pond water quality provided by this study, are useful for improving pond management practices to reduce the risks of exceeding DO thresholds. In particular, fish farmers should: (1) take care to avoid over-feeding and manage water and sediments to prevent excessive accumulation of organic matter and waste at the bottom of ponds, which can influence other water quality parameters; and, (2) use aeration and mechanical mixing interventions at critical times to reduce stress on fish from low DO concentrations, and thus avoid risks of mass mortality events.

Citation: Patcharawalai Sriyasak, Chanagun Chitmanat, Niwooti Whangchai, Jongkon Promya, Louis Lebel, Effect of water de-stratification on dissolved oxygen and ammonia in tilapia ponds in Northern Thailand, DOI: 10.1007/s40071-015-0113-y

INTRODUCTION

Nile tilapia culture in earthen ponds in Thailand is expanding: between 2003 and 2013 culture area and production increased by 12 and 46 %, respectively (DOF 2013). This is due to high consumer demand, improvements in seed production, including, the introduction of sex-reversal technologies (Bhujel 2013), and the availability of pellet feeds; thus allowing producers to achieve high fish growth rates and yields (Abdel-Tawwab 2012; Bhujel 2013). At the same time, fish farmers are facing increasing problems of maintaining adequate water quality in fish ponds (Sriyasak et al. 2014).

Water quality in a pond influences feeding, growth, disease burdens, and survival rates (Chainark and Boyd 2010; Pandit and Nakamura 2010). Water quality is controlled by a complex interplay of many factors, including weather conditions (Ahmed et al. 2011). For example, dissolved oxygen (DO) is related to phytoplankton production and respiration (Smith and Piedrahita 1988); nitrogen waste such as ammonia is related to the amount of organic matter inputs and ammonium excretion by fish (Gross et al. 1999; Hargreaves 1998); and, water temperature and thermal stratification are controlled by sunlight and air temperature (Egna and Boyd 1997; Losordo and Piedrahita 1991).

Adequate DO concentrations in a fish pond are essential for maintaining optimal growth (Chang and Ouyang 1988). Ponds with high nutrient input have high phytoplankton concentrations, and as a consequence, have high DO concentrations in the afternoon due to phytoplankton photosynthesis (Diana et al. 1991), and low DO concentrations in the early morning due to phytoplankton, water column, and sediment respiration (Chang 1989). Oxygen concentrations lower than 3 mg/L reduce feed intake and growth in Nile tilapia (Duy et al. 2008).

Ammonia is the first form of nitrogen released when organic matter decomposes, and can exist in two forms: un-ionized ammonia (NH_3) or ionized ammonia (NH_4^+) (Durborow et al. 1992; Gandhi 2012; Gross et al. 1999). Un-ionized ammonia (NH_3) is toxic to fish, and becomes even more toxic as water temperature and pH increase (EPA 2013; Hargreaves and Tucker 2004). Acute toxicity of ammonia is due to its effect on the central nervous system (Randall and Tsui2002); ammonia concentrations of 7.40 mg/L have been shown to cause mass mortality in tilapia fingerling within 24 h (Benli and Koksal 2005). Fish exposed to toxic levels of ammonia cannot excrete ammonia efficiently; as a result, ammonia levels in blood and tissues increase along with pH levels, therefore affecting enzyme activity (Lawson 1995). This can lead to poor feed conversion, slower growth rates, and reduced resistance to diseases (Gandhi 2012; Hargreaves and Tucker 2004).

Shallow aquaculture ponds undergo diel cycles of thermal and chemical vertical stratification (Losordo and Piedrahita 1991). Stratification of water in ponds arises due to the differences in water density, as the absorption of solar radiation decreases with increasing depth of the water column (Chowdhury et al. 2014; Klimaszyk and Tomaz 2007; Losordo and Piedrahita 1991). Warm surface water also has higher DO levels from photosynthesis, while cooler bottom water has lower DO, plus accumulated nitrogenous waste from uneaten feed and feces that drop down to the sediment (Gandhi 2012). De-stratification, or the remixing of water, occurs as air temperature drops and the surface layer cools down, therefore becoming denser than water in the lower layers. The surface water then sinks and forces water at the bottom to rise, thus turning over the layers (Chowdhury et al. 2014; Szyper and Lin 1990). After water mixing, DO in the surface layer falls, and the concentration of nitrogenous waste like ammonia increases. Stratification with depth in ponds, followed by rapid turnover or exchange of surface and bottom water layers, can expose fish to dangerously low dissolved oxygen levels, even near the surface. Conversely, mixing helps to increase oxygen levels near the bottom.

For Nile tilapia cultured in earthen ponds in Thailand, tilapia mortality from disease outbreak frequently occurs when DO levels are low and ammonia concentrations are high (Chitmanat2013). Fish culture in pellet feed based and poultry or livestock integrated systems appear to be more impacted by periods of prolonged cloud cover and heavy rain than subsistence culture system (Pimolrat et al. 2013). Fish in ponds that are exposed to hot weather for several days, and then treated to a sudden period of heavy rain or cloud cover, also tend to be vulnerable to diseases due to low oxygen condition and high organic wastes (Srisapoome and Areechon2014); this is because rain encourages water mixing between the surface and bottom layers.

Stratification and water turnover is known to occur in tilapia fish ponds in Northern Thailand (Sriyasak et al. 2013). The objective of this study was to measure the effect of water de-stratification on concentrations of DO and ammonia in fish ponds that are managed under different culture systems during different seasons in Northern Thailand. As tilapias are known to be sensitive to episodes of low concentrations of DO and high concentrations of ammonia, a better understanding of the effects of de-stratification on pond water quality should be useful for managing water- and climate-related risks to aquaculture production.

METHODS

Study Site

This study made observations and measurements in 15 tilapia earthen ponds located in five provinces in Northern Thailand: Chiangrai, Chiangmai, Phayao, Lampang, and Nakhon Sawan (Fig. 1). Pond sizes ranged from 0.16 to 0.64 ha, with depths of 0.8–2.0 m. Ponds were grouped according to culture system: commercial (where prepared pellet feed was regularly provided and crops tended to be harvested once); integrated (where fish and livestock were raised in the same area and the wastes from the latter were dumped into ponds); and subsistence (where feeding was sporadic and fish were harvested continuously for consumption and market). The characteristics of each culture system are summarized in Table 1.

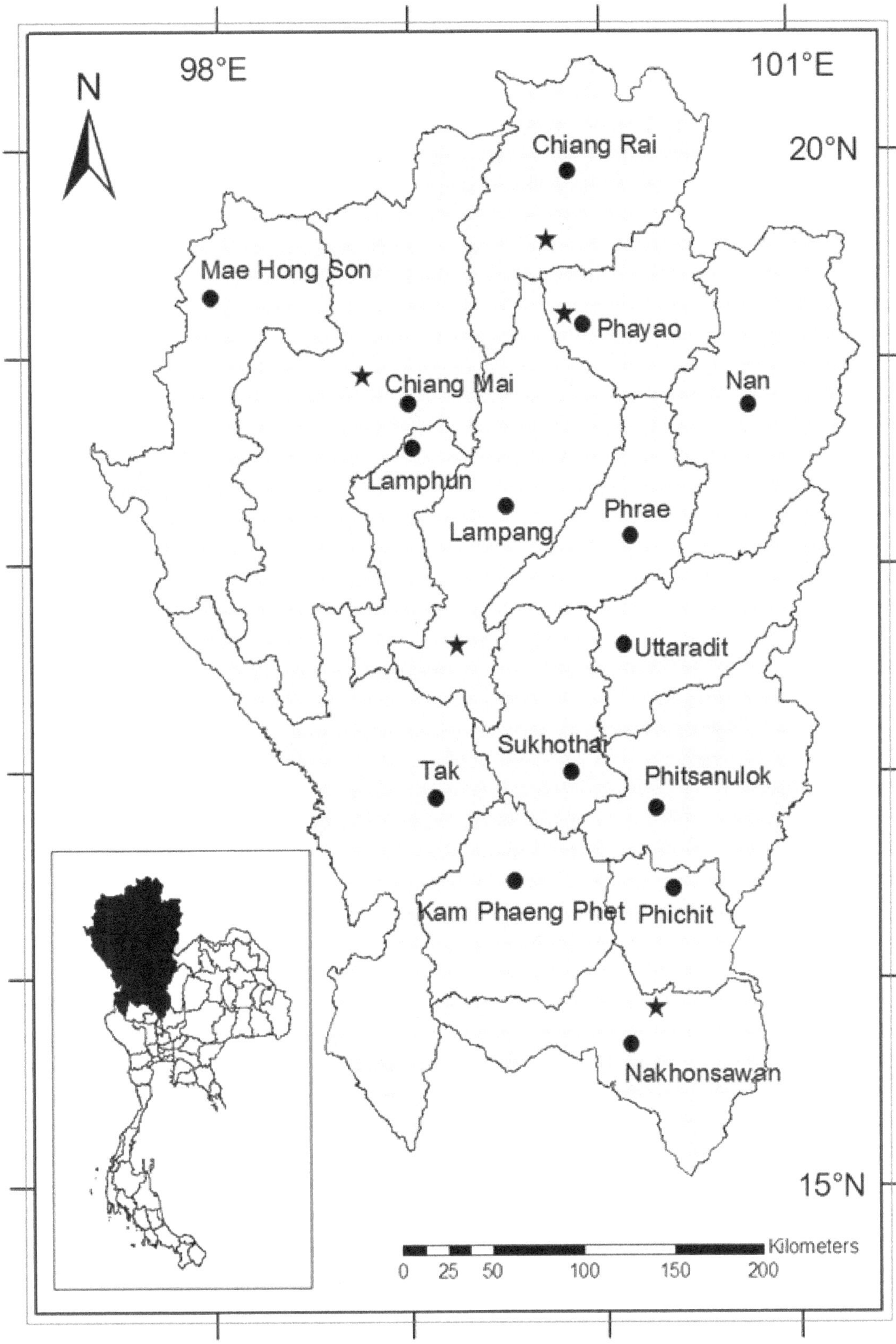
98°E
101°E
N
20°N
15°N
Chiang Rai
Mae Hong Son
Phayao
Chiang Mai
Nan
Lamphun
Lampang
Phrae
Uttaradit
Sukhothai
Tak
Phitsanulok
Kam Phaeng Phet
Phichit
Nakhonsawan
Kilometers
0 25 50 100 150 200

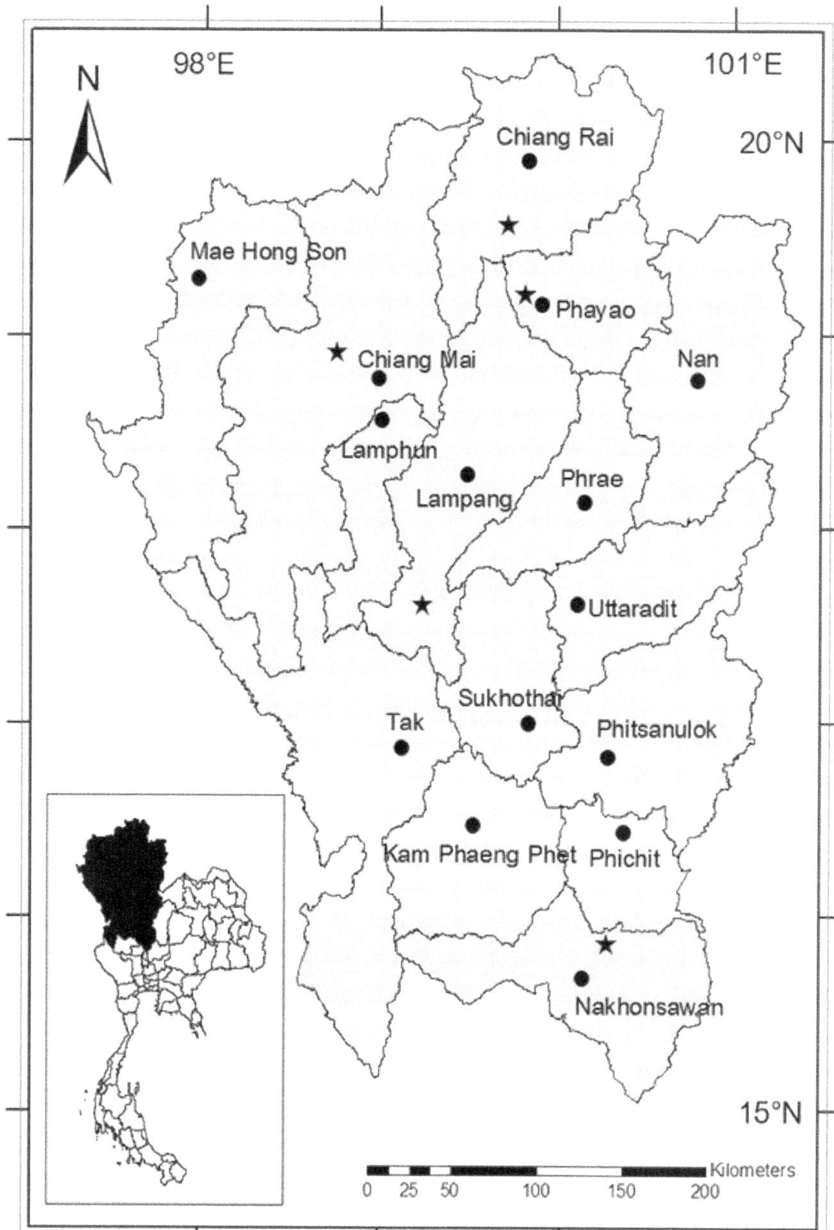

Figure 1: Map of study area in Northern Thailand. *Stars* indicate farm locations, *dots* the city district in each province.

Table 1: Major features of ponds managed under different culture systems that were monitored in the study

Pond feature	Culture system		
	Integrated	**Commercial**	**Subsistence**
Number of ponds	3	6	6
Pond area (m^2)	2600 ± 1830	3400 ± 1050	460 ± 340
Pond depth (m)	1.2 ± 0.4	1.2 ± 0.1	1.0 ± 0.2
Stocking rate (fish/m^2)	2.7 ± 0.5	2.8 ± 0.4	0.8 ± 0.3
Culture period (months)	5.8 ± 0.7	6.5 ± 0.5	14.0 ± 3.1
Feed type	Pellet feed	Pellet feed	Pellet feed, vegetable waste and food scarps
Manure inputs	Yes	No	No
Water renewal rate	Seldom	10 % per day	Seldom
Production	All harvest sold	All harvest sold	Harvest consumed and sold

Mean ± SD

Water Parameters

Altogether, 660 samples of fish pond water were collected for water quality analysis (15 ponds × 11 months × 4 samples) between May 2013 and 2014. Samples were classified into three seasons for analysis: the "dry" season (November–February) when air temperatures and precipitation are lowest; the pre-monsoon "hot" season (March–May) when temperatures are highest; and the monsoon "wet" season (June–October) when precipitation is highest (Thai Meteorological Department 2012). Ambient air temperature, water temperature, pH, turbidity and DO were monitored at 2 h intervals, over a 24 h period, in each season, at every 20 cm depth, with a multimeter (TOA DKK WQC-22A model, Japan). Moreover, water samples were collected 20 cm below the surface and 20 cm above the pond bottom during stratified and de-stratified conditions. Chemical analyses were carried out for total ammonia-nitrogen (TAN) and chlorophyll-a according to standard methods (APHA 1980).

Toxic un-ionized ammonia (NH_3) concentration was calculated from the equation:

$$NH_3 = (a)TAN$$

where, (a)(a) is the mole fraction of un-ionized ammonia in aqueous solution at different pH values and temperatures (Emerson et al. 1975).

The maximum difference between surface and bottom temperatures of the pond was used as an indicator of stratification (Stefan et al. 1996). The de-stratification time was defined as when there was no detectable difference (less than 0.1 °C) between surface and bottom water temperatures. Data on precipitation were taken from the nearest weather station to each farm (Thai Meteorological Department 2014).

Statistical Analysis

A factor analysis using principal component analysis (PCA) with a varimax rotation was first carried out to help understand the patterns of co-variation among the many water quality variables. Separate ANOVA models were then used to analyze the combined effects of culture system, season, and sampling depth factors on maximum water temperature stratification, minimum DO, DO stratification, and mean TAN. TAN values were log-transformed prior to analysis to homogenize variances, and then analyzed using repeated measures ANOVA for concentrations before and after stratification. Pond depth (<1 and >1 m) was included as a separate predictor as it could potentially influence thermal stratification; but as it was later found not to be significant it was thus excluded from the final models.

RESULTS

Principal Component Analysis

Logically related individual water quality measures were strongly correlated with the same factor (Table 2). Thus, temperature variables covaried and all loaded on Factor 1, whereas TAN variables were correlated with Factor 3, and DO variables with Factor 4. Factor 2 could be interpreted as high plankton and organic input conditions, as both chlorophyll-a and turbidity variables were positively correlated with the factor, while pH was negatively correlated. Based on the clear separation of outcome variables of interest in this exploratory analysis, it seems justified to proceed to analyzing the effects of season, culture system, and sampling depth on the selected temperature, DO and TAN variables separately.

Table 2: Factor loading, eigenvalues, and percentage of variance from the principal component analysis

Variables	Factor 1	Factor 2	Factor 3	Factor 4
Air temperature	**0.91**	0.12		−0.15
Surface water temperature	**0.97**	0.13		
Bottom water temperature	**0.92**		0.18	
DO surface		0.22		**0.83**
DO bottom	−0.19	−0.36		**0.73**
TAN before stratification			**0.96**	
TAN after stratification		0.21	**0.92**	
pH		−0.72	−0.17	0.43
Turbidity		**0.88**		0.14
Chlorophyll-a	0.20	**0.85**	0.14	
Eigenvalue	3.54	2.00	1.46	1.34
Percentage of variance	35.4	20.0	14.6	13.4
Cumulative (%)	35.4	55.3	70.0	83.4

Values larger than 0.7 are in bold

Weather, Water Stratification and de-Stratification in Fish Ponds

Averaged over 24 h, air temperatures, as well as surface and bottom water temperatures, at pond sites on sampling dates, were significantly lower in the dry season than in the hot or wet seasons as expected (Table 3). Maximum water surface temperature usually occurred at 16:00 h, and minimum at 06:00 h. The highest water temperature observed was 38.0 °C in the hot season, and the lowest was 18.3 °C during the dry season. Daily air temperature range was lower in the wet season than in the other two seasons. Average mean precipitation also varied among seasons as expected (Table 3).

Table 3: Air and water temperatures as well as daily precipitation (Mean ± SD) at fish ponds in three different seasons; different superscript letters in the same rows indicate a significant difference between means at $P < 0.05$ (Tukey's HSD)

Variables	Season		
	Hot (March–May)	Wet (June–October)	Dry (November–February)
Mean maximum air temperature (°C)	34.3 ± 6.9^{a}	30.5 ± 8.8^{b}	30.8 ± 6.6^{b}
Daily air temperature range (°C)	12.5 ± 3.7^{a}	8.1 ± 2.8^{b}	14.4 ± 4.2^{a}
Air temperature (°C)	28.2 ± 2.9^{a}	27.7 ± 2.7^{a}	22.8 ± 3.9^{b}
Surface water temperature (°C)	29.6 ± 2.6^{a}	28.7 ± 2.3^{a}	24.2 ± 3.4^{b}
Bottom water temperature (°C)	29.1 ± 2.7^{a}	28.3 ± 2.1^{a}	23.7 ± 3.4^{b}
Daily precipitation (mm/day)	2.9 ± 1.7^{b}	6.5 ± 2.2^{a}	0.8 ± 1.0^{c}

The maximum difference between surface and bottom water temperatures always occurred in measurements made in the afternoon, and ranged between 0 and 7.1 °C. The degree of stratification varied among seasons and culture systems with a significant interaction between the two factors ($P < 0.05$, Fig. 2). In the wet season, culture systems behaved alike, whereas in the hot and dry seasons, integrated ponds showed relatively larger thermal stratification (Fig. 2).

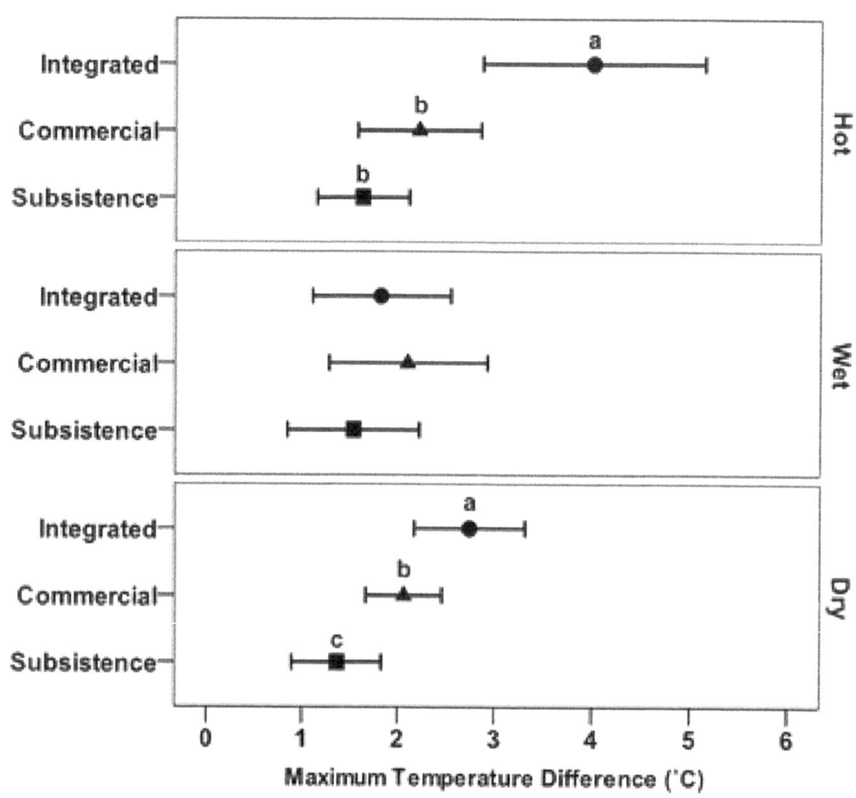

Figure 2: Maximum difference between surface and bottom water temperatures in different culture systems and seasons. *Plotted symbols* are means and bars 95 % confidence intervals (CI) for those means.

Stratification usually began from 14:00 to 16:00 h, and became less apparent during the night, as the upper layer cooled and water began to mix. Thermal de-stratification in the wet season occurred between 18:00 and 22:00 h; in the dry season between 20:00 and 24:00 h; and in the hot season it occurred between 22:00 and 04:00 h. Water de-stratification in the wet season in all culture system types occurred earlier than in the dry and hot seasons; possibly because cold water from rain decreased surface water temperature, in turn, encouraging de-stratification. To support this argument, Fig. 3 shows the thermal stratification on three separate observation days, one of which happen to rain heavily at about 16:00 h (Fig. 3c); 2 h later, complete stratification could be observed (Fig. 3c), two or more hours earlier than usual. Thermal stratification was sometimes substantial (Fig. 3b), and at other times less so (Fig. 3a).

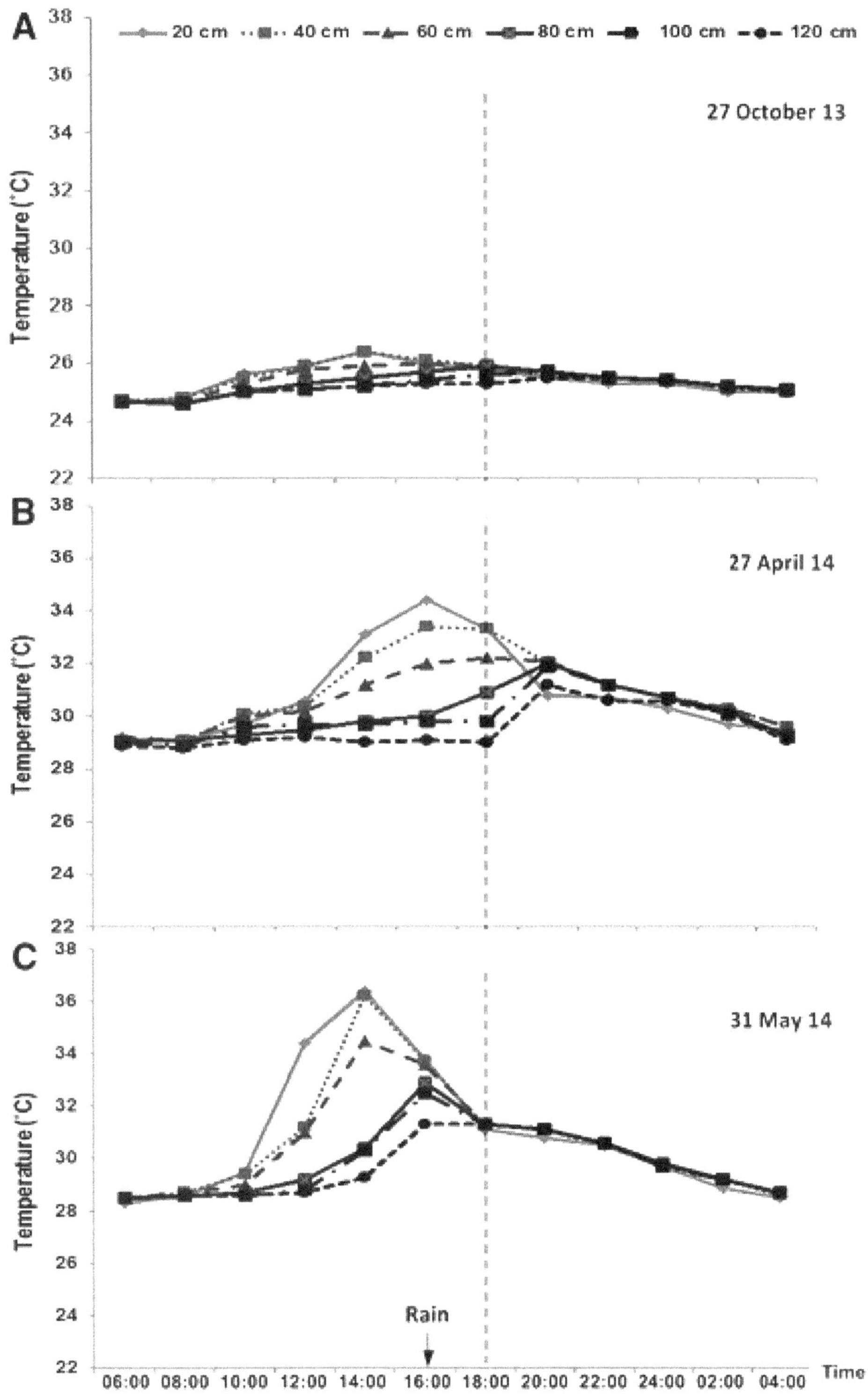

Figure 3: Thermal stratification of pond waters on three illustrative dates with small maximum temperature difference (**a**); large maximum temperature difference (**b**); and large temperature difference and heavy afternoon rainfall (**c**).

In a few instances however, no stratification was observed: 11 occasions in the dry season, 7 occasions in the hot season, and 4 occasions in the wet season. The reasons for these instances include: heavy rain during the day ($n = 6$); high cloud cover ($n = 8$); exchanging of pond water ($n = 4$); and the use of aerators ($n = 4$). These events were too rare for powerful statistical comparisons to be made.

Effect of water de-stratification on dissolved oxygen in fish ponds

Thermal stratification in ponds (Fig. 3) was accompanied by DO stratification (Fig. 4). The highest DO was found in the surface water layer during the afternoon, which began to decrease after sunset. The maximum DO recorded at the surface layer in the afternoon was 22.7 mg/L; this was in an integrated pond system in May. Oxygen oversaturation in the late afternoon was found to occur in both integrated and commercial ponds. On the other hand, under-saturation of oxygen in the late afternoon was observed in some subsistence ponds.

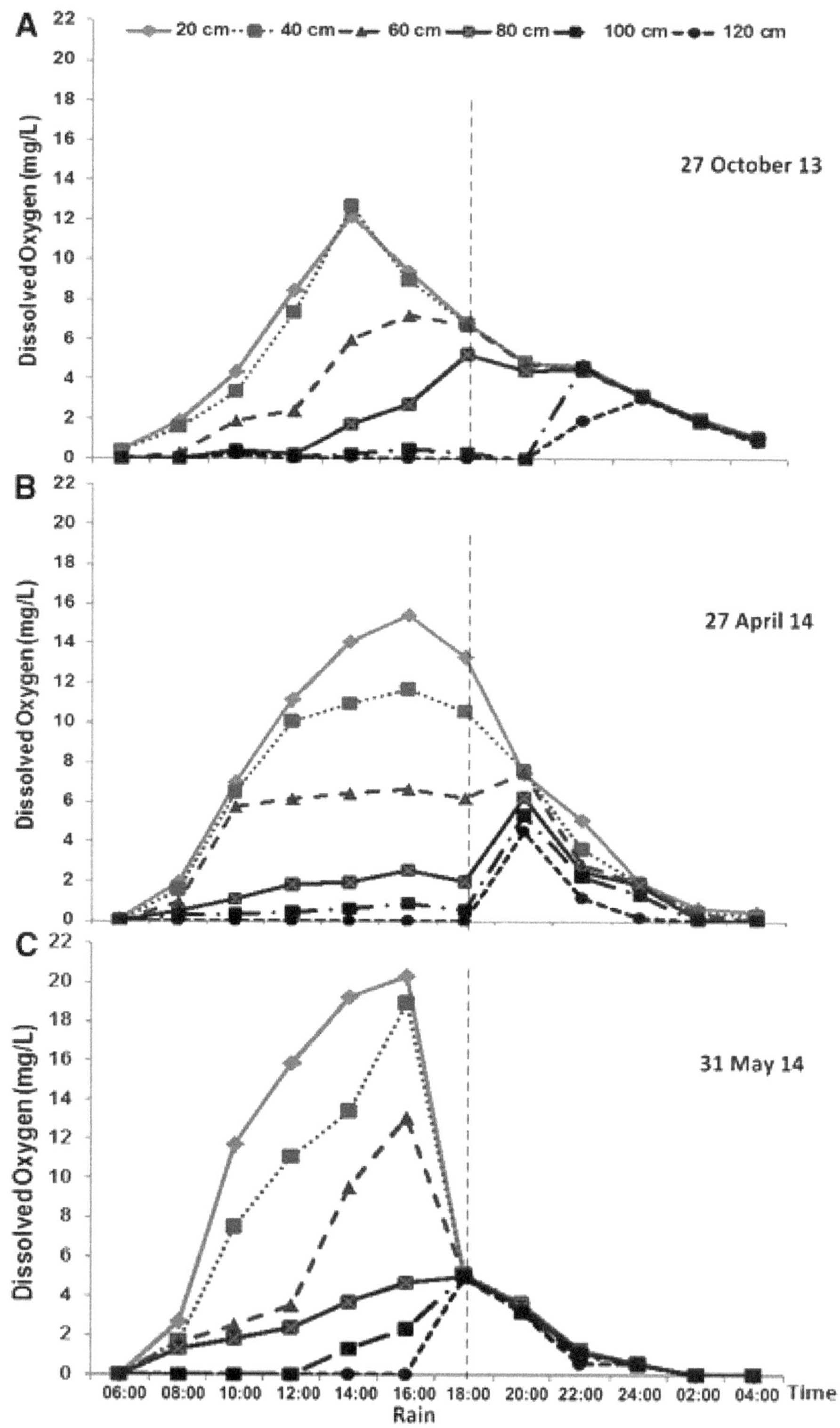

Figure 4: Diel stratification of dissolved oxygen of pond waters on three illustrative dates (as in Fig. 3) with small maximum temperature difference (a); large maximum temperature difference (b); and large temperature difference and heavy afternoon rainfall (c).

These differences in extremes among culture systems are consistent with measures of chlorophyll-a and turbidity. The mean chlorophyll-a concentration of integrated ponds (551 ± 300 μg/L) was significantly higher than commercial ponds (227 ± 208 μg/L) or subsistence ponds (36 ± 42 μg/L). Likewise, the turbidity in integrated ponds (153 ± 82 NTU) was significantly higher than commercial ponds (88 ± 52 NTU) or subsistence ponds (36 ± 26 NTU).

Water de-stratification resulted in decreased DO concentration at the surface, and increased DO at the bottom of the pond (Fig. 4). Mean DO concentration of integrated and commercial ponds were lower than 1 mg/L, from 02:00 to 06:00 h, in 68 % of observations; whereas the subsistence systems were always over 1 mg/L during the same periods.

ANOVA was used to analyze the influence season and culture system had on average minimum DO concentrations over a 24-h period. Average minimum DO concentration was low in integrated ponds, and was especially lower than the other two culture systems during the dry season (Fig. 5). Average 24-h DO concentrations were lower at the bottom of the pond (2.8 ± 2.3 mg/L) than at the surface (5.6 ± 2.3 mg/L, $P < 0.001$). Average 24-h DO concentrations were higher in integrated (6.5 ± 6.0 mg/L) and commercial ponds (6.1 ± 4.5 mg/L), than in subsistence ponds (4.6 ± 3.2 mg/L, $P < 0.05$).

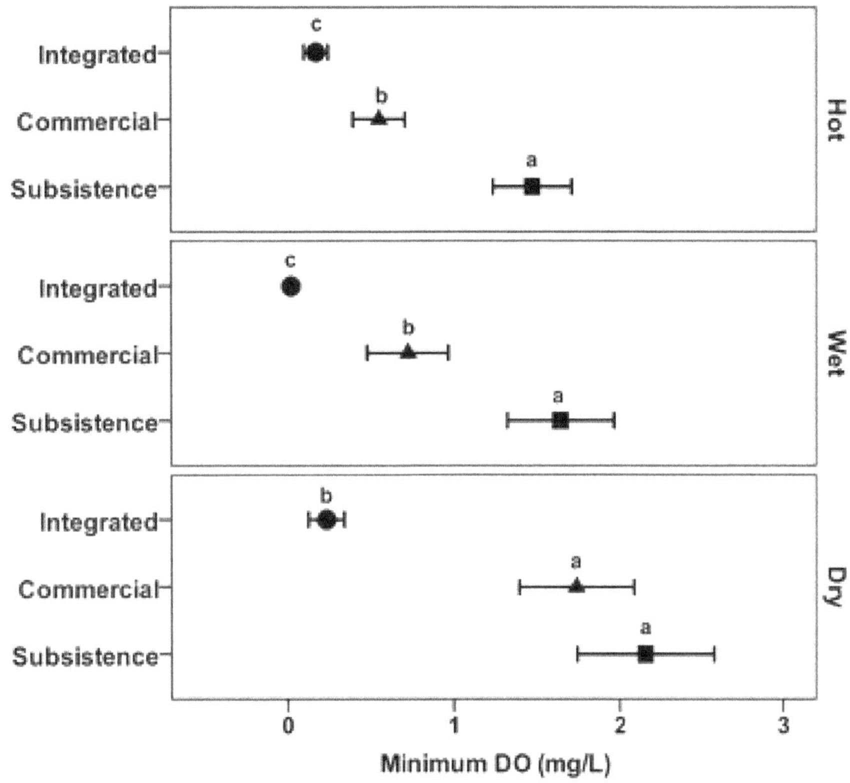

Figure 5: Minimum DO concentration in different culture systems and seasons. *Plotted symbols*are means and bars 95 % confidence intervals (CI) for those means.

Mean maximum DO difference between top and bottom water layers varied significantly among culture systems, but was unaffected by season (Fig. 6). Integrated culture system showed the greatest DO stratification, whereas subsistence system showed the least (Fig. 6).

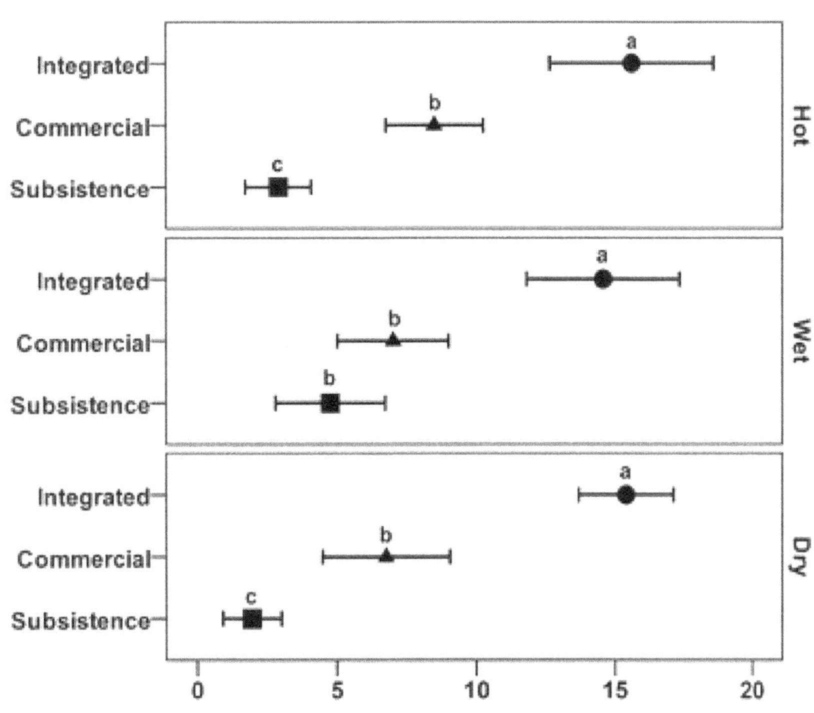

Figure 6: Difference between surface and bottom DO concentrations in the afternoon in different culture systems and seasons. *Plotted symbols* are means and bars 95 % confidence intervals (CI) for those means

Effect of Water de-Stratification on total Ammonia in Fish Ponds

Repeated measures, multi-factor ANOVA was used to analyze the influence of depth of water (surface and bottom), season, and fish culture system, on TAN before and after de-stratification. The analysis showed that fish culture systems and depth of pond had significant effects on TAN (Fig. 7). TAN concentrations in tilapia ponds at the surface were lower than at the bottom of the pond ($P < 0.05$). After de-stratification, TAN at the surface water layer clearly increased (Fig. 7). Water de-stratification resulted in increased TAN at the surface layer. TAN of the integrated system was significantly higher than in commercial and subsistence systems (Tukey's HSD,$P < 0.05$). However, season had no significant effects on TAN ($P > 0.2$). The average levels of un-ionized ammonia (NH_3) concentration of integrated and commercial pond systems were 0.012 ± 0.01 mg/L, whilst in subsistence system it was 0.005 ± 0.01 mg/L.

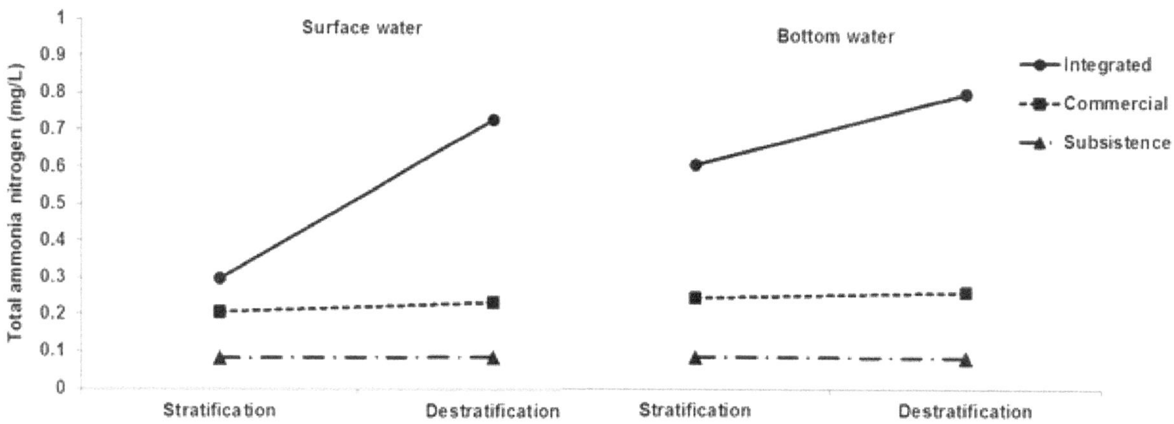

Figure 7: Total ammonia nitrogen of different culture systems at surface and bottom pond before and after de-stratification.

DISCUSSION

The largest differences in temperature between surface and bottom layer were observed in integrated ponds during the hot season; whereas differences in the dry season were modest and in the wet season largely absent (Fig. 2). Four factors help explain these finding. First, the high load of nutrient and fertilizer inputs in integrated ponds caused phytoplankton blooms (Little and Edwards 2004; Szyper and Lin 1990). In this study, concentrations of Chlorophyll-a and overall turbidity, were much higher in integrated ponds than the other two systems. Consequently, high biomass of plankton or suspended particles in the surface layer limits light penetration to the lower water layers (Litchman et al. 2004; Tadesse et al. 2004). Second, is that solar radiation in the hot season tends to be higher than in the other seasons of Thailand (Janjai and Wattan 2011). Factors one and two explain the high potential for thermal stratification to occur during the hot season. A third factor, is that the air temperature range within a 24-h period tended to be greater in the hot and dry seasons than in the wet season (Table 3), which, if wind-mixing was limited, would generate thermal gradients (Losordo and Piedrahita 1991). A fourth factor, is heavy day time rainfall in the wet season; which could help explain why thermal stratification sometimes does not occur in this season (Fig. 2).

Thermal stratification usually peaked in the afternoon and collapsed in the evening (Fig. 3). Water de-stratification in the wet season occurred earlier than in the dry and hot season, probably due to mixing effects from rain (Diana et al. 1991). Seasonal differences in typical wind conditions might also have played a role (Diana et al. 1991; Culberson and Piedrahita1996). Thermal de-stratification effects DO concentrations; decreasing them in surface waters while increasing them in the bottom layers (Boyd et al. 1978; Chang 1989). Water mixing is an important factor to consider therefore, as it effects fish respiration. On the other hand, the increase of DO that occurs at the bottom water layers as a result of mixing may be significant to improving the aerobic conditions at the bottom of ponds (Chang and Ouyang 1988). The use of aerators or pond mixers to assist in water mixing during the afternoons can help to reduce stratification in fish ponds; in addition, introduce more oxygen that can be used during the night, when DO concentrations otherwise might reach critically low levels (Chowdhury et al.2014; Szyper and Lin 1990).

Oxygen stratification had a similar diel pattern with thermal stratification. The highest DO level was found in surface water level in the afternoon, and decreased in the evening (Fig. 4). Photosynthetically active solar radiation (PAR) drives photosynthesis (Romaire and Boyd1979). Therefore, DO concentration generally increases with increasing PAR in the morning (Tadesse et al. 2004). In the afternoon, DO concentrations continue to increase even after PAR decreases; this observation is consistent with previous studies that suggest there is a time lag in photosynthesis response to changes in light intensity (Cloot 1994; Szyper et al. 1992).

The degree of DO stratification varied significantly among culture systems: integrated culture systems showed the greatest DO stratification, and subsistence systems the least (Fig. 6). In integrated and commercial ponds, oxygen oversaturation occurred during the late afternoon, while the bottom layer was often depleted of any oxygen. A likely explanation is that high input nutrient ponds have much higher phytoplankton biomass than the lower input systems, and this is thus reflected in DO concentrations. The bottom layers had low photosynthesis activity due to the abundance of phytoplankton in the upper layers, which prevented light from penetrating into the deeper layers (Diehl 2002). Respiration by phytoplankton, settling detritus and the sediments may have also contributed to the depletion of oxygen in the bottom layers (Chang1989; Hargreaves 1998).

In two-thirds of the early morning observations in integrated and commercial ponds, DO levels were lower than 1 mg/L. The critical threshold value for respiration for Nile tilapia is a DO concentration of 0.8 mg/L, at 26 °C (Duy et al. 2008). Be that as it may, tilapia can survive under DO concentrations of lower than 1 mg/L, by drawing air directly from the atmosphere at the pond surface. Mortality can occur when DO concentrations become lower than 2 mg/L for more than 2–3 days consecutively (Chervinski 1982). On days with dense cloud cover, low light intensities affect the oxygen budget of a pond; as a result, the oxygen production and the oxygen reserve for use at night declines (Kepenyes and Varadi 1984). One commercial pond in Nakhon Sawan experienced a mass mortality event, losing about 400 kg of full-sized fish (each weighing 700–800 g) during a week of cloudy and rainy days, when DO levels remained consistently low (DO concentration at night to early morning was about 0.1 mg/L). The average minimum DO concentrations in integrated and commercial culture systems were lower than 1 mg/L (Fig. 5), confirming that both types of culture systems are prone to risk of DO depletion (Sriyasak et al.2013). The reason for this is because these systems have high fish stocking densities and high nutrient inputs (Table 1) which stimulate high phytoplankton biomass.

Fish culture system and depth of pond had significant effects on TAN. In the bottom layer, TAN concentration was higher than surface water. This was likely due to the accumulation of nitrogen in the sediments (Hargreaves 1998; Kaggwa et al. 2010). Water de-stratification resulted in TAN concentration increasing at the surface layer; consequently, the diffusion of ammonia into the water column created a nutrient source to support the proliferation of phytoplankton (Hargreaves and Tucker 2004). TAN levels of integrated ponds were higher than commercial and subsistence systems. Ammonia of integrated and commercial culture systems, which had high feeding and fish stocking densities, were mainly from uneaten feed, feces, and fish excretes (Durborow et al. 1992; Gross et al. 1999). Average un-ionized ammonia concentration of all culture systems was lower than the proposed threshold toxicity levels for tilapia (less than 0.5 mg/L) (Hargreaves and Tucker 2004). The level of un-ionized ammonia which have an impact on growth rate and can kill fish over a few days ranged between 0.2 and 0.6 mg/L (Durborow et al. 1992; El-Shebly and Gad 2011; Joel and Amajuoyi 2010), while the acute toxicity (24 h) of Nile tilapia fingerling is 7.4 mg/L (Benli and Koksal 2005). The suggested safe level for chronic exposure (over 2–3 week) to un-ionized ammonia concentration is lower than 0.05 mg/L (Lawson 1995; MacIntyre et al. 2008). Ammonia levels measured in this present study were therefore not considered toxic to the fish; and with continued good water management practices in place, it should not be a major problem in future.

The complex interactions among factors as found in this study, suggest that there should be value in adapting existing pond aquaculture models (e.g. Culberson and Piedrahita 1996) to explore in more detail the dynamic consequences of weather and climate in different culture systems on thermal and DO stratification.

CONCLUSIONS

Thermal water stratification results in low DO levels in fish ponds. De-stratification reduces DO near the surface and increases DO at the bottom layer. Stratification interventions like mixing water or using an aerator should be employed to reduce stratification, and further increase oxygen levels at the bottom of the pond. Integrated and commercial culture systems are more prone to risk from very low DO (<1 mg/L), at around 02:00–06:00 h, than the subsistence system. In both systems, organic matter inputs from feed and waste are high, resulting in high concentrations of phytoplankton; which consume a lot of oxygen at night. De-stratification also causes ammonia concentrations in the surface water layer to increase. Ammonia concentrations in all ponds observed in this study however, were lower than the toxic threshold levels for tilapia. This study found widespread evidence indicating that integrated and commercial culture systems are especially prone to low levels of DO concentrations at night; approaching or exceeding thresholds likely to impact fish health and survival. Pond management practices should be adjusted to reduce the risks of exceeding DO thresholds. In particular, fish farmers should: (1) take care to avoid over-feeding; manage water and sediments to prevent excessive accumulation of organic matter and waste at the bottom of ponds, which can influence other water quality parameters; and, (2) use aeration and mechanical mixing interventions at critical times to reduce stress on fish from low DO concentrations, and thus avoid risks of mass mortality events.

Compliance with ethical standards

REFERENCES

1. Abdel-Tawwab M (2012) Effects of dietary protein levels and rearing density on growth performance and stress response of Nile tilapia, *Oreochromis niloticus* (L.). Int Aquat Res 4:3
2. Ahmed Z, Hisham A, Rahman A (2011) Ecomonitoring of climate impact on earthen pond water quality in El-Fayoum, Egypt. Int Res J Microbiol 2:442–454
3. APHA (1980) Standard method for the examination of water and wastewater. American public health association, Washington DC
4. Benli KA, Koksal G (2005) The acute toxicity of ammonia on tilapia (*Oreochromis niloticus* L.) larvae and fingerlings. Turk J Vet Anim Sci 29:339–344
5. Bhujel RC (2013) On-farm feed management practices for Nile tilapia (*Oreochromis niloticus*) in Thailand. In: Hasan MR, New MB (ed) On-farm feeding and feed management in aquaculture. FAO Fisheries and Aquaculture Technical Paper No. 583. Rome, FAO
6. Boyd CE, Romaire RP, Johnson E (1978) Predicting early morning dissolved oxygen concentration in channel catfish ponds. Trans Am Fish Soc 107:484–492. doi:10.1577/1548-8659(1978)107<484:PEMDOC>2.0.CO;2
7. Chainark S, Boyd CE (2010) Water and sediment quality, phytoplankton communities and channel catfish

production in sodium nitrate-treated ponds. J Appl Aquac 22:171–185. doi:10.1080/10454431003736524

8. Chang WYB (1989) Estimates of hypolimnetic oxygen deficits in ponds. Aquac Fish Manag 20:167–172
9. Chang WYB, Ouyang H (1988) Dynamics of dissolved oxygen and vertical circulation in fish ponds. Aquaculture 74:263–276. doi:10.1016/0044-8486(88)90370-5
10. Chervinski J (1982) Environmental physiology of tilapias. In: Roger S, Pullin RSV, Lowe-McConell RH (ed) ICLARM Conference: topic in the Biology and Culture of Tilapias, Manila, September 1980. Worldfish, Bellagio, pp 119–128
11. Chitmanat C (2013) Nile tilapia disease. Chiangmai Vet J 11:75–86
12. Chowdhury AS, Hasan K, Alam K (2014) The use of an aeration system to prevent thermal stratification of water bodies: pond, lake and water supply reservoir. Appl Ecol Environ Sci 2:1–7. doi:10.12691/aees-2-1-1
13. Cloot A (1994) Effect of light intensity variations on the rate of photosynthesis of algae: a dynamical approach. Math Comput Model 19(9):23–33. doi:10.1016/0895-7177(94)90038-8
14. Culberson SD, Piedrahita RH (1996) Aquaculture pond ecosystem model: temperature and dissolved oxygen prediction-mechanism and application. Ecol Model 89:231–258. doi:10.1016/0304-3800(95)00140-9
15. Diana SJ, Lin KC, Schneeberger JP (1991) Relationships among nutrient inputs, water nutrient concentrations, primary production, and yield of *Oreochromis niloticus* in ponds. Aquculture 92:323–341. doi:10.1016/0044-8486(91)90038-9
16. Diehl S (2002) Phytoplankton, light, and nutrients in a gradient of mixing depth: theory. Ecology 83:386–398. doi:10.1890/0012-9658(2002)083
17. DOF (2013) Report of tilapia production in the first quarter of 2013 [in Thai].http://fishco.fisheries.go.th/fisheconomic/Doc/fishnews_164.pdf. Accessed 10 Jan 2014
18. Durborow MR, Crosby MD, Brunson WM (1992) Ammonia in fish ponds. Southern Regional Aquaculture Center, Stoneville
19. Duy AT, Scharama J, Dam AV, Verreth AJ (2008) Effects of oxygen concentration and body weight on maximum feed intake, growth and hematological parameters of Nile tilapia,*Oreochromis niloticus*. Aquculture 275:152–162. doi:10.1016/j.aquaculture.2007.12.024
20. Egna HS, Boyd CE (1997) Dynamics of pond aquaculture. CRC Press, New York
21. El-Shebly AA, Gad MH (2011) Effect of chronic ammonia exposure on growth performance, serum growth hormone (GH) levels and gill histology of Nile tilapia (*Oreochromis niloticus*). J Microbiol Biotechnol Res 1(4):183–197
22. Emerson K, Russo RC, Lund RE, Thurston RV (1975) Aqueous ammonia equilibrium calculations: effect of pH and temperature. J Fish Res Board Can 32:2379–2383. doi:10.1139/f75-274
23. EPA (2013) Aquatic life ambient water quality criteria for ammonia—freshwater 2013. Office of Science and Technology, Washington, DC
24. Gandhi KT (2012) A study of water quality parameters to better manage our ponds or lakes. Int J Late Res Sci Technol 1:359–363
25. Gross A, Boyd CE, Wood CW (1999) Ammonia volatilization from freshwater fish ponds. J Environ Qual 28:793–797. doi:10.2134/jeq1999.00472425002800030009x
26. Hargreaves J (1998) Nitrogen biogeochemistry of aquaculture ponds. Aquaculture 166:181–212. doi:10.1016/S0044-8486(98)00298-1
27. Hargreaves JA, Tucker CS (2004) Managing ammonia in fish ponds. Southern Regional Aquaculture Center, Stoneville
28. Janjai S, Wattan R (2011) Development of a model for the estimation of photosynthetically active radiation from geostationary satellite data in a tropical environment. Rem Sens Environ 115:1680–1693. doi:10.1016/j.rse.2011.02.026
29. Joel O, Amajuoyi CS (2010) Determination of the concentration of ammonia that could have lethal effect on fish pond. J Eng Appl Sci 5:1–4
30. Kaggwa R, Dam AA, Kipkemboi J, Denny P (2010) Evaluation of nitrogen cycling and fish production in seasonal ponds ('Fingerpond') in lake victoria wetlands, East Africa using a dynamic simulation model. Aquac Res 42:74–90. doi:10.1111/j.1365-2109.2010.02563.x
31. Kepenyes J, Varadi L (1984) Aeration and oxygenation in aquaculture. In: FAO (ed) Inland aquaculture engineering. FAO, Rome
32. Klimaszyk P, Tomaz J (2007) Seasonal changeability of thermal and oxygen conditions in polyhumic lakes: Glodne Jeziorko III and Czarny Dol (northern Poland). Limnol Rev 7:199–203
33. Lawson TB (1995) Fundamentals of aquaculture engineering. Chapman & Hall, New York
34. Litchman E, Klausmeier CA, Bossard P (2004) Phytoplankton nutrient competition under dynamic light regimes. Limnol Oceanogr 49((4, part 2)):1457–1462. doi:10.4319/lo.2004.49.4_part_2.1457
35. Little DC, Edwards P (2004) Impact of nutrition and season on pond culture performance of mono-sex and mixed-sex Nile tilapia (*Oreochromis niloticus*). Aquaculture 232:279–292. doi:10.1016/S0044-8486(03)00482-4
36. Losordo TM, Piedrahita RH (1991) Modelling temperature variation and thermal stratification in shallow aquaculture ponds. Ecol Model 54:189–226. doi:10.1016/0304-3800(91)90076-D
37. MacIntyre CM, Ellis T, North BP, Turnbull JF (2008) The influences of water quality on the welfare of farmed rainbow trout: a review. In: Branson EJ (ed) Fish welfare. Blackwell Publishing Ltd., Oxford
38. Pandit NP, Nakamura M (2010) Effect of high temperature on survival, growth and feed conversion ratio of Nile Tilapia, *Oreochromis niloticus*. Our Nat 8:219–224. doi:10.3126/on.v8i1.4331

39. Pimolrat P, Whangchai N, Chitmanat C, Promya J, Lebel L (2013) Survey of climate-related risks to Tilapia pond farms in northern Thailand. Int J Geosci 4:54–59. doi:10.4236/ijg.2013.45B004

40. Randall DJ, Tsui TKN (2002) Ammonia toxicity in fish. Mar Pollut Bull 45:17–23. doi:10.1016/S0025-326X(02)00227-8

41. Romaire PR, Boyd EC (1979) Effects of solar radiation on the dynamics of dissolved oxygen in channel catfish ponds. Trans Am Fish Soc 108(5):473–478. doi:10.1577/1548-8659(1979)108<473:EOSROT>2.0

42. Smith DW, Piedrahita RH (1988) The relation between phytoplankton and dissolved oxygen in fish ponds. Aquculture 68:249–265. doi:10.1016/0044-8486(88)90357-2

43. Srisapoome P, Areechon N (2014) Serious disease outbreaks in Nile tilapia cultured in Thailand.http://www.rdi.ku.ac.th/kasetresearch52/05-animal/prapansak_Nile/animal_00.html. Accessed 24 Dec 2014

44. Sriyasak P, Chitmanat C, Whangchai N, Promya J, Lebel L (2013) Effects of temperature upon water turnover in fish ponds in northern Thailand. Int J Geosci 4:18–23. doi:10.4236/ijg.2013.45B004

45. Sriyasak P, Whangchai N, Chitmanat C, Promya J, Lebel L (2014) Impacts of climate and season on water quality in aquaculture ponds. KKU Res J 19(5):743–751

46. Stefan HG, Hondzo M, Fang X, Eaton JG, McCormick JH (1996) Simulated long-term temperature and dissolved oxygen characteristics of lakes in the north-central United States and associated fish habitat limits. Limnol Oceanogr 41:1124–1135. doi:10.4319/lo.1996.41.5.1124

47. Szyper PJ, Lin CK (1990) Techniques for assessment of stratification and effects of mechanical mixing in tropical fish ponds. Aquac Eng 9:151–165. doi:10.1016/0144-8609(90)90002-H

48. Szyper PJ, Rosenfeld ZJ, Piedrahita HR, Giovannini P (1992) Diel cycles of planktonic respiration rates in briefly incubated water samples from a fertile earthen pond. Limnol Oceanogr 37(6):1193–1201. doi:10.4319/lo.1992.37.6.1193

49. Tadesse I, Green FB, Puhakka JA (2004) Seasonal and diurnal variations of temperature, pH and dissolved oxygen in advanced integrated wastewater pond systems treating tannery effluent. Water Res 38:645–654. doi:10.1016/j.watres.2003.10.006

50. Thai Meteorological Department (2012) The weather of Thailand.http://www.tmd.go.th/info/info.php?FileID=22. Accessed 9 Dec 2014

51. Thai Meteorological Department (2014) The weather data of Thailand.http://119.46.126.50/web/aws/aws_alphanumeric.asp. Accessed 12 Feb 2015

CHAPTER 11

HOW EFFECTIVE ARE CREATED OR RESTORED FRESHWATER WETLANDS FOR NITROGEN AND PHOSPHORUS REMOVAL? A SYSTEMATIC REVIEW

Magnus Land[1], Wilhelm Granéli[2,3,5], Anders Grimvall[4], Carl Christian Hoffmann[5], William J. Mitsch[3], Karin S. Tonderski[6] and Jos T. A. Verhoeven[7]

[1]Mistra EviEM, Stockholm Environment Institute, 10451 Stockholm, Sweden

[2] Department of Biology, Lund University, 22362 Lund, Sweden.

[3] Everglades Wetland Research Park, Florida Gulf Coast University, 4940 Bayshore Drive, Naples 34112, FL, USA. [4] Swedish Institute for the Marine Environment, Box 260, 405 30 Gothenburg, Sweden.

[5] Department of Bioscience, Aarhus University, 8600 Silkeborg, Denmark.

[6] Department of Physics, Chemistry and Biology, Linköping University, 581 83 Linköping, Sweden.

[7] Department of Biology, Utrecht University, 3584 CA Utrecht, Netherlands.

ABSTRACT

Background

Eutrophication of aquatic environments is a major environmental problem in large parts of the world. In Europe, EU legislation (the Water Framework Directive and the Marine Strategy Framework Directive), international conventions (OSPAR, HELCOM) and national environmental objectives emphasize the need to reduce the input of nutrients to freshwater and marine environments. A widely used method to achieve this is to allow water to pass through a created or restored wetland. However, the large variation in measured nutrient removal rates in such wetlands calls for a systematic review.

Methods

Searches for primary studies were performed in electronic databases and on the internet. One author performed the screening of all retrieved articles at the title and abstract level. To check that the screening was consistent and complied with the agreed inclusion/exclusion criteria, subsets of 100 articles were screened by the other authors. When screening at full-text level the articles were evenly distributed among the authors. Kappa tests were used to evaluate screening consistency. Relevant articles remaining after screening were critically appraised and assigned to three quality categories, from two of which data were extracted. Quantitative synthesis consists of meta-analyses and response surface analyses. Regressions were performed using generalized additive models that can handle nonlinear relationships and interaction effects.

Citation: Magnus Land, Wilhelm Granéli, Anders Grimvall, Carl Christian Hoffmann, William J. Mitsch, Karin S. Tonderski, Jos T. A. Verhoeven, How effective are created or restored freshwater wetlands for nitrogen and phosphorus removal? A systematic review, doi:10.1186/s13750-016-0060-0

Results

Searches generated 5853 unique records. After screening on relevance and critical appraisal, 93 articles including 203 wetlands were used for data extraction. Most of the wetlands were situated in Europe and North America. The removal rate of both total nitrogen (TN) and total phosphorus (TP) is highly dependent on the loading rate. Significant relationships were also found for annual average air temperature (T) and wetland area (A). Median removal rates of TN and TP were 93 and 1.2 g m^{-2} $year^{-1}$, respectively. Removal efficiency for TN was significantly correlated with hydrologic loading rate (HLR) and T, and the median was 37 %, with a 95 % confidence interval of 29–44 %. Removal efficiency for TP was significantly correlated with inlet TP concentration, HLR, T, and A. Median TP removal efficiency was 46 % with a 95 % confidence interval of 37–55 %. Although there are small differences in average values between the two quality categories, the variation is considerably smaller among high quality studies compared to studies with lower quality. This suggests that part of the large variation between studies may be explained by less rigorous study designs.

Conclusions

On average, created and restored wetlands significantly reduce the transport of TN and TP in treated wastewater and urban and agricultural runoff, and may thus be effective in efforts to counteract eutrophication. However, restored wetlands on former farmland were significantly less efficient than other wetlands at TP removal. In addition, wetlands with precipitation-driven HLRs and/or hydrologic pulsing show significantly lower TP removal efficiencies compared to wetlands with controlled HLRs. Loading rate (inlet concentrations × hydraulic loading rates) needs to be carefully estimated as part of the wetland design. More research is needed on the effects of hydrologic pulsing on wetlands. There is also a lack of evidence for long-term (>20 years) performance of wetlands.

BACKGROUND

In Europe, like in many other parts of the world, nutrient enrichment of water bodies is a major environmental problem [1]. Several EU directives emphasize the need to reduce the input of nutrients to both freshwater and marine ecosystems (e.g., the Water Framework Directive, the Marine Strategy Framework Directive and the Nitrate Directive). This is also an important part of the Helsinki Commission (HELCOM) Baltic Sea Action Plan, which contains several suggested measures targeting nutrient losses from agricultural land. Wetland creation is one of these, as it is known that the biogeochemical transformations that occur in wetlands generally result in a reduction in the nutrient content of the water flow.

Wetland creation has been practiced, e.g., in Sweden, where wetlands have been constructed and restored on a fairly large scale since the 1990s—initially focused on nitrogen removal and biodiversity enhancement. Nitrogen was usually assumed to limit primary production in marine ecosystems [2, 3], and also in the brackish-water Baltic Sea [4] which catchment includes most of Sweden. However, this is a disputed assumption, and some scientists have the opinion that phosphorus is ultimately limiting production in the Baltic Sea [5–7]. In freshwater bodies eutrophication is usually thought to be controlled by phosphorus inputs only [6], although this is also somewhat controversial and some scientists argue that nitrogen inputs to lakes have to be reduced as well [5]. Thus, singling out one nutrient or the other as limiting in the marine and freshwater system, respectively, may simplify reality too far. In many systems both nitrogen and phosphorus are limiting depending on time of year and location [8]. Therefore, whether the major concern is the marine environment or freshwater ecosystems, quantifying the effect of measures to remove both nitrogen and phosphorus from water is very relevant.

Created wetlands can be of different types, and are usually classified as free water surface constructed wetlands (FWS), horizontal subsurface flow constructed wetlands (HSF), and vertical flow constructed wetlands (VF) [9]. FWS wetlands are usually between 0.1 and 2 m deep, with a plant community that can be composed of algae, and submersed, floating or emergent wetland plants. HSF constructed wetlands are typically designed with a permeable filter material ("soil") planted with emergent wetland plants. Water flows horizontally in and beneath their rhizosphere, which creates a mix of saturated anaerobic, and unsaturated aerobic zones. A VF wetland is similarly constructed, but water is applied on the surface of the filtering material, and percolates through the rhizosphere. This results in a typically unsaturated aerated "soil". When wetlands are restored, interventions are typically made to recreate previously drained, or by other means altered, natural wetlands. There are numerous studies of the physical and biogeochemical processes involved in the removal of both nitrogen and phosphorus. These are therefore relatively well known. When comparing different studies, the removal efficiency varies considerably, which makes it difficult to assess the extent to which wetland creation is an efficient measure to reduce eutrophication. Several previous studies [10–18] (cited in the protocol for this systematic review [19]) have indicated that removal differences between wetlands are related to a number of different factors such as inflow concentration, load variations, hydraulic retention time, temperature, hydraulic efficiency, and type of wetland. This calls for a systematic review of removal rates and how they are influenced by the wetland characteristics, loading differences, and environmental factors.

In Sweden, thousands of hectares of wetlands have been financed through various governmental funds and created as a means of reaching the Environmental objectives "Thriving wetlands" and "Zero eutrophication", and to meet the commitments made in the Baltic Sea Action Plan (BSAP) [20]. The nutrient removal effects of those investments have been estimated using both statistical models and a catchment hydro-chemical model [21–24]. In general, surprisingly low mean removal rates have been reported, and this is mainly attributed to the fact that many wetlands were created as part of biodiversity conservation efforts and thus receive low nutrient loads.

In a recent evaluation, Weisner et al. [25] assessed the nutrient removal effects of around 5300 ha wetlands created within the Rural Development Program 2007–2013. Previously used calculation models were modified based on new monitoring data, and it was estimated that those wetlands would remove 0.35–0.6 g m^{-2} $year^{-1}$ of phosphorus and 3.2–4.6 g m^{-2} $year^{-1}$ of nitrogen. If only looking at the group of wetlands that were created with nutrient removal as the main goal, the removal was about ten times higher or 3–4.5 g m^{-2} $year^{-1}$ for phosphorus and 30–34 g m^{-2} $year^{-1}$ for nitrogen. Based on the relatively small number of monitoring data sets that exist, the authors concluded that it would be possible to achieve a removal of around 10 g m^{-2} $year^{-1}$ of phosphorus and 100 g m^{-2} $year^{-1}$ of nitrogen in individual wetlands, provided they are located in optimal locations and with a design adapted to achieve high nutrient removal rates. For non-point source (mostly agricultural and urban runoff) nutrient removal in eastern USA, Mitsch et al. [26] suggested that sustainable removal rates would be 10–40 g m^{-2} $year^{-1}$for nitrogen and 0.5–5 g m^{-2} $year^{-1}$ for phosphorus, and based this on wetland studies in the last quarter of the 20th century.

The different models used in the cited evaluations are all based on assumed relationships between removal rates and loads. Although not directly comparable, the results clearly indicate that the removal rates for nitrogen and phosphorus in created or restored wetlands span over a wide range. Published research studies also show that even though there is a generally positive relationship between removal rate and loading rate, the variation is quite large between wetlands with similar loads. It is thus unclear to what degree created wetlands will contribute to fulfilling the different Swedish environmental goals related to eutrophication. In this context, the Swedish Board of Agriculture and the Swedish Agency for Marine and Water Management were interested in obtaining a comprehensive evaluation of measured retention rates in individual wetlands. They were also interested in obtaining a coherent picture of how different wetlands function in a variety of conditions to facilitate planning of more effective water pollution control. In addition, the wastewater treatment industries are important stakeholders, as they will most likely be expected to have to comply with even stricter regulations on nutrient emissions than today. Created wetlands may prove to be a cost-efficient polishing step to reach those regulation demands.

Objective of the review

The objective of this review is to quantify observed removal rates of nutrients in created or restored wetlands, and to examine the distribution of these rates and quantify the variation between different studies. The primary question this review seeks an answer to is *"How effective are created or restored freshwater wetlands for nitrogen removal and phosphorus retention?"* This question implicitly includes the relationship between removal and load, as effectiveness can be expressed both as removal rates (g m^{-2} $year^{-1}$) and relative removal (% of load). In this review, removal refers to a reduced amount of nitrogen and phosphorus in the water phase. Removal processes imply transformation of the nutrients to other forms; for nitrogen this often means nitrogen gas and a smaller proportion of the greenhouse gas nitrous oxide, which will be emitted to the atmosphere.

Secondary questions are related to how various effect modifiers, such as environmental conditions and wetland characteristics, influence the nutrient removal rates. For that reason this review covers a fairly wide range of climatic conditions, although the performance of wetlands in temperate and boreal regions is most relevant to the stakeholders in Sweden. The review will not engage in detailed investigations of various removal processes and mechanisms but rather treat each wetland as a "black box". This of course introduces some uncertainties. However, when assessing study quality, studies presenting complete nutrient budgets where removal by individual processes had been quantified were rated higher than studies merely providing inlet/outlet data. A fairly balanced budget may indicate that the numbers are reasonably accurate and/or that no major source or sink have been overlooked. The structure of the primary question is further discussed in the section Study Criteria, where information on relevant subjects, interventions, comparator and outcomes is given in more detail.

METHODS

Searches

Searches for literature were made in ten different literature databases. The searched fields and search date are shown for each database in Table 1. Specific search strings for each database. No particular time, document type or language constraints were applied. However, at a later stage it was decided that articles in Chinese should be excluded due to the lack of translation resources.

Table 1: Electronic databases used for searching

Database	Searched field	No of hits	Date
ISI web of science	Topic	3249	2013-02-26
Georef and geobase	Subject/title/abstract	2249	2013-02-26
Scopus	Title, abstract and keywords	2842	2013-02-26
Agricola	Keyword anywhere	524	2013-02-26
ASFA	All fields (no full text)	1933[a]	2013-02-27
Academic search	Title/abstract/subject/keyword	1131	2013-02-27
Biological abstracts	Keywords	401	2013-02-27
Wiley online library	Abstract or title or keywords	174	2013-02-27

Directory of open access journals	All fields	84[a]	2013-03-05
ScienceDirect	Abstract or title or keywords	876	2013-07-02

[a]After removal of duplicates

Grey literature was searched for using the search engine Google where simplified search strings were used. In addition to searches where English search terms were used, searches were also performed using Swedish, Danish, and Dutch search terms. Searches were performed 2014-03-03.

In addition, the websites of relevant specialist organisations (listed below) were also searched. Where possible the same search string as for Google was used (in relevant language). Generally, the first 100 hits were examined in the searches using Google and on specialist websites.

- Swedish Environmental Protection Agency (SEPA)
- Swedish Board of Agriculture
- The Swedish Agency for Marine and Water Management
- wedish directory of Master thesis (DiVA)
- South Florida Water Management District
- U.S. Environmental Protection Agency (EPA)
- North American Data Base (NADB)
- U.S. Department of Agriculture (USDA)
- Foundation for Applied Water Research (STOWA)
- Ekologgruppen i Landskrona AB
- Norwegian Institute for Agricultural and Environmental Research (Bioforsk)
- Danish Centre for Environment and Energy (DCE)
- European Environment Agency (EEA)
- Wetland Solutions Inc.
- Wetlands International
- Finnish Environment Institute (SYKE)
- Federal Environment Agency (UmweltBundesAmt, Germany)
- Stichting Toegepast Waterbeheer (STOWA, The Netherlands)

To test the comprehensiveness of the searches, bibliographies of review articles were examined and compared with the search results.

STUDY INCLUSION CRITERIA

In this review some constraints regarding the type of water entering the wetlands have been applied. Untreated wastewater was not considered since it is not permissible to discharge such water into the environment in most European countries. Industrial or agricultural wastewater can vary considerably in composition, and was therefore also excluded. Farmyard runoff was in most cases classified as agricultural wastewater, and thus excluded, since it is often mixed with untreated parlour washings and silage/farmyard manure effluents, among other things.

The inclusion criteria below were developed in collaboration with stakeholders during development of the review protocol [19].

- *Relevant subject*: Secondary or tertiary treated domestic wastewater, urban storm water, stream/river water, freshwater aquaculture effluents, and runoff from agricultural fields. To be able to make a distinction between treated and untreated wastewater, a guideline value of 100 mg/l was used for the highest acceptable concentration of BOD in the water entering the wetland.
- *Types of intervention*: Creation or restoration of wetlands. In this review, creation of a wetland refers to the construction of a wetland on a site that never was a wetland, regardless of the main purpose of the wetland. Sometimes the term Constructed wetland is used for a wetland with the specific aim of treating wastewater, storm water, acid mine drainage, or agricultural runoff. Such wetlands are thus regarded as Created wetlands. Created wetlands include both horizontal and vertical subsurface flow systems and free water surface systems. Restoration refers to recovery of ecological and hydrological processes as well as geomorphology in areas where natural wetlands previously have been drained or by other means altered. In this review the term restoration does not imply any specific purpose of the wetland. To be included the created or restored wetlands must host some type of vegetation.
- *Types of comparator*: No intervention (inlet conditions can serve as control).
- *Types of outcome*: Mass removal of total nitrogen (TN) or total phosphorus (TP) from the water body per unit wetland area and year. Removal efficiency of TN or TP (% of load).
- *Types of study*: The most common way to evaluate the overall retention rate in a wetland is to compare the nitrogen or phosphorus loads in the inlet and outlet water, respectively. Quite often the retention in wetlands is evaluated in experiments where effect modifiers such as loading rate or vegetation type are varied. This is a version of a control-impact (CI) study where inlet conditions serve as control. In rare cases, nutrient loads in a river or stream have been recorded both before and after the establishment of a wetland, which corresponds to a typical before–after (BA) study. Both types of studies are eligible.

The removal rates and efficiencies may show large seasonal variation. Therefore, to be included in the review, it is a prerequisite that the wetland is established in field conditions and exposed to the ambient climate. This means that laboratory and greenhouse studies were excluded and that each study must cover at least one complete annual cycle. Also, in order to reflect realistic conditions the wetland must be of a reasonable size. While typical microcosm studies were excluded, mesocosm studies were included since they potentially provide valuable information on the variability of the outcomes based on true replicates. A cut-off wetland size of 1 m^2 was applied.

This systematic review is focused on boreal and temperate regions, but for comparison sub-tropical regions have also been included. In the Köppen-Geiger climate classification system [27] this corresponds roughly to group D (snow climates), group C (warm temperate climates) and parts of group A (Equatorial climates with one dry season, i.e., As and Aw). Studies on wetlands located in other climates were excluded. Furthermore, to be included the studies must have taken all removal processes into account. Studies that only report results for selected processes were excluded.

Wetlands may be created or restored for purposes other than nutrient removal. Although some wetlands serve multiple purposes [28], the main purpose is in some cases to promote biodiversity or reduce flood risks. In this review wetlands have been considered regardless of the main purpose of the wetland, i.e., inclusion and exclusion was not based on the reasons for constructing or restoring the wetlands. However, the main purpose of the wetland was recorded during data extraction.

Screening Process

Articles found in the searches were checked for relevance at (1) title and abstract, and (2) full text levels. At the title and abstract level, the first author of this review performed the screening of all articles. To check that the screening was consistent and complied with the agreed inclusion/exclusion criteria, a subset of 106 articles was also screened by another author. A second subset of 106 articles was screened by two other authors, and a third subset of 107 articles was screened by a third pair of authors. In this way 319 articles were double-screened. Thus, we checked the consistency between the main screener and the other authors as well as between the authors within each screening pair. To evaluate the consistency Kappa tests were used.

Full-text articles were randomly and evenly distributed for screening among six authors. However, before full-text screening at full scale, three subsets of 34 articles in each were double-screened in the same manner as at the title and abstract level. Again, Kappa tests were used to test the consistency between the authors. Kappa values of 0.6 or higher were considered acceptable both at title and abstract screening and at full-text screening.

Potential effect modifiers and reasons for heterogeneity

The nutrient retention may vary considerably between different studies. The anticipated large variation is easy to understand in the light of the fact that the removal rate is a result of several independent processes. Nitrogen removal takes place through: (1) sedimentation and sediment accretion, (2) plant uptake, and (3) denitrification and volatilization. The processes involved in phosphorus removal are: (1) sedimentation and sediment accretion, (2) plant uptake (3) sorption and (4) precipitation/co-precipitation. The success of each of these mechanisms may depend on factors such as:

- Loading characteristics
    - Hydraulic loading rate (HLR)
    - Concentration and speciation of nitrogen and phosphorus at the inlet
- Wetland characteristics
    - Type of wetland
    - Size and shape (area, depth, length)
    - Flow pattern and hydraulic efficiency
    - Hydroperiod
    - Age
    - Sediment/soil type
    - Oxygen concentration and redox potential
    - Vegetation type and coverage
    - Fauna
    - Management methods and frequency
- Climate characteristics
    - Mean temperature
    - Ice coverage

Study Quality Assessment

Studies included after the full text screening were subject to critical appraisal. Bilotta et al. [29] suggested a systematic method for critical appraisal using an Environmental-Risk of Bias Tool for assessing the internal validity, and an Environmental-GRADE tool for assessing the overall quality of a study. Although the major part of our critical appraisal was performed before the article by Bilotta et al. [29] was published, we have applied a fundamentally similar approach. Concerning risk of bias the following conditions specific to this review should be noted. (1) Wetland studies included in the review have been designed in such a way that selection bias due to inadequate randomisation or inadequate allocation concealment is not an issue. However, another type of selection bias could potentially be introduced if samples were to be taken only at certain favourable (or unfavourable) conditions, e.g., at low hydrological loading rates or during growing seasons. To assess the risk for this type of selection bias the study length and sampling frequency were evaluated. Studies should cover complete annual cycles and, as a guideline value, include at least 12 sampling occasions. Furthermore, it has to be taken into account that the investigated wetlands do not form a random sample of a well-defined population of potential created wetlands. (2) Performance bias is mainly caused by exposure to factors other than the intervention and may be related to e.g., hydrological flow paths not accounted for or use of chemicals added to the water or soil in order to promote certain processes in the wetland. To assess the risk of bias caused by hydrological processes the hydrological mass balance was evaluated. (3) Detection bias is mainly related to sampling and analytical methods, and using different methods for treated and untreated water is most unusual. (4) Attrition bias may occur if there for some reason are fewer samples of either treated or untreated water compared to the other group. In some cases differing number of samples may however be justified. For example, during dry conditions it is not unusual that the hydraulic loading rate at the outlet is zero due to e.g., evapotranspiration while there still is an active water inflow to the wetland. This could potentially result in fewer samples of treated water than of untreated water. Provided that the water mass balance adds up reasonably well, attrition bias may not occur in such cases since the outcome measure is based on total nutrient mass transport during complete annual cycles. (5) The risk of selective reporting bias is fairly easy to assess by comparing e.g., study length with the number of years reported, or the measured outcomes with the reported outcomes.

When assessing the overall quality of the studies they were assigned to either of three quality categories: (1) Does not meet the quality criteria, (2) Acceptable, and (3) High standard. Studies in category 1 did not qualify for data extraction and quantitative synthesis. All studies relevant for this review (i.e., passed the screening at full text level) were observational studies, predominantly control-impact (CI) studies where inlet conditions served as control. Based on this study design, all studies were by default assumed to be of acceptable quality (assigned to category 2). A set of quality criteria (see Table 2) were then used to justify upgrading to category 3, downgrading to category 1, or keeping the study in category 2.

Table 2: Quality criteria and requirements for fulfillment

Quality criteria		Requirements
1	Moderate risk of bias	The study covered at least one complete annual cycle and included at least 12 sampling occasions evenly distributed in time The hydrologic data obtained in the study was adequate for calculating hydraulic loads at inlet and outlet There were no obvious deficiencies in the water balance
2	High sampling frequency	Samples were flow proportional composites or grab samples with a reasonably high sampling frequency (at least twice a month)
3	Hydraulic control	All major water flow paths were quantified and considered in the calculations of the outcomes
4	Removal processes quantified	At least one important removal process was quantified independently, and the result was reasonable in relation to the inlet/outlet difference
5	Effect modifiers	Sufficient information related to potential effect modifiers was reported
6	Replication	Replicate wetlands were studied, or the same wetland was studied for two or more complete annual cycles. If not reported it should be possible to calculate the variance of the replicates

True replication is unusual in studies of nutrient removal in wetlands. In most cases just one wetland, or a set of different wetlands, were studied. However, repeating the measurements in the same wetland during several complete years may be regarded as quasi-replication. We regarded this form of quasi-replication acceptable for meta-analysis.

The criteria for assigning the studies to either of the three quality categories are shown in Table 3.

Table 3: Quality categories and criteria for assignment

Quality categories	Criteria
Category 1	High risk of bias (quality criteria 1 is not fulfilled)
Category 2	Moderate risk of bias (quality criteria 1 is fulfilled)
Category 3	(1) Moderate risk of bias (quality criteria 1 is fulfilled) and (2) At least four of the six quality criteria listed in Table 2 are fulfilled and (3) Quality criteria 3 and 6 are fulfilled

Four reviewers performed the critical appraisal. To check the consistency between the reviewers a small number of articles were critically appraised by all reviewers. Differences were then discussed by all reviewers and, as a result of that discussion, the quality criteria were clarified. During the remainder of the critical appraisal the reviewers had an option to code studies as "uncertain". All studies with that code were then discussed at a meeting where consensus was reached.

Data Extraction Strategy

The outcomes evaluated in this review are the removal rate and removal efficiency of total nitrogen and phosphorus; typically the results are reported quantitatively as g m^{-2} $year^{-1}$ and as % of load, respectively. Results reported in other units were recalculated where possible. In cases where multiple-year studies just reported an overall average without the inter-annual variance, we extracted data for each sampling occasion and calculated annual values for each separate year and the inter-annual variance where possible.

In order to assess the quality of the studies and to be able to evaluate the importance of various effect modifiers, additional data was recorded as well (see Table 4). Not all studies provided information on all parameters shown in Table 4. For instance, a very small number of studies reported data on fauna in the wetland. It is however quite possible that nutrient cycling and retention are influenced by, e.g., benthic organisms through bioturbation [30] or by birds [31,32]. Also, planktivorous fish species can feed intensively on zooplankton and thereby protect phytoplankton from being grazed, leading to turbid water [33]. The parameters for which data was extracted from all included studies are indicated in Table 4 with underscored text.

Table 4: Type of extracted data (underscored parameters were recorded for all included studies)

Loading characteristics	Wetland characteristics	**Climate characteristics**	**Study design parameters**	**Results**
Type of water	Type of wetland	Annual mean air temperature [b]	Study length	Annual average removal rates (g m^{-2} $year^{-1}$)
Concentrations of TN and/or TP at inlet	Length, width, depth	Annual mean water temperature	Number of replicates	Annual average removal efficiency (% of load)
Speciation of N and/or P at inlet	Area	Ice coverage	Sampling frequency	Variance of annual removal rate
Average nutrient loading rates	Hydroperiod (d $year^{-1}$)	GPS coordinates	Flow proportional sampling (Y/N)	Variance of annual removal efficiency
Average hydraulic loading rates	Dominant plant species and coverage	Climate type [c]		
Average residence time (d)	Sediment/soil type			
Hydrologic regime	Land use history and wetland age			
pH	O_2 concentration and redox potential			
	Fauna			
	Management methods			
	Main purpose of wetland			
	Other characteristics[a]			

[a]Examples are, e.g., specific substrates, soil amendments, configuration of multiple wetlands, or other means of enhancing the outcome

[b]Extracted from WorldClim database (http://www.worldclim.org/)

[c]According to the Köppen-Geiger climate classification. This type of data was in most cases not provided by the primary studies and was therefore extracted from the map database referred to by [26]

The four reviewers who performed the critical appraisal extracted data related to wetland characteristics. A fifth reviewer extracted all other data. To make the data extraction as consistent as possible the data was entered into a pre-designed Excel spreadsheet. All reviewers tested the spreadsheet and after some minor modifications it was used throughout the entire data extraction process.

Data Synthesis and Presentation

The studies with true or temporal replication were subjected to meta-analyses. Log response ratios (ln R) where R = $Load_{out}$/$Load_{in}$ were used to quantify effect sizes, and random effects models [34] were used to calculate summary effects and uncertainty bounds of such effects. The between-study variance (τ^2) was estimated by calculation of T^2 using the DerSimonian and Laird method, and to estimate the ratio of true heterogeneity to total variance in observed effects, the I^2 statistic was calculated [35]. In subgroup analyses, separate estimates of τ^2 were made for each individual subgroup. The results were presented in forest plots and in tables where the log response ratios had been back-transformed and recalculated to median removal efficiencies and confidence intervals of median removal efficiencies.

To examine how the removal of nitrogen of phosphorus was influenced by effect modifiers all included studies, even those without replication, were subjected to response surface analyses using various regression models. The removal of nitrogen and phosphorus, expressed as removal efficiency and removal rate, were the primary target variables. The removal efficiency was defined as

$$Removal_{efficiency} = 100 \times \frac{Substance_{flow_{in}} - Substance_{flow_{out}}}{Substance_{flow_{in}}} = 100 \times \left(1 - \frac{Substance_{flow_{out}}}{Substance_{flow_{in}}}\right)$$

which is a monotonic function of the response ratio R.

The removal rate was defined as

$$Removal\_rate = Load\_in - Load\_out$$

i.e., the mass removed per unit time and per wetland area.

The hydraulic loading rate and concentrations of nitrogen and phosphorus in the inflow to the wetlands were considered to be the primary predictors or effect modifiers. Attention was also paid to type of wetland, type of inflow, climate zone, average air temperature, and wetland area. Cross-validations were not carried out because the number of wetlands included in our analysis turned out to be smaller than expected.

The relationships between the mean output-input ratio and potential predictors were assumed to have a multiplicative structure, e.g.,

$$Output\_input\_ratio = a(Hydraulic\_loading)\ b(Concentration\_in)\ c(Climate)$$

that after taking logarithms can be rewritten as a general additive model (*GAM*) of the form

$$LOG(Output\_input\_ratio) = f(LOG(Hydraulic\_loading)) + g(LOG(Concentration\_in)) + h(Climate)$$

where *a, b*, f and *g* are assumed to be smooth functions, and *c*(*Climate*) and *h*(*Climate*) are functions of climate zone indicators or average air temperature. The error terms on the log scale for different wetlands were assumed to be statistically independent and normally distributed with mean zero and constant variance.

The removal rate was assumed to be a smooth function of the magnitude of the hydraulic loading and the concentration of the chemical element or species under consideration. The basic model had an additive structure

$$Removal\_rate = f(LOG(hydraulic\_loading)) + g(LOG(inflow\_concentration)) + h(airtemperature)$$

Potential pairwise interaction effects of the predictors were taken into account by allowing thin plate splines (*TPS*) in the GAM models. Such splines encompass a very large class of smooth functions (response surfaces) that enable a very flexible description of both main effects and interaction effects of any pair of predictors. The error terms for different wetlands were assumed to be statistically independent and normally distributed with mean zero and constant variance.

GAM models with or without thin plate splines were fitted to the collected data using standard least squares algorithms in the software package SAS. Fitted values of the removal efficiency were obtained by first using *LOG*(*Output_input_ratio*) as target variable and then back-transforming the fitted values to the *Percentage_removed*. Because this transformation is nonlinear the back-transformed surfaces should be interpreted as estimates of the median removal efficiency. Fitted values of the removal rate were obtained by directly fitting GAM models to this target variable and various sets of effect modifiers.

Unless otherwise stated, the target variables as well as the predictors in the response surfaces analyses were temporal mean values for each of the studied wetlands. This simplified the modelling and justified the assumption that the underlying data were statistically independent. Separate analyses of studies with temporal replicates were carried out to reveal potential drawbacks of GAM modelling.

RESULTS

Review Descriptive Statistics

The searches were performed in February 2013. A flow chart of the screening process is shown in Fig. 1. The search in literature databases generated 13,463 records, of which 5853 records were unique.

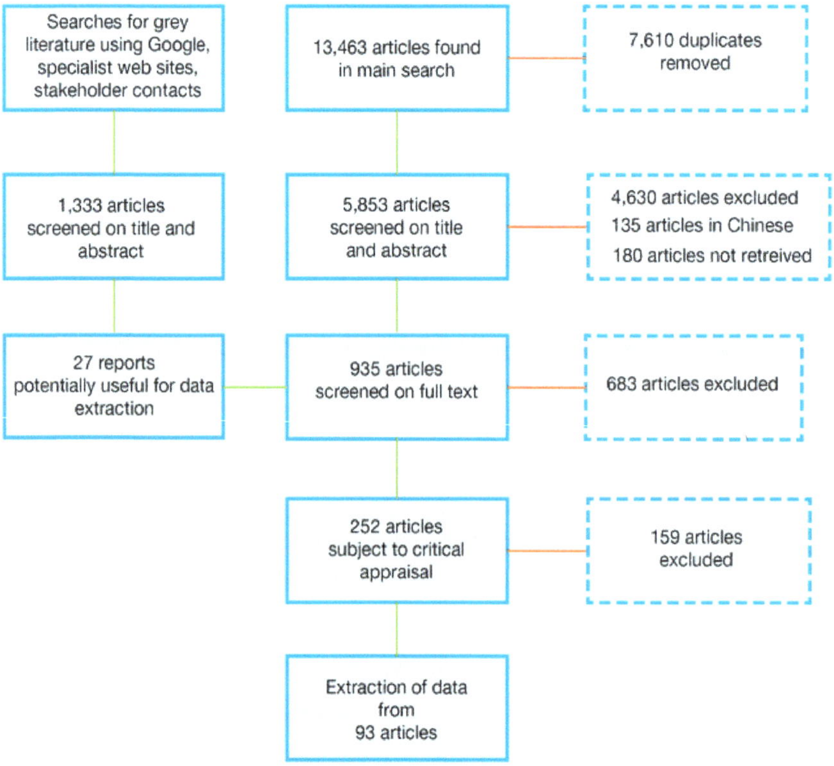

Figure 1: Chart of results from screening and critical appraisal.

At the title and abstract level 4630 articles were excluded while 1223 articles were included for full text screening. However, 135 of the included articles were in Chinese and had to be excluded due to lack of translation resources. Also, 180 of the articles could not be retrieved due to limitations in library resources. This resulted in 908 articles available for full text screening. Searches for grey literature using Google, specialist websites, and stakeholder contacts added another 27 reports that were screened at full text level. As a result of full text screening 685 articles were excluded. A list of these articles, with an indication of the reason for exclusion. Note however that only one reason for exclusion is shown for each article, although in many cases there were actually multiple reasons. The most common reason for exclusion was that the studied subject (type of water) did not conform to the inclusion criteria, followed by a lack of desired outcome data (Table 5). A total of 252 articles were subject to critical appraisal, and 93 of these passed to full data extraction.

Table 5: Number of articles (n) excluded at title and abstract screening and full text screening

Title and abstract screening	n	Full text screening	n
Exclude on intervention	1655	Exclude on subject	191
Exclude on outcome	1432	Exclude on outcome	164
Exclude on subject	719	Exclude on study length	143
Exclude on process study	467	Exclude on secondary data	81
Exclude on study length	156	Exclude on intervention	69
Exclude on secondary data	116	Exclude on process study	22
Exclude on location	75	Exclude on location	16
Exclude on no annual data	9	Abstract only	9
		Exclude on no annual data	4

The oldest article included in this review was published in 1981. The number of articles from each year was fairly constant between 1993 and 1999, after which it started to increase from 1 to 2 articles per year to around 5–9 articles per year (Fig. 2).

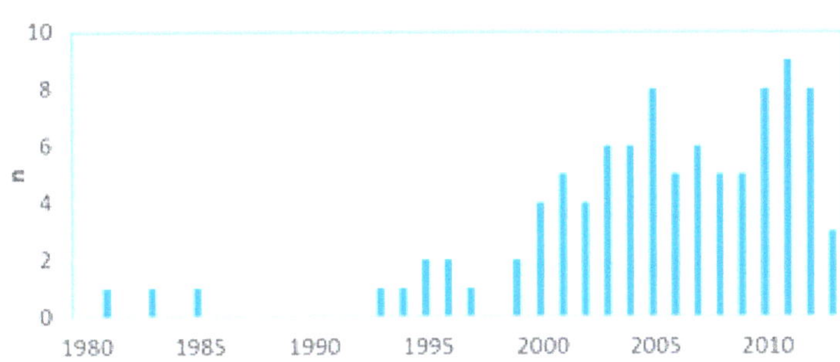

Figure 2: Distribution of included studies by publication year.

In total, 203 wetlands are included in this systematic review. Consecutive wetlands in series were treated as separate wetlands if removal rates and efficiencies were reported or could be calculated for each individual wetland. In other cases the train of wetlands was treated as one single wetland. Most of the wetlands are located in USA (n = 110) and Europe (n = 64). Seventeen wetlands are located in Sweden. The locations of the wetlands are shown in Fig. 3 the number of wetlands in each state or country is shown for nitrogen and phosphorus separately. The climate at these locations ranges from subtropical (climate zone Aw) in Florida to snow climate in parts of Scandinavia, northern USA and Canada (zone Df) and South Korea (zone Dwa).

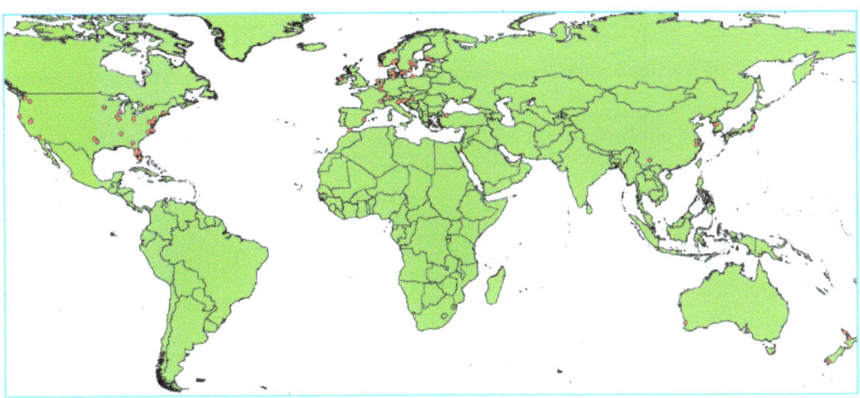

Figure 3: Location of included wetland studies.

The number of specific wetland types, inlet water types, water regimes, vegetation types, and climate zones are shown in Tables 6, 7, 8, 9 and 10. The most common wetland in this review is a free water surface (FWS) wetland with emergent vegetation treating agricultural runoff with a variable hydraulic loading rate. All of the included wetlands were primarily created or restored for the purpose of nutrient removal, although a small number of of them were multi-purpose wetlands.

Table 6: Number of included wetlands by wetland type.

Wetland type	Description	n
Free water surface	Wetland with visible water flowing horizontally	163
Horizontal subsurface flow	Wetland where water flows horizontally below the soil/substrate surface	17
Vertical subsurface down	Wetland where water flows vertically downwards through a soil/substrate	9
Combined horizontal	FWS wetland combined with other units such as sedimentation pond or overflow unit	8
Riparian	Any wetland at the interface between land and a river or stream	5
Vertical subsurface up	Wetland where water flows vertically upwards through a soil/substrate	1

Table 7: Number of included wetlands by inflow type

Inflow type	n
Agricultural runoff	100
River/lake water	38
Secondary domestic wastewater	29
Tertiary domestic wastewater	24
Urban storm water/runoff	7
Agricultural + urban runoff	4
Secondary domestic wastewater + urban runoff	1

Table 8: Number of included wetlands by water regime

Water regime	Description	n
Continuous, variable	Continuous flow with varying hydrologic loading rate (HLR)	72
Precipitation-driven	HLR driven by precipitation, continuous or intermittent but always variable	61
Constant	Continuous and constant HLR	43
Intermittent, constant	Periods with HLR = 0. When HLR > 0 it is constant and fixed at one value	14
Intermittent, variable	Periods with HLR = 0. When HLR > 0 it is variable	13

Table 9: Number of included wetlands by vegetation type

Vegetation type	n
Emergent	117
Mixed	28
Submerged	24
Unspecified	23
Trees	6
Filamentous algae	4
Floating	1

Table 10: Number of included wetlands in different climate zones

Climate zone	Description	Example location	n
Aw	Equatorial savannah with dry winter	Southern Florida	58
Cfa	Warm temperate climate, fully humid, hot summer	Northern Florida, Washington DC, Bucharest	28
Cfb	Warm temperate climate, fully humid, warm summer	London, Paris, Berlin, Southern Sweden	50
Csa	Warm temperate climate with dry summer, hot summer	Barcelona, Rome, Athens (Mediterranean)	20
Csb	Warm temperate climate with dry summer, warm summer	Ankara, Genoa, Seattle, San Francisco	10
Cwb	Warm temperate climate with dry winter, warm summer	Kunming	4
Dfa	Snow climate, fully humid, hot summer	Chicago, Boston	18
Dfb	Snow climate, fully humid, warm summer	Stockholm, Ottawa, Moscow	13
Dfc	Snow climate, fully humid, cool summer and cool winter	Northern Scandinavia, Anchorage	1
Dwa	Snow climate, dry winter, hot summer	Beijing, Pyongyang	1

The size of the included wetlands ranges from 1 to 10^7 m^2. As shown in Fig. 4, most of the studied wetlands were between 10^3 and 10^5 m^2, but smaller wetlands in the range 1–10 m^2 are also well represented.

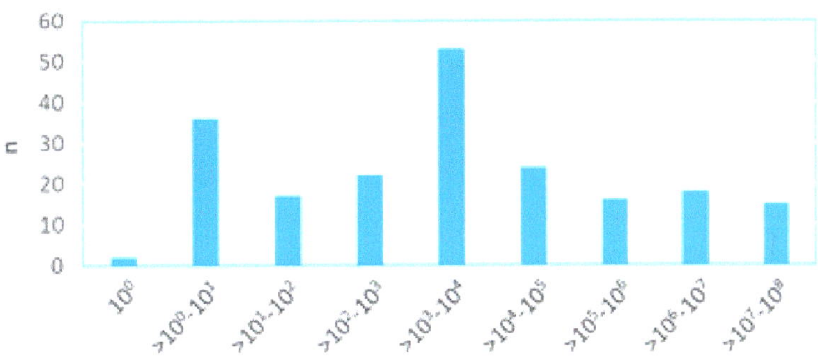

Figure 4: Size distribution (m^2) of wetlands included in the evaluation.

Study Quality Assessment

After full-text screening 252 articles remained for quality appraisal (Fig. 1). Some (15) articles described the same wetland study more than once, adding more data after the first study was published. In such cases the most recent and comprehensive article was chosen for quality assessment and potential data extraction.

Critical appraisal resulted in 143 articles in Category 1, i.e., articles that were not used for data extraction. From some of these articles it was not possible to calculate an annual total P and/or total N removal, expressed as mass/unit area/year. Even if removal rates were presented, or could be calculated, several articles were assigned to the lowest quality category due to deficiencies in the water budget. We deemed a high quality water budget (no large water flows unaccounted for) to be a necessary prerequisite for the study to be useful for data extraction. This is because calculations of nutrient removal are sensitive to uncertainties in water flow measurements. Another important reason to reject studies was measurements lasting less than a year, or more than a year, but without possibilities to break down measurements into a full year or several individual years. To be able to compare the annual efficiency of wetlands as nutrient traps, at least one annual cycle was demanded. The annual removal capacity is of main interest for stakeholders.

One of the reasons why quite a number of studies had to be excluded was the lack of a detailed description of the study design and sampling methodologies. Studies, which at first glance were expected to be suitable for extraction, proved to have insufficient detail for a proper quality appraisal and could not be included in the highest quality classes. Such shortcomings in methodological descriptions have been reported commonly in other systematic reviews and metadata analyses as well [36]. A more critical attitude of authors, reviewers and editors of scientific publications is desirable.

The difference between quality category 2 and 3 lies mainly in the demand for either two or more full years of measurements, or replicate wetlands, and that all major water flows should be quantified. For the studies in category 3 an annual mean value for nutrient removal, with standard deviation, is thus possible to calculate. A small number of studies (4) reported data from replicate wetlands, thus the majority of figures on variation in nutrient removal come from multiple year studies. This is a kind of pseudo replication, and variation in nutrient removal for the same wetland during different years may of course not reflect the true variation between replicate wetlands. On the other hand, compared to a 1-year study with true replicate wetlands, this pseudo replication is more likely to reflect changes in climatic conditions (e.g., temperature and precipitation), which may be equally important. Unfortunately, due to practical constraints, replication has only been done for mesocosm-size or experimental wetlands in wetland research parks. There were 41 studies of wetlands ≤ 10 m^2. For these, variation among replicates is likely to be smaller than for replicate "full-size" wetlands, i.e., wetlands restored or created not primarily for research purposes, but to reduce nutrient transport in a catchment. Such "full-size" wetlands are expected to produce more realistic figures on nutrient removal, but will lack replication. Some studies report major individual nutrient removal processes. If these match the total removal, based on a black box approach, then the quality of the study is strengthened. However, there were only a small number of studies (≈ 10) providing full nutrient budgets. Thus, from this review we cannot draw conclusions with respect to nutrient removal mechanisms, only the magnitude of nutrient removal, using a black box approach. Some indirect evidence on removal mechanisms may be indicated through statistical evaluations of effect modifiers, e.g., temperature and vegetation.

All studies included at the quality appraisal stage had at least one potential effect modifier reported, or such a modifier could be quantified from other sources (e.g., mean temperature from climate databases). However, many of the potential effect modifiers were reported in only a small number of articles, or were only semi quantitative. Thus a limited number of potential effect modifiers (c.f. Table 4) could be used in the final statistical meta-analysis and response surface analysis.

Of the 93 articles placed in category 2 and 3, 39 were in the highest quality category (3) and 54 in the second highest (2). Critical appraisal of individual studies, including assessments of internal validity (risk of bias) is reported.

Although there remained a reasonable number (39) of studies in quality category 3 with low susceptibility to bias (high internal validity), there is a risk of various types of bias of the body of evidence, affecting conclusions from this review. One of these is publication bias which may arise if results in a particular direction are less likely to be published. Figure 5 shows funnel plots for TN and TP. For TN there may be a small number of studies missing in the lower right corner, indicating that some studies showing net releases of TN from the wetland, albeit most probably insignificant, may have been overlooked or not reported. The distribution for TP is more symmetric but does also reveal a small number of studies showing very large effect sizes with small standard errors. Thus, in the present case publication bias seems not to be a major concern. This finding is not surprising since both positive and negative results are scientifically interesting, and also of importance for stakeholders.

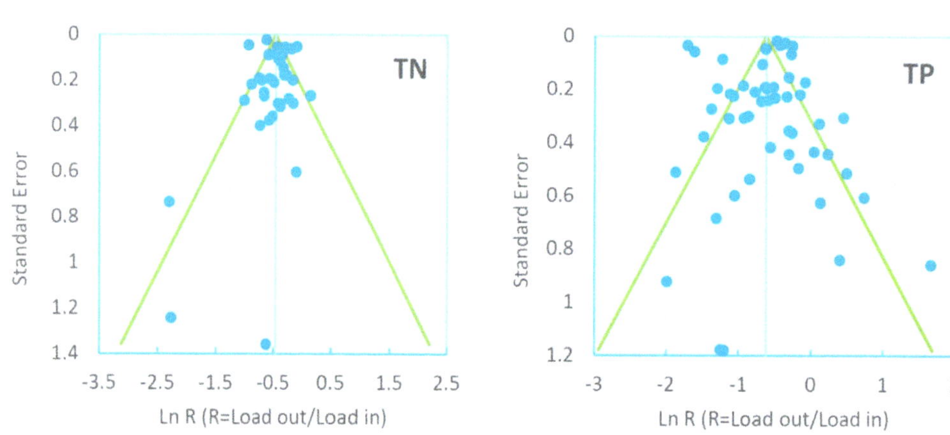

Figure 5: Funnel plots showing relations between effect sizes and standard errors for TN (*left*) and TP (*right*). An asymmetric distribution suggests the possibility of publication bias or a systematic difference between small and large studies (with large and small standard errors, respectively). In the absence of publication bias and systematic heterogeneity, 95 % of the data might be expected to lie within the *green funnel-shaped* delineation. The*blue vertical lines* indicate summary effect sizes.

Another source of bias affecting the external validity, or the generalizability of the included studies, is uneven geographical distribution of wetlands (Fig. 3). Indeed studies are clustered in two geographical regions: most studies are from Europe and North America. In so far as the bias is towards North Europe, especially Scandinavia, North America and areas with similar climate, geographical bias may, from a Swedish stakeholder point of view, not be a problem. However, a fairly large number of studies, especially for phosphorus, have been conducted in Florida and other states in Southern USA, which have a climate quite different from Sweden.

The size distribution of included wetlands may also be biased. Whereas most created or restored wetlands in Sweden range from 10^2 to 10^5 m^2 (Fig. 6), the wetlands included in this review show a much broader distribution with more of both smaller and larger wetlands. Almost 40 % of the included wetlands in this systematic review range between 1 and 10^2 m^2, and 17 % of the wetlands are between 10^5 and 10^7 m^2. The largest included wetland is 6.7×10^7 m^2.

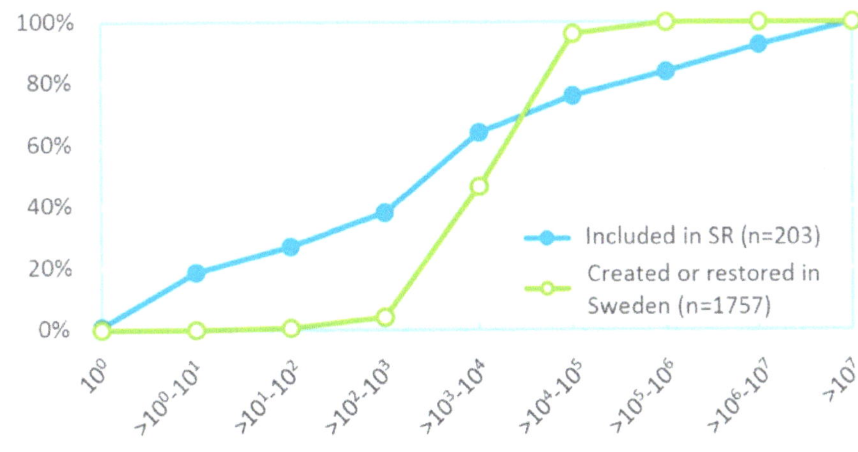

Figure 6: Cumulative frequencies of wetland areas (m^2) in this review (SR) and created or restored (with a known area) in Sweden before 2013 according to SMHI [20].

Most studies of nutrient removal in wetlands have been made during the years following wetland restoration or creation. Median wetland age at the start of study periods was 1 year for the included wetlands, whereas the median age at the end of the studies was 3 years. Thus our systematic review may be biased towards short-term nutrient removal effects. The performance of wetlands as nutrient traps after several decades (the expected minimum lifespan of a restored or created wetland) in comparison to the first years after restoration/construction has been studied in a very small number of cases. It might be that nutrient removal changes over time. It has, e.g., been reported that P removal decreases [17] and may even become negative with time. In a more recent study, Mitsch et al. [18] combined the data reported by Mitsch et al. [17] with data for two subsequent years and showed that although there was a significant declining trend for the entire period (1994–2010), there was also a significant improving trend at the end of the period (2003–2010). One explanation for this pattern could be that the wetlands were created on former agricultural soil and that it took a decade or so to wash the accumulated P out of the soil. Chen et al. [37] studied long-term (up to 17 years) TP removal in wetlands in Florida. The authors did not report any temporal trend but concluded that performance, in terms of outflow TP concentration and/or k value (first order removal constant, which could be interpreted as the settling rate for TP removal), depended primarily on HLR, inflow TP concentration, and TP loading rate. In addition, the impacts of these variables on P removal are often confounded by soil and vegetation conditions, regional rainfall, management activities, and other factors. In another study conducted in Florida, Moustafa et al. [38] showed that the TP removal efficiency remained relatively constant during their 9-year study. Also N removal may be expected to change, e.g., due to succession in aquatic vegetation and accumulation of organic matter in wetland sediments.

Given the large number of studies fulfilling our rigorous quality criteria, we feel confident that the review has answered the primary question about wetland effectiveness (see section "Objective of the review") in a general sense. However, the heterogeneity is high, so that it is not possible to use the graphs to estimate nutrient removal for an individual wetland. This was not unexpected since we have included studies from subtropical to cold climates, sizes of wetlands from one to more than 10^5 m^2, various types of created wetlands, wetlands receiving a wide variety of influent water quality, etc. On the other hand the statistically significant general trends found for all these different types of wetlands should represent a high quality outcome from the review.

With respect to the secondary questions about influence of effect modifiers, the results of the review for some effect modifiers are statistically solid, e.g., wetland size, hydraulic loading, temperature, water type, and substance concentration. However, for other potential effect modifiers (e.g., soil type, vegetation type and coverage, harvest, and fauna) there were not enough studies to make solid predictions (models).

We have chosen to include a rather wide variety of wetlands and water types. Thus the material is very heterogeneous, which increases variation. An alternative approach could have been to review only one wetland type and only one source of water in a more restricted geographical area, e.g., wetlands in agricultural areas in N Europe, receiving drainage from agricultural fields. This would probably have produced less variation, although the number of studies qualifying for the highest quality category would have been very small. Also, the possibility to generalize would have become reduced.

Narrative Synthesis

Most studies on nutrient removal in wetlands do not report any variance in annual removal rates or efficiencies. In some cases this is a consequence of the fact that only one wetland was studied and that the study only lasted for one year. There is thus neither any true replication nor any quasi-replication. In other cases the study of one wetland lasted for several years but only a long-term average was reported with no information about the inter-annual variance. Only four of the included studies investigated multiple wetlands that were similar enough to be regarded as replicates and also reported the results in such a way that it was possible to calculate the variance, while 60 wetlands were quasi-replicated through measurements for more than one year and reported in such a way that the inter-annual variance could be calculated. Total nitrogen was measured in two of the studies with replicated wetlands and TP was measured in three of them. Among the quasi-replicated wetlands, TN and TP were measured in 37 and 49 wetlands, respectively.

Removal of TN

The annual loading rates of TN in the included wetlands ranged from 2.1 to 2486 g m^{-2} $year^{-1}$, and averaged 505 g m^{-2} $year^{-1}$. The average removal rate of TN was 181 g m^{-2} $year^{-1}$, whereas the average removal efficiency was 39 %. Summary statistics for included wetlands are shown in Table 11. The ranges in loading and removal rates between wetlands are quite wide, and the distributions are skewed to the right, i.e., the median values are lower than the arithmetic means. The distribution of removal efficiencies is more likely to be normally distributed. Although there is no significant difference in average removal efficiencies between category 2 studies and category 3 studies, the variability is smaller among category 3 studies. It is worth noting that none of the wetlands among the category 3 studies had negative removal rates.

Table 11: Summary statistics for TN

Quality category	n[a]	TN loading rate (g m^{-2} $year^{-1}$)		TN removal rate (g m^{-2} $year^{-1}$)		TN removal Efficiency (%)	
		Av. ± 1SE (range)	Med	Av. ± 1SE (range)	Med.	Av. ± 1SE (range)	Med.
2 and 3	112	505 ± 579 (2.12–2486)	255	181 ± 251 (−0.3–1270)	93	39 ± 21 (−12.8–93)	38
2	88	578 ± 608 (2.12–2486)	378	207 ± 269 (−0.3–1270)	120	39 ± 23 (−12.8–93)	38
3	24	237 ± 353 (12.3–1424)	113	85 ± 128 (1.7–584)	43	39 ± 16 (14.8–87)	36

n number of wetlands, *Av.* arithmetic mean, *SE* standard error, *Med* median

[a]Some articles have reported results for more than one wetland

One included wetland showed a small (non-significant) negative TN removal rate [39]. This was a multi-purpose FWS wetland that had been restored on formerly drained cropland. Other studies, e.g., by Koskiaho et al. [40], have also shown negative TN removal rates but these were judged to be highly susceptible to bias and assigned to quality category 1.

Removal of TP

As in the case of TN, the spans in the loading and removal rates of TP are quite large (Table 12). The average loading rate and removal rate were 36 and 13 g m^{-2} $year^{-1}$, respectively. The average loading rate was considerably lower in category 3 studies compared to category 2 studies, even though the range was similar in both categories. There is no significant difference in average TP removal efficiencies between category 2 studies and category 3 studies but, as with TN, the variability between wetlands was smaller among the category 3 studies compared to the category 2 studies.

Table 12: Summary statistics for TP

Quality category	n[a]	TP loading rate (g m^{-2} $year^{-1}$)		TP removal rate (g m^{-2} $year^{-1}$)		TP removal efficiency (%)	
		Av. ± 1SE (range)	Median	Av. ± 1SE (range)	Median	Av. ± 1SE (range)	Median
2 and 3	146	36 ± 73 (0.03–373)	3.2	13 ± 38 (−16.8–240)	1.2	41 ± 52 (−422–99)	50
2	85	54 ± 83 (0.08–373)	14	19 ± 46 (−16.8–240)	2.9	39 ± 60 (−422–98)	51
3	61	10 ± 43 (0.03–307)	1.4	4 ± 18 (−7.3–133)	0.63	44 ± 38 (−104–99)	49

n number of wetlands, *Av.* arithmetic mean, *SE* standard error, *Med* median

[a]Some articles have reported results for more than one wetland

Negative removal rates of TP were reported in 17 of the 146 wetlands. Six of these were created on former cropland and one on previous cattle pasture [41]. The soil in such areas is usually rich in phosphorus that may be released after construction of the wetland. The other ten wetlands with reported negative TP removal rates were all FWS wetlands created 0–2 years before the start of the study, which means that release of P from initially P rich sediments could have contributed to the results. For example, Bass, Evans [42] argued that mineralization of phosphorus from formerly anoxic organic layers uncovered during the excavation could have caused the negative TP removal rate. Release of phosphorus associated with iron complexes under anaerobic conditions can also contribute to low or negative removal rates, as suggested by Healy, Cawley [43] as an explanation for the observed low TP removal rates.

QUANTITATIVE SYNTHESIS

Meta-analysis

Data from studies with replication were subjected to meta-analyses. The results obtained for individual wetlands are summarized in Figs. 7 and 8, which show forest plots of log response ratios (Ln (load out/load in)) for TN and TP, respectively.

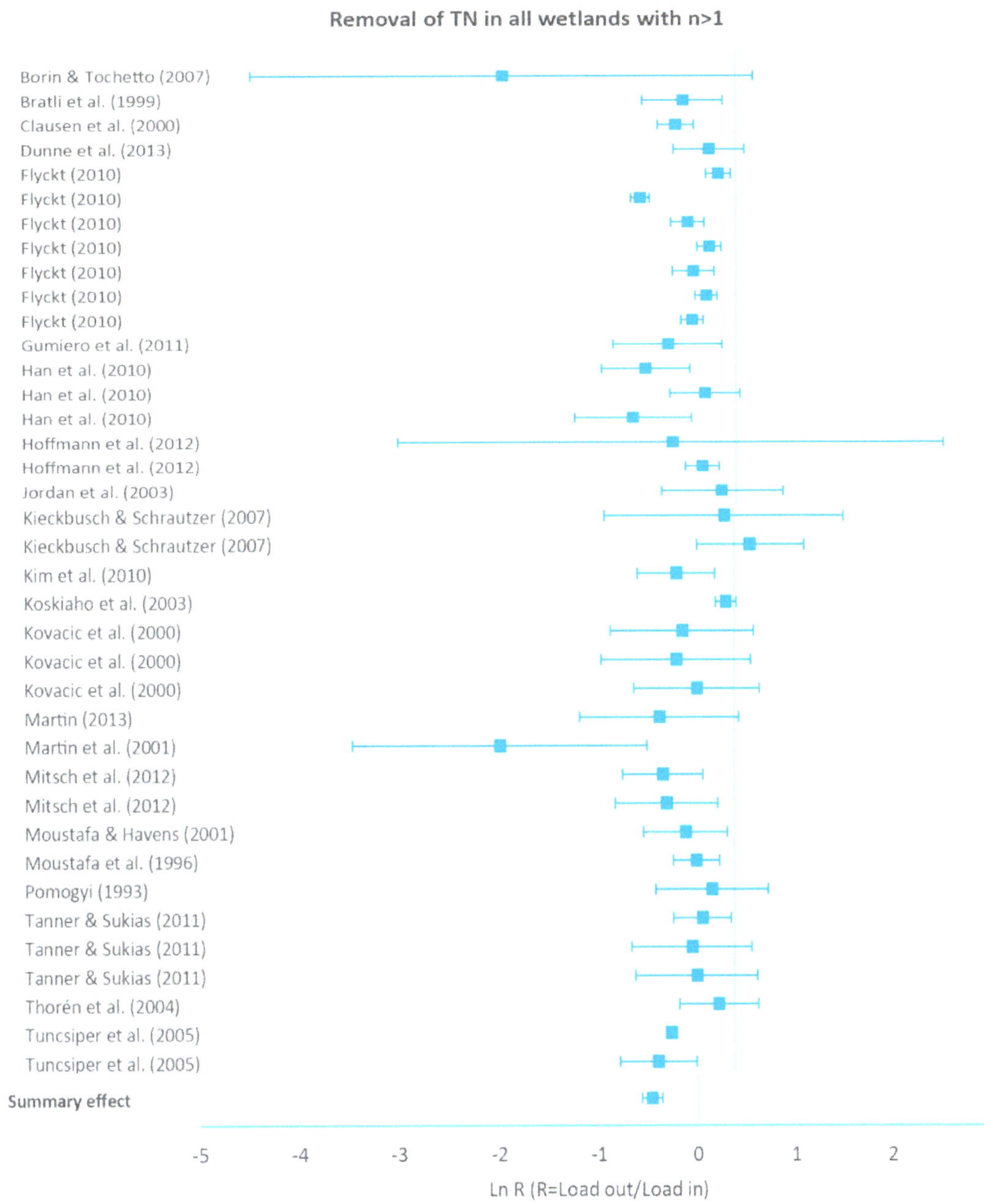

Figure 7: *Forest plot* showing average Effect sizes and 95 % confidence intervals.

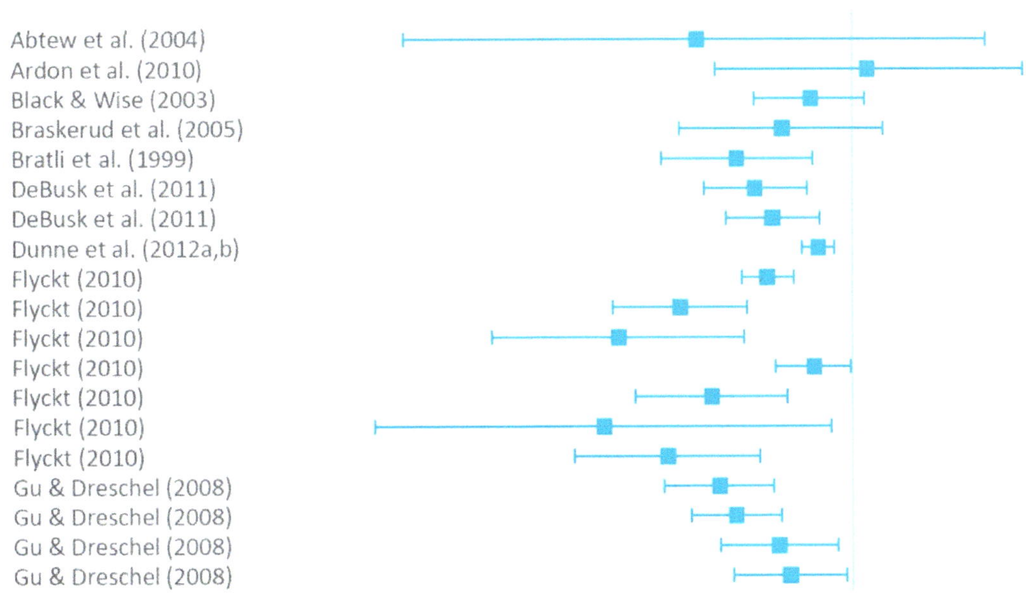

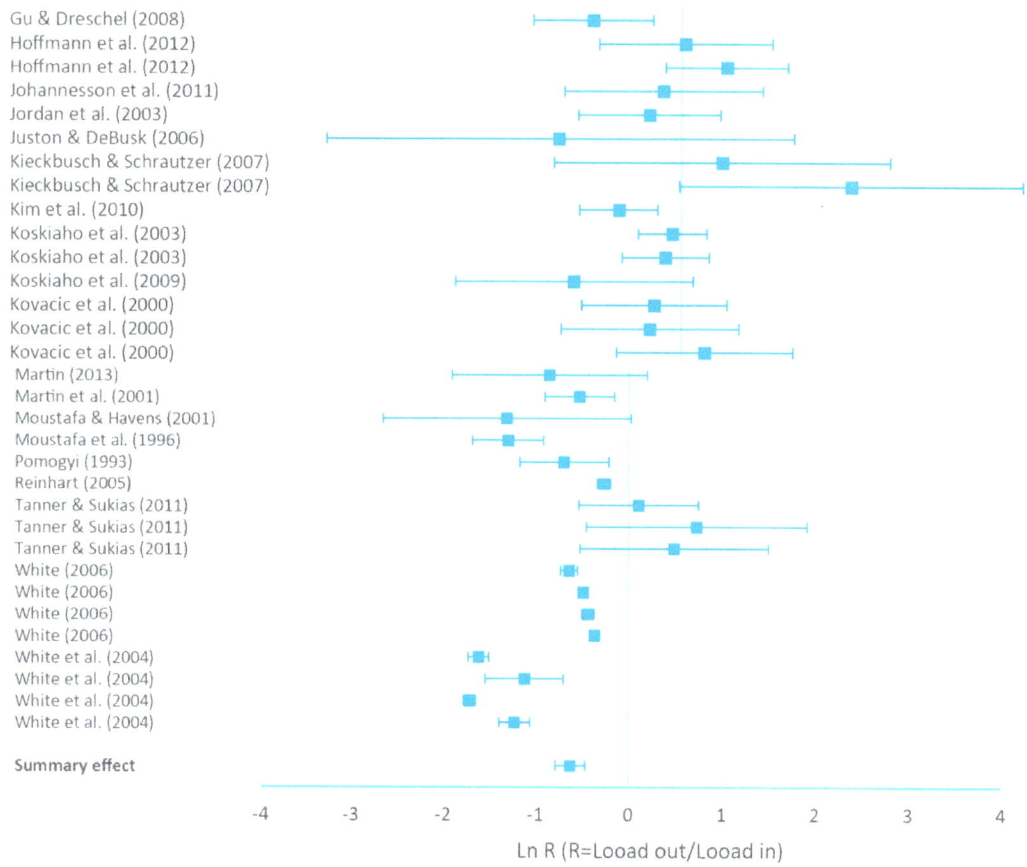

Figure 8: Forest plot showing average Effect sizes and 95 % confidence intervals.

The forest plot for TN removal shows an overall net removal with reasonably narrow confidence limits (Fig. 7). Only three out of 38 studies reported a strong variability, and it can be noted that two of them also had the highest removal rates. For 21 cases, the confidence limits indicate a statistically significant TN removal. For the remaining 17 cases, the confidence limits encompass the zero-effect line (including one case where the average indicate a release of TN). The overall average summary effect ± 1 SE is −0.46 ± 0.05. The between-study variance (T^2) was estimated to 0.06, and the I^2 statistic was 86 %. The heterogeneity of the evidence base may thus be regarded as high. The overall average summary effect represents a median TN removal ratio (R) of about 0.63. This means that the median TN load reduction, or removal efficiency, is 37 %, with a 95 % confidence interval of 29–44 %.

The forest plot for TP removal generally shows wider confidence intervals, and a higher number of cases were reported with an average net release rather than removal of TP (Fig. 8). For 29 out of 51 wetlands a significant net removal was reported. Among the remaining cases, 13 exhibited a non-significant net TP removal and nine a non-significant net TP release. The overall summary effect size is highly significant with rather narrow confidence limits. The average ± 1 SE is −0.62 ± 0.08, which is even lower than that for TN. Similar to TN, the heterogeneity among the studies is high also for TP. The estimated between-study variance (T^2) was 0.24, while the I^2 statistic was 97 %. The overall average summary effect represents a median TP removal ratio (R) of about 0.54, which can be recalculated to a median TP removal efficiency of 46 %, with a 95 % confidence interval of 37–55 %.

Summarizing, there is strong evidence that created wetlands generally remove TN and TP and that the overall removal efficiency is roughly 40 %. An annual release rather than removal of TN has been shown in a very small number of studies, whereas negative TP removal rates are less uncommon. Also, TP removal ratios generally show a larger variance.

The high heterogeneity of the evidence base calls for subgroup analyses that could potentially identify effect modifiers. Results of such subgroup analyses, i.e., where the studies were divided into different climate zones, wetland types, water types, hydrologic regimes etc., are shown in Figs. 9 and 10. The forest plot for TN removal with wetland studies grouped per climate zone shows that TN removal is significantly different from zero in all climate zones (Fig. 9a). The removal tends to be more efficient (i.e., the effect size tends to be more negative) for wetlands in groups with hot summers, although the 95 % confidence intervals overlap each other. It should also be noted that some groups include only a small number of studies, and hence the confidence intervals are quite broad. Subdivided by wetland type (Fig. 9b), the pattern is remarkably similar for the four types of wetlands: the summary effects all differ significantly from zero and the averages are relatively close to each other with confidence levels showing a strong overlap. Separated by water type, all four averages are again significantly different from zero, while river/lake water and agricultural runoff show quite similar averages and confidence limits (Fig. 9c). The secondarily treated domestic wastewater showed a higher removal efficiency as well as wider confidence interval.

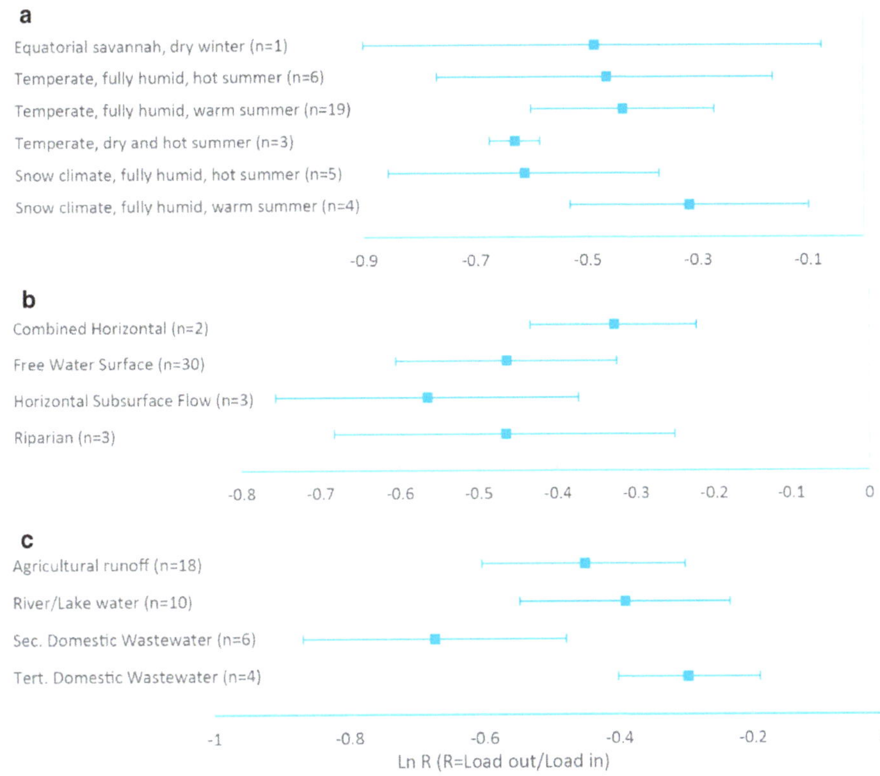

Figure 9: Summary effects for TN in wetland subgroups based on a climate zone, b wetland type, and c water type. *Error bars* show the 95 % confidence interval (where number of wetlands (n) is one it is based on the within study variance only).

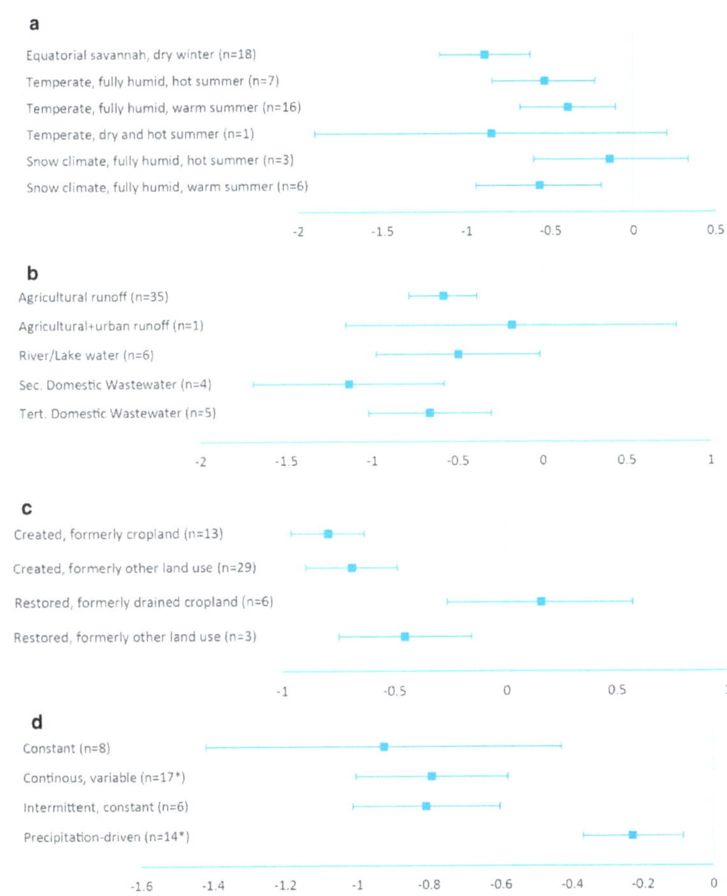

Figure 10: Summary effects for TP in wetland subgroups based on a climate zone, b water type, cwetland history, and d hydrologic regime. *Error bars* show the 95 % confidence interval (where number of wetlands (n) is one it is based on the within study variance only). *Restored wetlands on formerly drained cropland are not included (five precipitation-driven and one wetland with continuous flow and variable HLR).

The subgroup analyses demonstrate that wetlands have a robust capacity to remove TN from through-flowing water. Except for the observation that the removal of TN seems to be more efficient for secondarily treated wastewater than for tertiary treated wastewater, no significant effect modifier could be identified in the subgroup analysis. Numerical values of TN removal efficiency for each subgroup are shown in Table 13.

Table 13: Results of subgroup analyses for TN

	n[a]	Ln R ± 1 SE	R	Median removal efficiency (%)	95 % confidence interval
All wetlands	38	−0.46 ± 0.05	0.63	37	29–43
Climate zone					
Aw (equatorial savannah, dry winter)	1	−0.49 ± 0.21	0.61	39	7–60
Cfa (warm, fully humid, hot summer)	6	−0.47 ± 0.15	0.63	37	15–54
Cfb (warm, fully humid, warm summer)	19	−0.44 ± 0.08	0.64	36	23–46
Csa (warm, dry and hot summer)	3	−0.63 ± 0.02	0.53	47	44–50
Dfa (snow, fully humid, hot summer)	5	−0.61 ± 0.12	0.54	46	30–58
Dfb (Snow, fully humid, warm summer)	4	−0.31 ± 0.11	0.73	27	9–42
Wetland type					
Combined horizontal	2	−0.33 ± 0.05	0.72	28	19–36
Free water surface	30	−0.46 ± 0.07	0.63	37	27–46
Horizontal subsurface flow	3	−0.56 ± 0.1	0.57	43	31–54
Riparian	3	−0.47 ± 0.11	0.63	37	22–50
Water type					
Agricultural runoff	18	−0.45 ± 0.08	0.64	36	25–46
River/lake water	10	−0.39 ± 0.08	0.68	32	20–43
Sec. domestic Wastewater	6	−0.67 ± 0.1	0.51	49	38–59
Tert. domestic Wastewater	4	−0.29 ± 0.05	0.75	25	17–34
Wetland history					
Created, formerly cropland	3	−0.45 ± 0.05	0.64	36	29–43
Created, formerly other land use	25	−0.49 ± 0.06	0.61	39	30–46
Restored, formerly drained cropland	6	−0.33 ± 0.12	0.72	28	9–44
Restored, formerly other land use	3	−0.49 ± 0.36	0.61	39	−23–70
Water regime					
Continuous, variable	21	−0.54 ± 0.06	0.58	42	33–49
Intermittent, variable	1	−0.67 ± 0.27	0.51	49	12–70
Precipitation-driven	15	−0.43 ± 0.09	0.65	35	22–46
Intermittent, constant	1	−0.42 ± 0.1	0.66	34	19–47

R = load out/load in

[a]Where n = 1 the SE is based on the within study variance only

The forest plots for TP for the various wetland groupings show different patterns than those for TN. Two out of the six climate zones had wetlands with TP removal non-significantly different from zero, i.e., Mediterranean (Csa) and Snow climate with hot summers (Dfa) (Fig. 10a). However, the effect sizes in these climates are based on only 1 and 3 studies, respectively, and the 95 % confidence intervals are broad. Subdivided by water type, there was a TP removal significantly different from zero in wetlands receiving four out of five water types (Fig. 10b). The averages are not significantly different from each other, and the narrowest confidence interval was observed for agricultural runoff (also by far represented by the highest number of studies). The grouping based on wetland history shows a TP removal significantly different from zero in all cases except for restored wetlands on formerly drained cropland, for which there was an insignificant net release of phosphorus. Wetlands in all the other history groups had a significant removal, with no further differences among the types (Fig. 10c). Also, the grouping by water regime suggests that wetlands with precipitation-driven HLR are less effective than wetlands with other water regimes, although all subgroups significantly differed from zero. This is true also when the restored wetlands on former drained cropland are removed (Fig. 10d). If such wetlands are included the difference between precipitation-driven and other wetlands would appear to be even larger (data not shown). Inclusion or exclusion of restored wetlands on formerly drained cropland does not alter the general patterns shown in the other subgroup analyses. Numerical values of TP removal efficiency for each subgroup are shown in Table 14.

Table 14: Results of subgroup analyses for TP

	n[a]	Ln R ± 1 SE	R	Median removal efficiency (%)	95 % confidence interval
All wetlands	51	−0.62 ± 0.08	0.54	46	37–55
Climate zone					
Aw (equatorial savannah, dry winter)	18	−0.89 ± 0.14	0.41	59	46–69
Cfa (warm, fully humid, hot summer)	7	−0.53 ± 0.16	0.59	41	19–58
Cfb (warm, fully humid, warm summer)	16	−0.39 ± 0.15	0.68	32	9–50
Csa (warm, dry and hot summer)	1	−0.85 ± 0.54	0.43	57	−22 to 86
Dfa (snow, fully humid, hot summer)	3	−0.14 ± 0.24	0.87	13	−38 to 46
Dfb (snow, fully humid, warm summer)	6	−0.57 ± 0.19	0.57	43	17–62
Wetland type					
Combined horizontal	2	−1.22 ± 0.29	0.3	70	47–84
Free water surface	47	−0.64 ± 0.08	0.53	47	37–56
Riparian	2	0.31 ± 0.25	1.36	−36	−122 to 17
Water type					
Agricultural runoff	35	−0.58 ± 0.1	0.56	44	31–54
Agricultural + urban runoff	1	−0.18 ± 0.5	0.84	16	−120 to 69
River/lake water	6	−0.49 ± 0.24	0.61	39	1–63
Sec. domestic wastewater	4	−1.13 ± 0.28	0.32	68	43–82
Tert. domestic wastewater	5	−0.66 ± 0.18	0.52	48	26–64
Wetland history					
Created, formerly cropland	13	−0.79 ± 0.08	0.45	55	46–62
Created, formerly other land use	29	−0.69 ± 0.10	0.50	50	38–60
Restored, formerly drained cropland	6	0.15 ± 0.21	1.16	−16	−77 to 24
Restored, formerly other land use	3	−0.45 ± 0.15	0.64	36	14–53
Water regime[b]					
Constant	8	−0.92 ± 0.25	0.4	60	34–76
Continuous, variable	17	−0.79 ± 0.11	0.45	55	43–64
Intermittent, constant	6	−0.81 ± 0.1	0.44	56	45–64
Precipitation-driven	14	−0.23 ± 0.07	0.79	21	8–31

R = load out/load in

[a]Where n = 1 the SE is based on the within study variance only

[b]Restored wetlands on former drained cropland have been removed from these calculations

The subgroup analyses show that wetlands have a robust capacity to remove TP, although the 95 % confidence intervals of the means are generally wider, and there are more cases not significantly different from zero, than for TN. There is evidence that cases with a net TP release do occur, but mostly in wetlands that were restored on formerly drained cropland without excavating or isolating the soil. Water regime seems to be an additional significant effect modifier. Wetlands with a precipitation-driven HLR are less efficient than wetlands with a controlled HLR. Wetlands in warmer climates (tropical savanna and warm temperate) tend to have a more reliable TP removal than in colder climates, although the 95 % confidence intervals overlap each other.

Response Surface Analysis

In this section, the results obtained by response surface analyses are presented. This type of regression analyses was based on mean values per wetland study, and the response surfaces derived illustrate how estimates of median removal efficiency and median removal rate are influenced by various effect modifiers. Preliminary analyses demonstrated that, after both main effects and possible interaction effects of the two most relevant effect modifiers had been taken into account, introduction of additional explanatory variables or interaction usually had a rather low impact on the goodness-of-fit of the model. Moreover, we noticed that the statistical significance of individual components of complex response surface models could be strongly influenced by the inclusion/exclusion of a small number of observations representing rather unusual levels of the moderators. Therefore, we decided to focus on parsimonious models and present the results of a forward selection of explanatory variables and interactions.

TN removal efficiency (% load reduction) was significantly negatively related to hydraulic loading rate. According to a combined linear/spline model the linear component was strongly significant ($p < 0.0001$), whereas the non-linear component was less significant ($p = 0.034$). TN removal efficiency was also found to be positively correlated with annual average air temperature (Model 4). Other investigated predictors showed non-significant ($p > 0.05$) relationships to TN removal efficiency.

Using both hydraulic loading rate and air temperature as predictors in a general additive model (GAM) (Model 7) improved the model fit (reduced the deviance) and demonstrated that the linear response to air temperature was significant also in the presence of a function of hydraulic loading. The model fit was further improved when the one-dimensional splines in log hydraulic loading and air temperature, respectively, were substituted for a thin plate spline that allowed interaction effects between hydraulic loading and air temperature without changing the degrees of freedom of the model (Model 10). The fitted removal efficiency according to model 10 is shown in Fig. 11. Adding a cubic spline for wetland area resulted in an even better fit (Model 12) and demonstrated that the linear response to log wetland area was statistically significant also in the presence of a thin plate spline in log hydraulic loading and air temperature.

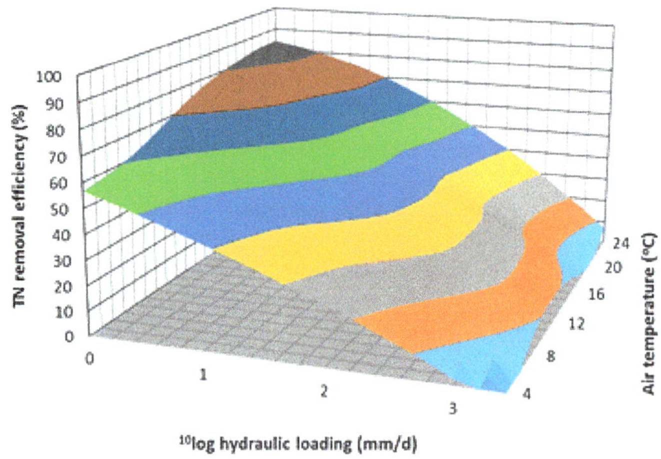

Figure 11: Median removal efficiency of TN (% of load) according to model 10.

The TN removal rate expressed as g m^{-2} day^{-1} was found to be positively correlated with the inflow concentration, with a steeper increase in removal rate at concentrations higher than about 18 mg/l (Model 14). The TN removal rate was also positively correlated with hydraulic loading, at least up to about 650 l m^{-2} day^{-1} at which level the TN removal rate started to slowly decline (Model 15). However, the non-linear component was not quite statistically significant ($p = 0.080$). Furthermore, the TN removal rate was negatively correlated with wetland area, but the decline in removal rate with wetland size appeared to be somewhat lower at areas above approximately 1 ha (Model 16).

When both hydraulic loading and TN concentration at inlet were used as predictors in a GAM the deviance was substantially reduced (Model 18), and a further reduction was achieved when the two one-dimensional splines were substituted for a thin plate spline allowing interaction effects without increasing the degrees of freedom of the model (Model 21). A plot of predicted removal rates according to this model is shown in Fig. 12, and the overall positive response to hydraulic loading and inflow concentration is clearly visible. When cubic splines for wetland area and temperature were added on top of the thin plate spline in inflow concentration and hydraulic loading the deviance was further reduced and the response to wetland size (log area) was statistically significant (Model 26).

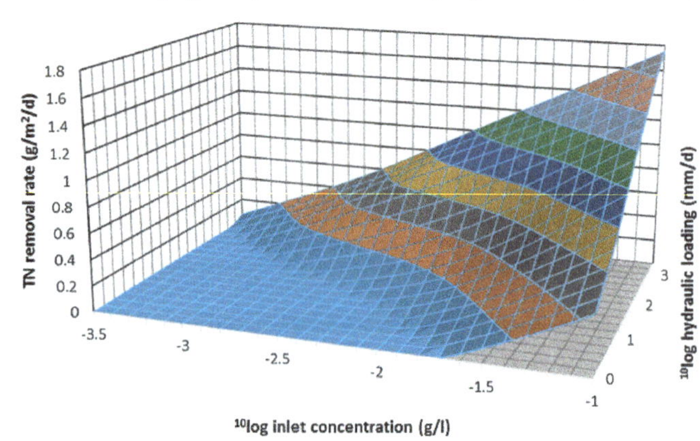

Figure 12: Median removal rate of TN (g m^{-2} day^{-1}) according to model 21. The surface has been truncated at 0 g m^{-2} day^{-1}.

According to combined linear/spline regression models, the removal efficiency of TP was influenced by all four of the investigated predictors, i.e., TP inlet concentration, hydraulic loading, wetland area, and air temperature (Models 27–30, respectively). More specifically, the response to the log-transformed values of inlet concentration, hydraulic loading, and wetland area had both linear and non-linear components, whereas the response to inlet concentration was primarily nonlinear. Closer examination of the removal efficiency indicated that it had a maximum for intermediate concentrations (0.05–0.5 mg/l) at wetland inlet.

When GAM models with two predictors were examined the best fit (lowest deviance) was obtained for a thin plate spline model with log inlet concentration and log hydraulic loading rate (Model 36; Fig. 13). However, the removal efficiency exhibited a substantial random variation and the deviance of model 36 was only slightly lower than that of model 28, which was the best one-dimensional spline model. Neither, was there any statistically significant linear response to air temperature or log area when model 36 was extended with one-dimensional splines in these variables.

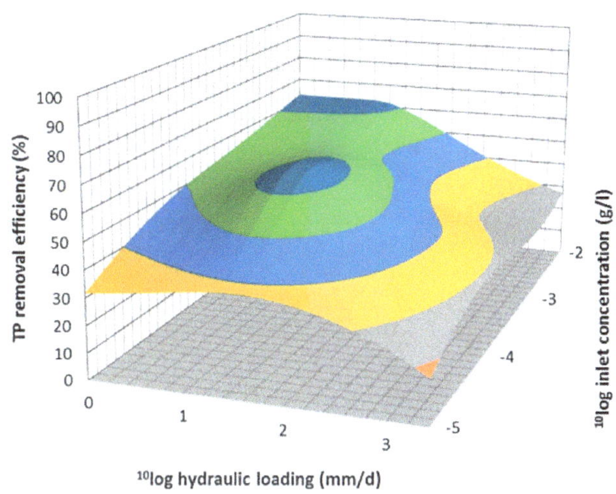

Figure 13: Median removal efficiency of TP (% of load) according to model 36.

The TP removal rate was positively correlated with TP concentration at inlet with a steeper increase in removal rate at concentrations above approximately 0.55 mg/l (model 40). In contrast, the TP removal rate was negatively correlated with wetland area, especially at areas below 2×10^4 m^2, above which the removal rate was fairly constant (model 42). A statistically significant spline function was found for air temperature (model 43), and a maximum in removal efficiency appeared at intermediate annual average temperatures (approximately 14–19 °C).

When both inlet concentration and hydraulic loading rate were used as predictors of removal rate and interaction effects between these predictors were taken into account using a thin plate spline function (model 47) an ordinary F-test indicated that the deviance was significantly lower than in the best one-dimensional spline model (model 40). Fitted TP removal rates according to model 47 are shown in Fig. 14. Further analysis showed that adding a one-dimensional spline for air temperature made the deviance even lower and that there was a statistically significant non-linear component of air temperature in this extended model (model 50).

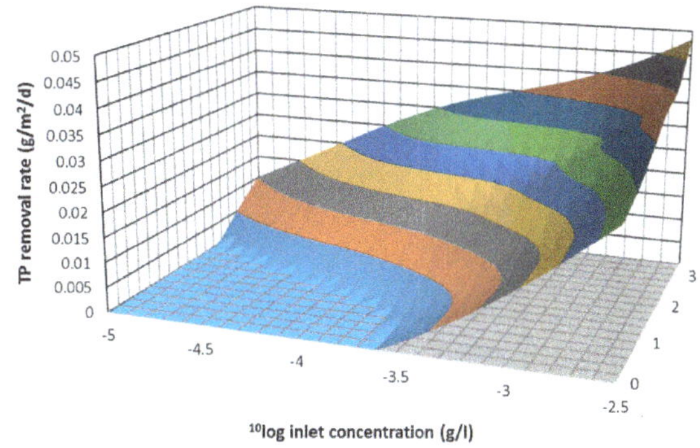

Figuer 14: Median removal rate of TP (g m^{-2} day^{-1}) according to model 47. The surface has been truncated at 0 g m^{-2} day^{-1}.

Figures 12 and 14 suggest that the removal rates are very low at low nutrient concentrations at the inlet and low HLRs. To obtain an appreciable removal rate either the inlet concentration or the HLR (or both) need to be increased. On the other hand, the HLR should be increased with some caution since the removal efficiency decreases with increasing HLR (Figs. 11, 13). When a wetland is being designed, a balance should thus be found between an HLR that is high enough to allow for a meaningful removal rate at a given inlet concentration, and an HLR that is low enough to keep the removal efficiency sufficiently high to make a significant difference to the total transport of nutrients. However, periods with none or very low removal rates will inevitably occur in wetlands with precipitation driven, intermittent, and to some extent variable continuous water flow. In all cases the removal rate increases with increasing concentration at inlet. The removal efficiency for TP generally increases with increasing inlet concentrations, primarily at low to intermediate HLRs, whereas the removal efficiency for TN is less influenced by the inlet concentration.

Because the number of wetlands included in our review is too small to allow sophisticated modelling of mean values per wetland, it is tempting to extract several effect sizes from the same study, for example by using data for individual years as model inputs. In principle, such data can be analysed by using hierarchical generalized linear models. However, the data collected in the present study were far from ideal for such models. First, less than half of the included wetland studies had any temporal replicates. Second, it was difficult to identify a suitable covariance structure for the study-specific random components; some of the studies exhibited a large inter-annual variation, whereas others exhibited a very small within-study variation.

DISCUSSION

Reasons for Heterogeneity and Review Limitations

Although the water quality criteria set for this review are narrowed to secondarily treated domestic wastewater, tertiary treated domestic wastewater, urban water runoff, agricultural drainage water, and river or lake water, these water types differ considerably in composition and thus also the processes needed for removal of nitrogen and phosphorus differs between the water types. It would have been desirable to look at the different

N and P species in connection with these removal and retention processes, as e.g., nitrate in connection with denitrification or particulate-P in connection with retention of phosphorus. This might have improved the analysis of water type divided into subgroups (Figs. 11, 13). However, in many cases the articles only reported TN and TP results, and therefore further analysis was impossible.

Especially for phosphorus, it might have been beneficial to look into different P-species because it has been documented that some created wetlands retain large amounts of particulate P (PP) when there are high loads of this species due to the upland characteristics [16]. Furthermore, phosphorus retention in wetlands is dependent on several factors, such as age, past history of P (i.e., P load, P fertilizer addition, P saturation in soil), redox conditions, climate conditions (e.g., frost–thaw), which are not always included in the information.

Excluded from this review due to limitations in sampling frequency are restored and created floodplains and other riparian areas subjected to flooding and inundation. In these areas sedimentation of suspended solids and particulate phosphorus is often high [15, 44–47] making sedimentation the most important process for retention of phosphorus in such wetlands.

This review is limited to only one function of created and restored wetlands—, namely their role in catchment nutrient losses. This function should be seen only as complementary to all the measures that can be undertaken to improve nutrient use and management on agricultural land and in fields. In reality, created and restored wetlands provide multiple ecosystem services, such as biodiversity enhancement, reservoirs for water and recreation that are difficult to value. Wetlands are also important ecosystems in the global cycles of greenhouse gases, as potential sources of CH_4, CO_2 and N_2O emissions and for their role in carbon sequestration and as possible sinks for N_2O in drainage water from farmland. These aspects are outside the scope of this review, but are potential subjects for other systematic reviews.

Hydrological processes and especially hydraulic loading are inadequately measured in many papers: 45 out of the 143 category 1 papers only included inlet measurements, had incomplete water balances, or lacked hydrological data, making it impossible or too uncertain to calculate mass balances. For created wetlands with a lining and constant load, it is relatively simple to set up a water balance. In contrast, for wetlands with varying or event driven hydraulic loading the measurements need to be much more comprehensive and require much higher temporal resolution to cover the variation. For restored wetlands it is difficult to set up and measure all variables in the water balance (surface water flow, groundwater flow, precipitation, evapotranspiration). In some wetland studies there are large deficits in the water balance due to unaccounted contributions from e.g., groundwater inflow and/or because all budget terms in the water balance have not been measured. Thus especially for restored or recreated wetlands it may be needed to carry out measurements with high spatial and temporal replication. For example, to calculate groundwater flow through a riparian meadow it is necessary to measure hydraulic potentials, hydraulic conductivity in different soil layers or at least to assign hydraulic conductivities according to soil profile descriptions.

In individual studies, measurement errors of the predictor loading rate may make the response variable removal efficiency correlated with the predictor irrespective of any underlying causality. This is one of the short-comings of treating a studied wetland as a black box (which most of the included studies do). From this point of view it would have been better if a larger number of studies used multiple measurements of the loading rate and removal processes (denitrification, plant uptake, chemical adsorption etc.). On the other hand, removal processes are variable in space and time, potentially leading to even higher experimental errors than input–output balance studies and difficulty to scale up for the whole wetland and an entire year. On the contrary, input and output fluxes can be measured with more precision because hydraulic flow measurements combined with frequent concentration measurements of N and P are relatively straightforward, and we have selected the studies for state-of the –art techniques in our critical appraisal. The input–output approach is therefore quite robust. Furthermore, while measurement errors in loading rates may be in the range 0–30 %, this review has included studies where the loading rates span three orders of magnitude for TN and four orders of magnitude for TP. Measurement errors in loading rates in individual studies should thus play a minor role for the correlation between loading rate and removal efficiency shown by the response surface analyses performed in this systematic review.

CONCLUSIONS

Implications for Policy/Practice

The objective of this review was to quantify observed retention rates of nutrients in created or restored wetlands and to quantify the variability between different studies. We also investigated importance of environmental conditions and wetland characteristics for nutrient removal rates. This study examined wetlands in a wide range of climatic conditions, although the performance of wetlands in temperate and boreal regions is emphasized. Overall, data from 203 wetlands has been used in our analyses.

- Our survey found that, on average, created or restored wetlands removed 184 g m^{-2} $year^{-1}$ of total nitrogen and 15 g m^{-2} $year^{-1}$ of total phosphorus. These average retentions are three to four times the retention rates suggested 15 years ago as sustainable for nonpoint source treatment wetlands (10–40 g m^{-2} $year^{-1}$ for nitrogen and 0.5–5 g m^{-2} $year^{-1}$ for phosphorus). However, the median values for TN and TP removal rates, respectively, are lower and in line with the earlier recommendations.
- Restored and created wetlands remain appropriate and potentially sustainable ecological engineering approaches for removing nutrients from treated wastewater and urban and agricultural runoff. Loading rates (inlet concentrations × hydraulic loading rates) need to be carefully estimated as part of the design of these wetlands. In general, high nutrient loading rates result in high removal rates (expressed in g m^{-2} $year^{-1}$). However, high hydraulic loading rates may result in reduced removal efficiency (expressed in %).

- Seventeen of 146 wetlands were shown as phosphorus sources; six of those had been restored on former drained cropland. The studies included in this systematic review suggest that TP removal is less efficient in such wetlands compared to other wetlands. Six out of nine restored wetlands on former drained cropland released more phosphorus than they received. However, one long-term study suggests that the performance of wetlands may be enhanced and that such wetlands can become significant phosphorus sinks after several years of operation.
- Water regime seems to be another factor that can influence phosphorus removal efficiency. Wetlands where the hydraulic loading rate is driven by precipitation show a lower phosphorus removal efficiency than wetlands with a controlled hydraulic loading rate.
- Removal efficiency of total nitrogen in wetlands was positively correlated with average annual air temperature and negatively correlated with hydraulic loading rate. The model fit was better if interaction effects between these variables were allowed. The total nitrogen removal rate was positively correlated with the inflow concentration and was also found to be positively correlated with hydraulic loading.
- The removal efficiency of total phosphorus was correlated with total phosphorus concentrations at the inlet, hydraulic loading, wetland area, and air temperature. The total phosphorus removal rate was positively correlated with concentration at inlet and hydraulic loading rate. In contrast, the total phosphorus removal rate was negatively correlated with wetland area, especially for wetlands smaller than 2×10^4 m^2.

Implications for Research

- Hydrological processes are inadequately measured in many papers: 45 of the papers excluded during critical appraisal only included inlet measurements, had incomplete water balances or lacked hydrological data, making it impossible or too uncertain to calculate mass balances.
- Only total nitrogen or total phosphorus was measured or reported in many studies. This prevented us from evaluating the influence of the speciation of these elements on the removal.
- Long-term performance of wetlands as nutrient sinks is poorly investigated.
- More research is needed on the effects of seasonality, particularly in wet/dry climates, and on hydrologic pulsing on wetlands used to treat agricultural and urban runoff. More research is also needed on the ecosystem services of carbon sequestration and flood mitigation that these wetlands could and do provide in addition to their primary role in water purification.

AUTHORS' CONTRIBUTIONS

The writing of this report was shared between KT (Introduction), ML (Methods and results), JV (Results), WG (Results), CCH (Discussion), and WM (Conclusions). Searches were performed by ML. Screening of literature was performed by ML, KT, JV, WG, CCH, and WM. Critical appraisal was undertaken by KT, JV, WG, CCH, and WM. Meta-analysis and response surface analysis were performed by ML and AG, respectively. All authors read and approved the final manuscript.

ACKNOWLEDGEMENTS

This systematic review was financed by the Mistra Council for Evidence-based Environmental Management (Mistra EviEM). EviEM is funded by the Swedish Foundation for Strategic Environmental Research (Mistra) and hosted by Stockholm Environment Institute.

REFERENCES

1. Smith VH. Eutrophication of freshwater and coastal marine ecosystems—a global problem. Environ Sci Pollut Res. 2003;10(2):126–39.
2. Nixon SW, Ammerman JW, Atkinson LP, Berounsky VM, Billen G, Boicourt WC, et al. The fate of nitrogen and phosphorus at the land sea margin of the North Atlantic Ocean. Biogeochemistry. 1996;35(1):141–80.
3. Ryther JH, Dunstan WM. Nitrogen, phosphorus, and eutrophication in coastal marine environment. Science. 1971;171(3975):1008.
4. Graneli E, Wallstrom K, Larsson U, Graneli W, Elmgren R. Nutrient limitation of primary production in the baltic sea area. Ambio. 1990;19(3):142–51.
5. Conley DJ, Paerl HW, Howarth RW, Boesch DF, Seitzinger SP, Havens KE, et al. Controlling eutrophication: nitrogen and Phosphorus. Science. 2009;323(5917):1014–5.
6. Schindler DW, Hecky RE, Findlay DL, Stainton MP, Parker BR, Paterson MJ, et al. Eutrophication of lakes cannot be controlled by reducing nitrogen input: results of a 37-year whole-ecosystem experiment. Proc Natl Acad Sci USA. 2008;105(32):11254–8.
7. Boesch D, Hecky R, O'Melia C, Schindler D, Seitzinger S. Eutrophication of Swedish Seas. Stockholm: Swedish Environmental Protection Agency, Report 5509; 2006.
8. Danielsson Å, Papush L, Rahm L. Alterations in nutrient limitations—scenarios of a changing Baltic Sea. J Mar Syst. 2008;73(3–4):263–83.
9. Kadlec RH, Wallace SD. Treatment wetlands. Boca Raton: CRC Press; 2009. p. 965.
10. Mitsch WJ, Day JW, Zhang L, Lane RR. Nitrate-nitrogen retention in wetlands in the Mississippi river basin. Ecol Eng. 2005;24(4):267–78.
11. Carleton JN, Grizzard TJ, Godrej AN, Post HE. Factors affecting the performance of stormwater treatment wetlands. Water Res. 2001;35(6):1552–62.
12. Vymazal J. Removal of nutrients in various types of constructed wetlands. Sci Total Environ. 2007;380(1–3):48–65.
13. Vymazal J, Kroepfelova L. Removal of nitrogen in constructed wetlands with horizontal sub-surface flow: a review. Wetlands. 2009;29(4):1114–24.
14. Kadlec RH. Nitrogen farming for pollution control. J Environ Sci Health Part A-Toxic/Hazard Subst Environ

Eng. 2005;40(6–7):1307–30.

15. Hoffmann CC, Kjaergaard C, Uusi-Kamppa J, Hansen HCB, Kronvang B. Phosphorus retention in riparian buffers: review of their efficiency. J Environ Qual. 2009;38(5):1942–55.

16. Braskerud BC, Tonderski KS, Wedding B, Bakke R, Blankenberg AGB, Ulen B, et al. Can constructed wetlands reduce the diffuse phosphorus loads to eutrophic water in cold temperate regions? J Environ Qual. 2005;34(6):2145–55.

17. Mitsch WJ, Zhang L, Stefanik KC, Nahlik AM, Anderson CJ, Bernal B, et al. Creating wetlands: primary succession, water quality changes, and self-design over 15 years. Bioscience. 2012;62(3):237–50.

18. Mitsch WJ, Zhang L, Waletzko E, Bernal B. Validation of the ecosystem services of created wetlands: two decades of plant succession, nutrient retention, and carbon sequestration in experimental riverine marshes. Ecol Eng. 2014;72:11–24.

19. Land M, Granéli W, Grimvall A, Hoffmann CC, Mitsch WJ, Tonderski KS, et al. How effective are created or restored freshwater wetlands for nitrogen and phosphorus removal? A systematic review protocol. Environ Evid. 2013;2(1):1–8.

20. SMHI Vattenwebb (in Swedish) [database on the internet]. Available from:http://vattenwebb.smhi.se/wetlands/. Accessed 25 Feb 2013.

21. Svensson JM, Strand J, Sahlén G, Weisner S. Utvärdering av våtmarker anlagda inom lokala investeringsprogram och med LBU-stöd avseende närsaltsretention och biologisk mångfald. Rikare mångfald och mindre kväve, Utvärdering av våtmarker skapade med stöd av lokala investeringsprogram och landsbygdsutvecklingsstöd (in Swedish). Naturvårdsverket, Rapport 5362; 2004.

22. Tonderski KS, Arheimer B, Pers CB. Modeling the impact of potential wetlands on phosphorus retention in a Swedish catchment. Ambio. 2005;34(7):544–51.

23. Brandt M, Arheimer B, Gustavsson H, Pers C, Rosberg J, Sundström M et al. Uppföljning av effekten av anlagda våtmarker i jordbrukslandskap. Belastning av kväve och fosfor (in Swedish). Naturvårdsverket, Rapport 6309; 2009.

24. Weisner S, Thiere G. Mindre fosfor och kväve från jordbrukslandskapet (in Swedish). Jordbruksverket, Rapport. 2010;2010:21.

25. Weisner S, Johannesson K, Tonderski K. Näringsavskiljning i anlagda våtmarker i jordbruket. Analys av mätresultat och effekter av landsbygdsprogrammet (in Swedish). Jordbruksverket, Rapport 2015:7.

26. Mitsch WJ, Horne AJ, Nairn RW. Nitrogen and phosphorus retention in wetlands-ecological approaches to solving excess nutrient problems. Ecol Eng. 2000;14(1–2):1–7.

27. Kottek M, Grieser J, Beck C, Rudolf B, Rubel F. World map of the Koppen-Geiger climate classification updated. Meteorol Z. 2006;15(3):259–63.

28. Thiere G, Milenkovski S, Lindgren PE, Sahlen G, Berglund O, Weisner SEB. Wetland creation in agricultural landscapes: biodiversity benefits on local and regional scales. Biol Conserv. 2009;142(5):964–73.

29. Bilotta GS, Milner AM, Boyd IL. Quality assessment tools for evidence from environmental science. Environ Evid. 2014;3(1):1–14.

30. Mermillod-Blondin F, Lemoine D, Boisson JC, Malet E, Montuelle B. Relative influences of submersed macrophytes and bioturbating fauna on biogeochemical processes and microbial activities in freshwater sediments. Freshw Biol. 2008;53(10):1969–82.

31. Hahn S, Bauer S, Klaassen M. Quantification of allochthonous nutrient input into freshwater bodies by herbivorous waterbirds. Freshw Biol. 2008;53(1):181–93.

32. Andersen DC, Sartoris JJ, Thullen JS, Reusch PG. The effects of bird use on nutrient removal in a constructed wastewater-treatment wetland. Wetlands. 2003;23(2):423–35.

33. Bernes C, Carpenter SR, Gårdmark A, Larsson P, Persson L, Skov C, et al. What is the influence of a reduction of planktivorous and benthivorous fish on water quality in temperate eutrophic lakes? A systematic review. Environ Evid. 2015;4(1):1–28.

34. Borenstein M, Hedges LV, Higgins JPT, Rothstein HR. Random-effects model. In: Introduction to meta-analysis. Chichester: Wiley; 2009. p. 69–75.

35. Borenstein M, Hedges LV, Higgins JPT, Rothstein HR. Identifying and quantifying heterogeneity. In: Introduction to meta-analysis. Chichester: Wiley; 2009. p. 107–25

36. Haddaway NR, Verhoeven JTA. Poor methodological detail precludes experimental repeatability and hampers synthesis in ecology. Ecol Evol. 2015;5(19):4451–4.

37. Chen HJ, Ivanoff D, Pietro K. Long-term phosphorus removal in the Everglades stormwater treatment areas of South Florida in the United States. Ecol Eng. 2015;79:158–68.

38. Moustafa MZ, Chimney MJ, Fontaine TD, Shih G, Davis S. The response of a freshwater wetland to long-term “low level” nutrient loads—marsh efficiency. Ecol Eng. 1996;7(1):15–33.

39. Kieckbusch JJ, Schrautzer J. Nitrogen and phosphorus dynamics of a re-wetted shallow-flooded peatland. Sci Total Environ. 2007;380(1–3):3–12.

40. Koskiaho J, Ekholm P, Räty M, Riihimäki J, Puustinen M. Retaining agricultural nutrients in constructed wetlands—experiences under boreal conditions. Ecol Eng. 2003;20(1):89.

41. Kovacic DA, David MB, Gentry LE, Starks KM, Cooke RA. Effectiveness of constructed wetlands in reducing nitrogen and phosphorus export from agricultural tile drainage. J Environ Qual. 2000;29(4):1262–74.

42. Bass KL, Evans RO. Water quality improvement by a small in-stream constructed wetland in North Carolina’s Coastal plain. Watershed Management and Operations Management 2000. Fort Collins: American Society of Civil Engineers; 2000.

43. Healy M, Cawley AM. Nutrient processing capacity of a constructed wetland in western Ireland. J Environ Qual. 2002;31(5):1739–47.

44. Walling DE. Linking land use, erosion and sediment yields in river basins. Hydrobiologia. 1999;410:223–40.

45. Walling DE, Owens PN. The role of overbank floodplain sedimentation in catchment contaminant budgets. Hydrobiologia. 2003;494(1–3):83–91.

46. Fink DF, Mitsch WJ. Hydrology and nutrient biogeochemistry in a created river diversion oxbow wetland. Ecol Eng. 2007;30(2):93–102.

47. Mitsch WJ, Zhang L, Fink DF, Hernandez ME, Altor AE, Tuttle CL, et al. Ecological engineering of floodplains. Ecohydrol Hydrobiol. 2008;8(2–4):139–47.

CHAPTER 12

EFFICIENCY OF THREE HALOPHYTE SPECIES IN REMOVING NUTRIENTS FROM SALINE WATER: A PILOT STUDY

H. J. De Lange and M. P. C. P. Paulissen

Alterra, Wageningen UR, 6700 Wageningen, The Netherlands

ABSTRACT

Saline wetlands may be well suited for purifying contaminated water from saline agriculture and aquaculture or from freshwater-based agriculture in areas subject to increased salinity. However, case studies on the nutrient removal efficiency of halophyte species are scarce, especially for temperate regions. Here we tested the nutrient removal efficiency and ability to store nutrients in aboveground and belowground biomass of three halophyte species, Aster tripolium, Bolboschoenus maritimus subsp. compactus, and Spartina anglica, in a greenhouse microcosm experiment at two salinity levels. Nutrient removal from water differed among the species: Spartina had the highest nitrogen removal, Bolboschoenus and Spartina had the highest phosphorus removal. The species also differed in the allocation of the nutrient uptake.Bolboschoenus had the highest absolute uptake of nitrogen and phosphorus in shoots, whereasSpartina had the highest uptake of nitrogen and phosphorus in roots. The applicability of these three species in constructed saline wetlands depends on the local salinity and water regime.

INTRODUCTION

Many constructed wetlands are designed to purify contaminated freshwater (e.g. Verhoeven and Meuleman 1999; Kadlec and Wallace 2009). Cases where the inflowing water is (highly) saline often originate from industrial sources and this water is typically also characterized by high concentrations of heavy metals or hydrocarbons (Kanagy et al. 2008; Calheiros et al. 2012; Carvalho et al. 2013; González-Alcaraz et al. 2013; Tao and Yu 2013). Our understanding of the ability of saline constructed wetlands to purify water from agricultural sources is mostly from traditional agricultural systems in arid regions (Lymbery et al. 2006), from aquaculture in (sub)tropical regions (Ye et al. 2001; Shpigel et al. 2013; Jesus et al. 2014), and the laboratory (Gregory et al. 2010, 2012). Larger scale field applications are available from tropical regions, and range from constructed recirculation aquaculture systems (Lin et al. 2010; Tilley et al.2002) to natural systems (Gautier et al. 2001; Gennaro et al. 2006). Examples of full-scale constructed wetlands in temperate regions purifying contaminated saline water from aquaculture sources are only recently becoming available (e.g. Webb et al. 2012, 2013; Quintã et al. 2015a, b). Yet, saline wetlands may be well suited for purifying contaminated water from saline agriculture or from freshwater-based agriculture in areas subject to increased salinity (e.g. Buhmann and Papenbrock 2013; De Lange et al. 2013). Such

Citation: H. J. De Lange, M. P. C. P. Paulissen, Efficiency of three halophyte species in removing nutrients from saline water: a pilot study, doi:10.1007/s11273-016-9489-8

saline constructed wetlands may help to prevent unwanted loading of vulnerable estuaries or coastal waters with nutrients, organic pollutants and/or pathogens from both point and diffuse sources. Salinity pressure is increasing in many low-lying coastal areas worldwide due to climate change and sea level rise (e.g. Rozema and Flowers 2008). In this perspective, saline treatment wetlands can be considered a logical and sustainable adaptation measure, with various additional advantages such as providing a semi-natural transition zone between intensively used agricultural areas and adjacent natural estuarine and coastal areas (Roncken et al. 2011). At the same time, the rehabilitation and/or enhancement of existing salt marshes will provide other essential ecosystem services, such as the stabilisation of fine sediments, providing a protective buffer between land and sea, and providing a habitat to a diversity of terrestrial and marine fauna (Laegdsgaard 2006).

The development of full-scale saline constructed wetlands is currently hampered, at least in the European context, by knowledge gaps about purification efficiencies of native salt-tolerant plants (e.g. Buhmann and Papenbrock 2013; De Lange et al. 2013). As reviewed by Flowers and Colmer (2008), halophytes are physiologically adapted to salinity stress, with some halophyte species having optimal growth rates under saline conditions and others under freshwater conditions. The zonation of halophyte species in a salt marsh reflects these differences in salt tolerance combined with tolerance to flooding (Rozema and Van Diggelen 1991). Higher tolerance to salt may co-occur with a lower tolerance to changes in nutrients, as shown for the halophyte *Bolboschoenus maritimus* compared with three freshwater species of *Bolboschoenus*, that may be attributed to evolutionary adaptation to salinity stress (Hroudová et al. 2014). For two other halophyte species, *Salicornia europaea* and *Aster tripolium*, it was shown that they can take up both organic and inorganic nitrogen (Quintã et al. 2015a, b), which is advantageous since salt marshes are often nitrogen-limited. This finding is interesting from the point of view of potential (organic) nutrient removal.

In order to compare nitrogen (N) and phosphorus (P) removal efficiencies of three plant species originating from a tidal brackish marsh in the Netherlands, we conducted a pilot greenhouse experiment using two salinity levels, representative of the tidal brackish marsh, and one nutrient level. The species tested were *A. tripolium*, *B. maritimus* subsp. *compactus*, and*Spartina anglica*. Choice for these species was based on general presence in saline areas in Western Europe, differences in salt tolerance (Rozema and Van Diggelen 1991; Buhmann and Papenbrock 2013), and differences in hydrodynamic tolerance. The focus of our study was on the nutrient uptake in above- and belowground plant material at the end of the experiment, relevant for situations where biomass can be harvested, and on the nutrient removal from the inflowing water during the experiment, relevant for water purification. The ultimate aim of the experiment was to make a contribution to the full-scale applicability of saline constructed wetlands.

Materials and Methods

The Plant Species

Aster tripolium is a biennial herbaceous plant, while *B. maritimus* subsp. *compactus* is a perennial graminoid herbaceous plant, and *Spartina anglica* is a perennial grass species.*Spartina* is commonly found in the transition zone from mud flat to salt marsh that is under tidal influence on a daily basis. It is relatively tolerant to prolonged inundation, e.g. at high tide. Compared to *Spartina*, *Aster* and *Bolboschoenus* are more typical of hydraulically less dynamic sites. They are common on Dutch salt marshes as well as in saline or brackish areas on the landward side of the dike.

Collection of Plants

We collected *A. tripolium*, *B. maritimus* subsp. *compactus* and *Spartina anglica* on 26 April 2013 in 'Het Verdronken Land van Saeftinghe' (51°34′-N, 4°18′-E), a saltmarsh reserve in the Scheldt estuary in the south-western Netherlands. In three sub-locations, each dominated by one of the three studied species, soil cores (diameter app. 20 cm, depth app. 20 cm) with living aboveground plant material were dug out using spades. For each species 24 cores were sampled, which was double the number needed in the experiment. The timing of the sampling was chosen to collect fresh sprouting plants, to study their performance during growth in the greenhouse. All sub-locations had brackish conditions and were situated in shallow depressions on the high inland end of the saltmarsh, with an average inundation frequency of approximately once or twice per month. Inundation is most frequent in the winter half-year. The sites where plant material was collected are grazed by cattle during the summer half-year (PepijnCalle, personal communication). The salinity of the Scheldt water varied between 10 and 20 PSU, and nutrient concentrations varied between 2 and 5 mg N/L, and 0.1–0.2 mg P/L.

The cores were put in 4 L plastic pots (diameter 18 cm, height 15 cm) and transported back to the greenhouse on the same day. Before putting the cores in the pots, the perforated bottoms of the pots were covered with a double layer of thinly perforated plastic foil to allow passage of effluent from the pots but minimise loss of soil material and roots growing out of the pots. To keep the soil cores moist during the acclimatization period, brackish surface water (salinity 10.8 PSU, 16.4 mS m^{-1}) was collected from two creeks near the sampling sites.

On 29 April 2013, standing dead litter from the previous growing season as well as other plant species present (notably *Puccinellia* sp.) were carefully removed from the soil cores. Each plastic pot was then placed in an individual metal pot lined with 40 × 60 cm plastic bags to collect effluent from the soil cores. 300 ml of the collected brackish ditch water was poured onto the soil cores of each pot. From 2 May 2013 onwards, a synthetic brackish solution (Coralsea, purchased from AquaHolland, Dordrecht, the Netherlands) was used for watering the pots during the remainder of the acclimatisation period (1/3× sea water strength, corresponding to ditch water salinity in the sites where the plant material was collected).

Experimental Set-up

After a 2 week acclimatisation period in the greenhouse, the experiment started on 8 May 2013. From the 24 available pots per species, we selected 12 pots where the plants had similar above ground biomass, similar total weight, similar number of sprouts, and had a healthy appearance. For each plant species we used two salinity levels and six replicate pots, resulting in

36 pots in total (three plant species × two salt concentrations × six replicates). The pots were placed on a large table in the greenhouse. For practical reasons, the experimental units were grouped according to salt concentration of the treatment solution. Within each of these salinity groups, the position of experimental units on the table in the greenhouse was randomised. To determine the plant conditions at the start of the experiment, of each plant species 4 additional pots were harvested to determine dry weight (DW) and nutrient concentration of shoots and roots at t = 0 (Supplemental Information, Table S-1). Climate conditions (light, temperature and relative humidity) in the greenhouse followed the natural fluctuations in Wageningen, as the greenhouse did not apply extra heating or light. Temperature and relative humidity (RH) were recorded every 5 min during the whole experiment inside the greenhouse. Average daily temperature during the experiment ranged from 8.4 to 25.4 °C with a median value of 15.7 °C. Average daily RH ranged from 48.5 to 90.4 % with a median value of 71.9 %.

All experimental units received the same nominal nutrient concentration of 15 mg N/L and 2.5 mg P/L. This concentration was chosen to be representative for saline aquaculture effluents and agricultural runoff (see De Lange et al. 2013) and not much higher than the natural range of concentrations in the Scheldt estuary. Treatment solutions were made of the synthetic Coralsea water (AquaHolland) with Pokon universal plant nutrient solution (PokonNaturado BV, Veenendaal, the Netherlands) added to achieve the required concentration. By using Pokon the presence of trace elements was ensured (K, Fe, Mn, B, Cu, Zn, and Mo) (cf. Keizer-Vlek et al.2014). Final pH of the used solution was >8, based on specifications of AquaHolland. Half of the experimental units received the nutrient solution at 1/10 x sea water strength, the other half at 1/2× sea water strength, further referred to as low and high salinity. This reflects the range of salinities occurring at this part of the Scheldt Estuary. Treatment solutions were applied twice per week at a dose of 250 ml per pot. The average salinity of the treatment solutions was 2.7 and 12.9 PSU for low and high salinity respectively.

Three weeks after the start of the experiment it was observed that some of the soil cores had become dry and hard with cracks in the clayey material mainly at the edges. After the second sampling point we applied extra rainwater to rewet the soil cores, including an additional weekly gift of 250 ml rainwater. Visual inspection during the experiment did not indicate a significant or lasting impact of the soil drought on plant health or survival. This was confirmed by the data on effluent water, the fraction of water collected as effluent remained relatively constant during the first weeks, and plant height showed a gradual increase during the experiment (see Supplemental Information, Figs. S-1, S-2).

Measurements

Once per week, the volume of the effluent collected was measured and the plastic bags were emptied and clean bags re-installed beneath each experimental unit. During the experiment, plant height was measured, photos of the plants were made, and water samples from the effluent were taken four times at 2 week intervals. The water samples were filtered (Whatman 595 1/2 filter paper) and stored at −20 °C prior to analysis. Concentrations of total dissolved N (NEN 6646 including internal extraction UV/persulfate) and total dissolved P (NEN6663 including internal extraction UV/persulfate) were determined using a Skalar SAN Plus continuous-flow analyser (Skalar Analytical BV, Breda, the Netherlands). Our focus in this pilot experiment was on the macro-nutrients N and P, though we recognize the importance of other elements in osmoregulation processes and possible uptake competition between salt ions and nutrient ions (e.g. Na^+ reducing K^+ uptake or Cl^- reducing NO^{3-}uptake). Since we used two fixed salinity levels, differences in nutrient uptake can be considered as the net functioning of the three species, including competition between salts and nutrients.

The experiment was ended after 9 weeks by harvesting the aboveground biomass at the soil surface of the pots containing plants. This is further referred to as 'shoots', and included both stems and leaves. Belowground biomass was harvested by rinsing away the soil particles, this is further referred to as 'roots'. All harvested plant material was dried at 70 °C for at least 72 h. After determining DW of the samples, the plant material was ground using ball mills and digested with H_2SO_4, salicylic acid, H_2O_2 and Se. Subsequently, the N and P concentrations of the diluted supernatant were determined using a Skalar SAN Plus continuous-flow analyser (Skalar Analytical BV, Breda, the Netherlands).

Statistical Analysis

The experiment was designed to use a two-factor ANOVA to test for differences between treatments, with plant species and salt concentration as factors. Prior to analysis, the data were checked for normal distribution and homogeneity of variances. If necessary log(x + 1)–transformation was applied to meet the ANOVA requirements. All statistical tests were done using Genstat 16th edition.

The ability of the plant-soil system to remove nutrients from inflowing water was calculated as concentration in the pouring solution minus the concentration in the effluent water, and referred to as delta N and delta P. Since this was measured four times during the experiment, the results were tested with a repeated measurements two-factor ANOVA. We expected this delta concentration to be positive due to plant uptake.

Dry weight (grams) and total N and P concentration (% of DW) of the roots and shoots at t = 0 (n = 4 per species) and t = end (n = 6 per salt concentration per species) were combined to calculate the absolute uptake in grams of N and P in shoots and roots during the experiment. Relative growth rate (RGR, d^{-1}) of the plants was calculated as (ln(final DW)–ln(initial DW))/(time in days), using the shoot DW (cf. Brown et al. 1999).

Data on N and P content in shoots and roots did not meet the conditions of homogeneity of variances and normal distribution of error terms, not even after log-transformation. Therefore we tested differences between treatments using the non-parametric Kruskal–Wallis test, followed by pair-wise comparisons using Mann–Whitney U tests.

Results and Discussion

Net nutrient removal: comparing concentrations in effluent water and in pouring solution

Delta N concentration differed between species ($p = 0.003$) with *Spartina* having significantly higher delta N than *Aster* (Fig. 1). There also was a significant time effect ($p = 0.041$), with the third sampling timepoint having significantly higher delta N than the first sampling timepoint. Finally, the high salinity pots had more variety in delta N at the different sampling points than the low salinity pots, resulting in significant time × salt interaction effect ($p = 0.002$) (Fig. 1). Delta P concentration also differed between species ($p = <0.001$) with *Spartina* and*Bolboschoenus* having significantly higher delta P than *Aster*. Again, there was a significant time effect ($p < 0.001$), with significantly increasing delta P for the first three sampling timepoints (Fig. 1). There was no significant effect of salt concentration on delta P.

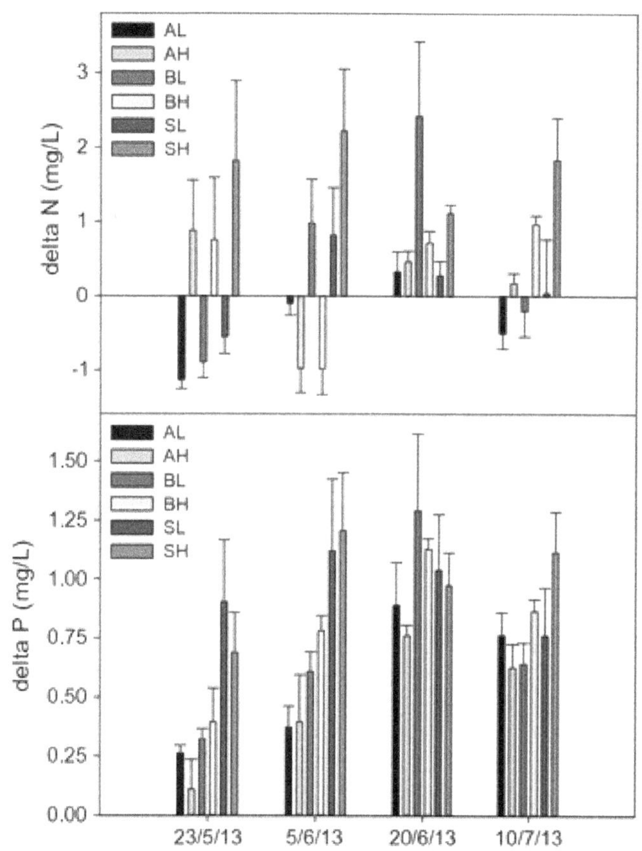

Figure 1: Delta total dissolved N concentration (*upper panel*) and delta total dissolved P concentration (*lower panel*) at the four sampling points in time, calculated as concentration in inflow minus percolated water, average ± 1 SE (n = 6). Treatments are abbreviated as follows: *AL Aster* low salinity, *AH Aster* high salinity, *BL Bolboschoenus*low salinity, *BH Bolboschoenus* high salinity, *SL Spartina* low salinity, *SH Spartina* high salinity. Positive values indicate net uptake in the pots.

For some samples, mostly in the low salinity treatment, the concentration of total N in the effluent water was higher than in the pouring water, resulting in a negative delta (see also Supplemental Information, fig. S-3). This might be caused by a complex of processes in the soil core causing net N release. This happened mostly during the first weeks of the experiment, during the second half of the experiment the plant-soil system retained more nitrogen. For total P, the delta concentration was positive for most replicate pots, indicating that the P is mostly retained in the plant and soil system (Fig. 1). Also, delta P is more stable than delta N during the experiment. The important difference between N and P behaviour in the plant-soil system is that N is more mobile and is subject to various microbial processes which makes it more prone to variation, whereas the behaviour of P is more governed by physico-chemical processes such as absorption to clay particles (Kadlec and Wallace 2009). The significant time effect for both delta N and delta P indicates that the plant-soil system needed to adapt to a new nutrient uptake equilibrium, which is in agreement with results of Lymbery et al. (2013). It also reflects that during the exponential growth phase, uptake of nutrients increased (see Supplemental Information Fig. S-4 for increase in plant height during the experiment). Based on these results, we expect that given a longer time period, delta N would become more stable as well, but with more fluctuations than delta P because of the microbial processes of N. However, given the pilot nature of our experimental setup, this should be further tested in more detailed experiments.

Delta N ranged from −1 to 2 mg N/L (Fig. 1), which corresponds to a maximal removal efficiency of 11 %. This is low compared with results from other greenhouse or microcosm studies testing different salt-tolerant species. These studies report removal efficiencies of around 70 % for NH_3-N (Klomjek and Nitisoravut 2005) or >95 % for total N (Brown et al.1999). At pilot-scale constructed wetland cells, removal efficiencies for total nitrogen are reported to be 98 % (Webb et al. 2012) and around 70 % (Lymbery et al. 2013).

Delta P ranged from 0.1 to 1.3 mg P/L (Fig. 1), which corresponds to a maximal removal efficiency of 35 %. This is in the same range as reported by Klomjek and Nitisoravut (2005) and Webb et al. (2012), but lower than reported by Brown et al. (1999), Lymbery et al. (2006) and Lymbery et al. (2013) who report removal efficiencies of >95 %, around 50 %, and around 80 % respectively.

Both Brown et al. (1999), Klomjek and Nitisoravut (2005) and Lymbery et al. (2013) report similar high removal efficiencies in the soil of control pots or in the unplanted wetland cells for both total N and total P, indicating that most of the removal action takes places in the soil core, and not by the plant. To better understand the processes responsible for nutrient removal and/or storage and their respective share in removal, microbial processes in the soil deserve further attention in follow-up experiments. Our pilot experiment only gives a first indication on the net outcome of these plant-soil processes.

Nutrient uptake in Plant Shoots and Roots

Above-and belowground biomass was harvested at the end of the experiment. The average shoot DW was not significantly different between species or salinity level, and was 20.0 ± 6.2, 18.5 ± 4.0 and 19.0 ± 6.2 g (average ± 1 SD) for *Aster*, *Bolboschoenus* and *Spartina* respectively. This absence of salinity effect on shoots was also found by Lymbery et al. (2013). However, in our experiment the RGR differed significantly between species ($p < 0.001$), but not between salt levels nor was there a significant interaction effect. RGR based on shoot DW was on average 0.026 ± 0.005, 0.042 ± 0.003, and 0.034 ± 0.005 d^{-1}(average ± 1 SD) for *Aster*, *Bolboschoenus*and *Spartina* respectively. This range of values was similar as reported by Brown et al. (1999), and higher than the values reported by Klomjek and Nitisoravut (2005). This supports that the conditions in our experiment were sufficient for healthy plant growth.

Root DW differed significantly between species ($p < 0.001$), but not between salt levels nor was there a significant interaction. Average root DW of *Aster* was significantly lower (58 ± 24 g) than of *Bolboschoenus* and *Spartina* (104 ± 45 and 119 ± 48 g, respectively, average ± 1 SD) (ANOVA $p < 0.001$).

At the end of the experiment, N concentration in shoots was highest in *Aster* and*Bolboschoenus*, while N concentration in roots was highest in *Aster* and *Spartina* (Kruskal–Wallis followed by Mann–Whitney U tests $p < 0.05$; Fig. 2). The N concentration in *Aster* shoots was slightly lower, and the concentration in *Aster* roots was much lower than reported values by Quintã et al. (2015b). This difference may be because our plants were larger and presumably in a different growth phase: 20 g DW in our experiment compared with 0.5–1.0 g DW reported by Quintã et al. (2015b), who grew their plants from seeds. P concentration in shoots was highest in *Aster* and lowest in *Spartina*, while P concentration in roots was highest in *Bolboschoenus*and lowest in *Spartina* (Kruskal–Wallis followed by Mann–Whitney U tests $p < 0.05$; Fig. 2). The nutrient concentration ratio between shoots and roots shows that N was more stored in the shoots, most strongly for *Bolboschoenus* (ratio > 1) whereas P was more stored in the roots, most strongly for *Bolboschoenus* and *Spartina* (ratio < 1) (Kruskal–Wallis followed by Mann–Whitney U tests $p < 0.05$; Fig. 2). Salinity level had no significant effect on the nutrient concentration in roots and shoots (Kruskal–Wallis, $p > 0.05$).

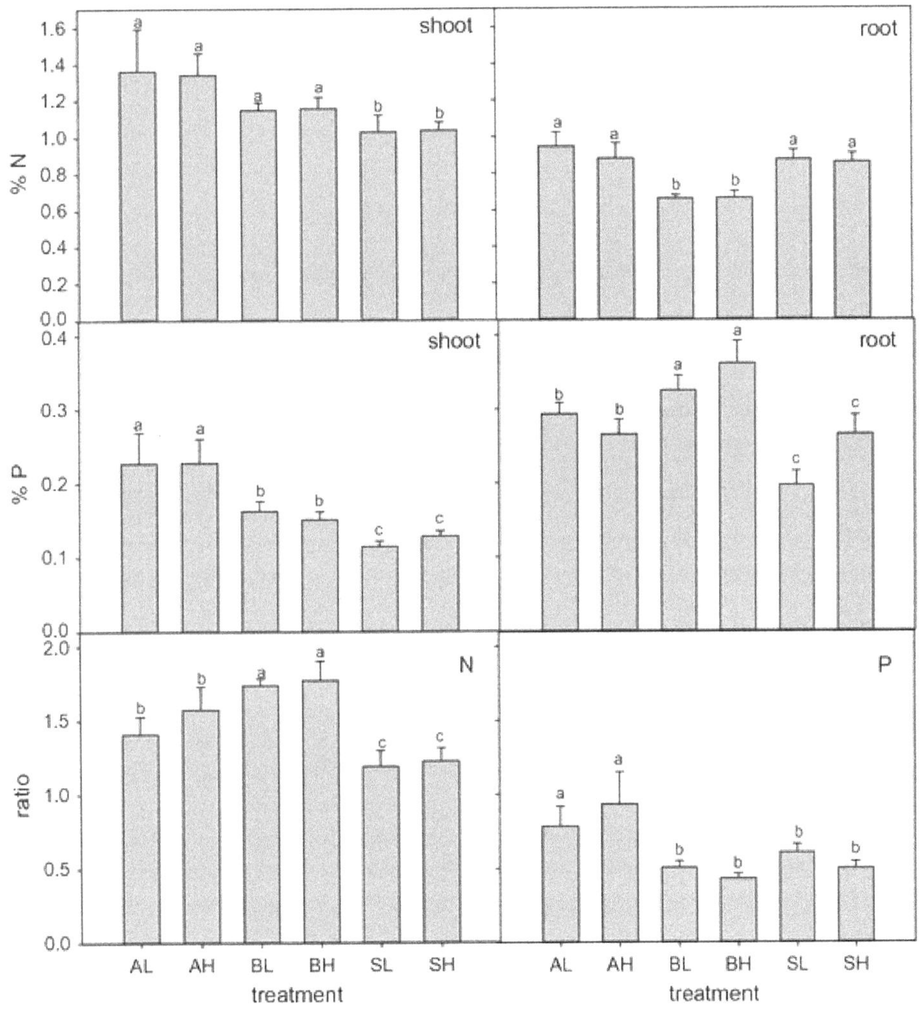

Figure 2: N concentration (*upper panels*) and P concentration (*middle panels*) for shoots (*left panels*) and roots (*right panels*), and shoot/root ratio of N and P (*lower panels*); average ± 1 SE (n = 6). Letters indicate homogeneous groups (Kruskal–Wallis followed by Mann–Whitney U tests $p < 0.05$). Treatment abbreviations as in Fig. 1.

The absolute uptake of nutrients during the experiment in plant shoots and roots was calculated using DWs and N and P concentration at start and end of the experiment (Fig. 3).*Bolboschoenus* had significantly higher shoot N uptake than *Aster*, with *Spartina* having intermediate values. There was no significant overall effect of salinity level. *Bolboschoenus* had significantly higher shoot P uptake than *Spartina*, with *Aster* having intermediate values. Again there was no significant overall effect of salinity level (Fig. 3), although *Bolboschoenus* had higher nutrient uptake in shoots at low salt concentrations (t test, $p = 0.055$ and $p = 0.002$ for N and P respectively).

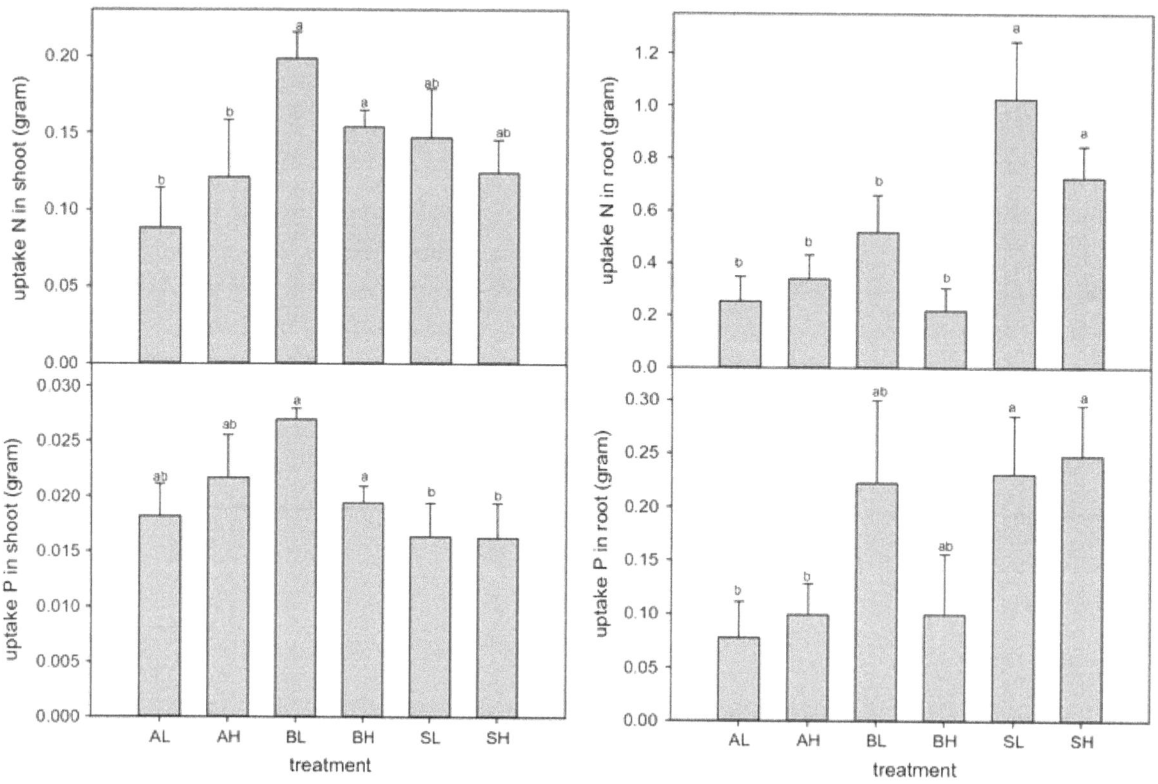

Figure 3: Uptake of N (*upper panels*) and P (*lower panels*) in aboveground plant biomass (*left side panels*) and belowground biomass (*right side panels*), average ± 1 SE (n = 6). Letters indicate homogeneous groups (ANOVA followed by post hoc LSD test). Treatment abbreviations as in Fig. 1.

N and P uptake in roots differed significantly between species, without a salinity effect, and was highest in *Spartina* and lowest in *Aster* (Fig. 3). In general, root DW was higher than shoot DW, and this resulted in a higher total amount of nutrients in roots compared with shoots. However, since in full-scale saline treatment wetlands aboveground biomass can be harvested more easily than belowground biomass, uptake of nutrients in shoots is more relevant when comparing performance of species for such cases.

Upscaling from Pilot Experiments

Although our experiment is on a pilot scale, the results of our experiment can be used in choosing the optimal plant species to test further for use in a saline constructed wetland. This choice depends on purpose and management options, salinity conditions and hydraulic regime. Some first conclusions can be drawn from our pilot experiments, but further upscaling experiments are needed to support our results when applied to field scale. If the focus is on nutrient removal from water, *Spartina* and *Bolboschoenus* achieve higher removal rates than*Aster*. If aboveground biomass can be harvested, *Bolboschoenus* has the advantage of the highest uptake rates for both N and P in aboveground biomass. This biomass could be used for other purposes, e.g. generating energy from biomass. However, the economic gain of this is still difficult to achieve (Meerburg et al. 2010). There was a limited effect of salinity. This might indicate that the two tested salinity levels are in the natural range of the three species. It was not our purpose to investigate the optimal salinity for each species, but for this pilot experiment we simply chose two salinities representative for the Western Scheldt estuary. Within this salinity range tested, *Aster* performed better under the high salinity conditions, so if the location or the effluent to be treated is more saline, *Aster* can be a good choice. *Bolboschoenus*and *Spartina* are comparable in removing nutrients from the water, and nutrient storage in the shoots. *Spartina* tolerates fluctuations in water level better than *Bolboschoenus*, so *Spartina*may be a good choice for lower-lying locations with large water level fluctuations.

The initial choice for these three species was made based on natural common occurrence in western Europe. The performance in the pilot experiment is promising enough to conduct further experiments with these three species, to investigate how the species will perform in terms of nutrient uptake and retention, and efficiency per unit area. Depending on the ecoregion, the search for better-performing could be extended to other species as well.

Conclusions

The main aim of this experiment was to compare different halophyte species of common occurrence in temperate Europe on nutrient removal in treatment water and nutrient uptake in plant biomass. Although our experiment treated the soil processes as a black box, the results show several outcomes that can be used in the future. The observed RGR of the three species showed that the experimental conditions were sufficiently good to support our conclusions. Nutrient removal from water differed between species: *Spartina* had the highest N removal,*Bolboschoenus* and *Spartina* had the highest P removal. *Bolboschoenus* had the highest absolute uptake of N and P in shoots, whereas *Spartina* had the highest uptake of N and P in roots. Salinity of the treatment water was not a significant factor, except for nutrient uptake in*Bolboschoenus* which was higher at low salinity. These results can be used in choosing the optimal plant species for use in a saline constructed wetland. This choice depends on site-specific characteristics and purpose of the constructed wetland, such as management type, salinity conditions and hydraulic regime.

ACKNOWLEDGMENTS

We thank Chiel Jacobusse of Stichting Het Zeeuwse Landschap for his help in obtaining permission to collect the plant material in the saltmarsh reserve. Pepijn Calle (Stichting Het Zeeuwse Landschap) kindly helped during fieldwork and provided information about the characteristics of the plant collection sites. We also thank Rinie Verwoert, Jolanda Bierman, Wim Lieftink and others at Unifarm (Wageningen University and Research Centre) for their advice and assistance before and during the experiment. Sija Stofberg (Wageningen University) assisted during the experiment and harvest. Jan van Walsem and Frits Gillissen (Wageningen University) carried out the nutrient analyses. We thank Monique Heijmans (Wageningen University) for constructive comments on an earlier version of the manuscript. Comments of two anonymous reviewers further improved the manuscript. The experiment was financed by the Dutch Ministry of Economic Affairs through the Kennisbasis programme (Project No. KB-14-005-034).

REFERENCES

1. Brown JJ, Glenn EP, Fitzsimmons KM, Smith SE (1999) Halophytes for the treatment of saline aquaculture effluent. Aquaculture 175(3–4):255–268. doi:10.1016/S0044-8486(99)00084-8
2. Buhmann A, Papenbrock J (2013) Biofiltering of aquaculture effluents by halophytic plants: basic principles, current uses and future perspectives. Environ Exp Bot 92:122–133. doi:10.1016/j.envexpbot.2012.07.005
3. Calheiros CSC, Quitério PVB, Silva G, Crispim LFC, Brix H, Moura SC, Castro PML (2012) Use of constructed wetland systems with *Arundo* and *Sarcocornia* for polishing high salinity tannery wastewater. J Environ Manag 95:66–71. doi:10.1016/j.jenvman.2011.10.003
4. Carvalho F, Prazeres AR, Rivas J (2013) Cheese whey wastewater: characterization and treatment. Sci Total Environ 445:385–396. doi:10.1016/j.scitotenv.2012.12.038
5. De Lange HJ, Paulissen MPCP, Slim PA (2013) Halophyte filters: the potential of constructed wetlands for application in saline aquaculture. Int J Phytoremediat 15:352–364. doi:10.1080/15226514.2012.702804
6. Flowers TJ, Colmer TD (2008) Salinity tolerance in halophytes. New Phytol 179(4):945–963
7. Gautier D, Amador J, Newmark F (2001) The use of mangrove wetland as a biofilter to treat shrimp pond effluents: preliminary results of an experiment on the Caribbean coast of Colombia. Aquacult Res 32(10):787–799. doi:10.1046/j.1365-2109.2001.00614.x
8. Gennaro P, Guidotti M, Funari E, Porrello S, Lenzi M (2006) Reduction of land based fish-farming impact by phytotreatment pond system in a marginal lagoon area. Aquaculture 256(1–4):246–254. doi:10.1016/j.aquaculture.2006.02.028
9. González-Alcaraz MN, Conesa HM, Álvarez-Rogel J (2013) Phytomanagement of strongly acidic, saline eutrophic wetlands polluted by mine wastes: the influence of liming and*Sarcocornia fruticosa* on metals mobility. Chemosphere 90:2512–2519. doi:10.1016/j.chemosphere.2012.10.083
10. Gregory SP, Shields RJ, Fletcher DJ, Gatland P, Dyson PJ (2010) Bacterial community responses to increasing ammonia concentrations in model recirculating vertical flow saline biofilters. Ecol Eng 36:1485–1491. doi:10.1016/j.ecoleng.2010.06.031
11. Gregory SP, Dyson PJ, Fletcher D, Gatland P, Shields RJ (2012) Nitrogen removal and changes to microbial communities in model flood/drain and submerged biofilters treating aquaculture wastewater. Aquacult Eng 50:37–45. doi:10.1016/j.aquaeng.2012.03.006
12. Hroudová Z, Zákravský P, Flegrová M (2014) The tolerance to salinity and nutrient supply in four European *Bolboschoenus* species (*B. maritimus*, *B. laticarpus*, *B. planiculmis* and *B. yagara*) affects their vulnerability or expansiveness. Aquat Bot 112:66–75
13. Jesus JM, Calheiros CSC, Castro PML, Borges MT (2014) Feasibility of *Typha latifolia* for high salinity effluent treatment in constructed wetlands for integration in resource management systems. Int J Phytoremediat 16(4):334–346. doi:10.1080/15226514.2013.773284
14. Kadlec RH, Wallace SD (2009) Treatment wetlands, 2nd edn. CRC Press, Boca Raton
15. Kanagy LE, Johnson BM, Castle JW, Rodgers JH Jr (2008) Design and performance of a pilot-scale constructed wetland treatment system for natural gas storage produced water. Bioresour Technol 99:1877–1885. doi:10.1016/j.biortech.2007.03.059
16. Keizer-Vlek HE, Verdonschot PFM, Verdonschot RCM, Dekkers D (2014) The contribution of plant uptake to nutrient removal by floating treatment wetlands. Ecol Eng 73:684–690. doi:10.1016/j.ecoleng.2014.09.081
17. Klomjek P, Nitisoravut S (2005) Constructed treatment wetland: a study of eight plant species under saline conditions. Chemosphere 58(5):585–593. doi:10.1016/j.chemosphere.2004.08.073
18. Laegdsgaard P (2006) Ecology, disturbance and restoration of coastal saltmarsh in Australia: a review. Wetl Ecol Manag 14(5):379–399
19. Lin YF, Jing SR, Lee DY, Chang YF, Sui HY (2010) Constructed wetlands for water pollution management

of aquaculture farms conducting earthen pond culture. Water Environ Res 82(8):759–768. doi:10.2175/106143010X12609736966685

20. Lymbery AJ, Doupé RG, Bennett T, Starcevich MR (2006) Efficacy of a subsurface-flow wetland using the estuarine sedge *Juncus kraussii* to treat effluent from inland saline aquaculture. Aquacult Eng 34:1–7. doi:10.1016/j.aquaeng.2005.03.004
21. Lymbery AJ, Kay GD, Doupé RG, Partridge GJ, Norman HC (2013) The potential of a salt-tolerant plant (*Distichlisspicata* cv. NyPa Forage) to treat effluent from inland saline aquaculture and provide livestock feed on salt-affected farmland. Sci Total Environ 445:192–201
22. Meerburg B, Vereijken P, de Visser W, Verhagen J, Korevaar H, Querner E, de Blaeij A, van der Werf A (2010) Surface water sanitation and biomass production in a large constructed wetland in the Netherlands. Wetl Ecol Manag 18(4):463–470. doi:10.1007/s11273-010-9179-x
23. Quintã R, Hill PW, Jones DL, Santos R, Thomas DN, LeVay L (2015a) Uptake of an amino acid (alanine) and its peptide (trialanine) by the saltmarsh halophytes *Salicornia europaea* and*Aster tripolium* and its potential role in ecosystem N cycling and marine aquaculture wastewater treatment. Ecol Eng 75:145–154
24. Quintã R, Santos R, Thomas DN, le Vay L (2015b) Growth and nitrogen uptake by *Salicornia europaea* and *Aster tripolium* in nutrient conditions typical of aquaculture wastewater. Chemosphere 120:414–421
25. Roncken PA, Stremke S, Paulissen MPCP (2011) Landscape machines: productive nature and the future sublime. J Landscape Archit (spring): 68–81. http://edepot.wur.nl/195288
26. Rozema J, Flowers T (2008) Crops for a salinized world. Science 322(5907):1478–1480. doi:10.1126/science.1168572
27. Rozema J, Van Diggelen J (1991) A comparative study of growth and photosynthesis of four halophytes in response to salinity. Acta Oecol 12(5):673–681
28. Shpigel M, Ben-Ezra D, Shauli L, Sagi M, Ventura Y, Samocha T, Lee JJ (2013) Constructed wetland with *Salicornia* as a biofilter for mariculture effluents. Aquaculture 412–413:52–63. doi:10.1016/j.aquaculture.2013.06.038
29. Tao R, Yu KW (2013) Nitrate addition has minimal effect on anaerobic biodegradation of benzene in coastal saline (salt), brackish and freshwater marsh sediments. Wetlands 33(4):759–767. doi:10.1007/s13157-013-0435-8
30. Tilley DR, Badrinarayanan H, Rosati R, Son J (2002) Constructed wetlands as recirculation filters in large-scale shrimp aquaculture. Aquacult Eng 26(2):81–109. doi:10.1016/S0144-8609(02)00010-9
31. Verhoeven JTA, Meuleman AFM (1999) Wetlands for wastewater treatment: opportunities and limitations. Ecol Eng 12:5–12. doi:10.1016/S0925-8574(98)00050-0
32. Webb JM, Quintã R, Papadimitriou S, Norman L, Rigby M, Thomas DN, Le Vay L (2012) Halophyte filter beds for treatment of saline wastewater from aquaculture. Water Res 46:5102–5114. doi:10.1016/j.watres.2012.06.034
33. Webb JM, Quintã R, Papadimitriou S, Norman L, Rigby M, Thomas DN, le Vay L (2013) The effect of halophyte planting density on the efficiency of constructed wetlands for the treatment of wastewater from marine aquaculture. Ecol Eng 61:145–153. doi:10.1016/j.ecoleng.2013.09.058
34. Ye Y, Tam NFY, Wong YS (2001) Livestock wastewater treatment by a mangrove pot-cultivation system and the effect of salinity on the nutrient removal efficiency. Mar Pollut Bull 42(6):513–521. doi:10.1016/S0025-326X(00)00196-X

CHAPTER 13

GROWTH AND NUTRIENT REMOVAL PROPERTIES OF THE DIATOMS, CHAETOCEROS CURVISETUS AND C. SIMPLEXUNDER DIFFERENT NITROGEN SOURCES

Panneerselvam Karthikeyan, Kuppusamy Manimaran, Pitchai Sampathkumar, Lakshmanan Rameshkumar

Centre of Advanced Study in Marine Biology, Faculty of Marine Sciences, Annamalai University, Parangipettai 608 502, Tamil Nadu, India

ABSTRACT

To investigate the suitability of the marine diatoms, *Chaetoceros curvisetus* and *C. simplex* for the removal of macronutrients from different wastewater, the growth and nitrate-phosphate removal properties were studied with nitrate, ammonium and urea nitrogen sources. Three separate experiments were conducted using modified F/2 medium with 12.35 mg L^{-1} total nitrogen and 1.12 mg L^{-1} total phosphorous (simulating the typical concentration of nitrogen and phosphorus in secondary effluent) as growth medium. The maximum cell densities of *C. curvisetus* and *C. simplex* were $7.16 \pm 0.34 \times 10^4$ cells mL^{-1} in NO−3NO3− and $3.88 \pm 0.32 \times 10^5$ cells mL^{-1} in urea, respectively. The maximum chlorophyll *a* per cell was 1.7 and 4.7 pg for *C. simplex* and *C. curvisetus,* cultured with urea and nitrate, respectively. The high protein contents of 4.7 pg $cell^{-1}$ in *C. simplex* with urea and 19.7 pg $cell^{-1}$ in *C. curvisetus*nitrate nitrogen sources were found. The higher cell density and protein content of both species from urea and nitrate nitrogen sources ($p < 0.05$) have shown that these were utilized by microalgae and were converted to protein. The *C. simplex* and *C. curvisetus* showed maximum removal efficiencies of nitrate by 97.86 and 91.62 % and phosphate by 98.5 and 100 %, respectively when urea used as nitrogen source than ammonia. The results indicated the *C. simplex* was more efficient than *C. curvisetus* and suitable for the removal of macronutrients when cultured with urea and nitrate nitrogen sources.

INTRODUCTION

The input of nutrients especially nitrogen and phosphorus by human activities increased into biochemical cycles through agricultural practices, urbanization, industrialization and other activities. The widespread use of nitrogen fertilizers, following discharges of agricultural and industrial effluents resulted in the pollution of aquatic environment and is the main cause of eutrophication (Karthikeyan et al. 2010; Manimaran et al. 2011). Nutrient enrichment or eutrophication of aquatic ecosystems can cause an increase in algae and aquatic plants, loss of component species and loss of ecosystem function. In connection with these problems, numerous studies have aimed on nitrogen and phosphorus removal from wastewater (Thakur and Kumar 1999). Most of these studies are based on biological processes and different combinations of anaerobic, aerobic and anoxic zones (Aslan and Kapdan 2006). It was

Citation: Panneerselvam Karthikeyan, Kuppusamy Manimaran, Pitchai Sampathkumar, Lakshmanan Rameshkumar, Growth and nutrient removal properties of the diatoms, Chaetoceros curvisetus and C. simplexunder different nitrogen sources DOI: 10.1007/s13201-012-0056-z reported that nitrate and phosphate could be effectively taken up by photosynthetic microorganisms namely, cyanobacteria and green algae. The phytoplankton has many advantages for application to nitrate and phosphate removal than macrophytes. It proliferates throughout the entire year with high growth rates when environmental conditions such as light, nutrients and water temperature are suitable (Alexander and Goldman 1994). Free-living and immobilized cells of microalgae have been extensively studied and used for the removal of the inorganic compounds from water (Garbisu et al. 1992; Urrutia et al. 1995; Gonzalez-Bashan et al. 2000; Tam and Wong 2000).

The advantages of using algae includes the low cost of the operation, the possibility of recycling assimilated nitrogen and phosphorus into algae biomass as a fertilizer, avoiding a sludge handling problem and the discharge of oxygenated effluent into the water body (Hu et al. 2000). In addition, the process has no carbon requirement for nitrogen and phosphorus removal, which is attractive for the treatment of secondary effluents (Aslan and Kapdan 2006). The concentrations of nitrate, phosphate in different sources (urea, nitrate and ammonia) in the secondary effluents vary depending on the wastewater characteristics (Aslan and Kapdan 2006). The total nitrogen concentration was reported as 788 mg L^{-1} in pretreated piggery wastewater by An et al. (2003) and around 1,450 mg L^{-1} by Olguin et al. (2003). The average values of $NH^{+}{}_{4}NH_{4}^{+}$and $PO{-}4PO_{4}{-}$ were 27.4 mg L^{-1} and 11.8 mg L^{-1}, respectively in the conventional secondary treatment effluent of domestic wastewater (Martinez et al. 2000) whereas they were 48 mg $NH^{+}{}_{4}NH_{4}^{+}L^{-1}$and 16 mg PO−4PO4− L^{-1} in the effluent of up-flow sludge blanket reactor, fed with domestic sewage (Van der Steen et al. 1999). The wastewater treatment with microalgae was reported in the use of high-rate algae ponds (HRAP) (Cromar et al. 1996; Deviller et al. 2004). Recently corrugated raceways (Olguin et al. 2003), a triangular photobioreactor (Dumas et al. 1998) and a tubular photobioreactor (Cattaneo et al. 2003) have been developed for nutrient removal. The microalgae *Chlorella* sp., *Scenedesmus* sp., *Spirulina* sp., *Nannochloropsis* sp. and *Phormidium*sp. have been widely investigated and used for nutrient removal from wastewater (Lee and Lee2001; Gonzales et al. 1997; Martinez et al. 1999; 2000; Olguin et al. 2003; Jimenez-Perez et al.2004; Laliberte et al. 1997; Dumas et al. 1998).

Many studies have focused on the objective of optimization of nitrogen and phosphorous removal and environmental factors influencing on microalgal growth, such as nutrient concentration (Aslan and Kapdan 2006), nutrient type (Hyenstrand et al. 2000), light availability (Janssen et al. 2000) and pH (Xin et al. 2010). Wastewater may contain different forms of nitrogen (nitrate, ammonium, urea, etc.), so it is important to know the effect of different nitrogen compounds on microalgal growth and nutrient removal (Xin et al. 2010). Among the phytoplankton, diatoms are contributing higher diversity and biomass production in marine environment and not well recognized for bioremediation purposes. The centric diatoms,*C. curvisetus* and *C. simplex* were studied extensively for their biochemical composition and used for larval diet in aquaculture practices. But the reports are very limited in the use of diatoms for wastewater treatment. Hence the present investigation focused on the objective of growth, biochemical composition and NP removal properties of these two marine diatoms with three different nitrogen sources viz. nitrate, urea and ammonia.

MATERIALS AND METHODS

Microalgae

The marine centric diatoms, *C. curvisetus* and *C. simplex* were isolated by serial dilutions in F/2 media from Vellar estuary, Southeast coast of India. The stock cultures were maintained in Guillard F/2 media following the methods described in Andersen (2005).

Experimental setup

The modified F/2 medium with 12.35 mg L^{-1} TN and 1.12 mg L^{-1} TP (simulating the typical concentration of nitrogen and phosphorus in secondary effluent) was used as growth medium. In all experiments, $K_2HPO_4{\cdot}3H_2O$ (5.6 mg L^{-1}) was used as phosphorus source. Besides nitrogen and phosphorus, the composition of other elements was the same with F/2 medium.

All the experiments were conducted in 5,000-ml Erlenmeyer flasks with 3,000-ml algal cultures in three replicates. The subcultures of *C. simplex* and *C. curvisetus* were maintained with three kinds of nitrogen sources, $NaNO_3$ (75 mg L^{-1} and 24.72 mg L^{-1} of $NaNO_3$ and TN, respectively), NH_4Cl (47 mg L^{-1} and 24.72 mg L^{-1} of NH_4Cl and TN, respectively) and urea (53 mg L^{-1} and 22.35 mg L^{-1} of urea and TN, respectively) for 7 days. The cell density was estimated at every 24 h intervals; the total nitrogen and phosphorous were analysed in culture medium, before inoculation of algal cells and after 7 days of culture. The protein and carbohydrate contents were estimated at the end of the experiment.

Growth Rate

The growth rate was calculated using the following formula (OECD 2002),

$$\mu = N_x - N_0/t_x - t_0$$

Where N_0 is the number of cells in time zero, N_x is the number of cells in time *x*, t_0 is the starting time (0) and t_x is the time *x* (in days).

Doubling Time

The doubling time was calculated by the following formula (Karthikeyan et al. 2011),

$$\text{Doubling time} = (N_0 \times 2)/(N_t) \times t$$

where, N_0 is the number of cells in time zero, N_t is the number of cell in time *t*, *t* is the time in hours. The results were presented in doubling time (DT) in hours.

Parameters Maintained

During isolation, stock culture maintenance and experiments were conducted; the parameters viz. temperature, light intensity, photoperiod and salinity were maintained as 24 ± 2 °C, 4500 ± 500 Lux, 14 ± 1:8 ± 1 h (light:dark) and 30 psu, respectively.

Analytical Methods

The total nitrogen and total phosphate were estimated at the beginning and the end of the experiment in cell-free media by centrifugation followed by Strickland and Parsons (1972). After 7 days of culture, the cells were isolated by centrifugation and the pellet used for further analysis of protein and carbohydrate. The protein was estimated by Folin's phenol reagent method (Lowry et al. 1951) and the carbohydrate was estimated by phenol sulphuric acid method (Dubois et al. 1956).

Data Analysis

Three replicates of each sample were used for statistical analysis and the values were reported as mean ± SD. One-way ANOVA and Duncan's multi-variant analysis were carried out using SPSS, version 16.0 software to study the difference between nitrogen sources and species.

RESULTS

The growth curves of *C. curvisetus* and *C. simplex* with different nitrogen sources are shown in Figure 1. The *C. curvisetus* was grown fastest in NO−3NO3− followed by urea whereas *C. simplex* was grown fastest in urea followed by NO−3NO3−. The organisms grew slowest on ammonia. The maximum cell density of 7.16 ± 0.34 × 10^4 cells mL^{-1} (growth rate 1.16 μ day^{-1}) and 3.88 ± 0.32 × 10^5 cells mL^{-1} (growth rate 0.59 μ day^{-1}) of *C. curvisetus* and *C. simplex* were found in NO−3NO3− and urea, respectively (Figure. 1 and Table 2). The maximum chlorophyll *a* per cell was 1.7 and 4.7 pg for *C. simplex* and *C. curvisetus* which were cultured with urea and nitrate, respectively. The *Chl a* of 3.9 pg $cell^{-1}$ was observed in urea used as nitrogen source in the medium for *C. curvisetus* and 1.6 pg $cell^{-1}$ for *C. simplex* in nitrate (Figure 5). The protein content of *C. simplex* did not showed any significant variations with urea and nitrate nitrogen sources (4.7 pg $cell^{-1}$ in urea and 4.6 pg $cell^{-1}$ in NO−3NO3−) where it was 3.4 pg $cell^{-1}$ in ammonia nitrogen. The maximum protein of 19.7 and 16.8 pg $cell^{-1}$ was found in *C. curvisetus* cultured with nitrate and urea, respectively. The carbohydrate contents of *C. simplex* were 0.86, 0.78 and 0.62 pg $cell^{-1}$, whereas it was 3.6, 4.5 and 0.4 pg $cell^{-1}$ in *C. curvisetus* with urea, nitrate and ammonia nitrogen, respectively (Figure. 6). The order of specific growth rate, chlorophyll, protein and carbohydrate contents with different nitrogen sources were NH+4NH4+ < urea < NO−3NO3− of *C. curvisetus* and NH+4NH4+ < NO−3NO3− < urea of *C. simplex,* respectively (Table 1; Figures. 2, 3, 4).

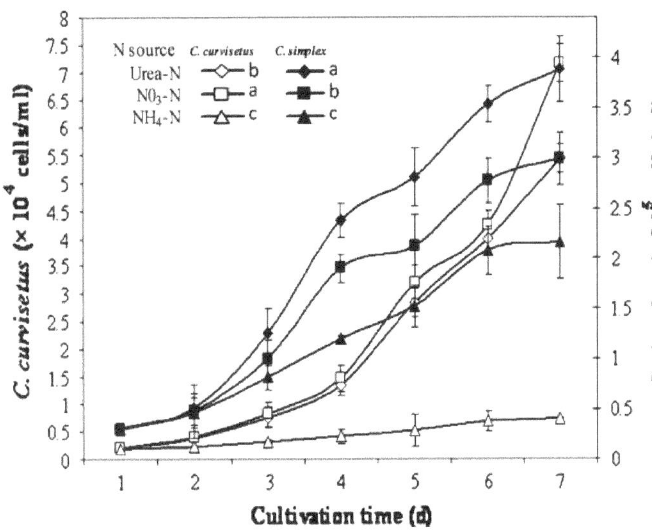

Figure. 1: Growth of *C. curvisetus* and *C. simplex* under three different nitrogen sources. Each value represents mean ± SD (n = 3). Each species with different *alphabet letters* are significantly different ($p \leq 0.05$)

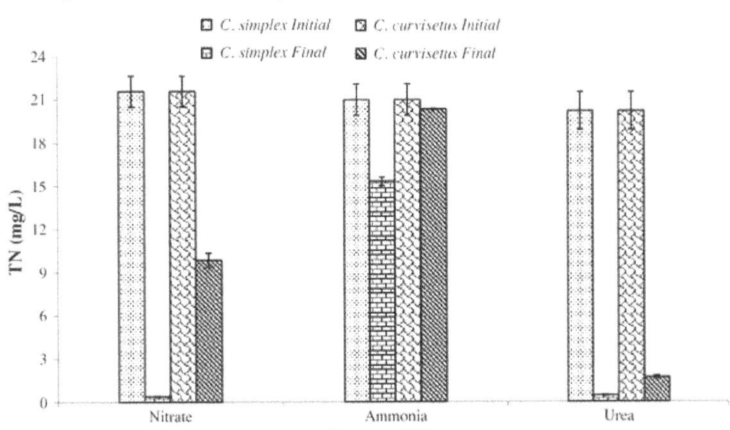

Figure. 2: Nitrogen removal properties of *C. simplex* and *C. curvisetus* under three different nitrogen sources. Each value represents mean ± SD (n = 3)

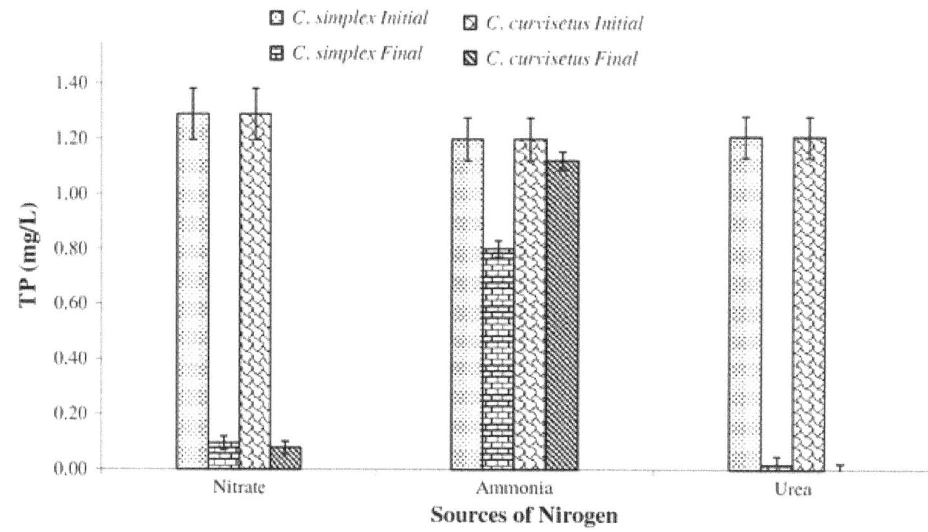

Figure. 3: Phosphate removal properties of *C. simplex* and *C. curvisetus* under three different nitrogen sources.

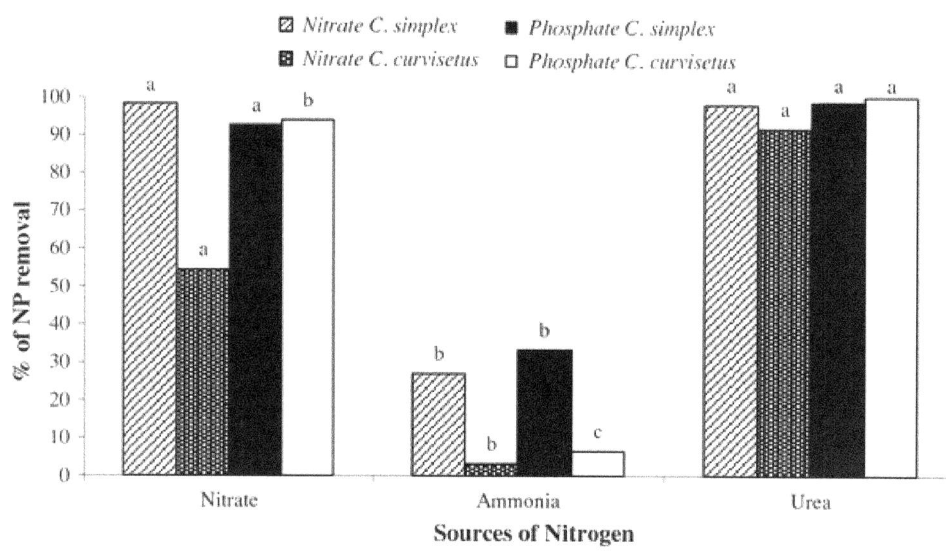

Figure. 4: Variations of Nitrate Phosphate removal of *C. simplex* and *C. curvisetus* under three different nitrogen sources. Each nitrogen sources (*bars*) with different *alphabet letters* of species are significantly different ($p \leq 0.05$).

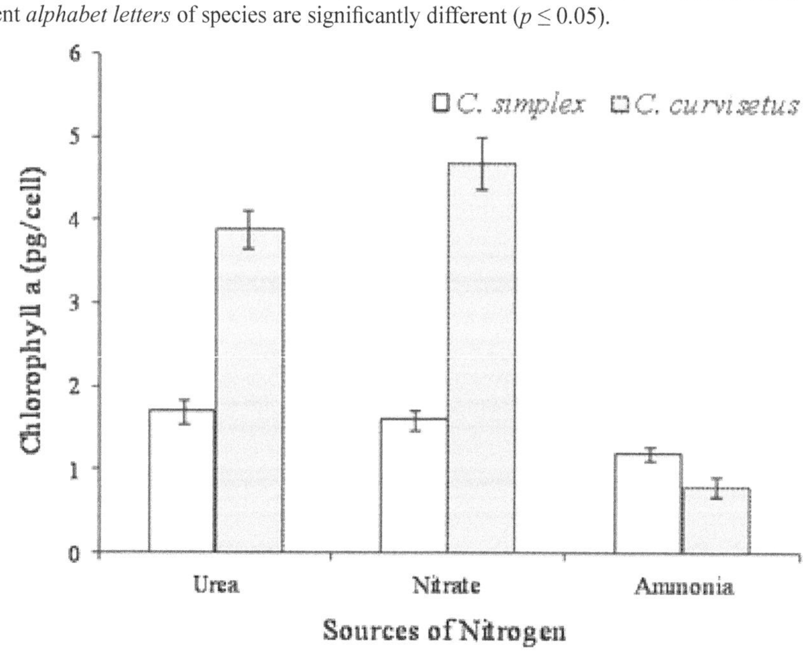

Figure. 5: Chlorophyll *a* concentration of *C. simplex* and *C. curvisetus* under three different nitrogen sources.

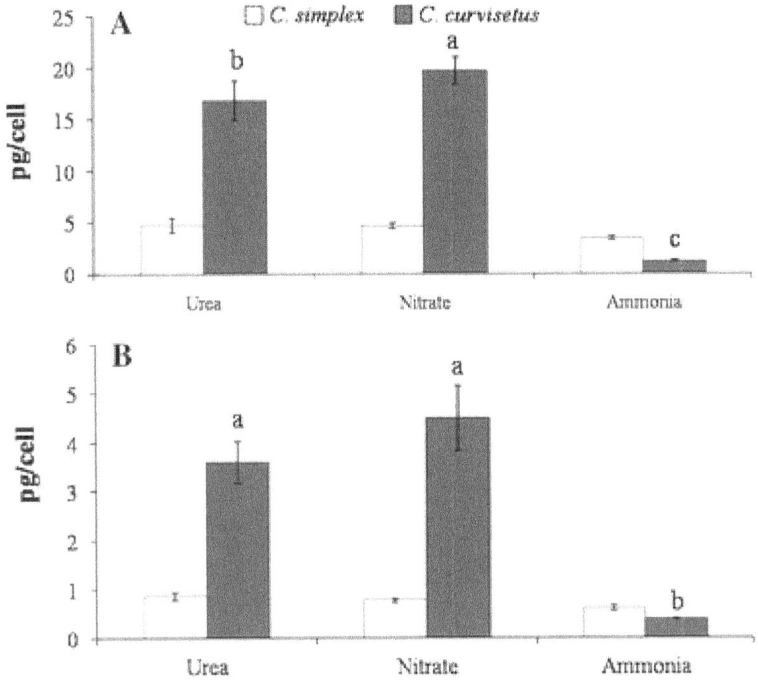

Figure. 6: Protein (**a**) and carbohydrate (**b**) contents of *C. simplex* and *C. curvisetus* under different nitrogen sources. Each nitrogen sources (*bars*) with different *alphabet letters* are significantly different ($p \leq 0.05$).

Table 1: The order of specific growth rate, N:P removal efficiency, protein and carbohydrate contents of the diatom with different nitrogen sources

	C. curvisetus	*C. simplex*
Growth rate	NH+4NH4+ < urea < NO−3NO3−	NH+4NH4+ < NO−3NO3− < urea
Chlorophyll	NH+4NH4+ < urea < NO−3NO3−	NH+4NH4+ < NO−3NO3− < urea
Nitrate removal	NH+4NH4+ < NO−3NO3− < urea	NH+4NH4+ < urea < NO−3NO3−
Protein content	NH+4NH4+ < urea < NO−3NO3−	NH+4NH4+ < NO−3NO3− < urea
Carbohydrate content	NH+4NH4+ < urea < NO−3NO3−	NH+4NH4+ < NO−3NO3− < urea

Table 2: Growth rate of *C. curvisetus* and *C. simplex* with different nitrogen sources

	C. curvisetus	*C. simplex*
Urea–N	0.87 ± 0.037^{b}	0.59 ± 0.012^{a}
NO−3NO3−–N	1.16 ± 0.045^{a}	0.45 ± 0.013^{b}
NH+4NH4+–N	0.09 ± 0.011^{c}	0.31 ± 0.01^{c}

Different superscripts in column are significantly different ($p \leq 0.05$)

The capacity of nitrate and phosphate removal differed between the species as well as among the nitrogen sources used (Figure. 4). The observations in the present study showed that both species performed well in the removal of both N and P when urea used as nitrogen source in culture medium. Next to urea, the nitrate was observed as a good source of nitrogen for each species in the removal of N and P. The removal of nitrate was 97.86 and 91.62 %; the phosphate was 98.5 and complete removal by *C. simplex* and *C. curvisetus,* respectively cultured with urea source for 7 days (Figure. 3). When compared to urea and nitrate the slow growth of the both species was observed with ammonia (Figure. 1). The order of nitrate and phosphate removal capacities with different nitrogen sources was NH+4NH4+ < NO−3NO3− < urea of *C. curvisetus* and NH+4NH4+ < urea < NO−3NO3− of *C. simplex* respectively (Table 1). The one-way ANOVA results that showed the nitrogen sources and the two species were significantly different for their growth, nitrate phosphate removal and biochemical compositions.

DISCUSSION

The microalgae could be used for the removal of phosphate and nitrate from wastewater. In this present study, the order of specific growth rate, chlorophyll, protein and carbohydrate contents with different nitrogen sources was NH+4NH4+ < urea < NO−3NO3− of *C. curvisetus* and NH+4NH4+ < NO−3NO3− < urea of *C. simplex,* respectively (Table 1). The order of growth rate and nutrient removal of *C. curvisetus* was similar to *Scenedesmus* sp. LX1 after 13 days of cultivation with nitrate, ammonium and urea as nitrogen sources (Xin et al. 2010). But the *C. simplex* was grown fastest with urea and the higher growth rate was observed in the present study from urea nitrogen (Figure.2). When urea used as a nitrogen source, the initial pH of the medium is slightly high compared to other nitrogen sources. This might be the reason for highest growth rate of *C. simplex* in urea (Garcia et al. 2006; Xin et al. 2010). Other than this reason, the *C. simplex* is very smaller in cell size and bio-volume than *C. curvisetus*. So the growth rate and doubling time of *C. simplex* were higher than *C. curvisetus* in this present study. The algal growth in the present study with ammonium nitrogen was significantly low. However, the ammonium is good nitrogen source for the phytoplankton growth, the unfavorable concentration causes growth inhibition due to toxicity and/or acidifying the culture media during algal growth (Garcia et al. 2006; Xin et al.2010).

After 7 days of cultivation with nitrate, ammonium and urea as nitrogen sources, TN removal efficiencies of *C. simplex* was significantly high with urea and nitrate nitrogen. But, *C. curvisetus* efficiently removed the TN from the media in the presence of nitrate nitrogen. The TP removal capacities were also observed in the same trend as TN removal. The TN and TP removal capacities of the diatoms in this study were slightly lower than *Scenedesmus* sp. LX1 after 13 days cultivation (Xin et al. 2010). However, the nutrient removal efficiencies of these two diatoms were higher than *Chlorella* spp. and *Phormidium bohneri* (Dumas et al. 1998; Lee and Lee 2001; Valderrama et al. 2002). The growth and nutrient uptake properties of these two species significantly varied because each species require unique suitable environmental conditions such as light intensity, temperature and salinity, etc. (Aslan and Kapdan 2006). From the available literatures, the different kinds of biomolecules have been produced in a single species of microalgae with changing environmental parameters.

Aslan and Kapdan (2006) found that the NH+4NH4+ was completely removed from the media by*Chlorella vulgaris* when the initial concentration was between 13.2 and 21.2 mg L^{-1}. Further, the removal efficiency ofNH+4NH4+ was decreased to less than 24 % when the NH+4NH4+ concentration was higher than 129 mg L^{-1}. In the present study, the initial NH+4NH4+ concentration was 47 mg $L^{-1}$and the nitrate removal was 26.9 and 3.1 % by *C. simplex* and *C. curvisetus,* respectively.

The chlorophyll content of *C. simplex* and *C. curvisetus* estimated in this study was slightly varied with cell density and nitrogen sources. The nutrient concentrations also influenced the chlorophyll *a* concentration and light intensity with major significance (Sciandra et al. 2000; Young and Beardall 2003; Lafarga-De la Cruz et al. 2006; Eker-Develi et al. 2006).

The protein and carbohydrate contents of *C. simplex* and *C. curvisetus* in this present study are under the range of reported amounts in other diatoms like *Chaetoceros calcitrans,C. gracilis, Nitzchia closterium, Phaeodactylum tricornutum, Skeletonema costatum, Thalassiosira pseudonana* and *Tetraselmis chui* by Brown (1991). So, the biomass produced after wastewater treatment could be used for many purposes after required precautions such as sterilization, etc.

CONCLUSIONS

Further optimization of other environmental parameters can improve the nutrient removal efficiency of *C. curvisetus.* If ammonium was the main nitrogen source in wastewater, a buffer solution should be added to keep pH above 7.5 for algal growth and nutrient removal.

ACKNOWLEDGMENTS

We are thankful to Dr. T. Balasubramanian, Dean, CAS in Marine Biology, Faculty of Marine Sciences, Annamalai University, Parangipettai and MoES-ICMAM to carry out the present work for providing facilities and fellowship to Mr. P. Karthikeyan, Mr. K. Manimaran.

REFERENCES

1. Alexander JH, Goldman CR (1994) Limnology, 2nd edn. Mcgraw-Hill, New York, pp 193–263
2. An JY, Sim SJ, Lee JS, Kim BK (2003) Hydrocarbon production from secondary treated piggery wastewater by the green algae. Botrycoccus braunii. J Appl Phycol 15:185–191
3. Andersen RA (2005) Algal culturing techniques. Elsevier Academic Press, Burlington. ISBN 0-12-088426-7
4. Aslan S, Kapdan IK (2006) Batch kinetics of nitrogen and phosphorus removal from synthetic wastewater by algae 'Dokuz eylu' University, Department of Environmental Sebnem Engineering, Kaynaklar Campus, 35160 Buca-IZMIR. Turk ecol eng 28:64–70
5. Brown MR (1991) The amino acid and sugar composition of 16 species of microalgae used in mariculture. Aquacult 145:79–99
6. Cattaneo C, Zolezzi M, Rovatti P, Fabiano M, Povero P, Rovatti M (2003) Growth and nutrient uptake of microalgae (*Nannochloropsis* sp.) in a tubular photo bioreactor. In: Pierucci S (ed) IcheaP-6 the six italian conference on chemical and process engineering florance, vol 1, pp 1677–1682
7. Cromar NJH, Fallowfield J, Martin NJ (1996) Influence of environmental parameters on biomass production and nutrient removal in a high rate algal pond operated by continuous culture. Water Sci Technol 34:133–140
8. Deviller G, Aliaume C, Nava MAF, Casellas C, Blancheton JP (2004) High-rate algal pond treatment for water reuse in an integrated marine fish recirculation system: effect on water quality and sea bass growth. Aquaculture 235:331–344
9. Dubois M, Gillies KA, Hamilton JK, Rebers F, Smith PA (1956) Calorimetric method for determination of sugars and related substances. Anal Chem 28:350–356
10. Dumas A, Laliberte G, Lessard P, Noue J (1998) Biotretament of fish farm effluents using the cyanobacterium *Phormidium bohneri*. Aquacult Eng 17:57–68
11. Eker-Develi E, Kideys AE, Tugrull S (2006) Effect of nutrients on culture dynamics of marine phytoplankton. Aquat Sci 68:28–39
12. Garbisu C, Hall DO, Serra JL (1992) Nitrate and nitrite uptake by free-living and immobilized N-starved cells of *Phormidium laminosum*. J Appl Phycol 4:139–148
13. Garcia J, Green B, Oswald W (2006) Long term diurnal

variations in contaminant removal in high rate ponds treating urban wastewater. Bioresour Technol 97:1709–1715

14. Gonzales LE, Canizares RO, Baena S (1997) Efficiency of ammonia and phosphorus removal from a Colombian agroindustrial wastewater by the microlagae *Chlorealla vulgaris* and*Scenedesmus dimorphus*. Bioresour Technol 60:259–262
15. Gonzalez-Bashan LE, Lebsky VK, Hernandez JP, Bustillos JJ, Bashan Y (2000) Changes in the metabolism of the microalga *Chlorella vulgaris* when coimmobilized in alginate with the nitrogen-fixing *Phyllobacterium myrsinacearum*. Can J Microbiol 46:653–659
16. Hu Q, Westerhoff P, Vermaas W (2000) Removal of nitrate from groundwater by cyanobacteria: quantitative assessment of factors influencing nitrate uptake. Appl Environ Microbiol 66(1):133–139
17. Hyenstrand P, Burkert U, Pettersson A, Blomqvist P (2000) Competition between the green alga *Scenedesmus* and the cyanobacterium *Synechococcus* under different modes of inorganic nitrogen supply. Hydrobiologia 435:91–98
18. Janssen M, Bresser LD, Baijens T, Tramper J, Mur LR, Snel JFH, Wijffels RH (2000) Scale-up aspects of photobioreacters: effects of mixing-induced light/dark cycles. J Appl Phycol 12(3–5):225–237
19. Jimenez-Perez MV, Sanches-Castillo P, Romera O, Fernandez-Moreno D, Perez-Martinez C (2004) Growth and nutrient removal in free and immobilized planktonic green algae isolated from pig manure. Enzyme Microbial Technol 34:392–398
20. Karthikeyan P, Jayasudha S, Sampathkumar P, Manimaran K, Santhoshkumar C, Ashokkumar S, Ashokprabu V (2010) Effect of industrial effluent on marine diatom. *Chaetoceros simplex*(Ostenfeld, 1901). J. Appl. Sci. Environ. Manage. 14(4):35–37
21. Karthikeyan P, Jayasudha S, Manimaran K, Sampathkumar P, Suriya S (2011) Effect of cadmium on marine diatom. *Chaetoceros curvisetus* with different ambient iron concentrations. Asian J of Biochem Pharm Res 2(1):240–248
22. Lafarga-De la Cruz F, Valenzuela-Espinoza E, Millan-Nunez R, Trees CC, Santamaria-del-Angel E, Nunez-Cebrero F (2006) Nutrient uptake, chlorophyll a and carbon fixation by *Rhodomonas*sp. (Cryptophyceae) cultured at different irradiance and nutrient concentrations. Aquacult Eng 35:51–60
23. Laliberte G, Lessard P, Noue J, Sylvestre S (1997) Effect of phosphorus addition on nutrient removal from wastewater with the cyanobacterium *Phormidium bohneri*. Bioresour Technol 59:227–233
24. Lee K, Lee CG (2001) Effect of light/dark cycles on wastewater treatments by microalgae. Biotechnol Bioprocess Eng 6:194–199
25. Lowry OH, Rosebrough NH, Farr AL, Randall RL (1951) Protein measurement with the Folin phenol reagent. J Biol Chem 193:265–275
26. Manimaran K, Karthikeyan P, Ashokkumar S, Ashok Prabu V, Sampathkumar P (2011) Effect of copper on growth and enzyme activity of marine diatom *Odontella mobiliensis*. Bull Environ Contam Toxicol 88(1):30–37
27. Martinez ME, Castillo JM, Yousfi EF (1999) Photoautotrophic consumption of phosphorus by*Scenedesmus obliquus* in a continuous culture. Influence of light intensity. Process Biochem 34:811–818
28. Martinez ME, Sanchez S, Jimenez JM, Yousfi FE, Munoz L (2000) Nitrogen and phosphorus removal from urban wastewater by the microalga Scenedesmus obliquus. 73(3):263–272
29. OECD (2002) Guidelines for the testing of chemicals, *Lemna* sp. Growth inhibition test draft guideline 221. Paris, France
30. Olguin EJ, Galicia S, Mercado G, Perez T (2003) Annual productivity of *Spirulina* (Arthrospira) and nutrient removal in a pig wastewater recycle process under tropical conditions. J Appl Phycol 15:249–255
31. Sciandra A, Lazzara L, Claustre H, Babin M (2000) Response of growth rate, pigment composition and optical properties of *Cryptomonassp.* to light and nitrogen stress. Mar Ecol Prog Ser 201:107–120
32. Strickland JDH, Parsons TR (1972) A practical handbook of seawater analysis. Bull Fish Res Board Can 167:311
33. Tam NFY, Wong YS (2000) Effect of immobilized bead concentrations on wastewater nutrient removal. Environ Pollut 107:145–151
34. Thakur A, Kumar HD (1999) Nitrate, ammonium, and phosphate uptake by the immobilized cells of *Dunaliella salina*. Bull Environ Contam Toxicol 62:70–78
35. Urrutia I, Serra JL, Llama MJ (1995) Nitrate removal from water by *Scenedesmus obliquus*immobilized in polymeric foams. Enzyme Microbial Technol 17:200–205
36. Valderrama LT, DelCampo CM, Rodriguez CM, De-Bashan LE, Bashan Y (2002) Treatment of recalcitrant wastewater from ethanol and citric acid production using the micro alga *Chlorella vulgaris* and macrophyte *Lemna minuscula*. Water Res 36:4185–4192
37. Van der Steen P, Brenner A, Van Buuren J, Oron G (1999) Post-treatment of UASB reactor effluent in an integrated duckweed and stabilization pond system. Water Res 33:615–620
38. Xin L, Hong-ying H, Ke G, Ying-xue S (2010) Effects of different nitrogen and phosphorus concentrations on the growth, nutrient uptake, and lipid accumulation of a freshwater micro alga *Scenedesmus* sp. Bioresour Technol 101:5494–5500
39. Young E, Beardall J (2003) Rapid ammonium and nitrate-induced perturbations to chl 'a' fluorescence in nitrogen-stressed *Dunaliella tertiolecta* (Chlorophyta). J Phycol 39:332–342

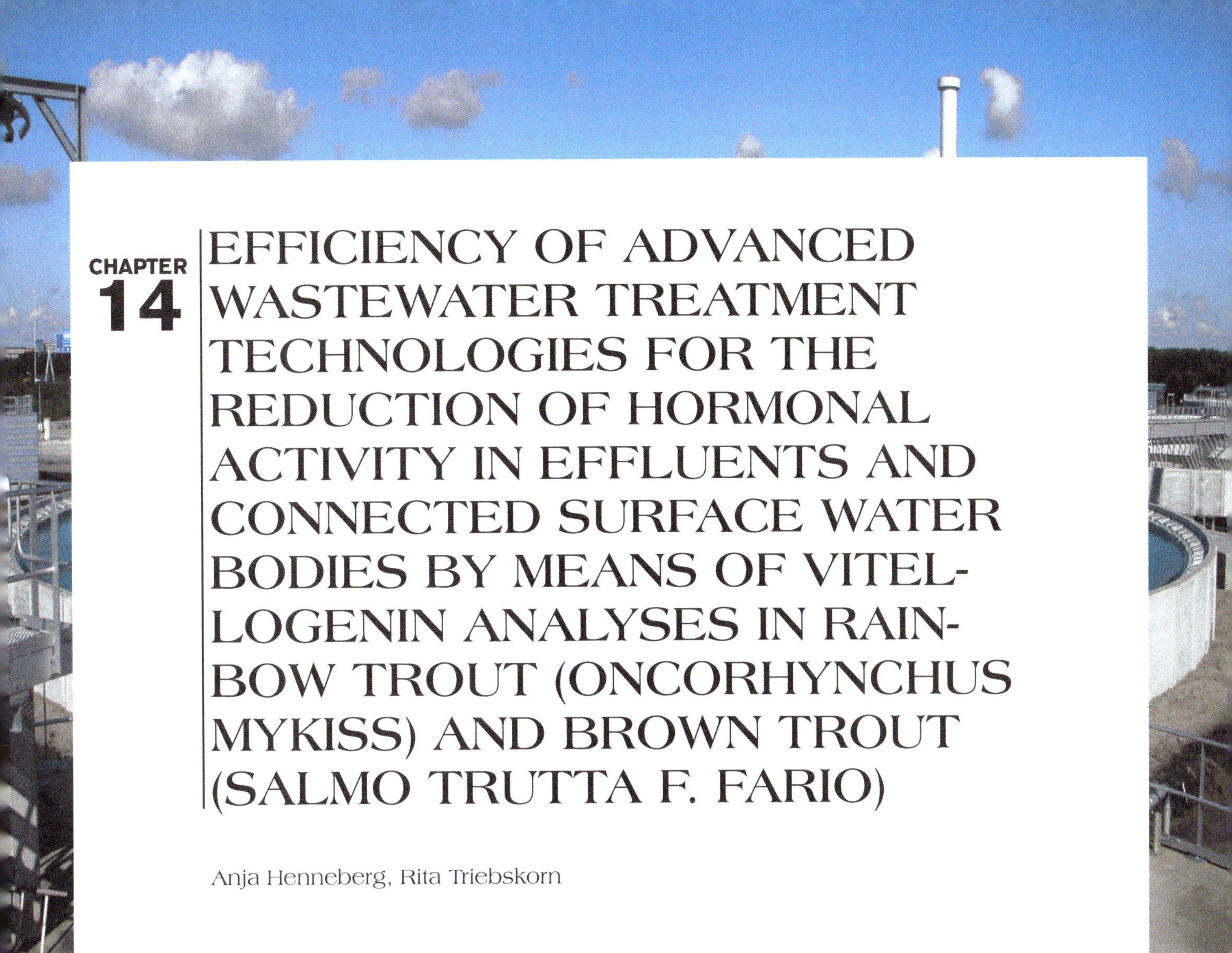

CHAPTER 14

EFFICIENCY OF ADVANCED WASTEWATER TREATMENT TECHNOLOGIES FOR THE REDUCTION OF HORMONAL ACTIVITY IN EFFLUENTS AND CONNECTED SURFACE WATER BODIES BY MEANS OF VITELLOGENIN ANALYSES IN RAINBOW TROUT (ONCORHYNCHUS MYKISS) AND BROWN TROUT (SALMO TRUTTA F. FARIO)

Anja Henneberg, Rita Triebskorn

Animal Physiological Ecology, *University of Tübingen*

ABSTRACT

Endocrine effects in the aquatic environment are in the focus of scientists and media along with debates on the necessity of further steps in wastewater treatment. In the present study VTG responses were compared to evaluate upgrades at wastewater treatment plants (WWTPs). We investigated several advanced sewage treatment technologies at two WWTPs connected to the Schussen, a tributary of Lake Constance, for the reduction of hormonal activity: (1) a powdered activated charcoal filter at the WWTP Langwiese; and (2) a combination of ozonation, sand filter, and granulated activated carbon filter at the WWTP Eriskirch. Rainbow trout and brown trout were either directly exposed to the effluents in aquaria or cages, or in a bypass system flown through by surface water of the Schussen. As a reference, trout were kept in bypass aquaria at the Argen River, which is less influenced by micropollutants. As a biomarker for estrogenicity, we analyzed the yolk precursor protein vitellogenin in immature rainbow trout and brown trout and in trout larvae (100 days post-fertilization) prior to and after the upgrade with the new technologies. Trout of different ages and species were used to detect differences in their sensitivity. At both bypass stations, larvae of brown trout showed significantly higher vitellogenin levels prior to the upgrade compared to negative control levels. Female brown trout exposed at the bypass station downstream of the WWTP showed decreased vitellogenin levels after the upgrade. In 1-year-old immature trout directly exposed to the respective effluents, no significant effects of the upgrades on vitellogenin levels were found. In general, larger effects were observed in brown trout than in rainbow trout, indicating that they are more sensitive test organisms.

BACKGROUND

Endocrine disruptors (EDs) are hormonally active chemicals which are able to influence the endocrine system of organisms by mimicking or repressing the body's own hormones. EDs are a very diverse group of chemicals including, for example, ingredients of personal care products, pharmaceuticals containing steroid hormones, pesticides, plasticizers, dioxins,

Citation: Anja Henneberg, Rita Triebskorn, Efficiency of advanced wastewater treatment technologies for the reduction of hormonal activity in effluents and connected surface water bodies by means of vitellogenin analyses in rainbow trout (Oncorhynchus mykiss) and brown trout (Salmo trutta f. fario), doi:10.1186/s12302-015-0056-3 furans, phenols, alkylphenols, polychlorinated biphenyls, and brominated flame retardants [1, 2]. Still more endocrine-active chemicals were identified over the last years. The priority list of the European Commission contains 564 chemicals that had been suggested by various organizations and published papers as being suspected EDs [1].

Because the aquatic environment is an important sink for natural and anthropogenic chemicals [3], the release of pollutants including EDs into surface waters via wastewater treatment plants (WWTPs) has come into the focus of scientists, authorities, and the public. Today, most wastewater is treated before it is released into bodies of water, but many studies show that not all hazardous chemicals, especially EDs, can be completely removed by routine wastewater treatment (see, e.g., [4]). Therefore, the discharge of wastewater treatment plants into recipient rivers is a main source for EDs to enter the aquatic environment. The level of pollution in rivers is particularly high if the catchment area is highly populated, has industry, or agriculture. Because wastewater can contribute up to 50 % and more of the flow of a river in months with low water [3], the released chemicals can play an important role for the occurring biota. For example, steroid estrogens, like the pharmaceutical ethinyl estradiol (EE2), are known to be extraordinarily active in fish at low to sub-ng/L concentrations [5, 6], and are found in many WWTP effluents at effect concentrations [7, 8].

This raises the question whether we should eliminate more pollutants, especially EDs, to improve wastewater quality. Whereas, for example, the Swiss Federal Government started projects introducing a tertiary treatment step at many of its WWTPs, the discussion whether additional wastewater treatment technologies are ecologically worthwhile is still ongoing [9].

The present study is part of the "SchussenAktiv*plus*" project in the Lake Constance area investigating differently sized WWTPs which were equipped with additional wastewater treatment techniques [10]. Two of them (WWTP Langwiese and WWTP Eriskirch) are in the focus of the present study. To characterize the efficiency of technologies newly introduced at these WWTPs, we investigated vitellogenin (VTG) in juvenile male and female trout as well as in trout larvae as a biomarker of estrogenicity [5, 11–16]. VTG is an egg yolk precursor protein which is normally only produced by female fish. It is estrogen-dependent and EDs can act on hepatic receptors to induce the synthesis of VTG in males and juveniles [11, 17]. We compared VTG levels of trout that were exposed (1) directly to the conventional and modified effluent in aquaria connected to the effluents; (2) upstream and downstream the effluent prior and after the WWTP upgrade; and (3) in bypass systems downstream the WWTP and at a reference river prior and after the WWTP upgrade. Figure 1 gives an overview of these three approaches. It also shows the two WWTPs with their new technologies and summarizes the exposure experiments in the years 2013 and 2014 (for detailed information see methods section).

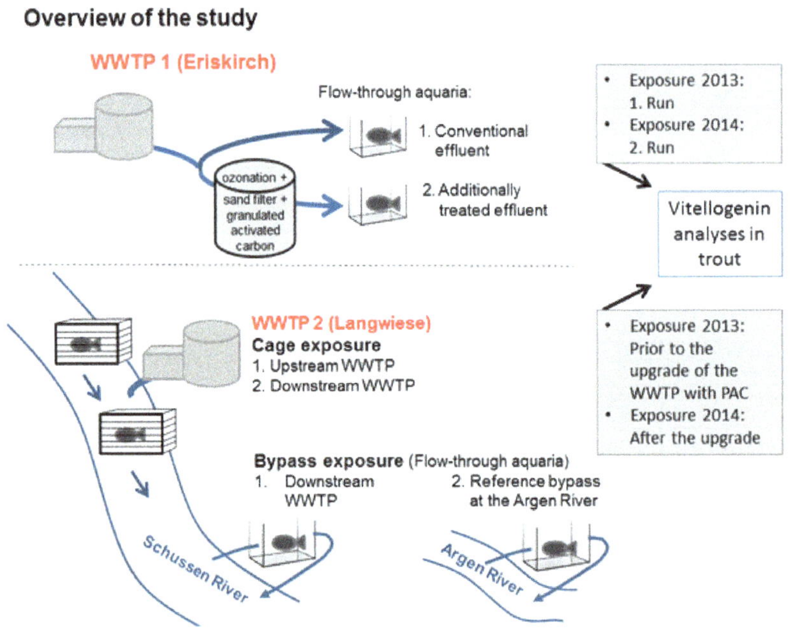

Figure. 1: Overview of the study design. *PAC* powdered activated charcoal.

Results and Discussion

Exposure Experiments at the WWTPs

In rainbow trout exposed at the conventional and modified effluent at the WWTP Eriskirch, VTG levels in females varied between treatments (conventional and additionally treated effluent) and years (exposure in 2013 and 2014), whereas VTG levels in males were constantly low or even non-detectable (significant differences could not be determined) in both years (Figure. 2).

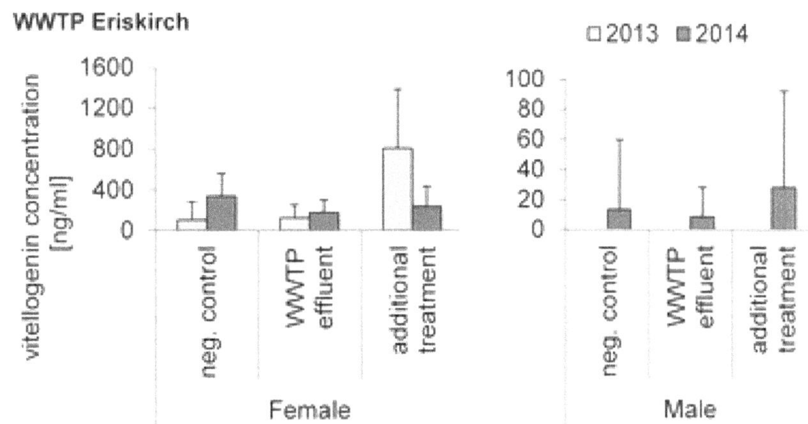

Figure. 2: Vitellogenin concentrations in blood plasma samples of rainbow trout exposed at the WWTP Eriskirch in aquaria connected to the conventional effluent or to the additionally treated effluent in 2013 and 2014; means and standard deviation (SD) are shown. Analyzed by Biosense rainbow trout vitellogenin ELISA kit. *N*-numbers 2013 females: negative control $n = 9$, WWTP effluent $n = 6$, additional treatment $n = 8$; males: negative control $n = 1$, WWTP effluent $n = 7$, additional treatment $n = 3$. No significant differences with Steel–Dwass test; $p > 0.05$. *N*-numbers 2014 females: negative control $n = 6$, WWTP effluent $n = 5$, additional treatment $n = 4$; males: negative control $n = 13$, WWTP effluent$n = 6$, additional treatment $n = 10$. No significant differences with Steel–Dwass test;$p > 0.05$. No significant differences between years; $p > 0.05$

The increased VTG level in females in 2013 after the exposure to additionally treated wastewater might be due to the altered composition of the effluent in 2013 compared to 2014 with more ozone used in 2013 compared to 2014. This possibly could have resulted in the formation of by-products with estrogenic activity [18]—thus leading to higher VTG levels in females. The lacking reactions in male fish, however, indicate that the wastewater at the WWTP Eriskirch was not highly estrogenic in general. In line with this, chemical analyses of the effluent showed only low concentrations of estrogen-active substances (Bisphenol A: 39–110 ng/L in the conventional effluent and 11-160 ng/L after the additional treatment) or concentrations below the detection limit (EE2 > 1 ng/L) [19].

In contrast to our results, a study with crucian carp showed that VTG levels in immature female and male carps were reduced when the wastewater was treated with ozone [20]; however, the carps already had higher VTG induction in the normal effluent compared with controls and our trout did not show higher VTG levels in the normal effluent compared with the negative controls.

Furthermore, we observed slightly higher VTG levels in females of our negative control in 2014 compared to 2013. Differences in VTG baseline levels in negative controls between the years 2013 and 2014 were probably due to the slower fish growth in the laboratory in 2013. In 2013, the mean weight of rainbow trout was 16.3 g ± 2.7 SD and, in 2014, the mean weight was 89 g ± 21.8 SD (Figure. 3). The brown trout showed similar results (Figure. 4). The gonadal development depends on the size of a fish. The bigger the fish the more developed are its gonads, and developed gonads are associated with higher VTG concentrations because the gonads induce the VTG synthesis in liver cells via hormones [21].

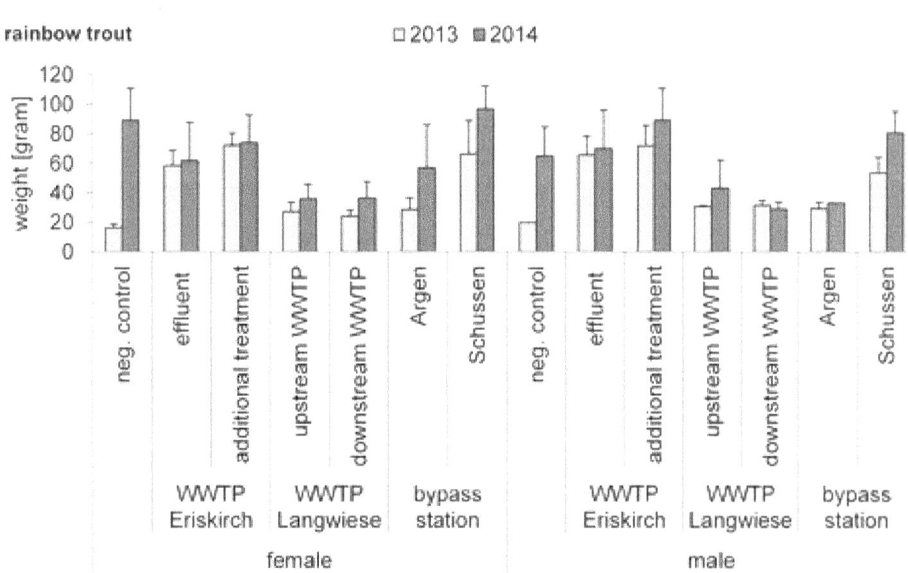

Figure. 3: Means of weight (gram) and SD of exposed rainbow trout in 2013 and 2014. For *n*-numbers see Table 1 and for significant differences see Table 2.

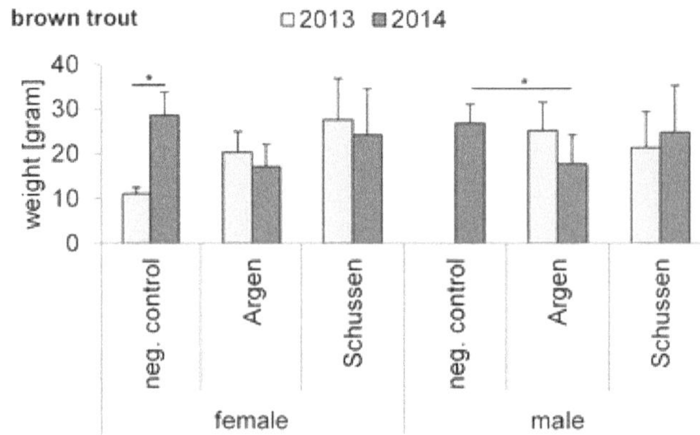

Figure. 4: Means of weight (gram) and SD of exposed brown trout in 2013 and 2014. For *n*-numbers see Table 3. Significant differences with the Tukey–Kramer HSD test: females: neg. control 2013—neg. control 2014 $p = 0.0226$ and males 2014: neg. control—Argen$p = 0.0498$ (*Asterisks* significant differences; $^{*}p < 0.05$)

Table 1: *N*-numbers of exposed rainbow trout in 2013 and 2014

Year	Treatment	*N*-numbers	
		Female	Male
Rainbow trout			
2013	Neg. control	9	1
WWTP Eriskirch	Effluent	6	7
	Additional	8	3
WWTP Langwiese	Upstream WWTP	15	2
	Downstream WWTP	7	4
Bypass	Argen	8	5
	Schussen	4	9
2014	Neg. control	6	13
WWTP Eriskirch	Effluent	5	6
	Additional	4	10
WWTP Langwiese	Upstream WWTP	9	11
	Downstream WWTP	13	8
Bypass	Argen	8	1
	Schussen	6	16

Table 2: Significant differences in weights of rainbow trout

Year	Treatment group	*p*value
Females		
2013	Neg. control 2013—additional treatment WWTP Eriskirch 2013	0.041
	Neg. control 2013—upstream WWTP Langwiese 2013	0.0080
2014	Downstream WWTP Langwiese 2014—bypass Schussen 2014	0.0465
2013 vs 2014	Neg. control 2013—upstream WWTP Langwiese 2014	0.0372
	Neg. control 2013—downstream WWTP Langwiese 2014	0.0081
	Upstream WWTP Langwiese 2013—additional treatment WWTP Eriskirch 2014	0.0092
	Upstream WWTP Langwiese 2013—effluent WWTP Eriskirch 2014	0.0451
	Upstream WWTP Langwiese 2013—neg. control 2014	0.035
	Upstream WWTP Langwiese 2013—bypass Schussen 2014	0.035
	Neg. control 2013—bypass Argen 2014	0.041
	Additional treatment WWTP Eriskirch 2013—upstream WWTP Langwiese 2014	0.0407
	Additional treatment WWTP Eriskirch 2013—downstream WWTP Langwiese 2014	0.0139
Males		

2014	Neg. control 2014—downstream WWTP Langwiese 2014	0.0138
	Additional treatment WWTP Eriskirch 2014—upstream WWTP Langwiese 2014	0.0411
	Additional treatment WWTP Eriskirch 2014—downstream WWTP Langwiese 2014	0.0299
	Bypass Schussen 2014—downstream WWTP Langwiese 2014	0.0076
	Bypass Schussen 2014—upstream WWTP Langwiese 2014	0.0201
2013 vs 2014	Bypass Schussen 2013—bypass Schussen 2014	0.0331

Data were logarithmised to get homoscedastic data and the Steel–Dwass test revealed the following p values

Table 3: *N*-numbers of exposed brown trout in 2013 and 2014

Year	Treatment	*N*-numbers	
		Female	**Male**
Brown trout			
2013	Neg. control	4	–
	Argen	4	4
	Schussen	3	4
2014	Neg. control	6	10
	Argen	6	13
	Schussen	9	5

The results of the caging experiments performed upstream and downstream of the WWTP Langwiese showed no evidence of estrogenic disruption in males, neither before nor after the upgrade (Figure. 5). Chemical analyses found no EE2 in the effluent (detection limit 1 ng/L), but in vitro tests revealed estrogenic potentials prior to the upgrade [22]. In females, slightly, but not significantly higher VTG levels were measured upstream the WWTP in both years. Lower values downstream might possibly be caused by the combined activity of estrogenicity, anti-estrogenicity, and androgenicity which were all detected in parallel in in vitro bio tests [22].

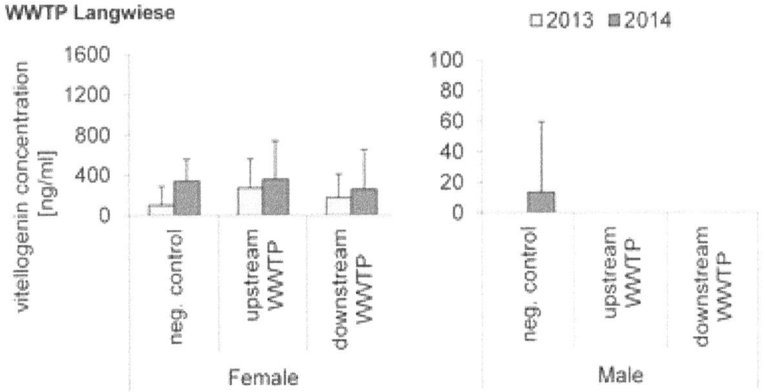

Figure. 5: Vitellogenin concentrations in blood plasma samples of rainbow trout exposed in 2013 and 2014 in cages upstream and downstream of the WWTP Langwiese; means and SD are shown. Analyzed with Biosense rainbow trout vitellogenin ELISA kit. *N*-numbers 2013 females: negative control $n = 9$, upstream WWTP $n = 15$, downstream WWTP $n = 7$; males: negative control $n = 1$, upstream WWTP $n = 2$, downstream WWTP $n = 4$. No significant differences with Steel–Dwass test; $p > 0.05$. *N*-numbers 2014 females: negative control $n = 6$, upstream WWTP $n = 9$, downstream WWTP $n = 13$; males: negative control $n = 13$, upstream WWTP $n = 11$, downstream WWTP $n = 8$. No significant differences with Steel–Dwass test; $p > 0.05$. No significant differences between years; $p > 0.05$

In summary, the results of our exposure experiments at the two WWTP effluents made evident that, in contrast to other studies which showed an induction of VTG by wastewater in juvenile, sexually immature, and male trout [11, 17, 23], even the conventional effluents of these WWTPs did not lead to increased VTG levels. This speaks for the high efficiency of the already established technologies at these two WWTPs, which, like most of the other larger WWTPs connected to tributaries of Lake Constance, are already equipped with a flocculation sand filter as a final cleaning step.

Exposure experiments at the bypass stations at the Schussen and the Argen River

Rainbow Trout

In the two bypass systems, at the Schussen downstream the WWTP Langwiese and at the reference river Argen, no VTG induction became evident in male fish, neither before nor after the upgrade of the WWTP Langwiese with the powdered activated charcoal filter (Figure. 6a). The VTG levels in females were highly variable; however, the highest percentages production in relation to the levels in the respective negative control fish were found in trout exposed at the bypass at the Schussen prior to the WWTP upgrade (Figure. 6b). Significant differences, however, did not occur.

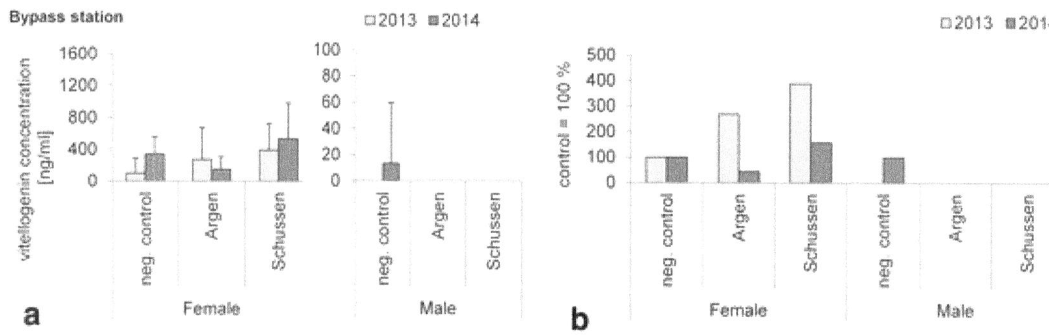

Figure. 6: a Vitellogenin concentrations in blood plasma samples of rainbow trout exposed in 2013 and 2014 at the bypass stations; means and SD are shown. Analyzed by Biosense rainbow trout vitellogenin ELISA kit. *N*-numbers 2013 females: negative control $n = 9$, Argen$n = 8$, Schussen $n = 4$; males: negative control $n = 1$, Argen $n = 5$, Schussen $n = 9$. No significant differences with Steel–Dwass test; $p > 0.05$. *N*-numbers 2014 females: negative control $n = 6$, Argen $n = 8$, Schussen $n = 6$; males: negative control $n = 13$, Argen$n = 1$, Schussen $n = 16$. No significant differences with Steel–Dwass test; $p > 0.05$. No significant differences between years; $p > 0.05$. **b** Values of **a** relative to negative control. Neg. control was set to 100 %

Juvenile rainbow trout which hatched at the bypass stations and were continuously exposed there afterwards showed neither before nor after the upgrade any induction of VTG. In contrast to that, Stalter et al. showed a significant increase in the VTG concentrations using yolk sac rainbow trout which were directly exposed to WWTP effluents for 60 days [17]. We used river water instead of effluent and the results of our other experiments revealed only a weak estrogenic pollution, explaining why we did not find increased VTG levels in juveniles.

These results for juvenile rainbow trout coincide with data for male fish, both indicating that neither at the Schussen downstream the WWTP nor at the Argen River are rainbow trout affected by estrogen disruptors.

Brown Trout

In 2013, prior to the upgrade, we found no significant differences in VTG levels in female and male brown trout exposed at the bypass stations (Figure. 7a). In 2014, after the upgrade, female brown trout showed significantly lower VTG levels at the Schussen (downstream WWTP Langwiese), whereas males showed no significant differences (Figure. 7a). Note that VTG levels of brown trout from different years cannot be compared because semi-quantitative VTG kits (semi-quantitative Salmonid (*Salmoniformes*) biomarker ELISA from Biosense) were used, implying that values are only comparable within one kit. This is the reason why the absolute values are also presented as relative values to the respective negative control levels in Figure. 7b. In 2014, VTG levels of females and males were lower at both rivers compared to negative control levels. Fish size did not vary strongly within treatment groups of each year (Figure. 4). Especially females showed no significant differences in their weights in 2014, hence we excluded differences in size as an explanation for differences in VTG levels (see Figure. 4). The fact that the VTG levels in females exposed at the Schussen were significantly lower than the negative control might be explained by the upgrade of the WWTP Langwiese. The additional treatment step at the WWTP Langwiese might have reduced estrogenic activities, and thereby unmasked anti-estrogenic activities which led to reduced VTG levels. Results by Stalter et al. indicated the importance of masking effects to evaluate wastewater [24]. Analyses of the same samples by in vitro yeast assays provided supporting results by showing elevated anti-estrogenicity and degraded estrogenic activities after the upgrade (not published).

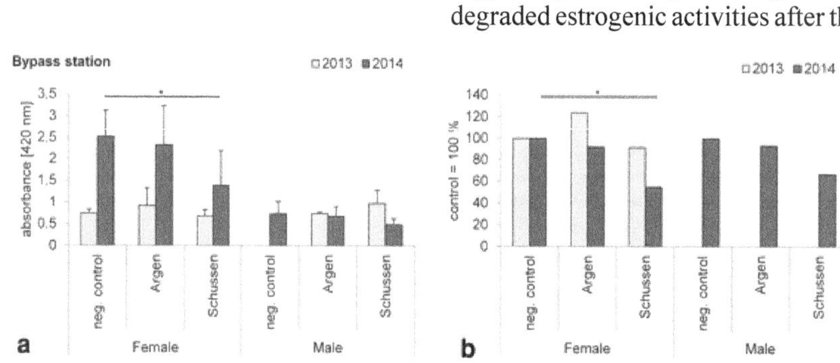

Figure. 7: a Absorbance measured in blood plasma samples of 1-year-old brown trout exposed at the bypass stations in 2013 and 2014; means and SD are shown. All samples of a group were analyzed within one semi-quantitative vitellogenin salmonid (Salmoniformes) biomarker ELISA kit (enzyme activity = color intensity is proportional to the concentration of vitellogenin in the sample). *N*-numbers 2013 females: negative control$n = 4$, Argen $n = 4$, Schussen $n = 3$; males: negative control $n = 0$, Argen $n = 4$, Schus-

sen$n = 4$. No significant differences; $p > 0.05$. N-numbers 2014 females: negative control$n = 6$, Argen $n = 6$, Schussen $n = 9$; males: negative control $n = 10$, Argen $n = 13$, Schussen $n = 5$. Significant differences with the Tukey–Kramer HSD test: females 2014 neg. control

Schussen p = 0.0231 (*Asterisks* significant differences; $*p < 0.05$). **b** Values of **a** relative to negative control.

Neg. control was set to 100 %. In 2013, no values could be given for males because of absence of males in the neg. control

At both bypass stations, brown trout larvae showed no increased VTG values after the upgrade of the WWTP Langwiese compared to the negative control levels (Figure. 8a). On the contrary, the levels are even lower than the negative control levels, which might again be related to unmasked anti-estrogenicity in 2014. These results differ from data we collected prior to the upgrade (Figure. 8, and see also Henneberg et al. [22]). In this previous study, brown trout showed significantly higher VTG levels at the Schussen bypass and at the Argen bypass compared to the negative control after the same exposure time. Estrogen-active compounds were likely causes for the increased VTG levels prior to the upgrade. Due to the fact that we did not observe differences in VTG levels after the upgrade at both bypass stations, we conjecture that it is mainly annual specific differences that caused these effects, and to a lesser degree the upgrade itself.

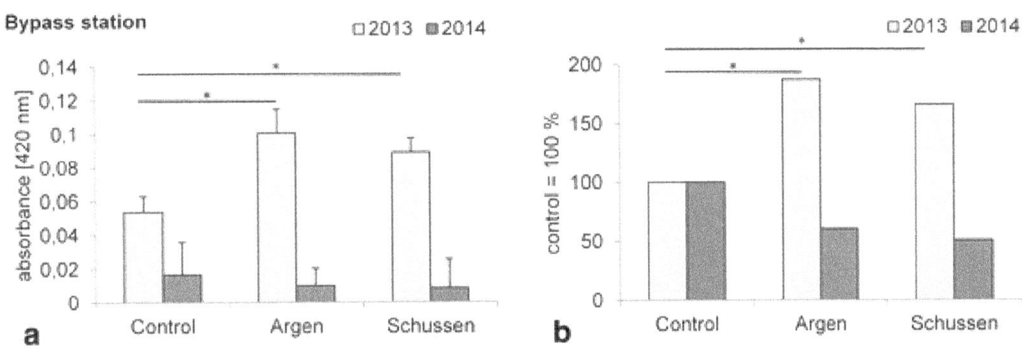

Figure. 8: a Absorbance measured in homogenates of juvenile brown trout exposed for 99 days post-fertilization at the bypass stations in 2014; means and SD are shown. All samples were analyzed within one semi-quantitative vitellogenin salmonid (Salmoniformes) biomarker ELISA kit (enzyme activity = color intensity is proportional to the concentration of vitellogenin in the sample). Each treatment $n = 12$. No significant differences with the Steel–Dwass test; $p > 0.05$. For better comparison, previous results from 2013 prior to the upgrade are also shown. These results were already published in PlosOne by Henneberg et al. 2014 [22]. **b** Values of **a** relative to negative control. Neg. control was set to 100 %.

In contrast to these results for brown trout, we observed no differences in VTG levels of juvenile and 1-year-old rainbow trout. Previous studies showed that brown trout are more sensitive to environmental stress than rainbow trout [25–27], and our results are in line with this observation. Bjerregaard et al. concluded that the sensitivity of brown trout to estrogens does not differ from the sensitivity of the majority of fish species; first- and second-year brown trout appear to be suitable monitoring organisms to demonstrate estrogenic effects in headwater streams [28]. Hence, our results indicate slight temporary estrogenic effects that might affect feral fish species. However, the differences in VTG levels of brown trout we observed were low, and we conclude that estrogenic effects in the two rivers investigated are generally low.

Organs of the trout we used in the present study were examined in a parallel study to assess their health status before and after the upgrade at the WWTP Langwiese. The results showed that the upgrade led to a better health status of trout and partly also of feral fish species. While this showed that the upgrade reduced toxic effects, the current study showed that estrogenic effects were only slightly reduced.

CONCLUSION

Overall, our VTG results showed no strong estrogenic effects of WWTP effluents at the Schussen River on trout. After the upgrade of WWTP Langwiese, juvenile and female brown trout showed significantly decreased VTG levels but especially the results for brown trout larvae indicated that annual variation might also play a major role. While rainbow trout showed no significant reduction in VTG levels, we found reduced VTG levels in brown trout, indicating that brown trout might respond more sensitively than rainbow trout.

Furthermore, we did not observe increased VTG levels in males in any experiment. Therefore, we classify the Schussen River as showing only low pollution with estrogens. In particular, neither effluents of the WWTP Langwiese nor effluents of the WWTP Eriskirch caused significantly higher VTG levels in trout, independently of additional wastewater treatment technologies.

METHODS

Test Organisms

For our investigations we used immature, 1-year-old brown trout (*Salmo trutta* f. *fario*) and rainbow trout (*Oncorhynchus mykiss*) delivered by the fish hatchery Lohmühle, Alpirsbach, Germany. We also obtained freshly fertilized trout eggs from there. For the experiments trout were transported from the hatchery to the exposure sites and directly released in cages or aquaria. The trout which grew up at the fish farm received a mixture of spring water with drinking water quality and stream water which originates in a water protection area (pH 7, nitrate <0.3 mg/L, nitrite <0.0033 mg/L) [29]. All fish were fed with food from the company BioMar, Denmark (INICO Plus for larvae and EFICO alpha for 1-year-old trout) in different particle sizes, depending on fish size. Trout in all our exposure experiments received the same amount of food, except for the negative control (rainbow trout) in 2014, which were sampled directly at the fish farm and for which the amount fed was not under our control.

Exposure Experiments at WWTPs and at Bypass Systems

As a model for a medium-sized WWTP with 40,000 population equivalents we chose the WWTP Eriskirch connected to the Schussen River in the Lake Constance catchment area, South Germany (Figure. 9). At this WWTP, a small-scale model installation was realized in 2013, which included different columns allowing cleaning of partial effluent flow by different combinations of ozonation, sand filtration, and granulated activated carbon filter. In 2013 and 2014, 1-year-old rainbow trout were exposed here in aquaria of which one was flown through by the conventional effluent and the second by the additionally treated effluent. In 2013, the additionally treated effluent was proportionately composed of wastewater treated by (1) ozonation + sand filter + granulated activated carbon, and (2) ozonation + granulated activated carbon. In 2014, the composition was changed as follows: (1) ozonation + sand filter; (2) ozonation + granulated activated carbon; and (3) only granulated activated carbon in the ratio 1:1:1. The aquarium with the regular effluent was aerated via a membrane pump to ensure sufficient oxygen concentrations for trout. Daylight was simulated by lamps using timer clocks, and the light/dark photoperiod was adapted to natural daylight. Fish were fed with equal amounts of food by an automatic feeder once a day.

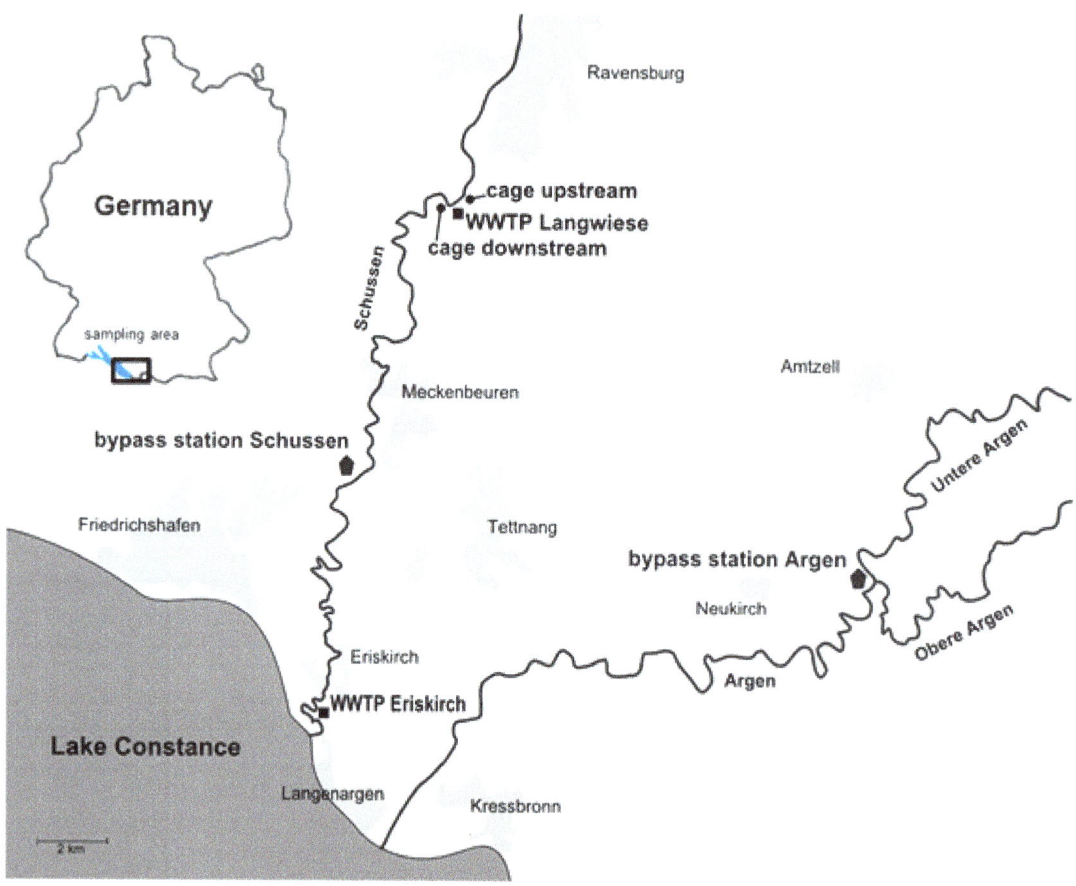

Figure. 9: Overview of sampling sites, bypass stations and examined WWTPs at the Schussen River and Argen River, Lake Constance, South Germany

As a model for a large WWTP with 170,000 population equivalents, the WWTP Langwiese was in the focus of our study, also situated at the Schussen River upstream of the WWTP Eriskirch (Figure. 8). At the WWTP Langwiese, an additional powdered activated carbon filter was put into operation after the biological treatment and before the final sand filter in September 2013. At that WWTP, we exposed 1-year-old rainbow trout in cages (for cage description see [30]) 100 m upstream the WWTP effluent and downstream of it (mixture of 50 % effluent and 50 % Schussen water) in the Schussen River. Trout were fed every second day with a comparable amount of food as the trout at the WWTP Eriskirch received in 2 days. The exposure experiments at the WWTP Langwiese were performed in spring 2013 prior to the upgrade with the powdered activated carbon filter and in spring 2014 after the upgrade.

In addition to the exposure experiments at the WWTPs, we used two bypass stations with 250 L aquaria continuously flown through by fresh river water (0.4 L/s): one setup was located downstream of the WWTP Langwiese at the Schussen River and one at the Argen River as a reference river less influenced by micropollutants [31] (see Figure. 9). Here, fertilized eggs and developing larvae as well as 1-year-old brown trout and rainbow trout were exposed (for a detailed description of the bypass station and trout exposure conditions see [22]).

As a negative control, we kept trout in 250 L aquaria under semi-flow-through conditions in climate chambers at the University of Tübingen. We used filtered tap water and exchanged a third of the water volume once a week. Water was aerated, temperature was kept at 6 °C, a stream pump (Co.: Tunze, Germany) guaranteed a constant stream, and a filter (Co.: JBL1500e) kept good water conditions. Temperature, ammonium and nitrite concentrations were controlled every other day (ammonium <0.05 mg/L, nitrite <0.01–0.05 mg/L). Light/dark photoperiod was adapted to natural daylight. The semi-static conditions implied that we could not feed fish as much as in the flow-through systems because we had to keep a good water quality. The poor growth of our negative control fish in 2013 was a main reason for us to change the negative control fish in 2014. For that, we sampled 1-year-old trout in 2014 directly at the fish farm where we bought all our trout. To ensure that the development status in all groups was comparable, we sampled the negative control fish at the fish farm shortly before sampling fish at the WWTPs.

To ensure that fish generally react to estrogenic substances by producing VTG, we exposed trout to EE_2 as a positive control. For this, fish were kept at same conditions as negative control fish in 2013, but EE_2 was added in concentrations which ranged from 5 to 20 ng/L. All trout exposed to EE_2 showed extreme higher VTG levels than the negative controls (see Table 4).

Table 4: Mean values and SD of exposure experiments with trout using EE_2 as positive control

	Females		**Males**	
	2013	**2014**	**2013**	**2014**
Brown trout				
Mean values (absorbance [420 nm])	326.33	21.77	500.5	4.86
SD	±70.11	±25.09	±114	±3.1
n-number	3	5	2	5
Rainbow trout				
Mean values (VTG [ng/ml])	2,699,183.9	3,812,659.5	3,362,476.1	3,830,203.9
SD	±3,074,723.7	±1,653,878.4		±1,867,237.4
n-number	8	5	1	7
Juvenile trout	**Rainbow trout (VTG [ng/ml])**	**Brown trout (absorbance [420 nm])**		
Mean values	2,030.54	0.0808		
SD	±2811.60	±0.0358		
n-number	8	9		

Exposure duration at WWTPs and at bypass systems

Prior to the upgrade at the WWTP Langwiese, we carried out one bypass exposure and one cage exposure experiment in the winter season 2012/2013. After the upgrade at the WWTP Langwiese, one bypass exposure and one cage exposure experiment were performed in the winter season 2013/2014. At the WWTP Eriskirch, we started the first exposure experiment in spring 2013 because the installation of the exposure aquaria was not completed until then. In the second year, 2014 (after the upgrade of the WWTP Langwiese), all exposure experiments started at the same time at all sites. Tables 5 and 6 summarize the time schedule for all exposure experiments, including exposure duration and exposure type.

Table 5: Time schedule for the exposure experiments performed at WWTPs and bypass stations with 1-year-old trout

Start of exposure	**End of exposure**	**Exposure duration (days)**	**Exposure type**	**Trout species**
Winter season 2012/2013 prior to the upgrade				
15 Nov 2012	24 Jan 2013	70	Laboratory neg. control + EE_2 control	Brown and rainbow trout
15 Nov 2012	17 Jan 2013	63	Cage exposure	Rainbow trout
15 Nov 2012	14 Feb 2013	91	Exposure in bypass systems	Brown and rainbow trout
6 Feb 2013	21 Mar 2013	43	Exposure at WWTP Eriskirch	Rainbow trout

Winter season 2013/2014 after the upgrade				
	29 Jan 2014	0	Neg. control from hatchery	Brown and rainbow trout
2 Dec 2013	23 Jan 2014	52	EE_2 control	Brown and rainbow trout
2 Dec 2013	4 Feb 2014	64	Cage exposure	Rainbow trout
2 Dec 2013	13 Feb 2014	73	Exposure at WWTP Eriskirch	Rainbow trout
2 Dec 2013	12 Mar 2014	100	Exposure in bypass systems	Brown and rainbow trout

Table 6: Time schedule for exposure experiments performed at the bypass stations with fresh fertilized trout eggs

Start of exposure	End of exposure	Exposure duration	Exposure type	Trout species
Winter season 2012/2013 prior to the upgrade				
07 Dec 2012	20 Mar 2013	103 days	Laboratory neg. control	Rainbow trout
07 Dec 2012	21 Mar 2013	104 days	Exposure in bypass systems	Rainbow trout
Results of exposure experiments using juvenile brown trout are published in Henneberg et al. [22].				
Winter season 2013/2014 after the upgrade				
24 Nov 2013	3 Mar 2014	99 days	Laboratory neg. control	Brown and rainbow trout
24 Nov 2013	4 Mar 2014	100 days	Exposure in bypass systems	Brown and rainbow trout
7 Mar 2014	28 Mar 2014	22 days	EE_2 control	Brown and rainbow trout

ETHIC STATEMENT

This study was carried out in strict accordance with German legislation (animal experiment permit nos. ZO 1/09 and ZP 1/12, District Magistracy of the State of Baden-Württemberg).

VITELLOGENIN DETECTION

Sampling

One-year-old brown trout and rainbow trout, sampled at each site, were killed with an overdose MS-222 (tricaine mesylate, Sigma-Aldrich, St. Louis, USA). Blood samples were taken immediately from the caudal vein by a sterile syringe, transferred in lithium-heparinized reaction tubes (Co. Sarstedt, Germany), and 4 TIU aprotinin (C. Roth, Germany) per mL blood was added. Samples were centrifuged (4 °C, 10 min, 2500 rpm Eppendorf 5810R) on-site and plasma samples were snap-frozen in liquid nitrogen. Thereafter, plasma aliquots were stored at −80 °C until we determined VTG levels. After taking the blood samples, the length and weight of each fish were measured, gonads were removed for histological examinations and fixed in 2 % glutaraldehyde dissolved in 0.1 M cacodylic acid.

Larvae were killed with an overdose MS-222 (tricaine mesylate, Sigma-Aldrich, St. Louis, USA), and the region between head and pectoral fin from each individual was placed in Eppendorf tubes, snap-frozen, and stored at −80 °C.

All the following steps were undertaken on ice. Homogenates of juvenile trout were prepared by adding homogenization buffer (4 times the sample weight; PBS + 2 TIU aprotinin, C. Roth, Germany), mixing with a plastic pestle, centrifuging (10 min, 4 °C, 20,000×*g* Eppendorf 5810R) [17] and storing the supernatants at −80 °C.

Vitellogenin ELISA

VTG levels of rainbow trout were measured using the rainbow trout (*Oncorhynchus mykiss*) vitellogenin ELISA kit (V01004402, Biosense Laboratories, Norway). For the analyses of the brown trout samples we used a semi-quantitative kit because the antibody of this kit shows a very good cross-reactivity against brown trout VTG (semi-quantitative vitellogenin Salmonid (*Salmoniformes*) biomarker ELISA kit (V01002402, Biosense Laboratories, Norway)). All steps were performed as described in the protocols. As recommended by the provider of the test kit, a minimum of 1:20 dilution was used and samples were tested in duplicates. The absorbance was measured by a microplate reader (Automated Microplate Reader Elx 8006, Bio-Tek Instruments, INC., USA).

The semi-quantitative ELISA test kit, which is recommended for VTG analyses of salmonids, was used for our brown trout samples. The enzyme activity (absorbance), which is measured by the assay, is proportional to the concentration of VTG in the sample. Purified VTG from Atlantic salmon (*Salmo salar*) was used as a positive control within every assay run. We analyzed all blood samples of females with one 96-well plate (neg. control, Bypass Schussen, Bypass Argen and, EE2 control), all samples of males on the next 96-well plate, etc. Hence, all these samples are comparable within their groups. All steps were performed as described in the protocols by the provider of the test kit.

Statistical Analyses

Statistical analyses were performed with JMP 10.0 (SAS Systems, USA). Data were tested for normality using the Shapiro–Wilk W test and for homogeneity of variance with the Levene test. If the data were normally distributed and the variance was homogeneous, the Tukey–Kramer HSD test was conducted. Otherwise, if the data were homoscedastic but not normally distributed, the Steel–Dwass test was used. If the data were normally distributed but not homoscedastic, the Welch's ANOVA was performed.

ACKNOWLEDGEMENTS

The technical help of S. Krais with laboratory analyses is highly acknowledged. Many thanks are due to M. Weyhmüller for the maintenance of the bypass systems as well as to the staff from the Department of Animal Physiological Ecology (A. Dietrich, M. Di Lellis, D. Maier, K. Peschke, A. and V. Scheil, P. Thellmann, and K. Vincze) for help with sampling.

FUNDING

We acknowledge support by the Deutsche Forschungsgemeinschaft and Open Access Publishing Fund of Tübingen University. The project SchussenAktivplus is funded by the Federal Ministry for Education and Research (BMBF) and cofounded by the Ministry of Environment Baden-Württemberg. In addition, Jedele & Partner GmbH, Ökonsult GbR, the city of Ravensburg, the AZV Mariatal and the AV Unteres Schussental financially contributed to the project. SchussenAktivplus is connected to the BMBF action plan "Sustainable water management (NaWaM)" and is integrated in the BMBF frame programme "Research for sustainable development FONA". Contract period: 1/2012 to 12/2014, Funding number: 02WRS1281A. All funders had no role in study design, data collection and analysis, decision to publish, or preparation of the manuscript. No competing interests between our study and the commercial funders exist and commercial founders do not alter our adherence to Environmental Sciences Europe policies on sharing data and materials.

REFERENCES

1. European Commission. Annex 1. Candidate list of 553 substances.http://ec.europa.eu/environment/chemicals/endocrine/strategy/substances_en.htm. Accessed 30 Mar 2015
2. Jobling S (1998) Natural and anthropogenic environmental oestrogens: the scientific basis for risk assessment. Pure Appl Chem 70(9):1805–1827
3. Sumpter JP (1998) Xenoendocrine disrupters—environmental impacts. Toxicol Lett 102:337–342. doi:10.1016/S0378-4274(98)00328-2
4. Bolong N, Ismail A, Salim MR, Matsuura T (2009) A review of the effects of emerging contaminants in wastewater and options for their removal. Desalination 239(1):229–246. doi:10.1016/j.desal.2008.03.020
5. Purdom CE, Hardiman PA, Bye VVJ, Eno NC, Tyler CR, Sumpter JP (1994) Estrogenic Effects of Effluents from Sewage Treatment Works. Chem Ecol 8(4):275–285. doi:10.1080/02757549408038554
6. Lange R, Hutchinson TH, Croudace CP, Siegmund F, Schweinfurth H, Hampe P et al (2001) Effects of the synthetic estrogen 17 alpha-ethinylestradiol on the life-cycle of the fathead minnow (Pimephales promelas). Environ Toxicol Chem 20(6):1216–1227. doi:10.1002/etc.5620200610
7. Routledge EJ, Sheahan D, Desbrow C, Brighty GC, Waldock M, Sumpter JP (1998) Identification of estrogenic chemicals in STW effluent. 2. In vivo responses in trout and roach. Environ Sci Technol 32(11):1559–1565. doi:10.1021/es970796a
8. Johnson AC, Dumont E, Williams RJ, Oldenkamp R, Cisowska I, Sumpter JP (2013) Do concentrations of ethinylestradiol, estradiol, and diclofenac in European rivers exceed proposed EU environmental quality standards? Environ Sci Technol 47(21):12297–12304. doi:10.1021/es4030035
9. Johnson AC, Sumpter JP (2015) Improving the quality of wastewater to tackle trace organic contaminants: think before you act! Environ Sci Technol. doi:10.1021/acs.est.5b00916
10. .Triebskorn R, Amler K, Blaha L, Gallert C, Giebner S, Güde H et al (2013) SchussenAktivplus: reduction of micropollutants and of potentially pathogenic bacteria for further water quality improvement of the river Schussen, a tributary of Lake Constance, Germany. Environ Sci Eur 25(1):1–9. doi:10.1186/2190-4715-25-2
11. .Kime DE, Nash JP, Scott AP (1999) Vitellogenesis as a biomarker of reproductive disruption by xenobiotics. Aquaculture 177(1–4):345–352. doi:10.1016/S0044-8486(99)00097-6
12. .Ackermann GE, Schwaiger J, Negele RD, Fent K (2002) Effects of long-term nonylphenol exposure on gonadal development and biomarkers of estrogenicity in juvenile rainbow trout (Oncorhynchus mykiss). Aquat Toxicol 60(3–4):203–221. doi:10.1016/S0166-445X(02)00003-6
13. Tyler CR, van Aerle R, Hutchinson TH, Maddix S, Trip H (1999) An in vivo testing system for endocrine disruptors in fish early life stages using induction of vitellogenin. Environ Toxicol Chem 18(2):337–347. doi:10.1002/etc.5620180234
14. Sumpter JP, Jobling S (1995) Vitellogenesis as a biomarker for estrogenic contamination of the aquatic environment. Environ Health Perspect 103(Suppl 7):173
15. OECD. Test No. 229: Fish Short Term Reproduction Assay. OECD Publishing
16. OECD. Test No. 230: 21-day Fish Assay. OECD Publishing
17. Stalter D, Magdeburg A, Weil M, Knacker T, Oehlmann J (2010) Toxication or detoxication? In vivo toxicity assessment of ozonation as advanced wastewater treatment with the rainbow trout. Water Res 44(2):439–448. doi:10.1016/j.watres.2009.07.025
18. Bila D, Montalvao AF, Azevedo DdA, Dezotti M (2007) Estrogenic activity removal of 17β-estradiol by ozonation and identification of by-products. Chemosphere 69(5):736–746
19. Scheurer M, Heß S, Lüddeke F, Sacher F, Güde H, Löffler H et al (2015) Removal of micropollutants, facultative pathogenic and antibiotic resistant bacteria in a full-scale retention soil filter receiving combined sewer overflow. Environ Sci Process Impacts 17(1):186–196. doi:10.1039/c4em00494a

20. An L, Hu J, Yang M (2008) Evaluation of estrogenicity of sewage effluent and reclaimed water using vitellogenin as a biomarker. Environ Toxicol Chem 27(1):154–158. doi:10.1897/07-096.1

21. Copeland PA, Sumpter JP, Walker TK, Croft M (1986) Vitellogenin levels in male and female rainbow trout (Salmo gairdneri richardson) at various stages of the reproductive cycle. Comp Biochem Physiol Part B Comp Biochem 83(2):487–493. doi:10.1016/0305-0491(86)90400-1

22. Henneberg A, Bender K, Blaha L, Giebner S, Kuch B, Köhler H-R et al (2014) Are in vitro methods for the detection of endocrine potentials in the aquatic environment predictive for in vivo effects? Outcomes of the Projects SchussenAktiv and SchussenAktivplus in the Lake Constance Area, Germany. PLoS One 9(6):e98307. doi:10.1371/journal.pone.0098307

23. Bjerregaard LB, Madsen AH, Korsgaard B, Bjerregaard P (2006) Gonad histology and vitellogenin concentrations in brown trout (Salmo trutta) from Danish streams impacted by sewage effluent. Ecotoxicology 15(3):315–327. doi:10.1007/s10646-006-0061-9

24. .Stalter D, Magdeburg A, Wagner M, Oehlmann J (2011) Ozonation and activated carbon treatment of sewage effluents: removal of endocrine activity and cytotoxicity. Water Res 45(3):1015–1024. doi:10.1016/j.watres.2010.10.008

25. Schmidt H, Bernet D, Wahli T, Meier W, Burkhardt-Holm P (1999) Active biomonitoring with brown trout and rainbow trout in diluted sewage plant effluents. J Fish Biol 54(3):585–596. doi:10.1111/j.1095-8649.1999.tb00637.x

26. Pickering A, Pottinger T, Carragher J (1989) Differences in the sensitivity of brown trout, Salmo trutta L., and rainbow trout, Salmo gairdneri Richardson, to physiological doses of cortisol. J Fish Biol 34(5):757–768. doi:10.1111/j.1095-8649.1989.tb03355.x

27. Schneeberger HU (1995) Abklärungen zum Gesundheitszustand von Regenbogenforelle (Oncorhynchus mykiss), Bachforelle (Salmo trutta fario) und Groppe (Cottus gobio) im Liechtensteiner-, Werdenberger- und Rheintaler-Binnenkanal [Inaugural Dissertation]: University of Bern, Bern

28. Bjerregaard P, Hansen PR, Larsen KJ, Erratico C, Korsgaard B, Holbech H (2008) Vitellogenin as a biomarker for estrogenic effects in brown trout, *Salmo trutta*: laboratory and field investigations. Environ Toxicol Chem 27(11):2387–2396. doi:10.1897/08-148.1

29. Schindler J (2015) Besatzforellen. http://www.forellenzucht-lohmuehle.de/besatzforellen.html. Accessed 11 Aug 2015

30. Vincze K, Scheil V, Kuch B, Köhler HR, Triebskorn R (2015) Impact of wastewater on fish health: a case study at the Neckar River (Southern Germany) using biomarkers in caged brown trout as assessment tools. Environ Sci Pollut Res. doi:10.1007/s11356-015-4398-6

31. Triebskorn R, Hetzenauer H (2012) Micropollutants in three tributaries of Lake Constance, Argen, Schussen and Seefelder Aach: a literature review. Environ Sci Eur 24(1):1–24.

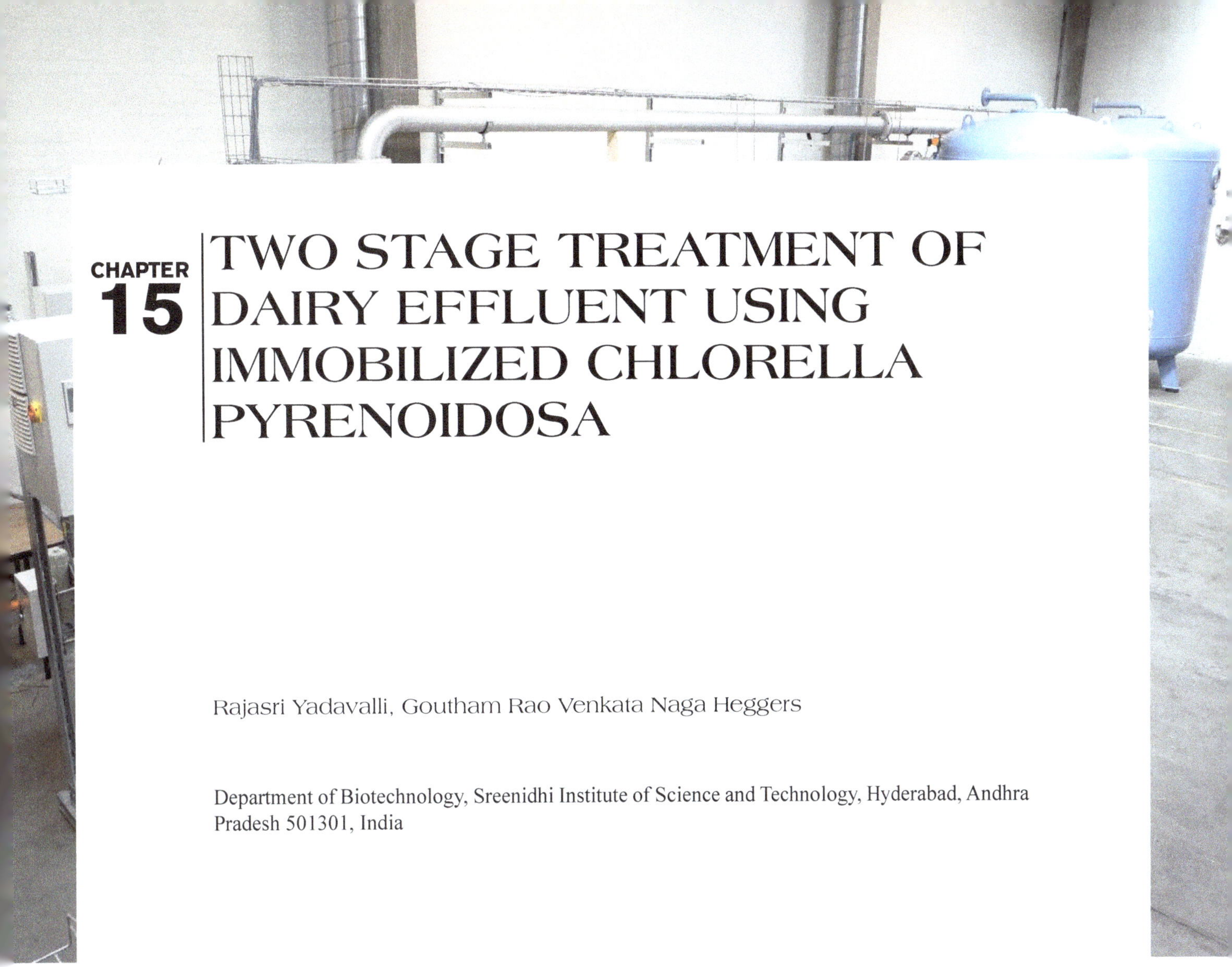

CHAPTER 15

TWO STAGE TREATMENT OF DAIRY EFFLUENT USING IMMOBILIZED CHLORELLA PYRENOIDOSA

Rajasri Yadavalli, Goutham Rao Venkata Naga Heggers

Department of Biotechnology, Sreenidhi Institute of Science and Technology, Hyderabad, Andhra Pradesh 501301, India

ABSTRACT

Background

Dairy effluents contains high organic load and unscrupulous discharge of these effluents into aquatic bodies is a matter of serious concern besides deteriorating their water quality. Whilst physico-chemical treatment is the common mode of treatment, immobilized microalgae can be potentially employed to treat high organic content which offer numerous benefits along with waste water treatment.

Methods

A novel low cost two stage treatment was employed for the complete treatment of dairy effluent. The first stage consists of **treating the diary effluent in a photo bioreactor (1 L) using immobilized** *Chlorella pyrenoidosa* **while the second stage involves a two column sand bed filtration technique.**

Results

Whilst NH_4^+-N was completely removed, a 98% removal of PO_4^{3-}-P was achieved within 96 h of two stage purification processes. The filtrate was tested for toxicity and no mortality was observed in the zebra fish which was used as a model at the end of 96 h bioassay. Moreover, a significant decrease in biological oxygen demand and chemical oxygen demand was achieved by this novel method. Also the biomass separated was tested as a bio fertilizer to the rice seeds and a 30% increase in terms of length of root and shoot was observed after the addition of biomass to the rice plants.

Citation: Rajasri Yadavalli, Goutham Rao Venkata Naga Heggers, Two stage treatment of dairy effluent using immobilized Chlorella pyrenoidosa, doi:10.1186/2052-336X-11-36

Conclusions

We conclude that the two stage treatment of dairy effluent is highly effective in removal of BOD and COD besides nutrients like nitrates and phosphates. The treatment also helps in discharging treated waste water safely into the receiving water bodies since it is non toxic for aquatic life. Further, the algal biomass separated after first stage of treatment was highly capable of increasing the growth of rice plants because of nitrogen fixation ability of the green alga and offers a great potential as a bio fertilizer.

BACKGROUND

Agro-food industries like the dairy and brewery sectors produce effluents with high organic load. These effluents contain high BOD, COD and nutrients such as phosphates, ammonia and/or nitrate, pose a serious threat to the water quality and should be appropriately treated before discharging into the environment [1]. There are several physical, chemical and biological methods for the treatment of effluents derived from the agro-food industries. Biological methods are proved to be much ecofriendly when compared with other types of treatments [2,3].

Research suggests that N- and P-rich wastewater is also considered as valuable substrate for cultivation of microalgae which is a type of biological treatment [4]. Microalgae are mostly suspension-type microorganisms and efficient solar driven cell factories that can be useful as treating agents for wastewater. The cultivation of algae in wastewaters offers the combined advantages of treating the wastewaters and simultaneously producing algal biomass, which can further be exploited for protein complements and food additives for aquaculture, animal and human feed, energies such as biogas and fuels and bio fertilizer [5]. Little information is available on the use of *Chlorella pyrenoidosa* as a potential agent for treatment of various types of waste waters among the several microalgae used to treat effluents [6].

Separation of algal biomass from the treated water discharge is the key for the success of waste water treatment. Numerous efforts have been devoted to developing a suitable technology for harvesting microalgae. The use of industrial filtration and centrifugation is not cost effective for wastewater treatment. In this context, immobilization of algal cells has been proposed for overcoming the harvest problem as well as retaining the high-value algal biomass for further processing [7].

Among all unicellular algal species, *Chlorella* is a common and effective species for the immobilization and nutrient removal purposes [8]. Uptake rate of phosphorous was low when compared with Nitrogen for *Chlorella sp* from waste waters [9, 10]. There are various methods of immobilization out of which calcium alginate is a very simple and cost-effective method, to entrap microbial cells as alginate beads.

Many green algae are also capable of using atmospheric dinitrogen (N_2) as the source of nitrogen. This important characteristic of nitrogen fixation has enhanced the agro production when green alga is used in agriculture [11]. Many studies have been reported on the use of dried green alga to inoculate soils as a means of aiding fertility. Algalization is the term applied to the use of a defined mixture of algal species to inoculate soil. A 15-20% increase in rice grain yield was observed because of algalization [12, 13].

Reduction of BOD and COD is one of the major challenges in wastewater treatment. If effluent with high BOD is discharged into environment, it will increase the bacterial growth in the water while reducing the dissolved oxygen levels in the water that may prove lethal to aquatic life. Much like BOD, a high COD can also deprive aquatic organisms of oxygen needed for survival [14].

In this study, we made an attempt to treat the diary effluent by a novel, low cost photobioreactor using immobilized *C. pyrenoidosa* and evaluate the potential of dairy effluent as a nutrient medium for microalgae growth. The efficacy of sand bed filtration to reduce/remove some of the organic compounds present in the dairy effluent was assessed. The treated dairy effluent is used to check the survival of zebrafish (*Danio rario*) whose gene sequence is identical with humans to assess the biosafety to aquatic life. In addition, the biomass produced is analyzed for its composition, and amenable to be used as fertilizer.

MATERIALS AND METHODS

Microalgae

Chlorella pyrenoidosa sp. (NCIM NO: 2738) was obtained from National Centre for Industrial Microorganisms (NCIM), Pune, India. Stock culture of *Chlorella pyrenoidosa* was grown photoautotrophically in BG11 media at 28°C under continuous light illumination in four 100 ml borosil flasks. Each liter of the BG11 medium contained $NaNO_3$-1.5 g, K_2HPO_4-0.04 g, $MgSO_4{\bullet}7H_2O$-0.075 g, $CaCl_2{\bullet}$2H2O-0.036 g, Citricacid-0.006 g, $NaCO_3$-0.02 g, H_3BO_3-0.00286 g, $MnCl_2{\bullet}4H_2O$-0.00181 g, $ZnSO_4{\bullet}$7H2O-0.00022 g, $Na_2MoO_4{\bullet}2H_2O$-0.00039 g, $CuSO_4{\bullet}5H_2O$-0.00008 g, $Co(NO_3)_2{\bullet}6H_2O$-0.00005 g, $(NH_4)_6Mo_7O_{24}.4H_2O$-0.003 g, Na_2EDTA-0.00001 g. The inoculum was prepared by transferring the cells from stock culture, and incubated aseptically in a 1000 ml flask containing 700 ml of fresh BG11 media under continuous illumination of 34 μmol m^{-2} s^{-1} at 28°C for four days on an orbital shaker set at 120 rpm. A 4 day old culture was used as inoculum at 10% volume for the preparation of immobilized microalgal beads. Gel entrapment is done using 2% Sodium alginate and 0.1 M Calcium chloride. The average diameter of the algal beads inoculated was 0.556 m. Untreated dairy effluent is collected from Vyshnavi dairy, Khammam, Andhrapradesh, India.

Fish Collection

Specimens of zebra fish (4–5 cm long; weight 1.25±0.25 g) were used for toxicity tests. These were procured from a local aquarium shop, Hyderabad were collected from. The fish were transported from the farm in oxygenated polythene bags to the laboratory and immediately transferred into 20 L capacity containing well –aerated unchlorinated ground water. The fish were allowed to acclimatize for 15 days before the experiments. They were fed with rice bran during the acclimation period. Only healthy and active fishes were used for experiments. The temperature, pH, dissolved oxygen, used for acclimatization were 29°C, 7.2±0.2, 5.0-6.5 mg/L respectively.

Photo bioreactor

The immobilized beads are weighed approximately to 50 gm and inoculated in 10 one litre inverted polycarbonate bottles set on a wooden rack (Figure 1). A light intensity of 55 μmol m^{-2} s^{-1} was maintained at 28°C for four days. Artificial air supply is provided through air pumps at 1 L min^{-1} for all 10 bottles continuously to ensure the suspension of immobilized beads in dairy water and thorough contact of beads with the nutrients present in dairy effluent. The filtered dairy water is analyzed for its biochemical composition.

Figure 1: Photo bioreactors with immobilized micro algal beads. Inverted polycarbonate bottles on a wooden rack with immobilized micro algal beads for treatment of dairy effluent.

Sand Bed Filtration

Figure 2: Low cost two column sand bed filtration. First column filled with stones and sand and second column spread with charcoal.

A two column sand bed is prepared for the filtration of algal treated dairy water. The first column contains 3 layers of small, medium and large stones each 4 cm and again each layer is covered by fine sand of about 2 cm in height (Figure 2). In the second column of sand bed filtration a charcoal bed is spread. The treated water from the first stage slowly drips into the second stage and the treated water will be collected finally into a borosil bottle after passing through charcoal bed. The treated dairy effluent after sand bed filtration was again analyzed and compared with control.

Estimation of Lipid, Carbohydrate and Protein

The carbohydrate content was analyzed based on the procedure published by Miao and Wu [15]. In short, 0.1 g dried algal pellet was acidified by adding 20 ml 2.5 M HCl. The acidified solution was then hydrolyzed at 100°C for 30 min and neutralized to pH 7. The volume was adjusted to 100 ml. The filtered sample was subjected to 3,5-dinitrosalicylic acid (DNS) assay [16].

Lipid extraction was performed in a way that was similar to the procedures reported by earlier studies [17, 18]. Briefly, 1.0 g dried cells were ground in a mortar and pestle. The dried powder is then transferred to a 50 ml glass centrifuge tube. Chloroform was then added to the tube to make the chloroform/ methanol (2:1, v/v). The tube was vortexed for 5 min and was allowed to stand for 24 h. After that, the tube was centrifuged at 4,000 g for 15 min to remove the algal solids. The supernatant was collected and the solvent was vaporized using Rotovap at 65°C. Oil left in the flask without solvent was weighed to calculate lipid content.

Protein estimation is done using Lowry's method [19]. Briefly, 0.1 g algal powder is taken and 2 ml of 5% TCA is added. The mixture is centrifuged at 2000 rpm for 10 min. Solution containing pellet is dissolved in 1.5 ml of 0.1 N NaOH. From the above prepared solution 0.2 ml is taken into 20 ml test tube and 5 ml of alkaline copper solution is added. After incubation at room temperature for 15 min, 0.5 ml of Folin-Ciocalten reagent is added. Again the mixture is incubated for 30 min and OD values were taken at 540 nm.

RESULTS AND DISCUSSION

Biochemical Analysis of Algal Biomass

After four days of the growth period the algal beads were separated from the dairy water effluent using nylon filter cloth (Figure 3). Earlier studies mentioned that even at high concentrations (up to 900 mg/L) of phosphates, immobilized beads tend to be stable in acidic pH (5.5-6.5) ranges up to 4 days [20]. Hence the bead stability was maintained throughout the experiment without cell leakage. The fully bulged algal beads of diameter 1.078 m were analyzed for lipid, carbohydrate and protein content (Table 1). The high protein content of algal biomass was supported by Rodulfo in his earlier studies [21]. The treated water is compared with the control (untreated dairy effluent) after each stage treatment process. A clear liquid without turbidity can be observed after two stage treatment process of dairy effluent.

Figure 3: Wet algal beads separated after four days of cultivation. The fully bulged algal beads of diameter 1.078 m were filtered after four days of cultivation in dairy effluent. The initial bead size before inoculation into photo bioreactor was 0.556 m.

Table 1: Algal biomass analysis

S.No	Biochemical parameter	Percentage
1	Lipid content	10.5
2	Carbohydrate	28
3	Protein	44

The results were an average of three experiments conducted.

DAIRY WATER ANALYSIS

Nitrogen and phosphate removal

The phosphates are reduced by 97% and 99% after algal bead treatment and sand bed filtration respectively (Table 2). The NH_4^+-N was almost completely reduced after two stage filtration process. There was a significant decrease of N and P which may be because of algal uptake and adsorption on alginate gels were found to be the major processes involved in the removal of N and phosphate in the present study. Previous studies have also reported that air-stripping of ammonia is a possible mechanism for N removal in an intensively aerated micro algal system with alkaline pH resulting from photosynthetic activity and aeration [22–24].

Table 2: Water analysis of three different samples

S.No	Parameter	Control (untreated dairy effluent)	Dairy effluent after algal bead treatment	Dairy effluent after sand bed filtration	Acceptable limits as per APPCB
1	BOD(3 @ 27°C)	1200 mg/L	146 mg/L	32 mg/L	<100 mg/L
2	COD	2900 mg/L	448 mg/L	106 mg/L	<250 mg/L
3	NH_4^+-N	35 mg/L	0.6 mg/L	0.54 mg/L	<50 mg/L
4	Phosphates	30 mg/L	0.87 mg/L	0.39 mg/L	--
5	pH	6.5	8.01	8.52	5.5-9.0

Sampling and analysis were done as per standard methods prescribed by BIS.

Effect on pH

The untreated dairy effluent has an acidic pH value initially but after algal beads treatment and sand bed filtration the treated effluent is shifted to alkaline pH range. NH_4^+-N could be lost via ammonia volatilization while PO_4^{3-}-P was removed by chemical precipitation, because of which alkaline pH was recorded in the two stage treatment system.

BOD and COD removal

Waste water of dairy industry contains large quantities of milk constituents such as casein, lactose, fat, inorganic salts. All these components contribute largely towards their high biochemical oxygen demand. High BOD and COD values lead to the deprival of oxygen for aquatic life in water [25]. BOD and COD have decreased up to 88% and 85% respectively after algal bead treatment (Table 2). The same were further reduced by 98% and 96% after sand bed filtration which was in acceptable levels prescribed by Andhra Pradesh Pollution Control Board (APPCB), Andhra Pradesh, India. Previous studies also showed that algal uptake had little effect on the removal of COD when compared with that of BOD values which is in agreement with our present study [26, 27]. Dairy water after algal beads treatment and sand bed filtration each one liter, were used for performing the mortality studies on twenty *Danio rario* (zebra fish). The results were analyzed in comparison with untreated dairy effluent. It was observed that the mortality was 10% after first stage of treatment and zebra fish survival percentage increased to 100% (Table 3) after second stage treatment which proved that the two stage treated water is non toxic to aquatic life.

Table 3: Relative toxic potential of dairy effluent to *Danio rario*

S.No	Sample	Survival percent				Toxic potential
		24 h	48 h	72 h	96 h	
1	Untreated Dairy effluent	0	0	0	0	Highly Toxic
2	Dairy effluent after algal bead treatment	100	90	90	90	Acceptable
3	Dairy effluent after Sand bed filtration	100	100	100	100	Non-toxic

Algal biomass as bio fertilizer

The algal biomass is also tested for its effectiveness as biofertiliser. A control is taken without adding biofertiliser to the rice seedlings. Another sample is taken with the same number of rice seedlings and 20 gms of dried algal beads were added as bio fertilizer. After 10 days of growth the average length of root and shoot were measured and recorded (Table 4).

Table 4: Lengths of root and shoot in rice plant

S.No	Parameter	Control (without bio fertilizer)	Test (with bio fertilizer)
1	Root	2 cm	4 cm
2	Shoot	10.5 cm	15 cm

Green alga can photosynthesize and fix nitrogen, and these abilities, together with great adaptability to various soil types, make them ubiquitous. Ammonia can be taken up by green alga through passive diffusion or as ammonium (NH_4^+) by a specific uptake system [11].

CONCLUSION

The present study demonstrated that two stage treatment was very effective in removing NH_4^+-N and phosphate from wastewater. A complete removal of NH_4^+-N (100%) and 99% reduction of PO_4^{3-}-P was achieved within 96 h of treatment. The completely treated dairy effluent was proved to be non toxic for the zebra fish because of the drastic decrease in BOD and COD values (98% and 96% respectively). The algal biomass was higher in protein content (44%) and when the same was used as bio fertilizer for rice seeds a 35% increase in growth of the rice plant was observed. Hence the process proved to be useful in treating wastewater for aquatic life and as a bio fertilizer.

ACKNOWLEGEMENTS

The authors thank the Management of Sreenidhi Institute of Science and Technology (SNIST) for their financial support in carrying out this in-house funded project. Special thanks to Vyshanavi DairyPvt Ltd., Khammam, Andhrapradesh, India for providing the dairy effluent for conducting the experiments. The authors also wish to show their gratitude towards Dr. V. Sesha Srinivas, Head department of Biotechnology SNIST, Dr. P. Ravindrababu, Associate Professor, Department of Biotechnology, SNIST and Mr. G. Praveen Kumar for their constant support during the project.

REFERENCES

1. Mulbry WW, Wilkie AC: Growth of benthic freshwater algae on dairy manures. *J Appl Phycol* 2001, 13: 301–306. 10.1023/A:1017545116317
2. Silambarasan T, Vikramathithan M, Dhandapani R, DJ M k, Kalaichelvan PT: Biological treatment of dairy effluent by microalgae. *World J Sci Technol* 2012,2(7):132–134.
3. Van-Coillie R, De La Noue J, Thellen C, Pouliot Y: Tertiary domestic wastewater treatment by Scenedesmus sp.Pilot scale culture. *Rev Sci Eau* 1990,3(4):441–456.
4. Voltolina D, Gómez-Villa H, Correa G: Nitrogen removal and recycling by Scenedesmus obliquus in semicontinuous cultures using artificial wastewater and simulated light and temperature cycle. *Bioresour Technol* 2005, 96: 359–362. 10.1016/j.biortech.2004.04.004
5. Cohen Z: *Chemicals from Microalgae*. Florida, US: Taylor & Francis Ltd: CRC Press; 1999.
6. Raposo MFdJ, Oliveira SE, Castro PM, Bandarra NM, Morais RM: On the utilization of microalgae for brewery effluent treatment and possible applications of the produced biomass. *J Inst Brew* 2010,116(3):285–292. 10.1002/j.2050-0416.2010.tb00433.x
7. Nirupama M: Immobilization of Microalgae. In *Methods in Biotechnology: Immobilization of Enzymes and Cells*. Second edition. Edited by: Guisan JM. Totowa, New Jersey: Humana Press Inc; 2006:373–391.
8. Robinson PK, Reeve JO, Goulding KH: Kinetics of phosphorus uptake by immobilized *Chlorella* . *Biotechnol Let* 1988, 10: 17–20. 10.1007/BF01030017
9. Megharaj M, Peardon HW, Venkateswarlu K: Removal of nitrogen and phosphorus by immobilized cells *Chlorella vulgaris* and *Scenedesmus bijugatus* isolated from soil. *Enz Microb Technol* 1992, 14: 656–658. 10.1016/0141-0229(92)90042-M
10. Tam NFY, Lau PS, Wong YS: Wastewater inorganic N and P removal by immobilized *Chlorella vulgaris* . *Wat Sci Technol* 1994, 30: 369–374.
11. Upasana M, Sunil P: Cyanobacteria: a potential biofertilizer for rice. *Resonance*2004,9(6):6–10. 10.1007/BF02839213
12. Fogg GE: Nitrogen fixation by photosynthetic organisms. *Annu Rev Plant Physiol*1956, 7: 51–70. 10.1146/annurev.pp.07.060156.000411
13. Venkataraman GS: *Algal biofertilizers and rice cultivation*. New Delhi, India: Today and Tommorrow Printers and Publishers; 1972.
14. *Description of commonly considered water quality constituents: Watershed Protection Plan Development Guidebook 0041ppendix B*. Georgia; 2001.
15. Miao X, Wu Q: High yield bio-oil production from fast pyrolysis by metabolic controlling of *Chlorella protothecoides* . *J Biotechnol* 2004, 110: 85–93. 10.1016/j.jbiotec.2004.01.013
16. Miao XL, Wu QY, Wu GF, Zhao NM: Sucrose accumulation in salt-stressed cells of agp gene deletion-mutant in cyanobacterium *Synechocystis sp* . PCC 6803. *FEMS Microbiol Lett* 2003, 218: 71–77. 10.1111/j.1574-6968.2003.tb11500.x
17. Zhu M, Zhou PP, Yu LJ: Extraction of lipids from *Mortierella alpina* and enrichment of arachidonic acid from the fungal lipids. *Bioresour Technol* 2002,84: 93–95. 10.1016/S0960-8524(02)00028-7
18. Shen Y, Yuan W, Pei Z, Mao E: Culture of microalga *Botryococcus* in livestock wastewater. *Trans ASABE* 2008, 51: 1395–1400.
19. Lowry O, Rosebrough N, Farr A, Randall R: Protein measurement with the folin phenol reagent. *J Biol Chem* 1951,193(1):265–275.
20. Dainty AL, Goulding KH, Robinson PK, Trevan MD: Stability of alginate immobilized algal cells routine cultivation of organism. *Biotechnol Bioeng* 1986, 28: 210–216. 10.1002/bit.260280210
21. Rodulfo BR: Culture and utilization of freshwater algae as protein source. In*Proceedings of the Symposium on Culture and Utilization of Algae in Southeast Asia*. Tigbauan, Iloilo, Philippines; 1981. [http://hdl.handle.net/10862/180]
22. McCarthy JJ, Taylor WR, Taft JL: Nitrogenous nutrition of the plankton in the Chesapeake Bay. I. Nutrient availability and phytoplankton preferences. *Limnol Oceanogr* 1977, 22: 996–1011. 10.4319/lo.1977.22.6.0996
23. Xianhai Z, Danquah M, Zheng C, Potumarthi R, Xiao DC, Lu Y: NaCS-PDMDAAC immobilized autotrophic cultivation of *Chlorella* sp. for wastewater nitrogen and phosphate removal. *Chem Eng J* 2012, 187: 185–192.
24. Tam NFY, Wong YS: Effect of immobilized microalgal bead concentrations on wastewater nutrient removal. *Environ Pollut* 2000,107(1):145–151. 10.1016/S0269-7491(99)00118-9
25. Hur J, Lee B-M, Lee T-H, Park D-H: Estimation of biological oxygen demand and chemical oxygen demand for combined sewer systems using synchronous fluorescence spectra. *Sensors* 2010, 10: 2460–2471. 10.3390/s100402460

26. Rajasri Y, Rao CS, Ramgopal Rao S, Ravichandra P: Dairy effluent treatment and lipids production by chlorella pyrenoidosa and euglena gracilis: study on open and closed systems. *Asia P J Chem Eng* Article In press

27. Lau PS, Tam NFY, Wong YS: Effect of algal density on nutrient removal from primary settled wastewater. *Environ Pollut* 1995, 89: 59–66. 10.1016/0269-7491(94)00044-E

CHAPTER 16

IDENTIFICATION OF BIOFLOC MICROSCOPIC COMPOSITION AS THE NATURAL BIOREMEDIATION IN ZERO WATER EXCHANGE OF PACIFIC WHITE SHRIMP, PENAEUS VANNAMEI, CULTURE IN CLOSED HATCHERY SYSTEM

Hidayah Manan[1] Julia Hwei Zhong Moh[1] Nor Azman Kasan[2] Suhaimi Suratman[3] Mhd Ikhwanuddin[1]

[1] Institute of Tropical Aquaculture, Universiti Malaysia Terengganu, 21030 Kuala Nerus, Terengganu, Malaysia

[2] School of Fisheries and Aquaculture Sciences, Universiti Malaysia Terengganu, 21030 Kuala Nerus, Terengganu, Malaysia

[3] Institute of Oceanography and Environment, Universiti Malaysia Terengganu, 21030 Kuala Nerus, Terengganu, Malaysia

ABSTRACT

Study on the microscopic composition of biofloc in closed hatchery culture system was carried out to determine the interaction between the aggregation flocs in the bioremediation process for the decomposition and degradation of organic matter loaded in the shrimp culture tanks. The study was done for 105 days of culture period in zero water exchange. All of the organic loaded in the culture tanks identified comes from the shrimp feces, uneaten fed, and the decomposed macro- and microorganisms died in the culture tanks. All of the microscopic organisms in the biofloc were identified using Advance microscopes *Nikon 80i*. From the present study, there were abundances and high varieties of phytoplankton, zooplankton, protozoa, nematodes and algae species identified as aggregates together in the flocs accumulation. All of these microscopic organisms identified implemented the symbiotic process together for food supply, become the algae grazer, act as natural water stabilizer in regulating the nutrients in culture tank and serve as decomposer for dead organic matter in the water environment. Heterotrophic bacteria identified from *Pseudomonas* and *Aeromonas* family consumed the organic matter loaded at the bottom of culture tank and converted items through chemical process as useful protein food to be consumed back by the shrimp. Overall it can be concluded that the biofloc organisms identified really contributed as natural bioremediation agents in zero water exchange culture system to ensure the water quality in the optimal condition until the end of culture period.

INTRODUCTION

Bioremediation is a process where microorganisms were stimulated with nutrients and other chemicals to enable them to wipe out contaminants in the targeted area and break down the hazardous substances into less toxic or non-toxic substances (Das 2014). Microorganisms are considered as the first living organisms to have evolved and are adaptive with the ecological changes. Nowadays, the use of microorganisms such as bacteria as biodegradation and bioremediation agent has come to attention because of its ability to reduce hazard, success in degrading natural and synthetic substances and accumulating toxic compound (Karigar and Rao 2011). According to Das et al. (2006), microorganisms are responsible for carbon fixation, nitrogen fixation, methane metabolism and sulfur metabolism,

Citation: Hidayah Manan, Julia Hwei Zhong Moh, Nor Azman Kasan, Suhaimi Suratman, Mhd Ikhwanuddin, Identification of biofloc microscopic composition as the natural bioremediation in zero water exchange of Pacific white shrimp, Penaeus vannamei, culture in closed hatchery system, DOI 10.1007/s13201-016-0421-4 thus controlling the biogeochemical cycle. Microorganisms are able to produce diverse metabolic enzymes that can assist for safe removal of contaminants either by direct destruction or converting to safer or less toxic intermediate (Dash and Das 2012). Any microorganisms used as bioremediation has to possess resistant genotype for the particular pollutant and because of it, microorganisms possess certain unique characteristic which make them suitable for bioremediation processes (Stelting et al. 2010).

Biofloc technology (BFT) is a promising technology which promotes the retention of waste and its conversion to biofloc as natural food for shrimp in the aquaculture system (Panigrahi et al.2014). Biofloc consists of microorganisms such as heterotrophic bacteria, algae (dinoflagellates and diatoms), fungi, ciliates, flagellates, rotifers, nematodes, metazoans and detritus that conglomerate together and perform symbiotic processes to maintain the water quality, maintain bio-security, support high density of shrimp culture and reduce water exchange in the aquaculture system. In the biofloc technology (BFT) application, protein is utilized as a feed for the shrimp when the heterotrophic microbe in the biofloc converts the nitrogenous waste in the culture tank from the uneaten feed into protein. Development of dense heterotrophic bacterial community rather than algae dominated will overcome the waste generated in the aquaculture system through in situ bioremediation (Panigrahi et al. 2014). By addition of cheap carbohydrate sources such as molasses or tapioca usually in ration around C: N 12–15:1 in the water column, biofloc will convert the toxic nutrients in the water to beneficial food sources for shrimp consumption. Avnimelech (2009) found out that in high stocking density and zero or minimal water exchange the additional carbon source encourage the development of heterotrophic bacteria in the pond or tank. Schneider et al. (2005) discovered that addition of organic nitrogenous waste, ammonium will be converted into bacterial biomass if C:N ratio is balanced at ratio 10–15:1. Usually in BFT, heterotrophic bacteria are more dominant than nitrifying bacteria because of their higher growth rate and microbial biomass yield per substrate, thus making many fold increase of heterotrophic bacteria (Hargreaves 2006). Identification of the microscopic biofloc composition can help in better understanding the application of biofloc. From the identification of each class of organisms' function (phytoplankton as primary producer, zooplankton as the algae grazer, bacteria and protozoa as organic matter decomposer) that occurs in the zero water exchange culture system, the interaction happening between the organisms in the biofloc system can be understood. Because of the potential of biofloc technology for bioremediation in the aquaculture system, present study was conducted to identify the biofloc microscopic composition and to determine the biofloc performance as the natural bioremediation agent for removal of organic matter loader in the culture tank in zero water exchange system.

METHODOLOGY

Experimental Design

Rounded tank with capacity 8 ton (height = 1.2 m and diameter = 3.3 m) was stocked with Pacific White shrimp, *Penaeus vannamei*, postlarvae at PL10 with density of 100 PL per cubic meter, m^3. Six tanks were used in the experiment for treatments (T1, T2 and T3 and for control tanks, C1, C2 and C3). Molasses as carbohydrate or carbon sources at ratio C:N 10:1 were transferred to the treatment culture tank after being fermented for 24 h to boost the breakdown process by the bacteria or the microorganisms for biofloc formulation. Shrimp were cultured for 105 days until reaching harvested size at PL115. During the culture period, the microorganisms in the culture tank were sampled every week to identify the microorganism's composition in the biofloc aggregation in the closed and zero water exchange system. All water parameters were checked weekly for pH, salinity, DO, TDS, and temperature using YSI multi-probe YSI 556 and nutrients (ammonia, nitrite, nitrate) were analyzed with spectrophotometer by ammonia–salicylate method (Standard Method 8155), nitrite diazotization method (Standard Method 8507) and nitrate–cadmium reduction method (Standard Method 8192) of (DR/2400 Procedure manual 2002).

Sample Collection

3 L of water sample from treatment tank was filtered using plankton net 20 μm for microscopic plankton identification. For bacterial identification, sample water was pipetted out using micropipette and serial diluted until 10^{-5} for bacteria analysis. Sample for plankton analysis was left 24 h for the substrate to settle at the bottom and concentrated to 10 ml of water sample and then preserved with 10 % formalin. All water samples were taken back to laboratory for further analysis.

Microbial Identification

Bacteria were isolated using trypticase soy agar (TSA) and selective agar thiosulphate–citrate–bile salts–sucrose agar (TCBS) for isolating *Vibrio* sp. The colony-forming unit (CFU) from fifth time serial dilution (10^{-5}) was selected for colony counting. Gram staining also was done to identify Gram-positive and -negative bacteria. Catalase test was done to identify Gram-positive bacilli. API kit (*Biomerieux*) API20E and API 20NE were used to identify Gram-negative bacteria. Incubation box, tray and lid were prepared for the strip preparation. For the inoculum preparation, an ampule of API NaCl 0.85 % (2 ml) was selected and 1–4 colonies of bacteria were picked up using inoculation loop from the agar plate and then suspension was prepared with the turbidity equivalent to 0.5 McFarland. For the API 20NE strip inoculation, test nitrate reduction (NO_3) and *p*-nitrophenyl-β-D-galactopyranoside hydrolysis (PNPG) were inoculated by distributing the saline suspension into the tubes using Pasteur pipette. API AUX medium was added to approximately 200 μl of the remaining suspension into the ampule and was homogenized well. Tubes and cupules of test glucose fermentation (GLU) and Phenyl-acetate assimilation test (PAC) were filled with the suspension. Mineral oil was added to the cupules of 3 tests (GLU), arginine hydrolysis (ADH), and urea hydrolysis (URE) until convex meniscuses formed. The incubation box was closed and incubated at 29 ± 2 °C for 24 h. After the incubation period, the strips were read by referring to the reading table. The reactions for (GLU, ADH, URE, aesculin hydrolysis test (ESC)), gelatine hydrolysis (GEL) and (PNPG) were recorded on result sheet. For NO_3 test, 1 drop of NIT 1 and 1 drop of NIT 2 reagents were added to NO_3cupule. For tryptophan deaminase test (TRP), 1 drop of JAMES reagent was added and immediate reaction took place. NIT 1, NIT 2 and JAMES reagents were removed using pipette and test NO_3 and TRP were covered with mineral oil. Kit was reincubated at 29 ± 2 °C for 24 h and all tests were read again except for NO_3, TRP and underlined GLU, which were only read once at 24 h. Identification is obtained with the numerical profile. Database (V6.0) in the API web index was used by entering the seven digit numerical profile in the identification software for species identification.

For API20E strip inoculation, bacterial suspension were distributed into the tubes with pipette for citrate assimilation test (CIT), Voges–Proskauer (VP) test for acetyl methyl carbinol detection, gelatine hydrolysis (GEL) test by filled in both tube and cupules, and for test ADH, lysine decarboxylase test (LDC), Ornithine decarboxylase test (ODC) and production of hydrogen sulfide test (H_2S) and urea hydrolysis (URE) filled with mineral oil in the cupules. The incubation box was closed and incubated at 36 ± 2 °C for 18–24 h. The strip was read by referring the reading table after incubation period. For Tryptophan deaminase test (TDA), 1 drop of TDA reagent was added, for Indole production test (IND) 1 drop of JAMES reagent was added and for VP test, VP 1 and VP 2 reagents were added. Identification is obtained with numerical profile nine digit using the database (v4.1) in the API web index for species identification.

PLANKTON MICROSCOPIC IDENTIFICATION

Advance microscope *Nikon 80i* was used for biofloc microscopic identification and for plankton length and size measurements. Qualitative and quantitative analyses of phytoplankton and zooplankton were done by Lackey's method. Compound microscope was used for phytoplankton counting. The cover slip was placed over a drop of water in the slide and whole of cover slip was examined by parallel overlapping strips to count all the organisms in the drop. About 22 strips were examined in each drop. Number of subsamples to be taken depended on examining 2–3 successive subsamples without addition of an encounter species when compared to the examined subsamples in the same sample (American Public Health Association APHA 1989).

Calculation formula:

Density (cells l−1)=C×AtAs×S×V×volume of concentrated sample(ml)volume of actual water filteredDensity (cells l−1)=C×AtAs×S×V×volume of concentrated sample(ml) volume of actual water filtered where*C* Number of organisms counted, A_t Area of cover slip (22 mm×22 mm), *S* Number of strip counted, A_s Area of strip (22 mm×1 mm), *V* Volume of sample under the cover slip.

RESULTS

There were various types of microscopic organisms identified from bioflocculation. All of the microscopic organisms identified come from different classes of phytoplankton algae and also numerous of algae grazer such as rotifer and nematode also the protozoa, *Vorticella* sp. (Fig. 1). From Gram staining, rod-shaped Gram-positive bacteria were also identified, which are *Bacillus*sp., from the positive result of catalase test. From the API kit analysis, species of bacteria identified come from heterotrophic bacteria (*Aeromonas hydrophila*, *Pseudomonas aeruginosa*) and also anaerobic bacteria, *Vibrio* sp. (ex. *V. fluvialis*). For the water parameter results, mean dissolved oxygen, DO, was 6.67 ± 0.97 mg l^{-1} (5.9–9.53 mg l^{-1}; n = 12), mean temperature 28 ± 0.30 °C (26–28 °C; n = 12), mean pH 7.36 ± 0.49 (6.1–8.2; n = 12), mean salinity 33.66 ± 1.45 ppt (31–36 ppt; n = 12) and mean total dissolved solid, TDS 33.52 ± 1.33 (31.5–35.5 mg l^{-1}; n = 12). The bioremediation process was successfully carried out by the microorganism in the biofloc as the nutrients ammonia, NH_3, drops from 8.0 to 0.3 mg l^{-1}, nitrite drops from 0.8 to 0.5, and nitrate drops from 15.3 to 5.7 mg l^{-1} during the culture period (Fig. 2). Species *Pseudomonas* sp. and *Aeromonas* sp. were identified to be dominant from the colony-forming unit (CFU) counting. *Vibrio* sp. was also identified as aggregates in the bioflocculation (Table 1). The percentage of bacteria identified dominantly come from heterotrophic bacterial species (ex. *Aeromonas hydrophyila*) and also from *Vibrio* spp. (*V. alginolyticus* and *V. fluvialis*) (Fig. 3). There were a lot of and varieties of microscopic organism compositions identified from bioflocculation in treatment tank 1 (Table 2) treatment tank 2 (Table 3) and treatment tank 3 (Table 4). The density of microorganism was also identified based on the Day of Culture (DOC) of shrimp, density of microorganism in DOC11 shrimp (Fig. 4), DOC17 (Fig. 5), DOC30 (Fig. 6), DOC58 (Fig. 7), DOC65 (Fig. 8), DOC76 (Fig. 9), and for DOC93 (Fig. 10).

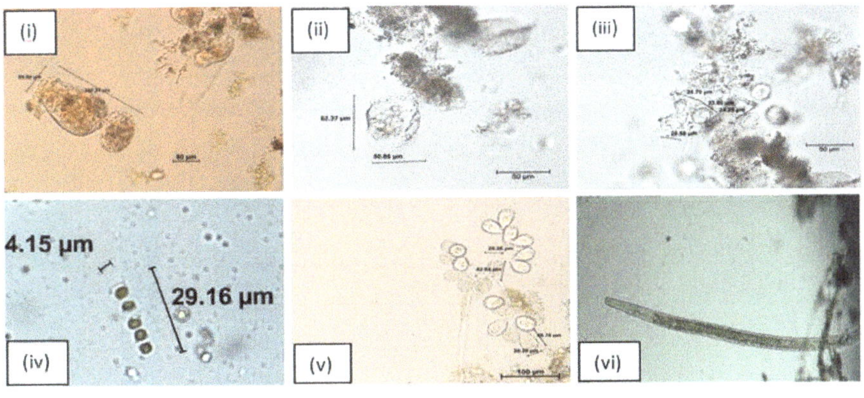

Figure 1: Microscopic organisms identified in the floc: a rotifer; b euplotes, the ciliate protozoa; c*Alexandrium* sp. of dinoflagellates; d *chaetoceros* diatom; e *vorticella* the protozoa and **f**nematode. All observed under ×400 magnifications.

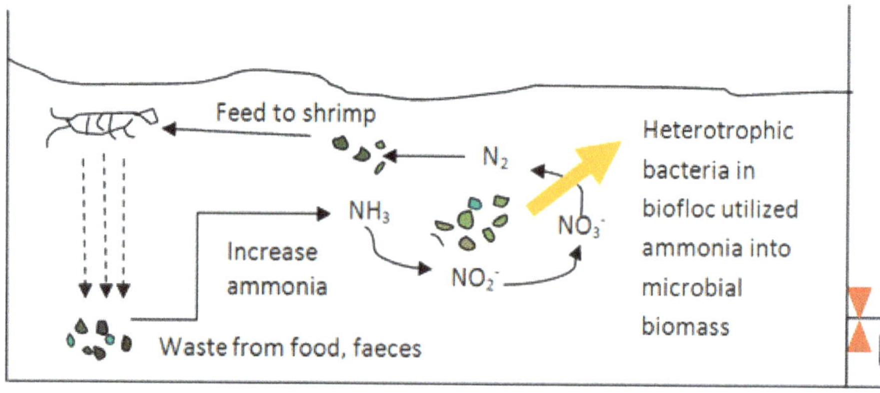

Figure 2: The bioremediation process was successfully carried out by microorganisms in the biofloc (bacteria, algae, plankton) to breakdown the hazardous nutrients in the treatments culture tank into the non-toxic substances and can be consumed back as additional protein feed diet for shrimp consumption.

Table 1: Species of Gram-negative bacteria identified using API kit 20E and 20NE *biomerieux*

Tank	Colony	CFU count	Species identify	% id	Id status	Code	API
Tank 1	3 colony	5×10^{-5}	*Vibrio fluvialis*	98.3	Good id	204650***	20E
		8×10^{-5}	*Vibrio alginolyticus*	82.6	Good id	7010***	20NE
		23×10^{-5}	*Aeromonas hydrophila*	98.9	Good id	704612***	20E
Tank 2	2 colony	13×10^{-5}	*Aeromonas hydrophila*	98.4	Good id	704612***	20E
		34×10^{-5}	*Aeromonas salmonicida*	99.9	Very good id	1550***	20E
Tank 3	3 colony	38×10^{-5}	*Aeromonas salmonicida*	99.9	Very good id	1550***	20NE
		16×10^{-5}	*Pseudomonas aeruginosa*	98.9	Good id	220200***	20E
		3×10^{-5}	*Vibrio alginolyticus*	86.3	Good id	7434***	20NE

*** Profile coding of the analytical profile index

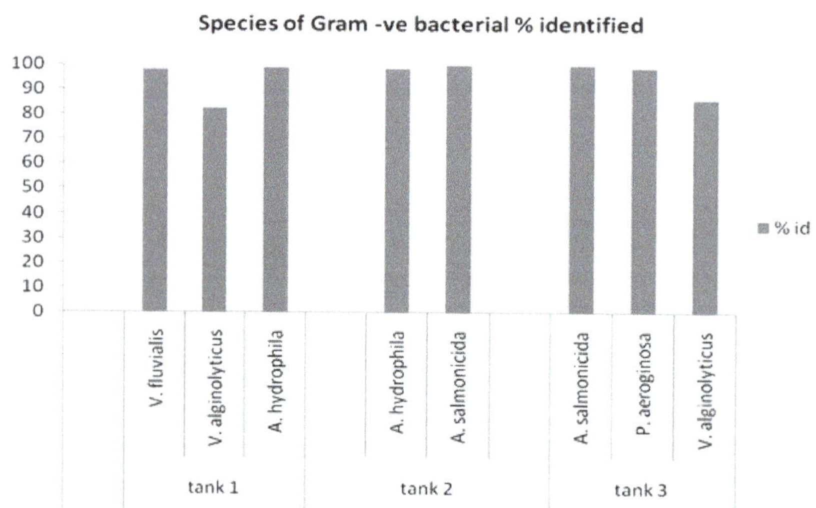

Figure 3: Percentage of bacterial species identified accumulated together in the biofloc in treatment tank 1 (T1), tank 2 (T2) and tank 3 (T3).

Table 2: Type of microorganisms identified as aggregates in the biofloc in Tank 1

Plankton	Tank	Phylum	Class	Genera
Phytoplankton	Tank 1	Ohcrophyta	Bacillariophyceae	*Nitzschia*
				Navicula
				Licmophora
				Amphora
				Cymbella

				Radiolarian
				Cyclotella
				Leptocylindrus
			Coscinodiscophyceae	*Cosinodiscus*
	Tank 1	Cynophyta	Cynophyceae	*Gomphoperia*
				Oscillatoria
	Tank 1	Chlorophyta	Chlorophyceae	*Borodinellopsis*
				Chlorella
				Chlamydomonas
				Tetraselmis
Zooplankton	Tank 1	Rotifera	Brachionidae	*Brachionus*
Protozoa	Tank 1	Ciliophora	Euplotidae	*Euplotes*
		Ciliophora	Vorticellidae	*Vorticella*
		Ciliophora	Parameciidae	*Paramecium*
		Euglenoidea	Euglenaceae	*Euglena*
		Ciliophora	Ciliatea	*Cilliate*
Nematode	Tank 1	Nematoda		*Nematode*

Table 3: Type of microorganisms identified as aggregates in the biofloc in Tank 2

Plankton	**Tank**	**Phylum**	**Class**	**Genera**
Phytoplankton	Tank 2	Ohcrophyta	Bacillariophyceae	*Nitzschia*
				Leptocylindrus
				Navicula
				Cyclotella
				Melosira
				Licmophora
				Cymbella
			Coscinodiscophyceae	*Cosinodiscus*
		Cynophyta	Cynophyceae	*Oscillatoria*
				Gomphosperia
				Gloeocapsa
		Chlorophyta	Chlorophyceae	*Chlorella*
				Borodinellopsis
				Scenedesmus
				Tetraselmis
		Dinophyta	Dinophyceae	*Protoperidinium*
				Alexandrium
Zooplankton	Tank 2	Arthropoda	Copepoda	*Copepod*
		Rotifera	Brachionidae	*Branchius*
Protozoa	Tank 2	Sarcomastigophora	Radiolaria	*Radiolarian*
		Ciliophora	Euplotidae	*Euplotes*
		Ciliophora	Parameciidae	*Paramecium*
		Ciliophora	Ciliatea	*Cilliate*
		Euglenoidea	Euglenaceae	*Euglena*
		Ciliophora	Vorticellidae	*Vorticella*
Nematode	Tank 2	Nematoda		*Nematode*

Table 4: Type of microorganisms identified as aggregates in the biofloc in Tank3

Plankton	Tank	Phylum	Class	Genera
Phytoplankton	Tank 3	Ohcrophyta	Bacillariophyceae	*Nitzschia*
				Leptocylindrus
				Cyclotella
				Licmophora
				Navicula
			Coscinodiscophyceae	*Cosinodiscus*
		Cynophyta	Cynophyceae	*Oscillatoria*
				Gomphosperia
				Gloeocapsa
		Chlorophyta	Chlorophyceae	*Chlorella*
				Borodinellopsis
		Dinophyta	Dinophyceae	*Protoperidinium*
				Alexandrium
Zooplankton	Tank 3	Arthropoda	Copepoda	*Copepod*
		Rotifera	Brachionidae	*Branchius*
		Gastrotricha	Chaetonotida	*Gastrotrich*
Protozoa	Tank 3	Euglenoidea	Euglenaceae	*Euglena*
		Ciliophora	Vorticellidae	*Vorticella*
		Ciliophora	Euplotidae	*Euplotes*
		Ciliophora	Parameciidae	*Paramecium*
		Ciliophora	Ciliatea	*Cilliate*
		Sarcomastigophora	Radiolaria	*Radiolarian*
Nematode	Tank 3	Nematoda		*Nematode*

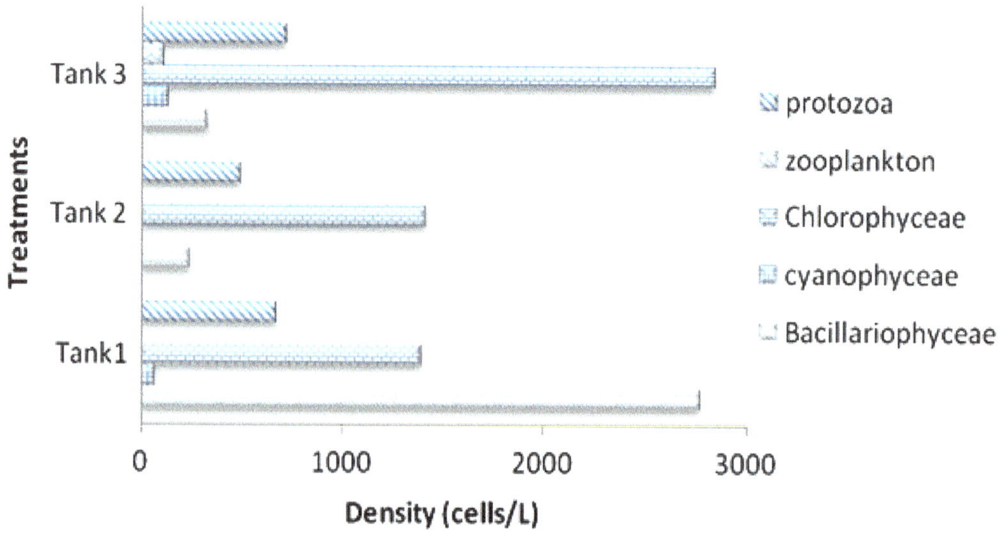

Figure 4: Density of microorganisms composition identified from the bioflocculation in the water column of treatment culture tanks for Day of culture, DOC11.

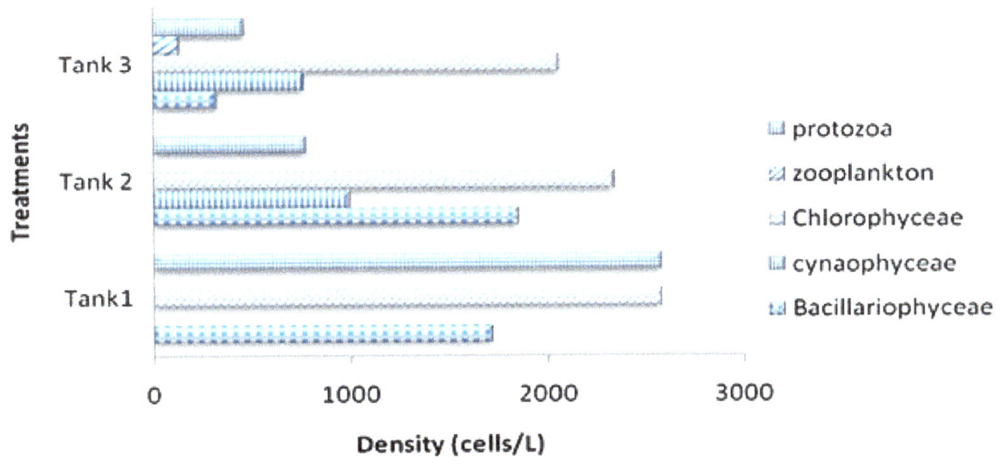

Figure 5: Density of microorganisms composition identified from bioflocculation in the water column of treatment culture tanks for Day of culture, DOC17.

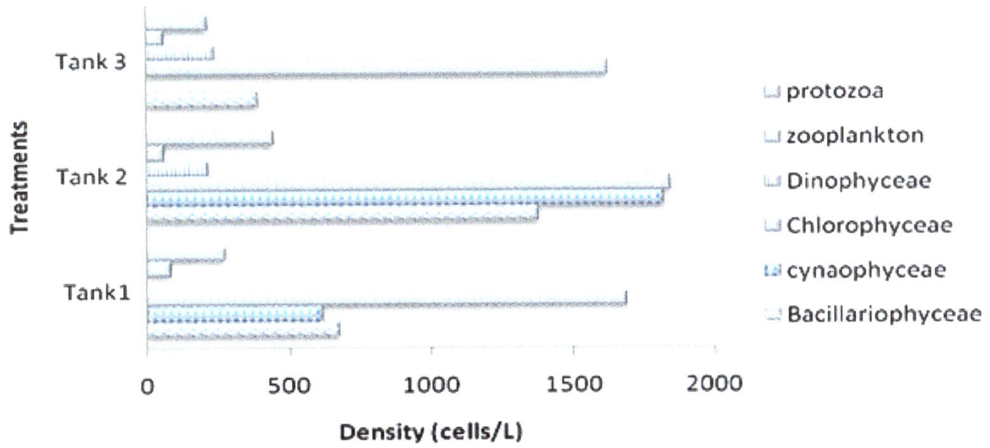

Figure 6: Density of microorganism composition identified from the bioflocculation in the water column of treatment culture tanks for Day of culture, DOC30.

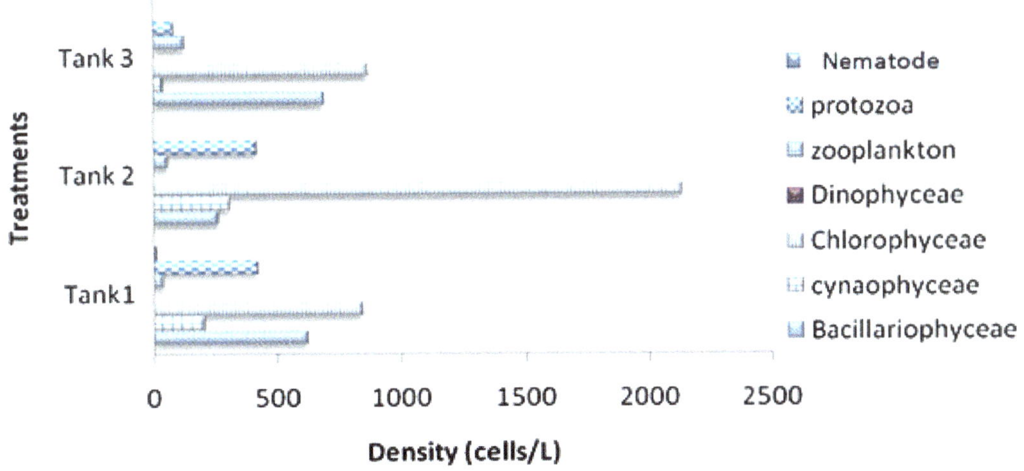

Figure 7: Density of microorganism composition identified from the bioflocculation in the water column of treatment culture tanks for Day of culture, DOC58.

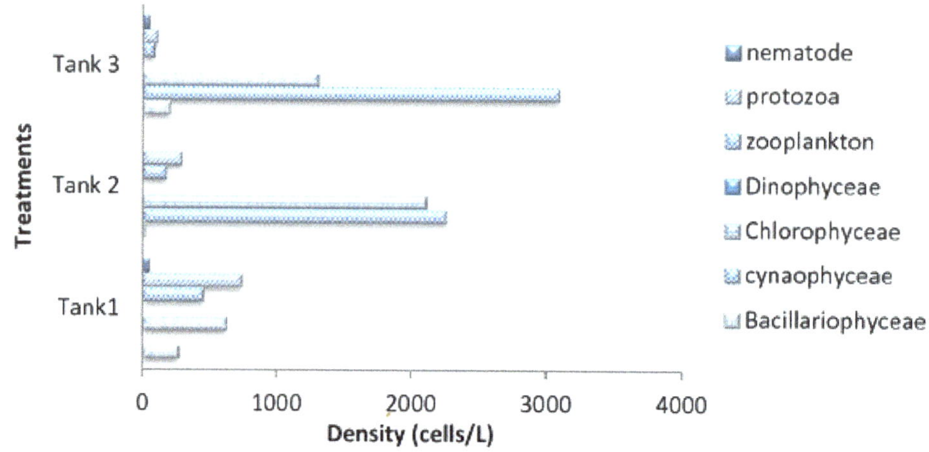

Figure 8: Density of microorganism composition identified from the bioflocculation in the water column of treatment culture tanks for Day of culture, DOC65.

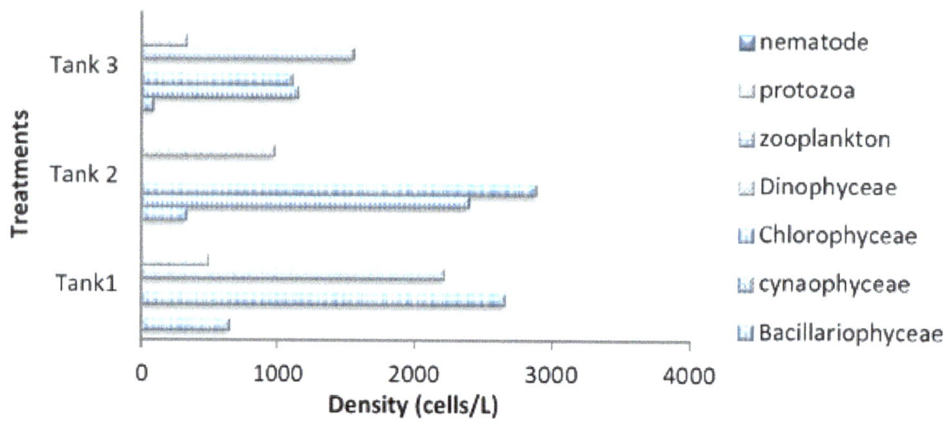

Figure 9: Density of microorganism composition identified from the bioflocculation in the water column of treatment culture tanks for Day of culture, DOC76.

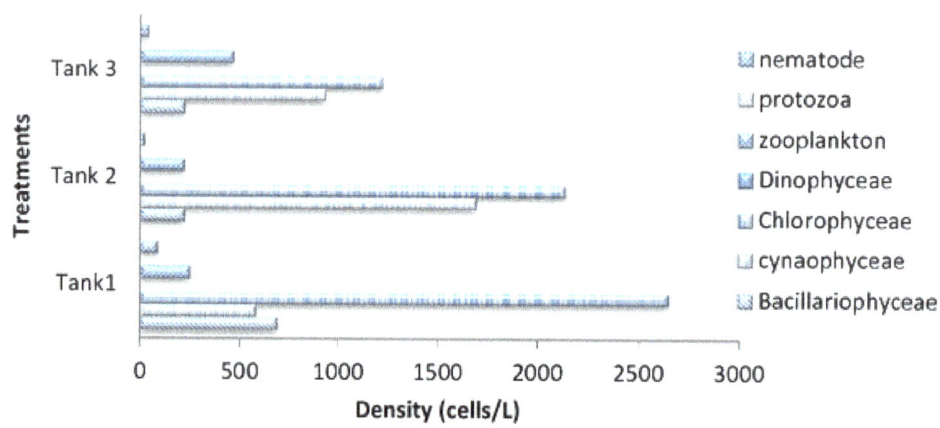

Figure 10: Density of microorganism composition identified from the bioflocculation in the water column of treatment culture tanks for Day of culture, DOC93

DISCUSSIONS

From te identifications done on the biofloc aggregations, the microorganisms can be divided into five overlapping groups which are: floc-forming organisms, saprophytes (organisms that obtain nutrients from dead organic matter), nitrifying bacteria, algae grazers, and pathogenic bacteria (*Vibrio* spp). All of these types of microorganisms have their own function and interaction between each other in the biofloc system to make the bioremediation process successfully happen. The organisms that forming the floc were identified come from some of algae and bacterial biomasses which they used in activated sludge form the organic matter come from the waste in the tank to secrete sticky EPS (Medina and Neis 2007). These EPS were known to have significant effect on the physiochemical properties of the microbial aggregates including structure, surface charge, flocculation, settling properties, dewatering and absorptive capacity Sheng et al. (2010). The floc itself also as the bioremediation agent is able to stick the detritus from the wastes together with other organisms such as protozoa and zooplankton during bioflocculation and bacteria in the floc take up ammonia in the water which mostly comes from the metabolic waste of shrimp and convert it into microbial protein. Hargreaves (2013) discovered that bioflocs are accumulation of algae, bacteria, protozoan, and other kinds of particulate organic matter such as feces and uneaten feed which are held together in a loose matrix of mucus secreted by bacteria and bound by filamentous algae or held by electrostatic attraction. During siphoning process, it can also be seen that the waste at the bottom of the tank was aggregated in small rounded shape, which means the biofloc microorganisms work in settling down the detritus and acted as the bioremediation agent in neutralizing the pollutant in the bottom of the tank and makes the condition of water optimum. For the saprophytes group or the organisms obtaining nutrients from the dead organic matter, heterotrophic bacteria is in this classification. Heterotrophic bacteria identified from the study are *Aeromonas* spp and*Pseudomonas* spp. Heterotrophic bacteria used the organic compound from the organic matter left in the tank as the sources of energy and food which contrast with the autotrophic organisms such as phytoplankton and algae. Protozoa identified in the biofloc treatment tank such as ciliate, vorticella, euplotes and paramecium also classified as saprophytic protozoa as absorb organic matter through their cell wall for food and takes 40 % of the nutrients for the production of protozoan biomass (Merriam Co 1913; Lal 2006). All of these saprophytic microorganisms acted as bioremediation agents in neutralizing the nutrients (ammonia, nitrite and nitrate) from the wastes of uneaten feed and shrimp's fecal secretion that produced ammonia product. The denitrifying bacteria, *Pseudomonas* sp., was identified in the bioflocculation that worked to convert the nitrates NO^{3-} into gaseous nitrogen; N_2 makes the water condition less toxic and maintains the water quality (Schramm et al. 1999). This bacteria also will convert the nitrates in the water to the beneficial protein for shrimp consumption besides getting the food from the pellet given. These are proved through Hargreaves's (2013) study in biofloc system, whereas some of the nitrogen is incorporated into the bacterial cells that become the main component of biofloc, shrimp consumption of this microbial protein will effect for a second time and contribute to shrimp growth. Zooplankton, protozoa and protozoa parasite are classified under the algae grazer category. The organisms identified in the bioflocculation were such as nematode, gastrotrich, euplotes protozoa, vorticella protozoa, ciliate, rotifer and copepod which in dense composition as higher food from algae and phytoplankton types were available in the biofloc treatment tank (Hargreaves 2013). During culture starting from DOC11 to DOC93 (Figs. 4, 5, 6, 7, 8, 9, 10), algae from three different classes indentified to dominate the biofloc culture treatment are Chlorophyceae (Green algae), Bacillariophyceae (Diatoms), Cyanophyceae (Blue green algae) and occasionally from Dinophyceae (Dinoflagellates) class. This finding is same as the studies by Galvez (2015) and Schrader et al. (2011) which found out that the most abundant algae in this biofloc study came from cyanobacteria class followed by Chlorophyta, Heterokontophyta, Euglenophyta and Dinophyta. The organisms classified under pathogenic group identified are protozoa that are harmful to the shrimp and the pathogenic bacteria *Vibrio* sp. such as *Vibrio alginolyticus* that can give infection to the shrimp as suggested by Wei and Wendy (2012). The *Vibrio* sp. was also identified in the biofloc aggregation but in less CFU number. These were not lethal to the shrimps as it is being controlled by the biofloc itself through higher diversity of phytoplankton and algae and also can compete with dominant number of heterotrophic bacteria. Refer to study done by Emerenciano et al. (2013) they discovered that the natural probiotic in the biofloc could internally or externally against the *Vibrio* sp. and ectoparasite from giving harmful to the shrimp. Competing with the dominant heterotrophic bacteria and nitrifying bacteria for the essential nutrients such as nitrogen also will limit the *Vibrio* sp. group from uncontrolled growth (Emerenciano et al. 2013).

CONCLUSION

Bio flocculation is a promising technology towards friendly aquaculture environment. In fact, it can supply additional diet for shrimp's consumption from the biofloculant of a variety microorganisms identified in the floc; biofloc also were recognized to be efficient and successful as a bioremediation and biodegradation agent for maintaining the water quality in the close aquaculture system with the zero water exchange.

ACKNOWLEDGMENTS

The author would like to thank private hatchery (AB Hatchery Sdn. Bhd) for PL shrimp supply for this research project. Authors also would like to thank Fisheries officer Mr Hanif, Assistant Science officer Mr. Sabri Muda and Research Officer Mr. Ikhwan Zakariah and all the hatchery officers and staffs of Institute of Tropical Aquaculture, Universiti Malaysia Terengganu for all the assistance and guidance throughout the research period.

REFERENCES

1. American Public Health Association. APHA (1989) Phytoplankton counting technique. In: Standard methods for the examination of water and wastewater, pp 10–23
2. Avnimelech Y (2009) Biofloc technology—a practical guide book. The world Aquaculture Society, EUA, BatonRouge, Louisiana
3. Das S (2014) Microbial biodegradation and bioremediation. Laboratory of Environmental Microbiology and Ecology (LenME). National Institute of Technology, Rourkela Odisha**(Elsevier Book First Edition)**
4. Das S, Lyla PS, Khan SA (2006) Marine microbial diversity and ecology: present status and future perspectives. Curr Sci 90(10):1325–1335
5. Dash HR, Das S (2012) Bioremediation of mercury and importance of bacterial mergenes. Int Biodeterior Biodegrad 75:207–213
6. Emerenciano M, Gaxiola G, Gerard C (2013) Biofloc Technology (BFT): a review for aquaculture application and animal food industry. Biomass Now-Cultiv Util. doi:10.57772/53902
7. Galvez AO (2015) Plankton communities in shrimp monoculture, integrated biofloc system. Global Aquaculture Advocate MAY/JUN 2015
8. Hargreaves JA (2006) Photosynthetic suspended-growth systems in aquaculture. Aquac Eng 34:344–363

9. Hargreaves JA (2013) Biofloc production systems for aquaculture. SRAC, Southern Regional Aquaculture Center Publication. No. 4503, pp 1–12

10. Karigar CS, Rao SS (2011) Role of microbial enzymes in the bioremediation of pollutants: a review. Enzym Res. doi:10.4061/2011/805187

11. Lal R (2006) Encyclopedia of soil science. In: Griffiths BS (ed) Nature. Scottish Crop Research Institute, Invergowrie. doi:10.1081/E-ESS-120042737

12. Medina M, Neis U (2007) Symbiotic algal bacterial wastewater treatment: effect of food to microorganism ratio and hydraulic retention time on the process performance. Water Sci Technol 55(11):165–171

13. Merriam Co G (1913) Protozoa, their role and impact on aquaculture. Webster Revised Unabridged Dictionary, Encyclopedia Dictionary (Aquaculture) published

14. Panigrahi A, Sundaram M, Ravichandran P, Gopal C, (2014) Microbial soup-Eco based approach for shrimp culture and management. CIBA (*Indian Council of Agricultural Research*) ENVIS Centre Newsletter vol 12:2. 2 Apr–Jun 2014

15. Schrader KK, Green BW, Perschbacher PW (2011) Development of phytoplankton communities and common off-flavors in a biofloctechnology system used for the culture of channel catfish (Ictaluruspunctatus). Aquac eng 45:118–126

16. Schramm A, de Beer DA, Van den Heuvel S, Ottengraf S (1999) Microscale distribution of populations and activities of *Nitrosospira* and *Nitrospira* spp. along a macroscale100 sustainable aquaculture technique gradient in a nitrifying bioreactor: quantification by in situ hybridization and the useofmicrosensors. Appl Environ Microbiol 65:3690–3696

17. Schneider O, Sereti V, Eding EH, Verreth JAJ (2005) Analysis of nutrient flows in integrated intensive aquaculture systems. Aquac Eng 32:379–401

18. Sheng G, Yu H, Li X (2010) Extracellular Polymeric Substances (EPS) of microbial aggregates in biological wastewater treatment systems: a review. Biotechnol Adv 28(6):882–894

19. Stelting S, Burns RG, Sunna A, Visnovsky G, Bunt C (2010) Immobilization of *Pseudomonas* sp. strain ADP: a stable inoculant for the bioremediation of atrazine. In: 19th World congress of soil science, soil solutions for a changing world, Brisbane, Australia

20. Wei LS, Wendy W (2012) Chracterization of *vibrio alginolyticus* isolated from white leg shrimp (Litopenaeus vannamei) with emphasis on its antibiogram and heavy metal resistance pattern. Verterinarski Arh 82(2):221–227

CHAPTER 17

THE USE OF NILE TILAPIA (OREOCHROMIS NILOTICUS) CULTIVATION WASTEWATER FOR THE PRODUCTION OF ROMAINE LETTUCE (LACTUCA SATIVA L. VAR. LONGIFOLIA) IN WATER RECIRCULATION SYSTEM

Hefni Effendi[1,4], Sri Wahyuningsih[2], Yusli Wardiatno[3]

[1]Center for Environmental Research, Bogor Agricultural University (IPB), Dramaga, Bogor 16680, Indonesia

[2] Graduate Program of Aquatic Resources Management, Bogor Agricultural University (IPB), Dramaga, Bogor 16680, Indonesia

[3] Department of Aquatic Resources Management, Faculty of Fisheries and Marine Sciences, Bogor Agricultural University (IPB), Dramaga, Bogor 16680, Indonesia

[4] Department of Aquatic Resources Management, Faculty of Fisheries and Marine Sciences, Bogor Agricultural University (IPB), Dramaga, Bogor 16680, Indonesia

ABSTRACT

In the recirculation aquaponic system (RAS), fish farming waste was utilized as a nutrient for plant, minimizing the water need, reducing the waste disposal into the environment, and producing the fish and plant as well. The study aimed to examine the growth of romaine lettuce (*Lactuca sativa* L. var. Longifolia) in aquaponic system without the addition of artificial nutrient. The nutrient relies solely on wastewater of nile tilapia (*Oreochromis niloticus*) cultivation circulated continuously on the aquaponic system. The results showed that tilapia weight reached 48.49 ± 3.92 g of T3 (tilapia, romaine lettuce, and inoculated bacteria), followed by T2 (tilapia and romaine lettuce) and T1 (tilapia) of 47.80 ± 1.97 and 45.89 ± 1.10 g after 35 days of experiment. Tilapia best performance in terms of growth and production occurred at T3 of 3.96 ± 0.44 g/day, 12.10 ± 0.63 %/day, 96.11 ± 1.44 % and 1.60 ± 0.07 for GR, SGR, SR, and FCR, respectively. It is also indicated by better water quality characteristic in this treatment. Romaine lettuce harvests of T2 and T3 showed no significant difference, with the final weight of 61.87 ± 5.59 and 57.74 ± 4.35 g. Overall, the integration of tilapia fish farming and romaine lettuce is potentially a promising aquaponic system for sustainable fish and horticulture plant production.

INTRODUCTION

Aquaculture has been a fast-growing industry because of significant increases in demand for fish and seafood throughout the world. It is growing more rapidly than any other segment of the animal culture industry (Qin et al. 2005). The development of this activity causes extremely high water needs, regardless of water scarcity (McMurtry et al. 1997), and accompanied by increasing environmental impacts (Endut et al. 2011). On the other hand, the clean water lack and land loss led to a decline in agricultural output which is the accommodation of human population growth (Fedoroff et al. 2010).

Discharge from tank effluents in aquaculture contain dissolved nutrient such as nitrogen (N) and phosphorus (P), specific organic and inorganic compounds, and total suspended solids (TSS). These constituents originate primarily from uneaten feed and metabolic wastes from the fish (Seawright et al. 1998; Piedrahita 2003; Sugiura et al. 2006). These waste will

Citation: Hefni Effendi, Sri Wahyuningsih, Yusli Wardiatno, The use of nile tilapia (Oreochromis niloticus) cultivation wastewater for the production of romaine lettuce (Lactuca sativa L. var. longifolia) in water recirculation system, DOI 10.1007/s13201-016-0418-z.

be accumulated in the culture system and exerting a negative feedback on fish growth and survival (Beveridge et al. 1997). From several types of nitrogen dissolved in the water, ammonia (NH_3) is most dangerous for fish, and most of tropical species are generally more sensitive to ammonia (Effendi et al. 2015c; Wang and Leung 2015). NH_3 causes decreasing growth due to reduced appetite and feed intake level (Hargreaves and Kucuk 2001). Ammonia in water is present in two forms, referred to as total ammonia nitrogen (TAN), and represented as non-ionized ammonia (NH_3) and ionized ammonia (NH_4^+) (Francis-Floyd et al. 1996; Losordo et al. 1998; Hargreaves and Kucuk 2001; Rahmani et al. 2004; Rijn et al. 2006; Titiresmi and Sopiah 2006).

Thus for further expansion of aquaculture activities, development and application of new technologies are required, by which water and nutrients can be recovered during cultivation process, so as to reduce the impact on environment (Hu et al. 2015). Recirculation aquaponic system (RAS) is a promising technology in the integration of fish and hydroponic plant production (Endut et al. 2010). RAS is designed to replace 5–10 % of the system volume daily with new fresh water (Masser et al. 1999). Aquaponic systems are recirculating aquaculture systems that integrate hydroponic production of plants and the aquaculture production of fish in recirculation systems (Diver 2006; Tyson et al. 2011; Rakocy et al. 2006; Endut et al. 2010; Roosta and Hamidpour 2011; Zheljazkov and Horgan 2011). In this system, ammonia is converted by nitrification bacteria and assimilated by the aquaponic plants (Tokuyama et al.2004). Compared with the conventional cultivation system, RAS is more profitable and able to provide additional benefits. Aquaponics can be used as a method or system to reduce water requirements, and fish and vegetables can be produced in a mutually benefit water-reuse (McMurtry et al. 1997; Simeonidou et al. 2012).

Plants that commonly used in aquaponic are water spinach (Endut et al. 2010, 2011; Effendi et al. 2015a), spinach (Shete et al. 2013), Lettuce (Simeonidou et al. 2012; Buzby and Lin 2014; Effendi et al. 2015b; Wahyuningsih et al. 2015), tomato (Roosta and Hamidpour 2011), cucumber (Tyson et al. 2008; Graber and Junge 2009), and pepper (Roosta and Mohsenian2012). Vegetable such as lettuce can be used in aquaponics system, because it can be harvested in a short time (3–4 weeks in the system), and relatively fewer problems with pests compared with fruiting plants, has low to medium nutritional requirements and is well adapted to aquaponic systems (Diver 2006, Rakocy et al. 2006). The type of romaine lettuce (*Lactucasativa* L. var. *Longifolia*) is green, and loved by consumer because its leaves are crispy (Zhan et al. 2012, 2013).

Nile tilapia (*Oreochromis niloticus*) is a type of fish used in the aquaponics system (Delis et al.2015; Liang and Chien 2013; Love et al. 2015; Wang et al. 2016). Nile tilapia has a good tolerance level to various environmental conditions, well-grown in aquaponic system using vegetables (Effendi et al. 2015c), and has a high economic value (Diver 2006). Therefore, this study aimed to examine the growth performance of romaine lettuce and nile tilapia in recirculation aquaponic system without the addition of artificial nutrient, and compare it with the system without the addition of plants.

MATERIALS AND METHODS

The study was conducted for 6 weeks on February–April 2015, in the Laboratory of Center for Environmental Research (PPLH-IPB), Indonesia. This research used recirculation aquaponic system (RAS), where the water was not changed during the experiment. RAS consisted of nine aquariums (80 × 40 × 40 cm^3) filled with 100 L of water for fish cultivation, tank (60 L) added biobal as bacteria surface media growth, and hydroponic subsystem for romaine lettuce cultivation. Each set of installation was equipped with a recirculation pump and aerator. The experiment consisted of three treatments with three replications, namely treatment without crops as a control (T1), treatment containing only romaine lettuce (T2), and treatment of romaine lettuce and inoculation with commercial bacteria (T3). The design of recirculation system can be seen in Fig. 1.

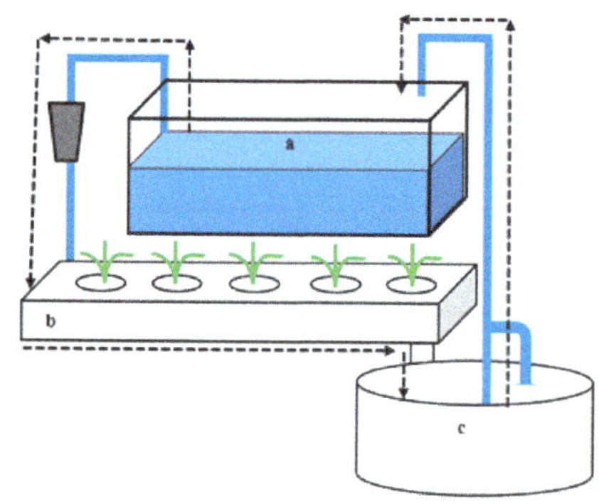

Figure 1: Aquaponic system design. a Aquarium for fish culture; b Gutter for romaine lettuce planting, c Tank; *dotted line with arrow* water current flow.

The water in the aquarium prior to usage was aerated for 1 week to dissolve oxygen in the water. Tilapia was stocked in each aquarium with the same density of 20 fishes per aquarium (average 20 g), sizes ranging from 9 to 10 cm. Fishes were acclimatized for 1 week before usage, in order to accumulate fish waste to enable the supply of nutrients for romaine lettuce. Fishes were maintained for 35 days and fed with pellet (commercial feed) as much as 3 % of body weight with 40 % feed protein content, three times a day (morning, noon, and afternoon). Water samples were taken every week. There was no water replacement during the experiment, except for the addition to replace the water lost due to evaporation.

Commercial bacteria (*Nitrobacter* sp. and *Nitrosomonas* sp. with a density of 10^6 CFU/mL) were added to T3 as much as 32 mL per week according to the instructions on the package. The addition of the bacteria into this system is intended to supply organic substances decomposing bacteria. The addition of the bacteria was expected to increase nitrification process, expected to reduce nitrogen content, especially in the form of NH_3 toxic to fish, and provided dissolved nutrients for plants.

2-week-old lettuce seedlings (average height 11 cm) were planted in small pots with planting distance of 20 cm (except T1). Lettuce introduction into hydroponics chamber was carried out 1 week after the fish entering into the aquarium. Each experimental plot was planted five lettuces with rockwool as planting medium. Romaine lettuce was planted by the nutrient film technique (NFT) system. During lettuce cultivation, there was no addition of nutrients because the nutrients relied only from tilapia fish farming waste.

The produced biomass of nile tilapia was harvested at the end of the experiment. Meanwhile, to assess the performance of nile tilapia in the system, growth and feed data were collected. Data included the number of fed, lengths and weights of fish measured every week. 50 % of fish were randomly selected to estimate the average weight, the fish production and adjustment of the amount of feed. Fish growth was monitored starting from stocking to the end of the experiment. Specific growth rate (SGR), growth rate (GR) and survival (SR) based on Zonneveld et al. (1991), and feed conversion (FCR) based on Ridha and Cruz (2001), were measured by the following equation:

$$GR = \frac{(W_t(g) - W_0(g))}{t(\text{day})} \quad (1)$$

$$SGR = \frac{(\ln W_t - \ln W_0)}{t(\text{day})} \times 100\,\% \quad (2)$$

where W_t and W_0 are the wet weight at time t and at time 0, and t is the observation time.

$$SR = \frac{N_0 - N_t}{N_0} \times 100\,\% \quad (3)$$

where N_0 and N_t are fish number at time 0 and at time t.

$$FCR = \frac{\text{Total weight of dry feed given (g)}}{\text{Total wet weight gain (g)}} \quad (4)$$

Romaine lettuce growth periods, height, leaf width, and number of leaves were measured every week. The produced biomass was harvested at the end of the experiment. Lettuce daily growth rate (DGR) (Ridha and Cruz 2001) and romaine lettuce relative growth rate (RGR) (Gaudet in Mitchell 1974) were calculated:

$$DGR = \frac{[H_t(\text{cm}) - H_0(\text{cm})]}{t(\text{day})} \quad (5)$$

where H_t and H_0 are romaine lettuce high at time t and at time 0, and t is the culture period.

$$RGR = \frac{[\ln W_t(g) - \ln W_0(g)]}{t(\text{day})} \quad (6)$$

where W_t and W_0 are wet weight of lettuce at time t and at time 0, and t is the culture period.

The frequency of sampling and observations was carried out every week for 35 days. Water samples were taken weekly from the tank, i.e., reservoir of water before entering into the aquarium. Water samples were analyzed for the content of TAN, nitrite (NO_2), nitrate (NO_3) and total bacteria. Dissolved oxygen and temperature were measured every week using DO meter TOA 14P, while the pH was measured using pH meter TI 9000. Water quality analysis referred to standard methods of APHA (2008).

Statistical analysis by ANOVA was carried out using statistical package for the social sciences (SPSS). If there were significant differences at significant level of 0.05, then Duncan multiple comparison test was used to compare the means to identify significant difference between the treatments.

RESULTS AND DISCUSSION

Water Quality

The quality of water at the beginning and end of the experiments is summarized in Table 1. These results were a characteristic quality of the water after treatment and before entering into the aquarium. Temperature and pH values fluctuated during the experiment and showed significant difference ($p < 0.05$). Water temperature for all treatments varied within a narrow range 29.37–29.80 °C at the beginning, and 28.47–29.20 °C at the end of the experiment. The average value of the temperature during the experiment is still within the normal range (average temperature 29 °C), but according to Colt (2006) optimal temperature for the life of tilapia is 28 °C. The pH values decreased during the experiment. At the beginning of the experiment, the pH value was still relatively high (>7), then declined until the end of the experiment (<7) for all treatments. The reduction in pH was likely caused by the respiration of fish and bacteria that produce CO_2. The presence of CO_2 will shift the equilibrium carbonate reaction, produces H^+ ions, and lowers the pH. Decrease in the pH was presumably also associated with the oxidation process undertaken by bacteria. According to Princic et al. (1998), in environments with high inputs such as ammonia from aquaculture wastewater, oxidation of this compound produces CO_2 and lowers the pH. DO levels were above 5 mg/L at the beginning of the experiment in all treatments, and ranged 4.82–4.98 mg/L at the end of the experiment. The concentration of dissolved oxygen indicated no significant difference among the treatments ($p > 0.05$). Dissolved oxygen becomes an important parameter, because it is needed in the process of oxidation of ammonia and becomes the major limiting factor for the survival of fish. The optimum DO concentration for optimum fish growth should be maintained above 5 mg/L (Masser et al. 1999; Colt 2006), and the DO concentration under 2 mg/L, ammonia and nitrite oxidation by nitrifying bacteria becomes inefficient anymore (Masser et al. 1999; Hargreaves2006).

Table 1: The quality of water at the beginning and end of the experiment

Parameters	At the beginning			At the end		
	Treatments			Treatments		
	T1	T2	T3	T1	T2	T3
Temperature (°C)	29.40	29.80	29.37	28.57	29.20	28.47
Dissolved oxygen (mg/L)	5.73	5.50	5.52	4.96	4.82	4.98

pH	8.15	8.14	8.16	6.99	6.38	6.17
NH_3 (mg/L)	0.40	0.45	0.39	0.04	0.007	0,004
NH_4 (mg/L)	3.17	3.98	3.60	4.72	3.33	3.10
NO_2 (mg/L)	0.03	0.03	0.03	0.93	0.67	0.47
NO_3 (mg/L)	0.95	0.97	0.91	1.52	1.12	1.25
Bacterial abundance (CFU/mL)	$2.25 \times 10^4 \pm 1.54 \times 10^4$	$5.90 \times 10^4 \pm 2.82 \times 10^4$	$1.11 \times 10^5 \pm 1.03 \times 10^5$	$4.00 \times 10^5 \pm 6.08 \times 10^4$	$5.46 \times 10^5 \pm 1.83 \times 10^4$	$1.15 \times 10^6 \pm 6.93 \times 10^5$

T1 control, *T2* romaine lettuce treatment, *T3* romaine lettuce and inoculation with commercial bacteria

Dissolved inorganic nutrients [ammonia (NH_3), ammonium (NH_4), nitrite (NO_2^-), and nitrate (NO_3^-)] were not significantly different for each treatment ($p > 0.05$). But T3 provided better results when compared with T1 (control). NH_3 concentration did not show any significant variation at the beginning among treatments, but showed reduction at the end of the experiment, and T3 showed value lower than T1 and T2. The concentration of NH_3 for fish growth should not exceed 0.05 mg/L (Losordo et al. 1998). It indicated that the value of ammonia at the end of the experiment for all treatments is still within safe limits for fish. NH_3concentration in this study was lower than the research results of Kamal (2006), using tilapia with bell pepper (*Capsicum annuum* L.) in aquaponic systems reaching 0.75 ± 0.03, 0.31 ± 0.03, and 0.41 ± 0.03 mg/L for treatment without plants, treatment of 15 plant per m^2, and the treatment of 10 plant per m^2, respectively. Different from the NH_3, NH_4 tended to increase during the experiment, and at the end of the experiment declined slightly in T2 and T3. Plants in T2 and T3 might absorb NH_4 as the major source of N for aquatic plants. Meanwhile no plant in T1 caused unutilized nutrient. Therefore, the NH_4 concentration in T1 was higher as compared with that in T2 and T3.

Nitrite (NO_2) is an unstable compound, and with enough oxygen will be easily oxidized to $NO_3$by nitrification bacteria. The concentration of the NO_2 tended to increase during the experiment, especially in T1. Although during the experiment NO_2 increased, the concentration was still safe for the life of nile tilapia. This is because oxygen supply is enough for NO_2oxidation process. Thus, NO_2 did not accumulate in the system. In the circulatory system, NO_2concentration should not exceed 10 mg/L for long periods and in most cases should remain under 1 mg/L (Losordo et al. 1998). On the other hand, NO_3 concentration also increased during the experiment, and the concentration at the end is greater than the beginning of the experiment. NO_3 is the source of N besides NH_4, and these results indicated that romaine lettuce prefers NH_4 than NO_3 as a source of N. NH_4 is a form of N that can be used directly by plants, while NO_3 should be converted in advance into another form. According to Xu et al. (1992), in the plant tissue, accumulation of NH_4 is higher than NO_3, especially when grown under high N conditions. NH_4 assimilation occurs relatively rapidly by plants and metabolic reactions are more efficient than NO_3. The low NO_3 removal by lettuce has been documented in other aquaponic systems (Lennard and Leonard 2006; Buzby and Lin 2014). During the experiment, the concentration of NO_3 was still supportive for the life of nile tilapia. According to Watson and Hill (2006), NO_3 should be maintained below 100 mg/L.

Bacterial abundance in each treatment was statistically different ($p < 0.05$). The abundance of bacteria in T3 was higher than in T1 and T2, from the beginning until the end of the experiment. Addition of commercial bacteria in T3 caused increasing bacterial population. The high bacterial density in T3 did not affect the high removal of inorganic nitrogen (especially NH_3) if compared with T1 and T2. According to Tyson et al. (2008), operating optimal conditions for the biofiltration process, especially reconciling pH for ammonia, was more important than nitrifying bacteria population in system.

Growth and Feeding of Tilapia

The average weight of tilapia increased from time to time for all treatments. Final weight of 48.49 ± 3.92 g maximum was reached at T3, followed by T2 (47.80 ± 1.97 g) and T1 (45.89 ± 1.10 g). Similarly, the tilapia fish biomass in the test system increased during maintenance. This could be seen from the growth of the fish biomass of 914.68 ± 53.09 g (T3), followed by T2 and T1 as much as 895.51 ± 104.23 g (T2) and 800.83 ± 21.10 g (T1), reaching twice of the initial biomass.

Average weight of tilapia was presented in Fig. 2. T3 growth was better than that of T2 and T1. During the experiment, the fishes were fed 3 % of body weight and the amount of feeding increased following the increase of fish weight each week. The average consumption of feed from the beginning to the end of the experiment ranged from 12.40 ± 0.14 to 29.10 ± 2.35 g/day.

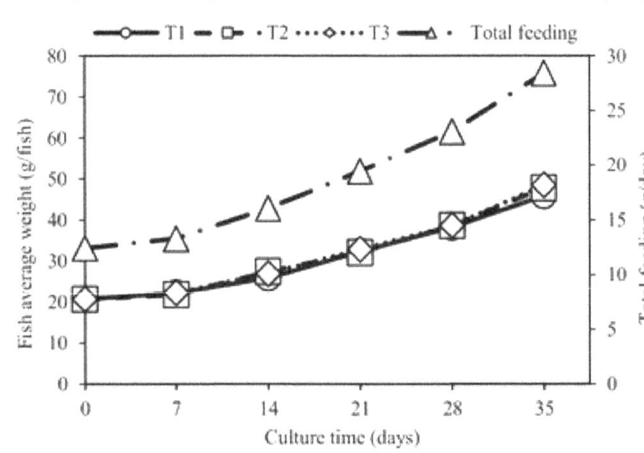

Figure 2: Average growth in weight and feeding for T1 control, T2 romaine lettuce, T3 romaine lettuce and inoculation with bacteria.

The GR and SGR rate in each treatment increased with time of experiment, and indicted no significant difference (*p*

> 0.05). The GR and SGR of nile tilapia can be seen in Fig. 3a, b, pointing out the same trend on each treatment; however, nile tilapia in T3 showed better growth than in T1 and T2. The highest GR at the end of experiment was 3.96 ± 0.44 g/day of T3, followed by T2 (3.88 ± 0.31 g/day), and T1 (3.59 ± 0.13 g/day). Likewise with the highest value of T3, SGR was 12.10 ± 0.63 %/day, followed by the T2 (11.97 ± 0.76 %/day), and T1 (11.37 ± 0.75 %/day). In the treatment without romaine lettuce and without inoculation of bacteria (T1), GR and SGR indicated the lowest value. This is in accordance with higher total ammonia (NH_3 and NH_4) concentration in T1 than that in T2 and T3. Exposure of nile tilapia to relatively high concentration of total ammonia in T1 (0.04 mg/L for NH_3 mg/L and 4.72 mg/L for NH_4 at the end of the experiment) indicated changes in the appetite, utilization of energy, and lower growth of tilapia. Ammonia concentration with a range of 0.02–0.07 mg/L has shown to slow growth and cause several tissue damage (Losordo et al. 1998). Reduced growth caused by ammonia exposure has been documented in other researches. Increasing ammonia concentration decreased growth of all species (juvenile hybrid striped bass, channel catfish, and blue tilapia). Ammonia exposure indicated reduction of fish appetite and reduction of digestibility of consumed feed. In addition, exposure to ammonia also caused liver glycogen depletion and consequent blood acidosis. This has contributed to increased susceptibility to hypoxia, histopathology effects mainly affecting the gills and liver function (Hargreaves and Kucuk 2001).

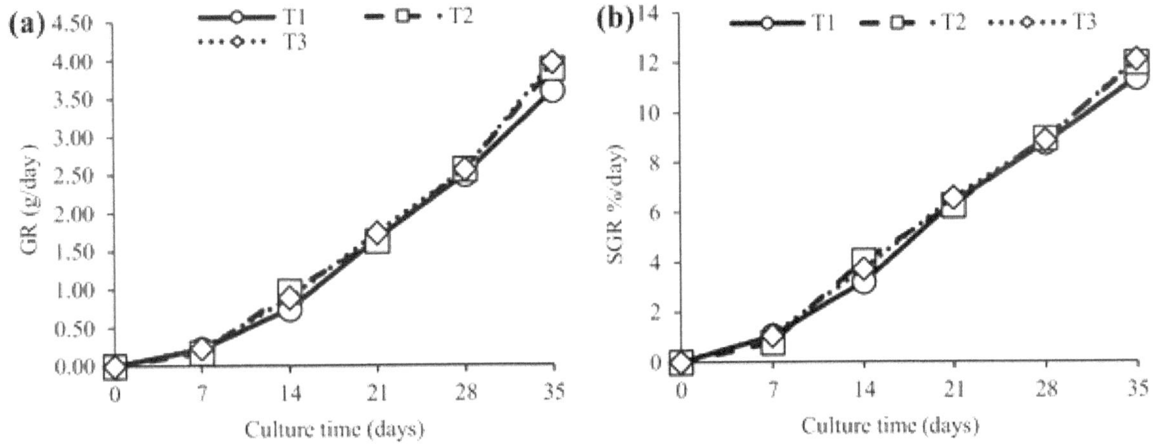

Figure 3: The rate of a GR and b SGR for T1 control, T2 romaine lettuce, T3 romaine lettuce and inoculation with commercial bacteria.

Meanwhile, the survival rate in T3 was higher than that in T1 and T2, and the statistical difference was significant ($p < 0.05$) (Table 2). Fish death at T1 occurred in the second week until the end of the experiment, the death in T2 occured in the second week, fourth week, and end of the experiment, meanwhile the death in T3 occured in the third week until the end of the experiment. Tilapia death that occured in this study (especially in T1) was allegedly caused by exposure of ammonia during the maintenance periods. The brief exposure levels of ammonia in this study only have an impact on growth, but the ammonia will cause lethal effects when exposed for a long time. Ammonia toxicity assumed due to NH_3 molecules has the ability to diffuse in the cell membrane (Colt 2006). Normally NH_3 is excreted by passive diffusion across the gill epithelium, from blood to water in response to an NH_3 gas partial pressure gradient maintained by combination of NH_3 with protons formed from the hydration of excreted CO_2 in the epithelial boundary layer. If the partial pressure of NH_3 in the epithelial boundary layer is greater than that in blood, excretion of NH_3 is inhibited, and NH_3 will diffuse from water into the blood (Hargreaves and Kucuk 2001).

Table 2: The average growth of nile tilapia (*Oreochromis niloticus*)

Growth indicator	Treatment		
	T1	T2	T3
Survival rate (SR) (%)	89.17 ± 0.96^a	94.44 ± 6.45^b	96.11 ± 1.44^b
Food conversion ratio (FCR)	2.02 ± 0.13^a	1.70 ± 0.40^a	1.60 ± 0.07^a

T1 control, *T2* romaine lettuce treatment, *T3* romaine lettuce and inoculation with commercial bacteria. Values with the same superscript letters are not significantly different at the test level of 5 %

The value of food conversion ratio (FCR) showed no significant difference among treatments ($p > 0.05$). FCR is the amount of feed (kg) which is given to produce 1 kg of fish meat, and the best FCR in this study belonged to T3, followed by T2 and T1 (Table 2). The lower value of FCR indicates less food needed to produce 1 kg of fish meat. FCR of T3 (1.60) means that for the production of 1 kg tilapia, it is required 1.60 kg of food. The higher ammonia concentration of T1 caused inefficient FCR, when compared with T2 and T3. Ammonia caused high FCR, as a result of some of the energy used to process ammonia excretion. According to Hargreaves and Kucuk (2001), high FCR in line with the increasing ammonia concentration is due to digestibility of dietary protein, and energy sources may have been affected. FCR for nile tilapia generally ranges from 1.4: 1 to 1.8: 1 (DeLong et al. 2009). The average of FCR value was comparable with the 1.69 (tilapia in the aquaponic system with prawn) and 1.72 (tilapia in the aquaponic system without prawn) reported by Sace and Fitzsimmons (2013), and better than 1.81 and 1.86 (tilapia and hydroponics bell pepper), and 2.2 (tilapia without plant) obtained by Kamal (2006).

Crop Yield

Romaine lettuce (*Lactuca sativa* L. var. Longifolia) could grow without the addition of extra nutrients, and only comes from tilapia fish farming waste. During the experiment, romaine lettuce grew rapidly and showed a positive response to nutrients derived from tilapia fish farming waste. Romaine lettuce growth was well marked with color of fresh green leaves and no signs of nutritional deficiency (Fig. 4a). Root system grew well with the length reaching 15.5–14.5 cm for T2 and T3 (Fig. 4b), and there was no significant difference. During the cultivation period, there was no dead of romaine lettuce. Plants from the type of lettuce are widely used as a plant cultivated along with fish farming, because these plants are able to adapt to the nutrients from the water. There was no significant difference between T2 and T3, but T2 has roots longer than T3. The longer roots provide an opportunity to absorb more nutrients, and provide a place for attachment surface for microbial communities. These results also explain the growth of romaine lettuce on T2 better than T3. Plant roots absorb the flowing nutrient in hydroponic subsystem, provide surface for nitrifying bacteria functioning for organic compound breakdown subsquently oxidation of ammonia to nitrate (Endut et al. 2010, 2011; Hu et al. 2015).

(a) (b)

Figure 4: The biomass of crops (a) and length of roots of romaine lettuce (b).

Results of romaine lettuce growth in terms of plant height are shown in Fig. 5. The height of romaine lettuce increased with time of experiment, and there was no significant difference. At the beginning of the experiment, plant height at T2 was 11.40 cm and at the end of the experiment reached 25.23 cm. Meanwhile, plant height at T3 at the beginning of the experiment was 11.60 cm, and reached 24.62 cm at the end of the experiment. The same growth pattern was shown in weight accretion of romaine lettuce. The increase of weight of romaine lettuce and other growth parameters can be seen in Table 3. The initial weight at T2 was 18.00 g, and then the final weight was 61.87 g. The weight of romaine lettuce at the beginning of the experiment at T3 was 20 g, and at the end of the experiment reached 57.74 g. Although the initial weight T3 is greater than T2, but at the end of the experiment the weight of romaine lettuce on a T3 is lower than T3. These results were likely caused by level of nutrients' absorption by the plants through the roots. Besides that the number of leaves at the beginning of the experiment was four strands either for T2 or T3, then the number of leave increased and reached 16 strands at the end of the trial for the T2 and T3.

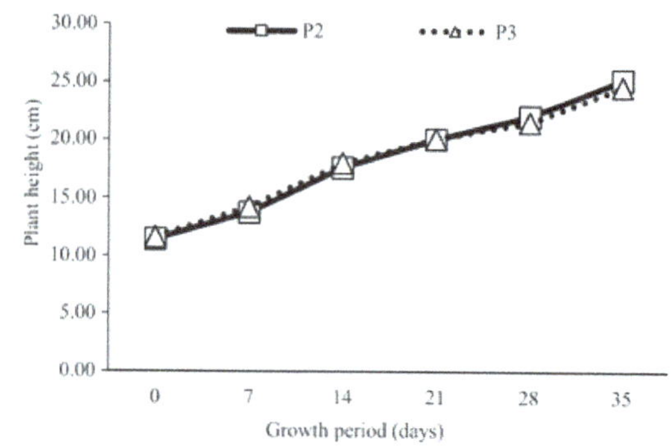

Figure 5: The height of romaine lettuce (*Lactuca sativa* L. var. Longifolia)

Table 3: The growth of romaine lettuce on aquaponic systems

Criteria	Treatments	
	T2	T3
Initial weight (g)	18.00 ± 0.17	20.00 ± 1.37
∑ Initial leaf (strand)	4 ± 0	4 ± 0
Final weight (g)	61.87 ± 5.59	57.74 ± 4.35
∑ Final leaf (strand)	16 ± 1	16 ± 1
DGR (cm/day)	0.39 ± 0.05[a]	0.37 ± 0.04[a]
RGR (g/day)	0.035 ± 0.002[a]	0.030 ± 0.002[a]

T2 romaine lettuce, T3 romaine lettuce and inoculation with bacteria. Values with the same superscript letters are not significantly different at the test level of 5 %.

The results of the final weight in this experiment were better than of 35.28 ± 0.80, 22.59 ± 0.3, and 18.32 ± 2.43 g of lettuce on the media followed by gravel and the control treatments by Sikawa and Yakupitiyage (2010), and this result was lower than the aquaponic research by Sace and Fitzsimmons (2013) using the lettuce, reaching 77.5 g (systems with lettuce and prawns) and 78.5 g (system lettuce without prawns), and other aquaponic experiments using lettuce with NFT system were 107.95 ± 2.20 g (Lennard and Leonard 2006).

Romaine lettuce growth in terms of DGR and RGR between T2 and T3 has no significant difference ($p > 0.05$) (Table 3). Romaine lettuce growth indicated the absorption of nutrients during the experiment. When plants are growing quickly, nutrient needs are high, and nutrient uptake will be greater (Buzby and Lin 2014). Lennard and Leonard (2006) research results also showed that lettuce grown in gravel as hydroponic media was efficient at removing nutrients. The RGR of romaine lettuce in this study (0.30–0.035) is better than the result of research by Buzby and Lin (2014), where RGR of lettuce on aquaponic system at the end of experiment only reached 0.01. Similarly, romaine lettuce growth in terms of DGR in this study is higher than result of Effendi et al.

(2015b) namely 0.04 cm/day, and lower than the research by Endut et al. (2011) namely 1.91 cm/day (water spinach) and 1.32 cm/day (mustard green).

CONCLUSION

Nile Tilapia and romaine lettuce can grow well together in aquaponic system in this study. Romaine lettuce growth in T2 (nile tilapia and romaine lettuce) and T3 (nile tilapia. romaine lettuce, and addition of bacteria) was well recorded by color of fresh green leaves, and there were no signs of nutritional deficiency. Romaine lettuce could utilize nutrient originating from fish culture. This nutrient was a result of decomposition of fish excreta and uneaten food. Best growth of nile tilapia was shown by T3 in terms of FCR, proportional to better water quality in this treatment. Thus, romaine lettuce could, to some extent, purify water quality of fish culture media to be reused in the fish culture. Aquaponic system (nile tilapia and romaine lettuce) without water exchange can maintain the quality of the water suitable for fish and plant growth, without having to replace the water regularly like in conventional cultivation system. Further research on optimization of the number of plant incorporated in fish culture in aquaponic system is required.

ACKNOWLEDGMENTS

We wish to thank the Ministry of Research, Technology, and High Education of Indonesia for providing financial support through BOPTN research scheme.

REFERENCES

1. American Public Health Association (APHA) (2008) Standard method for the examination of water and wastewater. Water Pollution Control Federation, Baltimore
2. Beveridge MCM, Philips MJ, Macintosh DJ (1997) Aquaculture and the environment: the supply of and demand for environmental goods and services by Asian aquaculture and the implications for sustainability. Aquacult Res 28:797–807
3. Buzby KM, Lin LS (2014) Scaling aquaponic systems: balancing plant uptake with fish output. Aquacult Eng 63:39–44. doi:10.1016/j.aquaeng.2014.09.002
4. Colt J (2006) Water quality requirements for reuse systems. Aquacult Eng 34:143–156. doi:10.1016/j.aquaeng.2005.08.011
5. Delis PC, Effendi H, Krisanti M, Hariyadi S (2015) Treatment of aquaculture wastewater using*Vetiveria zizanioides* (Liliopsida, Poaceae). AACL Bioflux 8:616–625
6. DeLong DP, Losordo TM, Rakocy JE (2009) Tank culture of tilapia. SRAC Publication 282:1–8
7. Diver S (2006) Aquaponics-integration of hydroponics with aquaculture. National Center for Appropriate Technology (NCAT), Butte, pp 1–28
8. Effendi H, Utomo BA, Darmawangsa GM (2015a) Phytoremediation of freshwater crayfish (*Cherax quadricarinatus*) culture wastewater with spinach (*Ipomoea aquatica*) in aquaponic system. AACL Bioflux 8:421–430
9. Effendi H, Utomo BA, Darmawangsa GM, Hanafiah DA (2015b) Wastewater treatment of freshwater crayfish (*Cherax quadricarinatus*) culture with lettuce (*Lactuca sativa*). Int J Appl Environ Sci 10(1):409–420
10. Effendi H, Delis PC, Krisanti M, Hariyadi S (2015c) The performance of nile tilapia (*Oreochromis niloticus*) and vetiver grass (*Vetiveria zizanioides*) concurrently cultivated in aquaponic system. Adv Environ Biol 9(24):382–388
11. Endut A, Jusoh A, Ali N, Nik WBW, Hassan A (2010) A study on the optimal hydraulic loading rate and plant ratios in recirculation aquaponic system. Bioresour Technol 101:1511–1517. doi:10.1016/j.biortech.2009.09.040
12. Endut A, Jusoh A, Ali N, Nik WBW (2011) Nutrient removal from aquaculture wastewater by vegetable production in aquaponic recirculation system. Desalin Water Treat 32:422–430. doi:10.5004/dwt.2011.2761
13. Fedoroff NV, Battisti DS, Beachy RN, Cooper PJM, Fischhoff DA, Hodges CN, Knauf VC, Lobell D, Mazur BJ, Molden D, Reynolds MP, Ronald PC, Rosegrant MW, Sanchez PA, Vonshak A, Zhu JK (2010) Radically rethinking agriculture for the 21st century. Sci 327:833–834. doi:10.1126/science.1186834
14. Francis-Floyd R, Watson C, Petty D, Pourder DB (1996) Ammonia in aquatic systems. FA16. Univ. Florida, Gainesville, pp 1–28
15. Graber A, Junge R (2009) Aquaponic systems: nutrient recycling from fish wastewater by vegetable production. Desalin 246:147–156. doi:10.1016/j.desal.2008.03.048
16. Hargreaves JA (2006) Photosynthetic suspended-growth systems in aquaculture. Aquacult Eng 34:344–363. doi:10.1016/j.aquaeng.2005.08.009
17. Hargreaves JA, Kucuk S (2001) Effects of diel un-ionized ammonia fluctuation on juvenile hybrid striped bass, channel catfish, and blue tilapia. Aquacult 195:163–181
18. Hu Z, Lee JW, Chandran K, Kim S, Brotto AC, Khanal SK (2015) Effect of plant species on nitrogen recovery in aquaponics. Bioresour Technol 188:92–98. doi:10.1016/j.biortech.2015.01.013
19. Kamal SM (2006) Aquaponic production of nile tilapia (*Oreochromis niloticus*) and bell peppers (*Capsicum annum* L.) in recirculating water system. Egypt J Aquat Biol Fish 10:85–97
20. Lennard WA, Leonard BV (2006) A comparison of three different hydroponic sub-systems (gravel bed, floating and nutrient film technique) in an aquaponic test system. Aquacult Int 14:539–550. doi:10.1007/s10499-006-9053-2
21. Liang JY, Chien YH (2013) Effects of feeding frequency and photoperiod on water quality and crop production in a tilapia-water spinach raft aquaponics system. Int Biodet Biodeg. doi:10.1016/j.ibiod.2013.03.029
22. Losordo TM, Masser MP, Rakocy JM (1998) Recirculating aquaculture tank production systems: an overview of critical considerations. SRAC Publication 451, USA.https://srac.tamu.edu/. Accessed 14 June 2015
23. Love DC, Uhl MS, Genello L (2015) Energy and water use of a small-scale raft aquaponics system in Baltimore,

Maryland, United States. Aquacult Eng 68:19–27. doi:10.1016/j.aquaeng.2015.07.003

24. Masser MP, Rackocy J, Losordo TM (1999) Recirculating aquaculture tank production systems: management of recirculating systems. SRAC Publication 452:1–12

25. McMurtry MR, Cure JD, Sanders DC, Hodson RG, Haning BC, Amand PCST (1997) Efficiency of water use of an integrated fish/vegetable co-culture system. J World Aquacult Soc 28:420–428

26. Mitchell DS (1974) The development of excessive populations of aquatic plants. In: Mitchell DS (ed) Aquatic vegetation and its use and control. UNESCO, Paris

27. Piedrahita RH (2003) Reducing the potential environmental impact of tank aquaculture effluents through intensification and recirculation. Aquacult 226:35–44. doi:10.1016/S0044-8486(03)00465-4

28. Princic A, Mahne I, Megusar F, Paul EA, Tiedje JM (1998) Effects of pH and oxygen and ammonium concentrations on the community structure of nitrifying bacteria from wastewater. Appl Environ Microbiol 64(10):3584–3590

29. Qin G, Liu CCK, Richman NH, Moncur JET (2005) Aquaculture wastewater treatment and reuse by wind-driven reverse osmosis membrane technology: a pilot study on Coconut Island, Hawaii. Aquacult Eng 32:365–378. doi:10.1016/j.aquaeng.2004.09.002

30. Rahmani AR, Mahvi AH, Mesdaghinia AR, Nasse S (2004) Investigation of ammonia removal from polluted waters by clinoptilolite zeolit. Int J Environ Sci Technol 1:125–133

31. Rakocy JE, Masser MP, Losordo TM (2006) Recirculating aquaculture tank production systems: aquaponics-integrating fish and plant culture (revision). SRAC Publication 454:1–16

32. Ridha MT, Cruz EM (2001) Effect of biofilter media on water quality and biological performance of the nile tilapia *Oreochromis niloticus* L. reared in a simple recirculating system. Aquacult Eng 24:157–166

33. Rijn JV, Tal Y, Schreier HJ (2006) Denitrification in recirculating systems: theory and applications. Aquacult Eng 34:364–376. doi:10.1016/j.aquaeng.2005.04.004

34. Roosta HR, Hamidpour M (2011) Effects of foliar application of some macro and micro nutrients on tomato plants in aquaponic and hydropinic systems. Sci Hortic 129:396–402. doi:10.1016/j.scienta.2011.04.006

35. Roosta HR, Mohsenian Y (2012) Effects of foliar spray of different Fe sources on pepper (*Capsicum annum* L.) plants in aquaponic system. Sci Hortic 146:182–191. doi:10.1016/j.scienta.2012.08.018

36. Sace CF, Fitzsimmons KM (2013) Recirculating aquaponic systems using nile tilapia (*Oreochromis* niloticus) and freshwater prawn (*Macrobrachium rosenbergii*) polyculture and the productivity of selected leafy vegetables. Merit Res J Bus Manag 1:11–29

37. Seawright DE, Stickney RR, Walker RB (1998) Nutrient dynamics in integrated aquaculture–hydroponics systems. Aquacult 160:215–237

38. Shete AP, Verma AK, Tandel RS, Prakash C, Tiwari VK, Hussain T (2013) Optimization of water circulation period for the culture of goldfish with spinach in aquaponic system. J Agricult Sci 5(4):26–30

39. Sikawa DC, Yakupitiyage A (2010) The hydroponic production of lettuce (*Lactuca sativa* L) by using hybrid catfish (*Clarias macrocephalus* x *C. gariepinus*) pond water: potentials and constraints. Agricult Water Manag 97:1317–1325. doi:10.1016/j.agwat.2010.03.013

40. Simeonidou M, Paschos I, Gouva E, Kolygas M, Perdikaris C (2012) Performance of a small-scale modular aquaponic system. AACL Bioflux 5:182–188

41. Sugiura SH, Marchant DD, Kelsey K, Wiggins T, Ferraris RP (2006) Effluent profile of commercially used low-phosphorus fish feeds. Environ Pollut 140:95–101. doi:10.1016/j.envpol.2005.06.020

42. Titiresmi Sopiah N (2006) Biofilter technology for ammonia wate treatment. J Teknol Lingkung 7:173–179 **(in Indonesian)**

43. Tokuyama T, Mine A, Kamiyama K, Yabe R, Satoh K, Matsumoto H, Takahashi R, Itonaga K (2004) *Nitrosomonas communis* strain YNSRA, an ammonia-oxidizing bacterium, isolated from the reed rhizoplane in an aquaponics plant. J Biosci Bioeng 98(4):309–312

44. Tyson RV, Simonne EH, Treadwell DD (2008) Reconciling pH for ammonia biofiltration and cucumber yield in a recirculating aquaponic system with perlite biofilters. Hort Sci 43:719–724

45. Tyson RV, Treadwell DD, Simonne EH (2011) Opportunities and challenges to sustainability in aquaponic systems (reviews). Hort Technol 21:6–13

46. Wahyuningsih S, Effendi H, Wardiatno Y (2015) Nitrogen removal of aquaculture wastewater in aquaponic recirculation system. AACL Bioflux 8:491–499

47. Wang Z, Leung KMY (2015) Effects of unionised ammonia on tropical freshwater organisms: implications on temperate-to-tropic extrapolation and water quality guidelines. Environ Pollut 205:240–249. doi:10.1016/j.envpol.2015.05.045

48. Wang CY, Chang CY, Chien YH, Lai HT (2016) The performance of coupling membrane filtration in recirculating aquaponic system for tilapia culture. Int Biodet Biodeg 107:21–30. doi:10.1016/j.ibiod.2015.10.016

49. Watson CA, Hill JE (2006) Design criteria for recirculating, marine ornamental production systems. Aquacult Eng 34:157–162. doi:10.1016/j.aquaeng.2005.07.002

50. Xu QF, Tsai CL, Tsai CY (1992) Interaction of potassium with the form and amount of nitrogen nutrition on growth and nitrogen uptake of maize. J Plant Nutr 15(1):23–33

51. Zhan L, Hu J, Pang L, Fan H (2012) Browning inhibition and quality preservation of fresh-cut romaine lettuce exposed to high intensity light. Innov Food Sci Emerg Technol 14:70–76. doi:10.1016/j.ifset.2012.02.004

52. Zhan L, Hu J, Ai Z, Pang L, Li Y, Zhu M (2013) Light exposure during storage preserving soluble sugar and l-ascorbic acid content of minimally processed romaine

lettuce (*Lactuca sativa* L.var. longifolia). Food Chem 136:273–278. doi:10.1016/j.foodchem.2012.07.123

53. Zheljazkov VD, Horgan TE (2011) Study on shrimp waste water and vermicompost as a nutrient source for bell peppers. Hort Sci 46:1493–1496

54. Zheljazkov VD, Zonneveld A, Huisman EA (1991) The principles of aquaculture. PT. Gramedia Pustaka, Jakarta **(in Indonesian)**

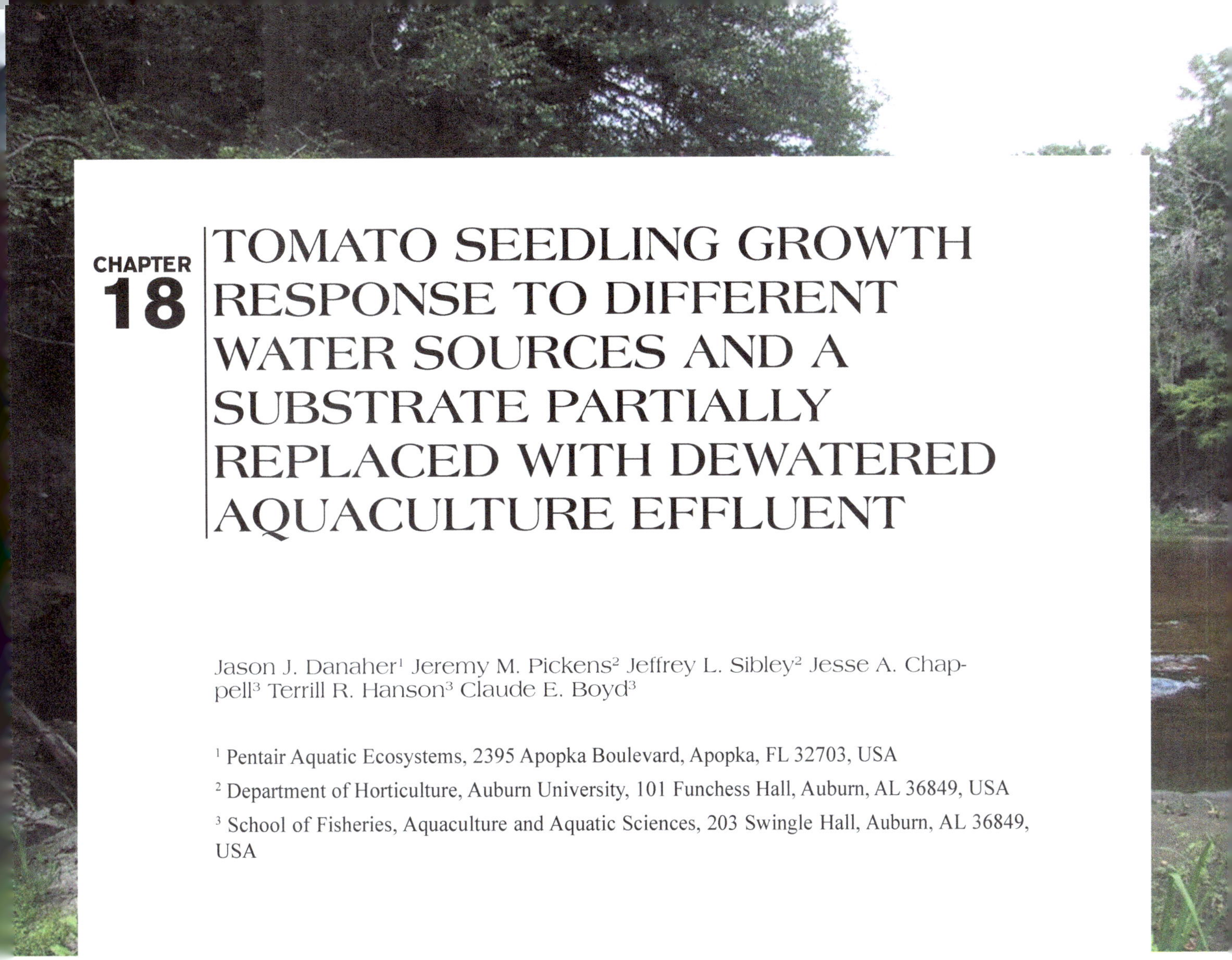

CHAPTER 18

TOMATO SEEDLING GROWTH RESPONSE TO DIFFERENT WATER SOURCES AND A SUBSTRATE PARTIALLY REPLACED WITH DEWATERED AQUACULTURE EFFLUENT

Jason J. Danaher[1] Jeremy M. Pickens[2] Jeffrey L. Sibley[2] Jesse A. Chappell[3] Terrill R. Hanson[3] Claude E. Boyd[3]

[1] Pentair Aquatic Ecosystems, 2395 Apopka Boulevard, Apopka, FL 32703, USA

[2] Department of Horticulture, Auburn University, 101 Funchess Hall, Auburn, AL 36849, USA

[3] School of Fisheries, Aquaculture and Aquatic Sciences, 203 Swingle Hall, Auburn, AL 36849, USA

ABSTRACT

Purpose

The experiment was performed to determine the effect a commercial potting mix partially replaced with dewatered aquaculture effluent had on tomato transplant growth.

Methods

The experiment was designed as a 2 × 3 factorial and evaluated two water sources (water-soluble, inorganic fertilizer or municipal water) and three soilless substrates with 0, 5 or 10 % dewatered aquaculture effluent (v/v) on substrate properties and tomato (*Solanum lycopersicum* Mill. 'Bolseno') transplant growth. The layout was a completely randomized design with twelve single-pot replications for each treatment.

Results

There was a substrate and water interaction affecting plant height, leaf dry matter (LDM), stem dry matter, root dry matter (RDM), and total dry matter (TDM). Tomato plants watered with inorganic fertilizer and grown in substrates replaced with 0 and 5 % dewatered aquaculture effluent had greater LDM, RDM, and TDM compared to plants watered with municipal water. However, tomato plant growth in substrate partially replaced with 10 % dewatered aquaculture effluent was similar irrespective of water source.

Citation: Danaher JJ, Pickens JM, Sibley JL, Chappell JA, Hanson TR, Boyd CE (2016). Tomato seedling growth response to different water sources and a substrate partially replaced with dewatered aquaculture effluent. International Journal of Recycling of Organic Waste in Agriculture, v. 5, Issue 1, 25-32. DOI: 10.1007/s40093-016-0114-x.

Conclusion

Substrates incorporated with 10 % aquaculture effluent provided optimal physical and chemical properties along with sufficient nutrients for tomato transplants without the need for commercial, inorganic fertilizer.

INTRODUCTION

Aquaculture is one of the fastest growing sectors of agriculture. A decline in capture fisheries, coupled with a strong consumer-driven demand for aquaculture products, has resulted in the adoption of intensive fish production facilities. Aquaculture farms have intensified and require solutions to manage discharged wastes. Recirculating aquaculture systems (RAS) have incorporated water treatment technology to manage toxic dissolved nutrients and solid waste. This technology allows the producer to maximize production per unit area and reuse limited freshwater resources. The RAS technology is used to produce popular food species like Nile tilapia (*Oreochromis niloticus*; Azim and Little 2008). To ensure water quality remains optimal for the fish, a RAS discharges concentrated organic matter and dissolved wastes daily. Even though the point of discharge is well defined, the concentrated organic matter and dissolved nutrients are still a liability for the producer. Effluent leaving production facilities is regulated by environmental agencies (Ebeling et al. 2005). Capturing and treating discharged wastes on-site would help intensive aquaculture production facilities address waste management regulations and prevent eutrophication.

While intensive aquaculture facilities are examining technologies to capture and treat discharged effluent, the horticulture industry is evaluating alternative soilless substrates for vegetable seedling production. Sphagnum peatmoss remains extremely important to the US greenhouse industry and is used as a primary component for greenhouse grown plants because of its desirable physical and chemical properties. Some regions of peatlands around the world face pressure from industrial and agricultural interests. These environmental impacts of peat harvesting have been debated in North America and Europe for a number of years. As a result, environmental agencies along with the peat industry have launched conservation efforts. In addition, transportation costs have escalated to ship peat affecting profitability for horticulture producers; therefore, the horticulture industry has evaluated alternative substrates to substitute peat for tomato seedling production. Alternative substrates investigated were coconut coir (Arenas et al. 2002), wood fiber (Gruda and Schnitzler 2004), rice hulls (Evans and Gachukia2004), spent mushroom waste (Eudoxie and Alexander 2011), swine waste (Ribeiro et al. 2007), municipal solid waste (Herrera et al. 2008; Kasmi et al. 2012), pulp mill sludge (Levy and Taylor2003), vermicompost (Atiyeh et al. 2000; Bachman and Metzger 2008) and green waste composts (Ceglie et al. 2011). However, few experiments investigated solid matter in aquaculture effluent as a substrate amendment for vegetable seedlings.

Boyd and Tucker (1998) reported only 25 to 30 % of the nitrogen input to an aquaculture production system is assimilated in the final fish biomass. Integrated agriculture systems could help improve use of farm inputs, like fish feed. Nitrogen (Rakocy et al. 2003) and phosphorus (Adler et al. 2003) concentrations in treated aquaculture effluent are suitable for plant production. Palada et al. (1999) reported field grown bell peppers (*Capsicum annuum* L.) provided aquaculture effluent grew well compared to other fertilizers evaluated. Nair (2006) reported coleus (*Coleus* × *hybridus* Blume) grew well in substrates containing vermicomposted aquaculture effluent. Danaher et al. (2011) reported composted aquaculture effluent was a good alternative substrate for tomato (*Solanum lycopersicum*). Tomato transplants also responded positively to substrate partially replaced with dewatered aquaculture effluent at 5–15 % container volume (Danaher et al. 2014). Pantanella et al. (2011) reported composted aquaculture effluent was a suitable substrate replacement for containerized lettuce (*Lactuca sativa* L.) and basil (*Ocimum basilicam* L.) production.

The objective of this experiment was to evaluate tomato plant growth in response to different combinations of substrate and water sources and determine if dewatered aquaculture effluent could partially replace a commercial substrate while providing adequate nutrients for plant growth.

MATERIALS AND METHODS

The discharged aquaculture effluent (AE) was collected from a 100-m^3 intensive, freshwater RAS producing Nile tilapia. A 3.5 m × 1.5 m woven geotextile bag (U.S. Fabrics, Inc., Cincinnati, OH) and polyacrylamide polymer Hyperfloc CE 854 (Hychem, Inc., Tampa, FL) were used to dewater discharged AE. After repeated fillings, the bag was allowed to dewater and was cut open to remove the solids. Solids were allowed to air-dry and then further processed with a hammer mill (Model 30; C.S. Bell Co., Tiffin, OH) to pass through 0.635-cm screen. Two substrates were prepared prior to the experiment by substituting (v/v) Fafard 3B (F3B) mix (Conrad Fafard, Inc., Agawam, MA) with 5 or 10 % AE. The F3B mix consisted of Canadian sphagnum peat moss (50 %), processed pine bark, perlite, vermiculite, starter nutrients, wetting agent and dolomitic limestone. A composite sample of the F3B and AE was taken and analyzed at the Auburn University Soils Laboratory for chemical properties (Table 1). Saturated media method was used to extract soluble salts and elements were determined simultaneously by inductively coupled plasma atomic emission spectrometry using a radial spectrometer (Vista-MPX; Varian Inc., Palo Alto, CA). The substrate pH was determined using a bench-top meter (Fisher Accumet Model 50, Fisher Scientific). Inorganic nitrogen was analyzed according to Sims et al. (1995) and total nitrogen and carbon according to methods described by Kirsten (1979).

Table 1: Chemical properties from composite sample of Fafard 3B mix (F3B) and dewatered aquaculture effluent (AE) used as soilless substrates in the tomato experiment

	Parameter[z]													
	pH	EC	NH_4-N	NO_3-N	Ca	Mg	P	K	S	Na	B	Fe	Mn	Zn
F3B	6.4	1.3	28.2	46.9	61.7	66.0	10.5	129.5	130.5	17.5	0.03	0.56	0.38	0.99
AE	6.6	2.2	129.8	0.8	83.6	51.6	70.2	204.5	111.9	82.6	0.13	0.70	0.21	0.83

[z]Electrical conductivity (EC), 1 mS cm^{-1} = 1 mmho cm^{-1}; macronutrients and micronutrients reported as mg kg^{-1}

Three replicates of each substrate were used to determine total porosity (TP), container capacity (CC), airspace (AS), and bulk density (BD) following procedures described by Bilderback et al. (1982). The BD (g cm^{-3}) was determined from 347.5 cm^{-3} substrate samples dried in a forced air oven at 70 °C for 72 h. Physical properties of substrates are presented in Table 2. Four replicates of each substrate were analyzed for particle size distribution (PSD) by passing a 100 g sample through 9.5, 6.35, 3.35, 2.36, 2.0, 1.4, 1.0, 0.5, and 0.25 mm sieves with particles passing through the 0.25 mm sieve collected in a pan. Sieves were shaken for 3 min with a Ro-Tap sieve shaker [278 oscillations/min, 159 taps/min (Ro-Tap RX-29; W.S.Tyler, Mentor, OH)]. The PSD for each substrate is presented in Table 3. The non-destructive Virginia Tech pour-through extraction method (Wright 1986) was used to determine substrate pH and electrical conductivity (EC) of substrate leachate using a bench-top multiparameter meter (Accumet Excel XL50; Fisher Scientific; Table 4).

Table 2: Physical properties of substrates substituted with different volumes of dewatered aquaculture effluent (AE) to grow tomato seedlings

Substrate	Total porosity (%)[z]	Container capacity (%)[y]	Air space (%)[x]	Bulk density (g cm^{-3})[w]
0 % AE	85.1a[v]	66.6a	18.5a	0.17c
5 % AE	83.7a	66.0a	17.7a	0.19b
10 % AE	80.5b	68.1a	12.4a	0.21a
Sufficiency range[w]	50–80	45–65	10–30	0.19–0.70

[z]Total porosity is container capacity + air space

[y]Container capacity is (wet weight − oven dry weight)/volume of the sample

[x]Air space is volume of water drained from the sample/volume of the sample

[w]Bulk density after forced air drying at 105 °C for 48 h; 1 °C = (1.8 × °C) + 32; 1 g cm^{-3} = 0.5780 oz $inch^{-3}$

[v]Means within columns followed by different letters were significant with Tukey's test ($P < 0.05$)

[u]Sufficiency ranges reported by Yeager et al. (2007)

Table 3: Particle size analysis as percent of sample weight for substrates substituted with different volumes of dewatered aquaculture effluent (AE)

	Substrate		
	0 % AE	5 % AE	10 % AE
Sieve opening (mm)			
9.50	1.4a[z]	1.4a	1.4a
6.35	7.3a	5.7ab	4.5b
3.35	14.7a	13.1ab	11.3b
2.36	9.1b	10.3a	9.8ab
2.00	3.7c	4.6b	5.4a
1.40	9.8c	11.4b	12.8a
1.00	12.2a	10.9c	11.6b
0.50	28.3a	20.3b	19.5b
0.25	11.6b	14.2a	13.7a
0.11	1.9c	7.6b	8.9a
0.05	0.1c	0.5b	1.1a
0.00	0.0a	0.0a	0.0a
Texture[y]			

Coarse	23.3a	20.2b	17.3c
Medium	34.8c	37.1b	49.5a
Fine	41.9a	42.7a	43.2a

Means within rows marked by a different letter were significant with Tukey's test ($P \leq 0.05$);$n = 3$

[y]Coarse > 3.35 mm; medium ≥ 1.00 mm and ≤3.35 mm; fine < 1.00 mm

Table 4: The pH and electrical conductivity (EC) of substrates substituted with different volumes of dewatered aquaculture effluent (AE) 4, 19, and 25 days after potting (DAP)

Main effect	pH			EC (mS cm^{-1})		
	4 DAP[z]	19 DAP	25 DAP	4 DAP	19 DAP	25 DAP
Substrate						
0 % AE	6.4c	7.1a	6.7a	1.0c	0.3b	0.2b
5 % AE	6.5b	6.4b	6.4b	1.6a	1.0b	0.4b
10 % AE	7.1a	6.4b	6.5ab	1.3b	2.8a	0.7a
Water						
Municipal	6.6a	6.7a	6.6a	1.3a	1.3a	0.4a
Fertilizer	6.6a	6.7a	6.5a	1.3a	1.4a	0.4a
Significance						
Substrate	***	***	***	***	***	***
Water	NS	NS	NS	NS	NS	NS
Substrate and water	NS	NS	NS	NS	NS	NS

[z]Mean separation of main effects within the same column followed by a different letter are significantly different by Tukey's test at * $P \leq 0.05$, ** 0.01, or *** 0.001; NS = non-significant

The experiment was performed in a double-layer, polyethylene-covered greenhouse at the E.W. Shell Fisheries Center, North Auburn Unit, in Auburn, Alabama from 22 October to 17 November 2012. The trial was designed as a 2 × 3 factorial evaluating two water sources (water-soluble, inorganic fertilizer or municipal water) and F3B mix substituted with different amounts (0, 5, or 10 %) of AE. The experiment was a completely randomized design with twelve single-pot replications for each treatment. Tomato seeds were germinated in a 288-cell flat containing F3B mix and on 22 October, one uniform transplant was transferred into each 473 cm^{-3} square (9.84 cm × 8.57 cm) plastic pot (Dillen™ Products, Middlefield, OH) containing the aforementioned substrates when the first true leaves developed. All pots were placed on raised benches and for the first week all pots were watered with municipal water as needed. Thereafter, pots were watered as needed according to treatment with either municipal water or fertigated, twice weekly, using a Dosatron® (Dosatron International, Inc., Clearwater, FL) injector at 100 mg·L^{-1} nitrogen with a water-soluble 20N–4.4P–16.6K fertilizer (SDT Industries, Inc., Winnsboro, LA) containing chelated micronutrients. All treatments were watered until substrate reached saturation (i.e., until water leached from the bottom of the pot). At 25 days after potting (DAP) leaf greenness was quantified for all plants using a chlorophyll meter (SPAD-502; Minolta Camera Company, Ramsey, NJ) and taking the average reading of four random leaflets per plant (Table 5). Plant height was measured 26 DAP (Table 5). Stem dry matter (SDM), leaf dry matter (LDM), root dry matter (RDM), and total plant dry matter (TDM) were measured 26 DAP (Table 6). The TDM included cotyledons.

Table 5: Effect of substrate and water interaction on tomato seedling SPAD readings 25 days after potting (DAP) and plant height 26 DAP in substrates substituted with different volumes of dewatered aquaculture effluent (AE)

Water source[y]	SPAD (cm)[z]		Plant height (cm)	
	Fertilizer[x]	Municipal	Fertilizer	Municipal
Substrate[w]				
0 % AE	42.2 Ab	34.7 Bb	19.2 Ab	15.1 Bb
5 % AE	42.2 Bb	43.8 Aa	23.1 Aa	22.2 Aa
10 % AE	44.9 Aa	44.0 Aa	19.7 Ab	20.3 Aa
Significance[v]				
Substrate	***		***	

Water	***	**
Substrate and water	***	***

[z]Leaf greenness of four recently mature leaves per plant was quantified with a chlorophyll meter (SPAD-502; Minolta Camera Company, Ramsey, NJ)

[y]Fertilizer = 20N–4.4P–16.6K; Municipal = Auburn, AL city water

[x]For each parameter values within column followed by different lower case letters are significantly different for pairwise comparisons of substrate within each level combination of water by Bonferroni adjusted α-level ($P \leq 0.05$)

[w]For each parameter values within row followed by different upper case letters are significantly different for pairwise comparisons of water within each level combination of substrate by Bonferroni adjusted -level ($P \leq 0.05$)

[v]NS = non-significant; * $P \leq 0.05$; ** 0.01, or *** 0.001 based on two-way analyses of variance

Table 6: Effect of substrate and water interaction on tomato seedling leaf dry matter, stem dry matter, root dry matter and total dry matter 26 d after potting in substrates substituted with different volumes of dewatered aquaculture effluent (AE)

Water source[y]	**Leaf dry matter (g)**		**Stem dry matter (g)**		**Root dry matter (g)**		**Total dry matter (g)**[z]	
	Fertilizer[x]	**Municipal**	**Fertilizer**	**Municipal**	**Fertilizer**	**Municipal**	**Fertilizer**	**Municipal**
Substrate[w]								
0 % AE	2.4 Ab	1.0 Bb	0.8 Ac	0.3 Bb	0.7 Ab	0.5 Bb	3.9 Ab	1.7 Bb
5 % AE	4.0 Aa	3.5 Ba	1.4 Aa	1.3 Aa	1.0 Aa	0.8 Ba	6.3 Aa	5.5 Ba
10 % AE	3.9 Aa	3.7 Aa	1.2 Ab	1.2 Aa	1.0 Aa	1.0 Aa	6.0 Aa	5.8 Aa
Significance[v]								
Substrate	***		***		***		***	
Water	***		***		**		***	
Substrate and water	***		***		*		***	

[z]Total dry matter includes cotyledons

[y]Fertilizer = 20N–4.4P–16.6K; Municipal = Auburn, AL city water

[x]For each parameter values within column followed by different lower case letters are significantly different for pairwise comparisons of substrate within each level combination of water by Bonferroni adjusted α-level ($P \leq 0.05$)

[w]For each parameter values within row followed by different upper case letters are significantly different for pairwise comparisons of water within each level combination of substrate by Bonferroni adjusted α-level ($P \leq 0.05$)

[v]NS = non-significant; * $P \leq 0.05$; ** 0.01; or *** 0.001 based on two-way analyses of variance

Two-way analysis of variance (ANOVA) was used to determine the main effect of substrate and water on tomato plant growth. If a significant interaction existed ($P \leq 0.05$), pairwise comparisons on the individual group means within each simple effect were conducted and means were separated using the Bonferroni adjusted α-level ($P \leq 0.05$). If no significant ($P > 0.05$) statistical interaction was identified, the main effects of substrate and water were analyzed separately and means were separated by Tukey's test ($P \leq 0.05$).

RESULTS AND DISCUSSION

The AS and CC were unaffected by increased proportions of AE (Table 2); however, the inclusion of AE at ≥5 % container volume resulted in greater substrate BD. Bulk density of soilless substrate typically increases as the amount of animal-based wastes used to amend the substrate increases (Atiyeh et al. 2001). The PSD provides some explanation for the difference in TP and BD among substrates (Table 3). As the amount of AE increased, the percent of coarse-sized particles decreased and the percent of medium-sized particles increased. This redistribution of particle size affected the physical properties of the substrates. Melgar-Ramirez and Pascual-Alex (2010) and Hicklenton et al. (2001) reported container substrate physical properties were influenced by increased levels of compost; nonetheless, the substrate physical properties in the present study were optimal (Yeager et al. 2007).

There was no substrate and water interaction for pH or EC of container leachate at 4, 19, or 25 DAP (Table 4). The main effect of water did not affect pH at 4, 19 or 25 DAP; but, the main effect of substrate directly affected both substrate leachate pH and EC values at 4, 19 and 25 DAP. Increased proportions of AE increased leachate pH at 4 DAP. Tyler et al. (1993) and Marble et al. (2010) also reported elevated pH of container substrates with increased amounts of composted poultry litter. In addition, Melgar-Ramirez and Pascual-Alex (2010) also reported a direct relationship between the quantity of vermicompost and substrate pH of container media. In the present study, the substrate leachate pH for AE amended substrates approached optimal ranges by 19 DAP and remained lower than 0 % AE for the remainder of the experiment.

Proportions of ≥5 % AE increased substrate leachate EC values at 4 DAP (Table 4). The substrate leachate EC value of 10 % AE remained greater than other substrates at 19 and 25 DAP, but 5 % AE was similar to 0 % AE at 19 and 25 DAP (Table 5). Increased substrate leachate EC values observed with 10 % AE may have resulted from the continuous physical breakdown and release of soluble salts. Smaller particle size could have accelerated leaching of salts through physical breakdown of the AE substrates. In addition, the nutrient content of AE, itself, may have also contributed to observed EC leachate values. Substrates did not exceed the threshold of 3.5 mS cm^{-1} reported by Wright (1986) for healthy seedling growth.

A substrate and water interaction affected plant SPAD readings at 25 DAP (Table 5). Fertigating tomato plants grown in substrates with 0 % AE improved SPAD readings 21.6 % compared to plants grown in the same substrate with municipal water. The commercial mix alone did not contain enough nutrients to sustain plant growth. Fertigating plants in 5 % AE slightly decreased SPAD readings 3.7 % compared to plants grown in the same substrate with municipal water. Water source did not affect SPAD reading of plants grown in 10 % AE. Fertigating tomato plants grown in 10 % AE improved SPAD readings 6.4 % compared to substrates with ≤5 % AE. Commercial mix partially replaced F3B with ≥5 % AE and watered with municipal water improved SPAD readings 26.5 % compared to plants grown in commercial mix alone.

There was a substrate and water interaction affecting plant height at 26 DAP (Table 5). Fertigating tomato plants grown in substrate with 0 % AE improved plant height 27.2 % compared to plants grown in the same substrate and municipal water, indicating the potting mix alone did not have sufficient nutrients to maintain plant growth for a 26-day growing period. The combination of fertilizer and 5 % AE increased plant height 19 % compared to the other substrates. Plant height improved approximately 41 % using ≥5 % AE and municipal water compared to substrate with 0 % AE and municipal water.

There was a substrate and water interaction affecting LDM, SDM, RDM, and TDM at 26 DAP. Fertigating tomato plants grown in substrate with 0 % AE improved LDM, SDM, RDM, and TDM 140, 167, 40, and 129 %, respectively, compared to plants grown with the same substrate with municipal water (Table 6). Fertigating tomato plants grown in 5 % AE improved LDM, SDM, RDM, and TDM 14, 8, 25, and 15 %, respectively, compared to plants grown with 5 % AE and municipal water (Table 6). This suggests these potting mixtures, alone, were unable to supply the plants with sufficient amounts of nutrients, but fertigating these substrates improved plant growth. When substrate was partially replaced with 10 % AE water source had no effect on plant growth indices, indicating this potting mix did have sufficient nutrients to maintain plant growth without the need for additional inorganic fertilizer.

Atiyeh et al. (2000) and Danaher et al. (2014) report substrates partially replaced with <20 % animal waste and supplied with all required mineral nutrients improved growth of tomato plants compared to traditional substrates. Their results correspond to the present experiment. But the present findings further advocate the potential benefit of AE alone as a nutrient source for tomato seedling production. In this experiment, different growth responses of tomato seedlings occurred when grown with specific combinations of amended F3B under fixed water regimes. Starter nutrients in the commercial mix were unable to supply sufficient amounts of nutrients for 26-day-old tomato plants requiring the addition of commercial fertilizer to improve plant growth. Although water source did make a difference in final growth indices for 5 % AE, the difference between plants receiving fertigation or municipal was small. Water source had no impact on plant growth 26 DAP for plants grown in 10 % AE, indicating this substrate could provide optimal physico-chemical parameters and sufficient nutrients for tomato plant growth without the need for fertigation.

The physical and chemical properties of different organic substrates are unique; therefore, each substrate should be characterized prior to partial replacement of container substrate to optimize plant growth. Atiyeh et al. (2001) and Subler et al. (1998) reported the incorporation of 5 and 10 %, respectively, vermicompost into container media improved growth of tomato transplants. Both Atiyeh et al. (2000) and Herrera et al. (2008) reported 30 % replacement of commercial mix with vermicompost and municipal solid waste, respectively, enhanced tomato plant growth while greater amounts decreased plant growth. In studies by Lazcano et al. (2009) and Danaher et al. (2011), substitution of ≥50 % container mix with alternative soilless substrates improved tomato transplant growth. Jahromi et al. (2012) reported commercial mix replaced with ≥60 % compost outperformed the peat-based substrate even though EC concentrations exceeded 3.5 mS cm^{-1}. Physical and chemical properties of organic wastes are non-uniform between animal species and consecutive batches at the same farm (García-Delgado et al. 2007; Naylor et al. 1999). Therefore, it is critical the horticulture producer evaluates the material before and after traditional substrates are partially replaced.

A simple economic analysis was done to determine the cost of obtaining the solid component created by the geotextile bag and polymer. It was based on an actual 5-month production period of Nile tilapia produced in the 100 m^3 biofloc system at the E.W. Shell Fisheries Center, North Auburn Unit, in Auburn, AL. The cost of the 4.6 × 3.1 m, 10 oz, non-woven geotextile bag was $171 and two, 19 liter buckets of liquifloc 1 % chitosan ($102 per bucket) were required to flocculate the discharged effluent. The total cost of materials was $375. A total of 1505 kg of tilapia diet (dry weight) was fed to the 100 m^3 biofloc system over the five-month production period and approximately 284 kg of solids (dry weight), or 19 % of the diet fed, was captured in the geotextile bag. This equates to $1.32 per kg for dewatered solid matter (dry weight).

CONCLUSIONS

Increasing the proportion of dewatered aquaculture effluent to 10 % container volume provided optimal physical and chemical properties for tomato plant growth, but different tomato growth responses occurred when grown with specific combinations of substrate and water source. In an integrated production system the dewatered effluent from the fish production system could be utilized as a resource for the production of tomato transplants without the need for additional inorganic fertilizer.

REFERENCES

1. Adler PR, Summerfelt ST, Glenn DM, Takeda F (2003) Mechanistic approach to phytoremediation of water. Ecol Eng 20:251–264. doi:10.1016/s0925-8574(03)00044-2CrossRef
2. Arenas M, Vavrina CS, Cornell JA, Hanlon EA, Hochmuth GJ (2002) Coir as an alternative to peat in media for

tomato transplant production. HortScience 37:309–312. Retrieved fromhttp://hortsci.ashspublications.org

3. Atiyeh RM, Subler S, Edwards CA, Bachman G, Metzger JD, Shuster W (2000) Effects of vermicomposts and composts on plant growth in horticultural container media and soil. Pedobiologia 44:579–590. doi:10.1078/s0031-4056(04)70073-6CrossRef

4. Atiyeh RM, Edwards CA, Subler S, Metzger JD (2001) Pig manure vermicompost as a component of a horticultural bedding plant medium: effects on physicochemical properties and plant growth. Bioresour Technol 78:11–20. doi:10.1016/s0960-8524(00)00172-3CrossRef

5. Azim ME, Little DC (2008) The biofloc technology (BFT) in indoor tanks: water quality, biofloc composition, and growth and welfare of Nile tilapia (*Oreochromis niloticus*). Aquaculture 283:29–35. doi:10.1016/j.aquaculture.2008.06.036CrossRef

6. Bachman GR, Metzger JD (2008) Growth of bedding plants in commercial potting substrate amended with vermicompost. Bioresour Technol 99:3155–3161. doi:10.1016/j.biortech.2007.05.069CrossRef

7. Bilderback TF, Fonteno WC, Johnson DR (1982) Physical properties of media composed of peanut hulls, pine bark and peat moss and their effects on azalea growth. J Am Soc Hortic Sci 107:522–525. Retrieved from http://journal.ashspublications.org

8. Boyd CE, Tucker CS (1998) Pond aquaculture water quality management. Kluwer Academic Publishers, Boston. doi:10.1007/978-1-4615-5407-3CrossRef

9. Ceglie FG, Elshafie H, Verrastro V, Tittarelli F (2011) Evaluation of olive pomace and green waste composts as peat substitutes for organic tomato seedling production. Compost Sci Util 19:293–300. doi:10.1080/1065657x.2011.10737011CrossRef

10. Danaher JJ, Pantanella E, Rakocy JE, Shultz RC, Bailey DS (2011) Dewatering and composting aquaculture waste as a growing medium in the nursery production of tomato plants. Acta Hortic 891:223–229. doi:10.17660/actahortic.2011.891.26CrossRef

11. Danaher JJ, Pickens JM, Sibley JL, Chappell JA, Hanson TR, Boyd CE (2014) Growth of tomato seedlings in commercial substrate amended with dewatered aquaculture effluent. Int J Veg Sci 20(4):340–353. doi:10.1080/19315260.2013.809622CrossRef

12. Ebeling JM, Rishel KL, Sibrell PL (2005) Screening and evaluation of polymers as flocculation aids for the treatment of aquacultural effluents. Aquac Eng 33:235–249. doi:10.1016/j.aquaeng.2005.02.001CrossRef

13. Eudoxie GD, Alexander IA (2011) Spent mushroom substrate as a transplant media replacement for commercial peat in tomato seedling production. J Agric Sci 3:41–49. doi:10.5539/jas.v3n4p41

14. Evans MR, Gachukia M (2004) Fresh parboiled rice hulls serve as an alternative to perlite in greenhouse crop substrates. HortScience 39:232–235. Retrieved fromhttp://hortsci.ashspublications.org/

15. García-Delgado M, Rodríguez-Cruz MS, Lorenzo LF, Arienzo M, Sánchez Martín MJ (2007) Seasonal and time variability of heavy metal content and of its chemical forms in sewage sludges from different wastewater treatment plants. Sci Total Environ 382:82–92. doi:10.1016/j.scitotenv.2007.04.009CrossRef

16. Gruda N, Schnitzler GH (2004) Suitability of wood fiber substrates for production of vegetable transplants II. The effect of wood fiber substrates and their volume weights on the growth of tomato transplants. Sci Hortic 100:333–340. doi:10.1016/j.scienta.2003.09.004CrossRef

17. Herrera F, Castillo JE, Chica AF, Lopez Bellido L (2008) Use of municipal solid waste compost (MSWC) as a growing medium in the nursery production of tomato plants. Bioresour Technol 99:287–296. doi:10.1016/j.biortech.2006.12.042CrossRef

18. Hicklenton PR, Rodd V, Warman PR (2001) The effectiveness and consistency of source separated municipal solid waste and bark composts as components of container grown media. Sci Hortic 91:365–378. doi:10.1016/s0304-4238(01)00251-5CrossRef

19. Jahromi MA, Aboutalebi A, Farahi MH (2012) Influence of different levels of garden compost (garden wastes and cow manure) on growth and stand establishment of tomato and cucumber in greenhouse condition. Afr J Biotechnol 11:9036–9039. doi:10.5897/ajb11.4139

20. Kasmi A, Latigui A, Metai K, Sahli B, Dilem A (2012) Use of sewage sludge and fiber palm co-compost as components of substrates *Lycopersicum esculentum* and *Cucumis melo* cultivated in soilless crop. Am J Plant Physiol 7:92–103. doi:10.3923/ajpp.2012.97.103

21. Kirsten WJ (1979) Automated methods for the determination of carbon, hydrogen, nitrogen, and sulfur alone in organic and inorganic materials. Anal Chem 51:1173–1179. doi:10.1021/ac50044a019CrossRef

22. Lazcano C, Arnold J, Tato A, Zaller JG, Domínguez J (2009) Compost and vermicompost as nursery pot components: effects on tomato plant growth and morphology. Span J Agr Res 7:944–951. doi:10.5424/sjar/2009074-1107CrossRef

23. Levy JS, Taylor BR (2003) Effects of pulp mill solids and three composts on early growth of tomatoes. Bioresour Technol 89:297–305. doi:10.1016/s0960-8524(03)00065-8CrossRef

24. Marble SC, Gilliam CH, Sibley JL, Fain GB, Torbert HA, Gallagher TV, Olive JW (2010) Evaluation of composted poultry litter as a substrate amendment for wholetree, clean chip residual, and pinebark for container grown woody nursery crops. J Environ Hortic 28:107–116. Retrieved from http://hriresearch.theknowledgecenter.com/

25. Melgar-Ramirez R, Pascual-Alex MI (2010) Characterization and use of a vegetable waste vermicompost as an alternative component in substrates for horticultural seedbeds. Span J Agric Res 8:1174–1182. doi:10.5424/sjar/2010084-1407CrossRef

26. Nair DNS (2006) Recycling aquacultural waste through horticultural greenhouse production as a resource recovery approach. Master's Thesis, Virginia Polytechnic Institute and State University. Retrieved from http://scholar.lib.vt.edu/

27. Naylor SJ, Moccia RD, Durant GM (1999) The chemical composition of settable solid fish waste (manure) from commercial rainbow trout farms in Ontario, Canada. N Am J Aquac 61:21–26. doi:10.1577/1548-8454(1999)061<0021:tccoss>2.0.co;2CrossRef

28. Palada MC, Cole WM, Crossman SMA (1999) Influence of effluents from intensive aquaculture and sludge on growth and yield of bell peppers. J Sustain Agric 14:85–103. doi:10.1300/j064v14n04_08CrossRef

29. Pantanella E, Danaher JJ, Rakocy JE, Shultz RC, Bailey DS (2011) Alternative media types for greenhouse seedling production of lettuce and basil. Acta Hortic 891:257–264. doi:10.17660/actahortic.2011.891.31CrossRef

30. Rakocy JE, Shultz RC, Bailey DS, Thoman ES (2003) Aquaponic production of tilapia and basil: comparing a batch and staggered cropping system. Acta Hortic 648:63–69. doi:10.17660/actahortic.2004.648.8

31. Ribeiro HM, Romero AM, Pereira H, Borges P, Cabral F, Vasconcelos E (2007) Evaluation of a compost obtained from forestry wastes and solid phase of pig slurry as a substrate for seedlings production. Bioresour Technol 98:3294–3297. doi:10.1016/j.biortech.2006.07.002CrossRef

32. Sims GK, Ellsworth TR, Mulvaney RL (1995) Microscale determination of inorganic nitrogen in water and soil extracts. Commun Soil Sci Plant Anal 26:303–316. doi:10.1080/00103629509369298CrossRef

33. Subler S, Edwards CA, Metzger J (1998) Comparing vermicomposts and composts. BioCycle 39:63–66. Retrieved from http://www.nal.usda.gov/

34. Tyler HH, Warren SL, Bilderback TE, Fonteno WC (1993) Composted turkey litter: I. Effect on chemical and physical properties of a pine bark substrate. J Environ Hortic 11:131–136. Retrieved from http://hriresearch.theknowledgecenter.com

35. Wright RD (1986) The pour-through nutrient extraction procedure. HortScience 21:227–229. Retrieved from http://hortsci.ashspublications.org

36. Yeager T, Bilderback T, Fare D, Gilliam C, Lea-Cox J, Niemiera A, Ruter J, Tilt K, Warren S, Whitwell T, Wright R (2007) Best management practices: guide for producing nursery crops, 2nd edn. Southern Nursery Association, Georgia. Retrieved from http://www.sna.org/

INDEX

T

U

W